Advances in Ophthalmic Optics Technology

Online at: https://doi.org/10.1088/978-0-7503-3263-7

Series Editor

R Barry Johnson, a Senior Research Professor at Alabama A&M University, has been involved for over 50 years in lens design, optical systems design, electro-optical systems engineering, and photonics. He has been a faculty member at three academic institutions engaged in optics education and research, has been employed by a number of companies, and has provided consulting services.

Dr Johnson is an IOP Fellow, an SPIE Fellow and Life Member, an OSA Fellow, and was the 1987 President of SPIE. He serves on the editorial board of *Infrared Physics & Technology* and *Advances in Optical Technologies.* Dr Johnson has been awarded many patents, has published numerous papers and several books and book chapters, and was awarded the 2012 OSA/SPIE Joseph W Goodman Book Writing Award for Lens Design Fundamentals (second edition). He is a perennial co-chair of the annual SPIE Current Developments in Lens Design and Optical Engineering Conference.

Foreword

Until the 1960s the field of optics was primarily concentrated in the classical areas of photography, cameras, binoculars, telescopes, spectrometers, colorimeters, radiometers, etc. In the late 1960s optics began to blossom with the advent of new types of infrared detector, liquid crystal display (LCDs), light emitting diode (LEDs), charge coupled device (CCDs), laser, holography, and fiber optics along with new optical materials, advances in optical and mechanical fabrication, new optical design programs, and many more technologies. With the development of the LED, LCD, CCD, and other electro-optical devices, the term 'photonics' came into vogue in the 1980s to describe the science of using light in the development of new technologies and the operation of a myriad of applications. Today optics and photonics are truly pervasive throughout society and new technologies are continuing to emerge. The objective of this series is to provide students, researchers, and those who enjoy self-education with a wide-ranging collection of books, each of which focuses on a topic relevant to the technologies and applications of optics and photonics. These books will provide knowledge to prepare the reader to be better able to participate in these exciting areas now and in the future. The title of this series is *Emerging Technologies in Optics and Photonics*, in which 'emerging' is taken to mean 'coming into existence', 'coming into maturity', and 'coming into prominence'. IOP Publishing and I hope that you will find this series of significant value to you and your career.

A full list of titles published in this series can be found here: https://iopscience.iop.org/bookListInfo/emerging-technologies-in-optics-and-photonics.

Advances in Ophthalmic Optics Technology

Edited by
Davies William de Lima Monteiro
Federal Univerisity of Minas Gerais (UFMG), Belo Horizonte, Brazil

Bruno Lovaglio Cançado Trindade
Feluma Medical School, Belo Horizonte, Brazil

IOP Publishing, Bristol, UK

ISBN 978-0-7503-3263-7 (ebook)
ISBN 978-0-7503-3261-3 (print)
ISBN 978-0-7503-3264-4 (myPrint)
ISBN 978-0-7503-3262-0 (mobi)

DOI 10.1088/978-0-7503-3263-7

Version: 20221201

IOP ebooks

British Library Cataloguing-in-Publication Data: A catalogue record for this book is available from the British Library.

Published by IOP Publishing, wholly owned by The Institute of Physics, London

IOP Publishing, No.2 The Distillery, Glassfields, Avon Street, Bristol, BS2 0GR, UK

US Office: IOP Publishing, Inc., 190 North Independence Mall West, Suite 601, Philadelphia, PA 19106, USA

The editors would like to dedicate this book to the many millions of people whose vision has been and will continue to be benefited from steady advances in ophthalmic optics.

Contents

Acknowledgement

Davies William de Lima Monteiro and Bruno Lovaglio Cançado Trindade would like to acknowledge Prof. Barry Johnson for his enthusiasm with the topic and for the invitation to put this book together. They would also like to thank Ms. Ashley Gasque and team for the editorial support, tolerance and understanding with the several requested extensions, especially when a substantial part of the work was happening during the pandemics. They also want to acknowledge the diligent involvement of all authors of chapters in this book, who made it possible to bring so many complementary topics together.

Additionally, Davies William de Lima Monteiro wants to acknowledge the support by the Department of Electrical Engineering (DEE) and the Graduate Program on Electrical Engineering (PPGEE) of the Federal University of Minas Gerais (UFMG); the support and hard work of all the students who have contributed to our joint advances in the realm of ophthalmic optics; and also the funding agencies CNPq (Brazilian National Research Council), FAPEMIG (Research Support Foundation of the State of Minas Gerais) and CAPES (Coordenação de Aperfeiçoamento de Pessoal de Nível Superior - Brasil; finance code 001).

Lucas S Batista would like to thank the support from UFMG and the Brazilian agencies FAPEMIG (Research Support Foundation of the State of Minas Gerais), CNPq (The National Council for Scientific and Technological Development), and CAPES (Coordination for the Improvement of Higher Education Personnel).

Diogo Ferraz Costa, Rodrigo de Abreu, Thiago Daniel Moura, Luiz Melk de Carvalho, Otávio Gomes de Oliveira, Jhonattan Córdoba Ramirez, Carlos Henrique Lauro and Pablo Nunes Agra Belmonte would like to thank the research funding agencies FAPEMIG, CAPES and CNPq as well as the Graduate Program on Electrical Engineering (PPGEE), Universidade Federal de Minas Gerais, Brazil.

Andrés Vásquez Quintero would like to thank Ghent University, VLAIO and Azalea Vision (Belgium) for their support and fruitful discussions that led to this work.

Ado Jorio, Alexandre S Barbosa, Emerson A Fonseca, Lucas Lafetá, Leandro M Malard, Rafael P Vieira and Renan Cunha acknowledge João Luiz Campos and Marco A Romano-Silva for important contributions. The work in chapter 14 was financed by FINEP 01.13.0330.00 and INCT-MM.

Editor biographies

Davies William de Lima Monteiro

Davies William de Lima Monteiro has been Full Professor of Electrical Engineering and Optics, Research Manager of Optical Metrology in Industry, and Founding member of the Brazilian Society for Optics and Photonics. He was also Director at the Brazilian Microelectronics Society and Member of the advisory board to Microelectronics: Brazilian National Research Council. As a leading expert in the field, he has written more than 150 scientific publications.

Bruno Lovaglio Cançado Trindade

Graduated in medicine in 2006 and concluded his Ophthalmology residency in 2011. PhD degree 2017. Dr Trindade is a Professor of Ophthalmology at the Feluma Medical School in Belo Horizonte, Brazil. He is also the head of the Cornea Department of the Feluma Eye Institute in the Belo Horizonte, Brazil.

List of contributors

Rodrigo de Abreu
Rua Cambuí, 40, 31840-120, Guarani, Belo Horizonte, MG, Brazil

Alexandre Barbosa
Av. Antonio Carlos 6627, Pampulha, 31270-901, Belo Horizonte, MG, Brazil

Lucas Batista
DEE/UFMG, Av. Antonio Carlos 6627, Pampulha, 31270-010, Belo Horizonte, MG, Brazil

Jacqueline Beltz Franzco
Royal Victorian Eye and Ear Hospital, 32 Gisborne Street, East Melbourne, Victoria 3002, Australia

Andrea Chiuchiarelli
DELT/UFMG, Av. Antonio Carlos 6627, Pampulha, 31270-010, Belo Horizonte, MG, Brazil

Jhonattan Cordoba Ramirez
DELT/UFMG, Av. Antonio Carlos 6627, Pampulha, 31270-010, Belo Horizonte, MG, Brazil

Glauber Coutinho Eliazar
Perfil Oftalmologia Personalizada, Rua Gonçalves Dias, 82, Santa Efigênia, 30350-090, Belo Horizonte, MG, Brazil

João Marcelo de Almeida Gusmão Lyra
Rua Eng Mário de Gusmão 90 Sala 712, Brazil

Davies William de Lima Monteiro
Department of Electrical Engineering, Graduate Program on Electrical Engineering, Universidade Federal de Minas Gerais (UFMG), Av. Antonio Carlos 6627, Pampulha, 31270-010, Belo Horizonte, MG, Brazil

Thiago Daniel de Oliveira Moura
Maukenbach, 15, 6242, Radfeld, Austria, Austria

Diogo Ferraz Costa
Rua Fidélis Martins, 303, apto. 202, Buritis, Belo Horizonte, MG, Brazil

Emerson A Fonseca
Physics Department, Instituto de Ciências Exatas, Universidade Federal de Minas Gerais, Av. Presidente Antônio Carlos, 6627, Pampulha, 31270-901, Belo Horizonte, MG, Brazil

Otávio Gomes de Oliveira
Rua São Simão 42/1106, Bairro Bom Pastor, Manhuaçu/MG, 36902-269, Brazil

Ado Jorio
Physics Department, Universidade Federal de Minas Gerais, Av. Antônio Carlos 6627, Pampulha, 31270-901, Belo Horizonte, MG, Brazil

Marjean T Kulp
The Ohio State University College of Optometry, 338 West 10th Avenue, Columbus, OH, 43210-1280, USA

Lucas Lafeta Prates da Fonseca
Physics Department, Instituto de Ciências Exatas, Universidade Federal de Minas Gerais, Av. Presidente Antônio Carlos, 6627, Pampulha, 31270-901, Belo Horizonte, MG, Brazil

Carlos Henrique Lauro
DEMEP Praça Frei Orlando, 170, Centro, São João del Rei/MG, 36307-352, Brazil

Mateus Lins dos Santos
Av. Poeta Mario Jorge Menezes Vieira, 1159, apt 401, Coroa do Meio, Aracaju/SE, Brazil

Bruno Lovaglio Cançado Trindade
Rua Manaus 595, 30.150-350, Sao Lucas, Belo Horizonte, MG, Brazil

Leandro Malard Moreira
Av. Antonio Carlos 6627, Pampulha, 31270-901, Belo Horizonte, MG, Brazil

Monica Matsumoto
Aeronautics Institute of Technology (ITA), Praça Marechal Eduardo Gomes, 50 - Vila das Acácias - São José dos Campos - SP - ZIP 12228-900, Brazil

Luiz Melk de Carvalho
10 Wells Court, Pevensey Garden, Worthing, BN11 5PE, UK

Pablo Nunes Agra Belmonte
Rua Virgínia 246 Carlos Prates, Belo Horizonte, MG, Brazil

Diego Oliveira Miranda
Rodovia MG 262, KM 10, S/N, Bairro, Sobradinho, Sabará, MG, 34590-390, Brazil

Rafael Pinto Vieira
Departamento de Bioquímica e Imunologia, Instituto de Ciências Biológicas, Universidade Federal de Minas Gerais Av. Antonio Carlos 6627, Pampulha, 31270-901, Belo Horizonte, MG, Brazil

Thomas W Raasch
The Ohio State University College of Optometry, 338 West 10th Avenue, Columbus, OH, 43210-1280, USA

Paulo Schor
Universidade Federal de São Paulo (UNIFESP), R. Botucatu, 806 - São Paulo, SP 04038-001, Brazil

Renan Souza Cunha
Physics Department, Instituto de Ciências Exatas, Universidade Federal de Minas Gerais, Av. Presidente Antônio Carlos, 6627, Pampulha, 31270-901, Belo Horizonte, MG, Brazil

Andrew J Toole
The Ohio State University College of Optometry, 1664 Neil Ave, Columbus, OH, 43201, USA

Andrés Vásquez Quintero
Technologiepark 126, iGent tower, 9052 Zwijnaarde, Belgium

Mariana Vaz Goulart
Aeronautics Institute of Technology (ITA), Praça Marechal Eduardo Gomes, 50 - Vila das Acácias - São José dos Campos - SP - ZIP 12228-900, Brazil

Marcus Henrique Victor Jr
Aeronautics Institute of Technology (ITA), Praça Marechal Eduardo Gomes, 50 - Vila das Acácias - São José dos Campos - SP - ZIP 12228-900, Brazil

Chapter 1

Introduction

Bruno Lovaglio Cançado Trindade and Davies William de Lima Monteiro

The eye is an ingenious and yet rather simple opto-mechanical system that images the world onto the retina to allow the magic of the most dominant of the human senses to occur, vision. Understanding the intrinsic dynamic mechanism of light management through the visual pathway requires knowledge of different disciplines, as anatomy, optics and mechanics. Vision is an involved process that comprises neural and optical domains, the latter being the focus of this book.

To compensate for refractive imperfections in the eye optics, a variety of ophthalmic lenses can be used to enable a better image quality on the retina. Spectacles and contact lenses are non-invasive clinical options for millions of people today. They have been available for centuries, witnessing many improvements in optical quality and materials, while more recent deployments also venture towards immersive augmented reality scenarios.

Intraocular lenses, on the other hand, are implanted into the eye in surgical procedures, of which cataract surgery is the most common one, which allows millions of people worldwide to have vision restored. During this fascinating surgery, this artificial lens replaces the clouded natural crystalline lens in a minimally invasive operation that is packed with cutting-edge technology. Although this type of prosthesis has undergone substantial evolution since its introduction a few decades ago, it has not yet achieved all the combined desired characteristics of an ideal lens, which should offer a broad range of sharp uncorrected vision, sustain a long-lasting clear capsule, be made of a biologically inert material and be able to be inserted into the eye through a very small incision. Design compromises have been adopted in the myriad of current off-the-shelf lenses, whose specific features, functional performance, materials and topologies result from collaborative work among physicists, ophthalmic surgeons, optical designers, computer scientists and engineers from various specialties. Modern ophthalmic surgeons need to thoroughly understand and propose the most suitable available model to each individual, which requires an overarching knowledge beyond clinical practice. Likewise, lens designers

doi:10.1088/978-0-7503-3263-7ch1

aware of the peculiarities of ophthalmic optics are geared up to deploy the most suitable physical concepts and engineering solutions to high performing lenses.

This book was conceived to be a comprehensive textbook that encompasses different aspects of modern ophthalmic lenses. There is a vast underpinning interdisciplinary work from initial requirements and design to fine tuning, manufacturing, performance testing and using these lenses in the human eye. In the following chapters, you will be guided through a detailed text on the different aspects of all these processes.

The chapters cover the basics of the human eye optics, types of refractive errors and correction, fundamental optics parameters, types of intraocular lenses, lens design and optimization methods. It also presents novel concepts on lens design such as intraocular lenses with extended depth of focus, electronically equipped lenses, mechanically adjustable and accommodative lenses, and the evaluation of health parameters and diseases through the eye.

This volume presents an independent generalist view for the specialists. It intends to bridge the gap between the clinical practice and the engineering of lens design and manufacturing, especially in the scope of intraocular lenses. The ophthalmic sector for this type of lens solely is a several billion-dollar market where ingenious solutions have been developed and adopted in the past few decades. Although there is an abundance of consolidated technologies, the tricks and trades of lens prototyping is mostly restricted to industrial classified procedures, and improvements are based on physical trials and the assessment of consultant experts under contract.

There are also brewing technologies such as smart contact and intraocular lenses, achromatic metalenses and refractive-index shaping on the horizon. However, it is worth repeating that solutions often come with trade-offs most patients and sometimes ophthalmologists are oblivious to. This text will help the medical doctor to better understand the potential and limitations of the lenses to be chosen for each patient. It will also assist both the experienced and the junior lens designer to better understand the constraints of ophthalmic lenses, in particular intraocular ones. And, finally, it will serve as a reference to any professional in the ophthalmic sector, from sales to manufacturing, to put the parts together.

We introduce selected aspects of ophthalmic optics that directly lead to understanding ametropias, visual quality, lens prescription and lens add-on features. For complex in-depth material, the reader is referred to more specialized literature. We hope you find useful information in the following chapters and that they can be used to expand ophthalmic lens science and technology even further in the coming years.

IOP Publishing

Advances in Ophthalmic Optics Technology

Davies William de Lima Monteiro and Bruno Lovaglio Cançado Trindade

Chapter 2

The human eye and refractive correction

Bruno Lovaglio Cançado Trindade, Glauber Coutinho Eliazar and Davies William de Lima Monteiro

The human eye is a remarkable organ that fundamentally captures light from a three-dimensional scene harboring several features such as optical projection, focus adjustment, aperture resizing, object scanning, image averaging, color modulation and intensity gain. To execute all these functions, many processes and components are involved and must work in a harmonic and precise balance. The internal structures of the eye have properties that contribute to focusing images and to converting their spatial intensity profile into neural impulses that enable vision to take place at the cortical level of the central neural system.

Many physical dimensions are required to evaluate the optical properties of the eye, including the corneal curvatures, the axial length and the anterior chamber depth. These dimensions vary in a limited range among individuals and can be measured using various pieces of equipment (see chapter 6). It is important to understand these limits and variations in order to plan any form of optical correction.

One of the unique features of the healthy human eye is the ability to quickly refocus light coming from objects at multiple longitudinal distances. This allows sharp vision across a dynamic and diverse environment. This process of accommodation requires many structures and different stimuli to work in conjunction in a precise arrangement.

Although the eye is a terrific piece of natural engineering, some errors may be present or might evolve to a level where they can compromise an acceptable visual outcome. Due to various reasons, the eye may have a refractive error and may not be capable of correctly focusing an image on the retina, resulting in a blurred or distorted object representation. Many alternatives have been developed to correct for these errors, some of which date back to centuries ago while others are still on the drawing board. Each of them needs to be carefully weighed before considering their use.

In this chapter we will detail the anatomical characteristics of the eye, the visual pathway, as well as the importance of each of the structures in the physiology of vision formation. We will describe each of the eye dimensions and their range of

doi:10.1088/978-0-7503-3263-7ch2

normal and pathological variations. We will then discuss the accommodation phenomenon and the impact of its decrease with age. And, finally, we will tackle the refractive errors and their clinical and surgical treatment options.

The full comprehension of these topics is essential to adequately understand the other chapters in this book and to develop the ability to further investigate and propose new treatments for the refractive errors.

2.1 Anatomy

The eye is responsible for more than 80% of the data communicated from the external environment to the human body [1]. Its complex structure is an anatomical window capable of capturing light stimuli that is interpreted as vision at the cerebral cortex level.

The eyeball has three concentric layers: an outer structure as a protective layer, an intermediate layer responsible for nutrition, and an inner one made up of nervous tissue. The outermost layer consists of the sclera posteriorly and the cornea anteriorly. The middle layer, the uvea, is subdivided into choroid, ciliary body and iris. The more internal neural layer is responsible for transforming the light stimuli into electric signals, which are transported by the optic nerve to the central nervous system.

2.1.1 The tear film

The entire ocular surface exposed to the external environment is covered by a tear film. It is a superficial lipid layer, produced by the meibomian glands, that covers an aqueous layer with diluted proteins and mucins. The mucous components interact directly with epithelial cells glycocalyx.

A tear film homeostasis is essential for corneal functioning, promoting lubrication, nutrition and antimicrobial protection. Besides, the air–tear film interface is the major refractive element of the eye due to their refractive index differences, providing a smooth optical surface. Thus, changes in the tear film directly affect the eye health of the patient and visual quality. Tear film thickness is approximately 3.4 μm and its basal volume 7.4 μl, which decreases with age.

2.1.2 The cornea

The cornea is a transparent tissue and because it is avascular, its oxygen nutrition comes dissolved in the tear film from the air and from palpebral conjunctival vessels. Other nutrients come from the aqueous humor. Its dimensions are about 12 mm on the horizontal meridian and 11 mm on the vertical meridian. The cornea is practically spherical on the central 4 mm diameter, and it becomes flatter as it moves towards the periphery, mainly in the nasal and upper quadrants. The posterior surface is more curved than the anterior one, hence it is thinner in the central region (~0.5 mm) and thicker in the periphery (~1.0 mm).

The cornea consists of five layers, where the epithelium is the most external, followed by the Bowman's layer, the stroma and more internally by the Descemet's membrane and the endothelium.

The epithelium is stratified nonkeratinized, with 4–6 layers of cells, leading to around 50 μm thickness. The most superficial cells are joined by tight junctions that guarantee a barrier to the diffusion of substances. Its surface is irregular due to the presence of microvilli and the tear film is responsible for making it optically smooth. The sources of new epithelial cells are limbus stem cells.

The Bowman's layer is a more resistant anterior stroma region unable to regenerate. Damage in this layer can originate scar tissue and make the region opalescent. The stroma constitutes 90% of the corneal thickness and is formed by keratocytes, ground substance and collagen lamellae (predominantly type I). These collagen lamellae are less compact posteriorly. Corneal transparency is possible due to uniformity in size and separation of collagen lamellae. In cases of edema, i.e. swelling, and separation of collagen fibrils, stromal clouding occurs.

The Descemet's membrane is the basal membrane of the corneal endothelium and increases in thickness with age. The innermost layer is the endothelium, a monolayer of hexagonal cells originated from the neural crest. Cell density is greater at the periphery than in the center and the number of cells gradually decreases with age. Endothelium cell division capacity is limited. In cases of cell loss, the cell density decreases, and the remaining cells migrate and increase in size. There is an intense metabolic activity and an active ionic transport responsible for removing water from the corneal stroma and maintaining stromal deturgescence and transparency. Injuries in this layer can lead to endothelial decompensation, stromal edema, loss of corneal transparency and visual impairment.

2.1.3 The anterior chamber

The anterior chamber is located between the cornea and the plane of the iris and pupil. The angle of the anterior chamber is located at the intersection of the cornea and the iris and has five important structures: Schwalbe line, Schlemm's canal and trabecular meshwork, scleral spur, anterior border of the ciliary body and peripheral iris. The aqueous humor that fills the anterior chamber is produced by the ciliary body in the posterior chamber, passes through the pupil and is drained by the trabecular meshwork and the Schlemm's canal. There is also an alternative drainage pathway known as the uveoscleral pathway (through the iris and ciliary body to the supracoroidal space) which corresponds to up to 50% of the aqueous outflow in young people.

2.1.4 The iris

The iris is constituted by connective tissue with vessels and nerves, dilator muscle, sphincter muscle and a double layer of pigmented epithelial cells posteriorly. There are melanocytes and pigment cells responsible for iris coloring. The muscle tissues have a neuroectoderm origin and guarantee iris mobility and pupil size change. There are both sympathetic and parasympathetic innervation. The dilator muscle contracts with adrenergic sympathetic stimulus ($\alpha 1$) while cholinergic parasympathetic stimulus has an inhibitory function. Conversely, the sphincter muscle is innervated by parasympathetic nerve fibers originated from the Edinger–Westphal nucleus and the oculomotor nerve is responsible for carrying these parasympathetic fibers. The sympathetic innervation helps to relax the sphincter.

2.1.5 The crystalline lens

This lens is a biconvex refractive structure located between the vitreous and the posterior chamber. It has a gradient index of refraction both along its longitudinal axis and from its center to its periphery. The lens contributes with one third of the eye dioptric power, approximately 20.00 D. At birth its equatorial diameter is 6.5 mm and the anteroposterior width is 3 mm. The former increases to 9–10 mm in the first 2–3 decades of life and stabilizes in adulthood. The latter, however, continues increasing throughout life, reaching 6 mm by the age of 80 years old. Concomitant to these changes, the radius of curvature decreases. There is no innervation or blood supply for the lens and its nutrition occurs by the contact with the aqueous and vitreous humors. Since its embryogenesis, the lens is surrounded by a capsule, which is a basal membrane.

2.1.6 The vitreous

The vitreous is a gel-like substance that occupies four-fifths of the volume of the globe. It is composed of 99% water in addition to mucopolysaccharide, hyaluronic acid, collagen (mainly type II) and hyalocytes. Its volume is close to 4.0 ml and occupies the posterior segment of the eyeball. The vitreous assists in lens, ciliary body and retina nutrition. It adheres to the peripheral retina at the vitreous base, at the optic nerve, at the perimacular region, along the retinal vessels, and at the periphery of the posterior lens capsule. The common complaints of floaters originate from small vitreous opacities that produce shadows on the retina and visual disturbance.

2.1.7 The retina

The retina and its underlying epithelial layer have a common embryological origin, the optic vesicle. Therefore, the retina has two parts: the neurosensory retina and the retinal pigment epithelium (RPE). The layers of the neurosensory retina, from inner to outer retina, are as follows:

- internal limiting membrane;
- nerve fiber layer;
- ganglion cell layer;
- inner plexiform layer;
- inner nuclear layer;
- middle limiting membrane;
- outer plexiform layer;
- outer nuclear layer;
- external limiting membrane;
- rod and cone inner segments;
- rod and cone outer segments.

The photoreceptor layer of the neurosensory retina are the rods and cones. There are approximately 100–125 million rods and 6–7 million cones, in a ratio of 20:1. Between the temporal vascular arcades, the retina is called macula. The fovea is the macula's central area, which contains a specialized region at its center known as foveola, whose diameter is about 250 μm. The photoreceptor layer of the foveola is entirely composed of a high density of cones, responsible for great visual acuity and color vision, compliant with direct photopic and mesopic illuminances. The cones are 2.5 μm wide in the fovea and become 10 μm at the periphery. Rods are more peripheral photoreceptors, not specialized in color vision but with a high light sensitivity, appropriate for scotopic conditions, and feature a diameter ranging from 3 μm to 10 μm, with the largest diameter corresponding to the more peripheral loci.

The retina neuronal interconnection elements are highly complex, with many types of bipolar, amacrine, and ganglion cells. The RPE is a monolayer of hexagonal cells with several functions like vitamin A metabolism, maintenance of the outer blood–ocular barrier, phagocytosis of the photoreceptor outer segments, absorption of light (reduction of scattering), active transport of materials into and out of the RPE and others.

2.1.8 The optic nerve

The optic nerve (cranial nerve II) consists of more than 1 million ganglion cells axons and extends toward the lateral geniculate nucleus. It has four topographic parts: intraocular, intraorbital, intracanalicular and intracranial.

2.1.9 The visual pathway

The visual path begins at photoreceptors and follows until the cerebral cortex. After the axons of the ganglion cells form the optics nerves, they proceed towards the central nervous system where some fibers decussate (cross to the other side) at the optic chiasm and continue as optics tracts until they reach the lateral geniculate nucleus. The first ganglion cell axons synapse happens at this point. Then, the fibers keep traveling towards the primary visual cortex (Brodmann's area 17).

2.2 Physiology

2.2.1 Retinal image formation

The transformation of the incoming light by photosensitive cells into an electrical impulse is called phototransduction. Rods are more sensitive than cones because they have a much larger membrane surface area. A single photon can stimulate a rod. Cones, although less sensitive, can adapt to various light intensities and respond more quickly to repetitive stimuli.

The photoreceptor cells have 1000 invaginations of their plasma membrane, forming discs with more than 1 million rhodopsin trans-membrane proteins each. Rhodopsin is responsible for capturing and amplifying light energy. One rhodopsin molecule activates 100 transducing molecules, amplifying the reaction.

This protein best absorbs light with wavelengths around 510 nm (green light) in rods. Blue and yellow beams are less well absorbed and it is insensitive to longer wavelengths (red light). In the dark, rhodopsin is inactive and the cell is depolarized releasing the neurotransmitter glutamate from its axonal terminal. With light activation, the cell membrane hyperpolarizes and glutamate release is inhibited.

Color vision in humans happens due to three different types of cones, i.e. three types of opsins:

- short-wavelength-sensitive cones (S cones): blue–yellow color vision. These cones detect color by comparing their signals with those of the M cones;
- middle-wavelength-sensitive cones (M cones): they detect high-resolution achromatic contrast (black and white);
- long-wavelength-sensitive cones (L cones): this type creates green–red color vision.

L and M cones are more numerous than S cones because both contribute to achromatic and chromatic contrast. Most color vision defects involve L and M cones and the red–green discrimination.

2.2.2 Visual field

The visual field is composed of information that is produced by different retinal locations. For example, the upper visual field is produced by information received from the lower retinal photoreceptors. In fact, the entire visual field can be mapped to a specific retinal topography. Knowing the anatomy of the visual pathway leads to understanding the formation of the visual field. Both eyes are responsible for the composition of the visual field perceived by our brains. The left part of the visual field is produced by information received from the temporal retina of the right eye and the nasal retina of the left eye. The right visual field is the opposite. This is why nerve fibers from the nasal retina of both eyes have to cross to the other side of the brain to form the corresponding temporal visual field (together with same-sided temporal fibers). The crossover region of the optic tract is called optical chiasm and, from that point onwards, visual information from both eyes travels together in both sides of the brain.

2.2.3 Cortical processing

The end of the visual pathway occurs when the light stimulus initiated in the retinal photoreceptors reaches the primary visual cortex (known variously as V1, striate cortex, or Brodmann area 17), located in the calcarine groove which divides the medial face of the occipital lobe.

Predominantly, human vision comes from the central retina, which occupies most of the cortical area. The central 10° alone is responsible for 50%–60% of the cortical activity, and if the macular area is considered, within 30°, the corresponding activity rises up to 80%.

After reaching the primary cortex, the visual electrical impulse is transmitted to secondary areas (V1 to V5) through a complex network of synapses. Each area is responsible for different visual interpretations. The parastriate cortex (also called V2, or Brodmann area 18) is contiguous and receives inputs from V1. Area V3 also receives direct inputs from V1 and sends efferent information to the basal ganglia (pulvinar) and the midbrain. Visual integration may occur in this region. Area V4 seems to be particularly sensitive to color, and damage to this area is probably responsible for most cerebral achromatopsia cases. Areas V4 and V5 are very sensitive to movement and direction. Neurons in area V5 encode the speed and direction of moving stimuli.

Didactically, visual processing pathways in the human brain are divided into occipitotemporal (ventral, or 'what') pathway, responsible for language processing, object identification, and also for projections to limbic structures; and the occipitoparietal (dorsal, or 'where') pathway that is responsible for visual-spatial analysis [1].

2.3 Anatomical dimensions

2.3.1 Corneal curvature

Keratometry is the measure of the corneal curvature. The standard manual keratometer views the cornea as a convex mirror and measures only a small central portion (3.2 mm diameter). The corneal radius of curvature is calculated based on the size of the reflected image. Both anterior and posterior corneal surfaces contribute to corneal power, but keratometry is based exclusively on the measurement of the anterior surface's radius of curvature. In lens-power calculation formulas, the central corneal power is a very important factor. Each diopter error in corneal power leads to a 1.0 diopter postoperative refractive error.

Keratometry and corneal topography can estimate corneal power, but neither of them measure corneal power directly. Corneal tomography which also measures the posterior corneal curvature can define the total corneal power without the need for estimates.

The cornea has an average anterior radius of curvature of 7.80 mm, with observed limits ranging from 7.00 mm to 8.65 mm; and an average posterior radius of curvature of 6.50 mm, ranging from 6.00 mm to 6.60 mm across the population [1]. Its refractive index is 1.376. The cornea produces most of the eye's refractive power contributing with an average of 43.25 D of the total 58.60 D refractive power required by an average eye.

There is a wide range of keratometry values in the population. Keratometry readings in candidates to keratorefractive surgery typically range from 38.0 D to 48.0 D. Excessively flat (<33.0 D) or excessively steep (>50.0 D) corneal powers decrease vision quality and increase aberrations.

2.3.2 Corneal thickness

Corneal thickness is the distance between the epithelial and endothelial corneal surfaces. In normal conditions, it is about 0.5 mm: epithelium (40–50 μm), Bowman's layer (8–15 μm), stroma (470–500 μm), Descemet's membrane (10–12 μm), and endothelium (4–6 μm).

2.3.3 Anterior and posterior chamber depths

The anterior chamber is the space between the iris and the cornea. Anterior chamber depth varies among individuals and between regional populations. It is deeper in aphakia, pseudophakia, and myopia and shallower in hyperopia. The anterior chamber is deepest centrally and it becomes narrower towards the iridocorneal angle.

The average volume of the anterior chamber is 220 μl and the average depth is 3.11 mm.

The posterior chamber is the anatomical portion of the eye posterior to the iris and anterior to the lens and vitreous face with an average volume of 60 μl.

The average distance from the apex of the cornea to the front surface of the lens is 3.6 mm in an emmetropic eye. Considering the thickness of the cornea is 0.5 mm, the depth of the anterior chamber is 3.1 mm, ranging from 1.5 mm to 4 mm.

2.3.4 Crystalline-lens dimensions (thickness and diameter)

The lens is a biconvex refractive structure with anteroposterior (from anterior to posterior capsule) and equatorial diameters (vertical orientation). The anteroposterior diameter increases throughout life, reaching 6 mm in the eighth decade of life. The equatorial diameter at birth is 6.5 mm and increases until adulthood to 9–10 mm.

The average natural lens diameter of the human eye is 9.57 mm (range: 9.2–9.95 mm). The mean capsule bag diameter increases slightly after lens extraction to 9.75 mm (range 9.6–9.85). The mean capsular bag diameter measured with a special device (capsular tension ring using the ring measurement system/viscoelastic) was 10.32 mm (range: 10.0 mm to 10.65 mm).

2.3.5 Vitreous cavity

The vitreous cavity is the largest compartment in the eye, contributing more than two-thirds of its volume (5–6 ml from 6.5 to 7.0 ml in an adult eye).

The vitreous cavity longitudinal diameter, from the posterior capsule to the fovea, is around 16.5 mm.

It is usually filled with humor vitreous, but in some situations after retinal surgery, this cavity could be filled with silicone oil. There are two major problems regarding accuracy in biometry.

The first is obtaining an accurate axial length measurement with the ultrasonic biometer. Ultrasound biometry measures the transit time of the ultrasound pulse with an estimated ultrasound speed through the various ocular media. When silicone oil fills the posterior segment, the speeds differ from the usual range (980 m s^{-1} for silicone oil and 1532 m s^{-1} for vitreous). The use of optical biometry to axial length measurement, which uses light and not ultrasound for measurements, partially solves this problem.

The second problem is that the oil filling the vitreous cavity reduces the optical power of the posterior surface of an implanted biconvex lens, as the silicone oil has a greater refractive index than the vitreous humor. This issue can be counterbalanced increasing the intraocular lens (IOL) power by 3–5 D. See chapter 6 for a more detailed discussion on ocular measurements.

2.3.6 Retinal thickness

The retinal thickness varies considerably according to its region. The retina is thickest in the papillomacular bundle near the optic nerve (0.23 mm) and thinnest in the foveola (0.10 mm) and ora serrata (0.11 mm).

The anteroposterior (AP) diameter of the adult eye can vary significantly. In normal eyes, the average axial length is approximately 23.5 mm, ranging from 19.5 mm to 26.5 mm. However, nanophthamic eyes can be as short as 14 mm and long pathological myopic eyes can exceed 32 mm.

2.4 Accommodation

Accommodation is a physiological phenomenon in which the natural lens increases its refractive power by decreasing the radius of curvature of the front and back surfaces and by promoting an anterior shift in its position. The purpose of this process is to promote focused vision to near objects. It is said that the radius of curvature of the anterior surface reduces 0.33 mm for each diopter of accommodation while that of the posterior one decreases 0.15 mm for the same amount of additional convergence [2].

The accommodation phenomenon is controlled by the ciliary muscles that transmit their contractile force to the lens capsule through the zonules. The ciliary muscle is a group of circumferent nonstriated muscle fibers located anteriorly in the eye behind the iris root under the sclera. These fibers are usually divided into three different layers: the longitudinal, the radial and the circular layers. Despite the different orientation of the anatomical fibers, the ciliary muscle behaves as a unit decreasing its annular diameter with muscle contraction. Variations of the ciliary muscle tension (and thus in its annular diameter) are what make possible the accommodation phenomenon. The zonules are made of fibrillin fibers and are responsible for keeping the lens in place much like the springs in a child's trampoline. These fibers connect the ciliary muscle to the anterior and posterior lens capsule and transmit the alteration of tensile force with changes in the ciliary muscle tone.

The classic accommodative theory was proposed by Helmholtz back in the 19th century [3]. In this theory, the contraction of the ciliary muscle fibers moves the muscle mass forward and inward allowing relaxation of the zonules. This promotes relaxation of the tension over the anterior and posterior capsule and allows the lens nucleus to become more spherical increasing dioptric power. On the other hand, Schachar *et al* have proposed a new theory in which the contraction of the ciliary muscle would cause tension in the equatorial region of the lens and relaxation in the anteriorly- and posteriorly-fixated zonules fibers [4]. This way, when accommodation occurred, the equatorial portion of the lens would move closer to the sclera. During subsequent studies, Glasser *et al* could demonstrate in monkeys a centripetal movement of the equatorial region of the lens during both centrally and pharmacologically-induced accommodation supporting the classical Helmholtz theory [5]. Complementary discussion on the accommodation theory is presented in chapter 13.

The accommodation reflex is usually accompanied by two other important processes. The first is pupil constriction (miosis), in an attempt to increase depth of focus, and the second is interocular convergence to give binocular near vision. The combination of these three processes is often called the near vision triad [6]. This reflex is triggered by the stimulation of the mid-brain-located Edinger–Westphal nucleus. This nucleus receives afferent input through the peristriate area 19 of the cortex and sends efferent stimulus through the third cranial nerve (oculomotor). The parasympathetic fibers of the third nerve go through the ciliary ganglion and then through the short ciliary nerves to promote miosis and ciliary muscle contraction (accommodation). The motor fibers of the third nerve will stimulate medial rectus muscle to converge the two eyes. The secondary visual cortex neurons are responsible for detecting the 'out-of-focus' image and starting the feedback loop to promote accommodation.

As stated, accommodation causes miosis through the near vision triad. Pupillary contraction limits light entrance in the eye, reduces optical aberrations caused by imperfections of the cornea and increases the depth of focus. This is one of the reasons why a well-lit room favors one's ability to read fine print material. The increase in the depth of focus reduces the demand to accommodate and this can be helpful, especially in older individuals with decreased accommodation amplitude [7].

The maximum amount that the natural lens can increase its vergence is called amplitude of accommodation. This limit is more or less constant between individuals and gradually decreases with age. It is thought that this is caused by the constant growth of the human lens through life. Due to the embryological characteristics of the lens, its growth pattern involves deposition of new lens fibers from the outside inwards, increasing its axial and equatorial diameters and compacting the central core (nucleus) that consists of older cells [8]. This way, the lens not only becomes bigger as the years go by, but it also gets stiffer. The ciliary muscle contraction is no longer able to change its shape and position as before and accommodation cannot be successfully accomplished. This process is called presbyopia and causes a blurred near image in an emmetropic eye.

The usual near activities are performed around 25 cm to 40 cm away and they require an accommodation of around 2.5–4 D, which can be easily accomplished by a healthy young eye. As a person reaches the fifth decade of life, the amplitude of accommodation starts to lose its ability to withstand these requirements and a progressive difficulty of reading fine prints settles. Presbyopia continues to progress until the individual reaches 55–60 years of age when the amplitude of accommodation is essentially zero.

Presbyopia is a condition that universally affects the population after reaching a certain age (usually around 40–45 years of age). It is thought that presbyopia afflicts almost two billion people worldwide and, as the population ages, this number will tend to increase over the next decades [9]. Initially, patients present asthenopic symptoms (eye discomfort when focusing for near) and try to compensate this by increasing the usual distance for reading fine prints. As it gets more significant, presbyopia requires the use of some form of optical correction. The usual methods can be divided into clinical and surgical alternatives. Bifocals or progressive spectacles are the most used method for adding convergence to the presbyopic eye. Multifocal contact lenses or monovision, a method in which one eye (usually the dominant eye) is corrected to see far objects and the other (in general the non-dominant one) to see near targets can also be used. Many corneal and intraocular surgeries have been proposed to treat this condition. Each of these methods will be extensively discussed later in this chapter.

The economic impact of presbyopia is very important since if affects people that are usually in the most productive years of their lives [6]. It also affects mental health since it is usually one of the first signs of the transition to a more mature life stage and onset of its eventual limitations. With the current social requirements of the western developed world, the inability to read fine print can have a real psychological impact in people's lives. This has been a pushing force towards new alternative treatments for presbyopia. Moreover, massive adoption of technological devices and displays and their applications has led people to be facing a near screen for a prolonged period of time. Computers, tablets and mobile phones are now part of a daily routine in modern society. The call for sharp near vision has probably never been so high. More on accommodation and accommodative intraocular lenses is presented in chapter 13.

The constant and gradual growth of the lens is also responsible for another important change. The lens fibers are metabolic active cells and their nutritional requirements are met via diffusion of the aqueous humor. As the lens gets larger with a progressively more compacted nucleus, the innermost fibers start to get metabolically deprived. This leads to glycation and precipitation of the crystallin which are the structural proteins found in these cells, causing them to lose transparency in a process that leads to the opacification of the lens. This opacification is known as cataract.

2.5 Ametropia

To produce a sharp image, light coming from the object in gaze must be properly refracted so the image is focused on the retina. This is naturally accomplished by the convergence power of two different lenses, namely the cornea and the crystalline lens. This system has a focal length in the vitreous that must equal the longitudinal axial distance between the posterior surface of the crystalline and the retina. When the object is infinitely far, the plane where the best focused image is projected coincides with the focal plane of this dual-optic system, which lies on the retina. This favorable condition is termed emmetropia.

The cornea is responsible for about 2/3 of the entire convergency requirements of the normal eye staring at a far object. The crystalline adds about 22 D of convergency in this unaccommodated state. However, when the object approaches the eye, if the system does not change, the image will look blurred on the retina because the best-focus plane for a close object lies further than the focal length, i.e. it moves beyond the retina. The closer the object, the farther the image plane. Conveniently, accommodation occurs and extra convergence is added by the crystalline, as explained in the previous section. This change shortens the focal length of the compound optic such that the image plane, which lies further than this length, now rests on the retina. The extension of this change varies according to the subject's age.

In general, there must be a match between the amount of convergence that the eye can reach and its axial length. Any unbalance produces an unfocused image on the retina and vision is compromised. This situation is referred to as ametropia. There are basically three different refractive errors that can be present in an ametropic eye: myopia, hyperopia and astigmatism. Each of these will be detailed next.

2.5.1 Myopia

Also known as shortsightedness or nearsightedness, this refractive error is caused by an excessive convergence of the cornea/lens for a specific axial length. In other words, the focal length of the system is shorter than the distance to the retina. Therefore, light coming from an infinitely far away object is focused before it reaches the retina. This produces a blurred image of distant objects (figure 2.1).

However, as light coming from near objects reaches the eye with a divergent pattern, the best-focus image plane lies beyond the focal length, therefore favoring a sharper image on the retina than that of a distant object. This way, myopic eyes usually have very good uncorrected near vision, hence the alternative names. There are different causes of myopia, whose identification is relevant as it may change the incidence, the risks and treatment options. The cornea curvature may be steeper than normal producing more convergent light rays. This is called keratometric myopia. Another common cause of myopia results from the change of the refractive index of the crystalline lens that can happen with age. The density of the nucleus usually increases with age and, more importantly, with the setting of a cataract. This leads to extra convergence power added to the lens. This is usually referred to as index myopia and may account for the improvement of uncorrected near vision in some older patients with mild to moderate nuclear cataract. However, the main type of myopia is due to an excessive growth of the eyeball. The retina is moved away with globe elongation and the convergence produced by the cornea leads to a focal length short of the retina. This growth usually happens during teenage years and corresponds to the usual progressing of myopia in these years. This type is referred to as axial myopia. Myopia has gained significant academic attention lately due to the increased prevalence in many countries, with many studies being performed to improve understanding about its origins and treatments [10–12]. Depending on the amount of globe growth, this can have an impact on the health of associated structures of the eyeball, the main one being the retina. As the globe grows, the

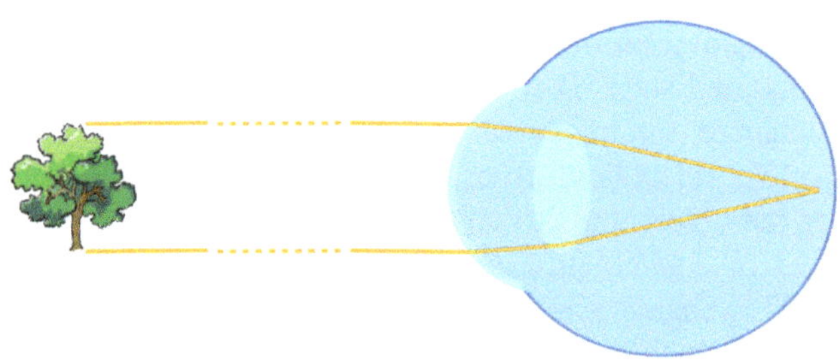

Figure 2.1. Note the light being focused before the retina.

entire ocular wall becomes thinner. The thinning of the sclera may increase its transparency and the adjacent choroidal blood vessels may produce a blueish hue to the white part of the globe typically seen in high myopes. However, the main problem of this abnormal globe elongation is the thinning of the retina. This may produce degenerations and holes especially in its periphery that can become problematic, leading to a condition known as retinal detachment. This is a sight-threatening complication that requires urgent surgery. Axial myopic eyes are especially prone to develop vitreo-retinal complications over the person's entire lifespan. Glaucoma can also be more common in these eyes requiring constant monitoring by a specialist. Myopia is usually classified as mild (<3 D), moderate (3–6 D) and high (>6 D). The term pathological myopia is sometimes used to describe eyes that have a very high myopic error (usually >10 D). These eyes are particularly at risk for the above-mentioned complications.

The incidence of myopia is dramatically increasing worldwide, especially in Asia [13]. This has led researchers to try to better understand the mechanisms involved in globe elongation and to propose mechanisms to avoid such an epidemic. The lack of violet light exposure has been described as one of the factors involved in the development of myopia [14]. Modern societies are living indoors much of the time [1]. Artificial LED lights usually lack the violet end of the spectrum. Moreover, the glasses used in many windows usually filter these shorter wavelengths leading to deprivation of violet light exposure. This has led to the encouragement of outdoors activities for children, especially in Asia. Muscarin receptors inhibition, using topical atropine, has also been shown to be effective in preventing myopia progression in children. The use of a low dose of daily atropine drops is now the standard of care for young children with myopia.

2.5.2 Hyperopia

Also known as hypermetropia or farsightedness, as opposed to myopia, this refractive error is caused by the insufficient convergence power by the cornea/lens for a specific axial length. It is mainly due to a smaller-than-normal eye anatomy causing light coming parallel to the eye to be focused after the retina, causing, in principle, a blurred image already for distant objects (figure 2.2).

This way, focusing near objects is particularly compromised, as their corresponding image planes are even further than the retina, and blurred near vision occurs. If it were not for accommodation, distant objects would always be blurred since, as

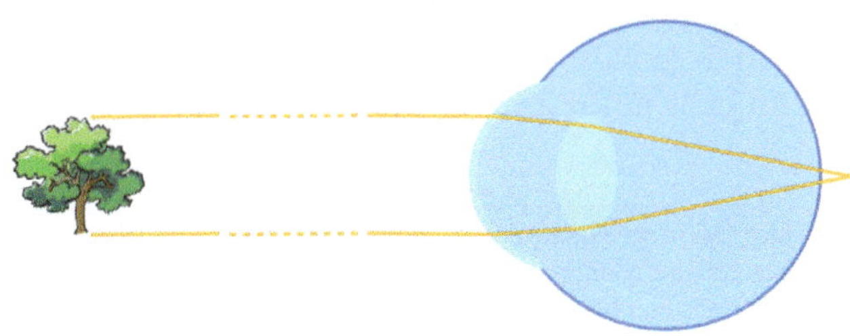

Figure 2.2. Note the light being focused after the retina.

defined, hyperopia presents with a far point (a virtual point) located behind the eye. Nevertheless, the additional convergence provided by accommodation may bring distant objects into focus and this explains the alternative naming of this refractive error. The amount of compensation provided by accommodation basically depends on the magnitude of hyperopia and the subject's age. As explained in the previous section, the amplitude of accommodation decreases with age. This way, for mild hyperopia in a young individual, symptoms of blurred vision may be absent. Hyperopia can thus be classified into three different magnitudes: manifest, latent and total hyperopia [15]. Manifest hyperopia is the amount of residual lack of convergence needed to bring distant objects into focus in an accommodating eye. Note that this can be negligible or absent even in the presence of a significant convergence deficit in a young eye. Latent hyperopia is the amount of hyperopia that is compensated by accommodation. This can be measured after cycloplegia and paralyzation of the ciliary muscles. Total hyperopia is the sum of the former two and represents the total deficit of convergence presented by the eye. As a subject ages, the manifest hyperopia tends to increase whereas the latent hyperopia decreases. As accommodation decreases to a clinical negligible level (around 70 years of age), the manifest becomes approximately the total hyperopia.

One special issue that arises with hyperopia relates to the fact that the accommodation triad consists of lens power increase, myosis and interocular convergence. This way, when a subject is forced to accommodate to see distant objects, the convergence produced by this accommodation may be enough to create strabismus, compromising binocular fused vision. There is a ratio between the accommodation convergence and accommodation (AC/A) that is somewhat constant for each individual. In the setting of an increased AC/A ratio or a high hyperopia, esotropia (convergent strabismus) will develop [16]. This is especially important in children since they have a larger accommodative amplitude, a smaller eye and are susceptible to amblyopia[1]. Hence, while the correction of myopia is based only on providing the best possible focused image, in hyperopia, refractive corrections must account not only for that but also for the correction of any imbalance in binocular vision.

2.5.3 Astigmatism

While the two former refractive errors, myopia and hyperopia, are known as spherical errors, this third refractive abnormality is caused by the rotationally asymmetric radii of curvature on a specific lens. In regular astigmatism, there is a steeper meridian that is opposite or 90° away from a flatter one. In contrast, irregular astigmatism does not respect this orthogonality.

[1] This is a condition in which vision development is affected mainly due to unbalance in vision between the two eyes. It can originate while cortical connections are still being created usually up until 6 years of age. The difference in image quality or position between the two eyes causes cortical connections to be created only for the better-seeing eye. Consequently, the worse seeing eye becomes a 'lazy eye' and cannot see well even after correcting the refractive error and/or the strabismus later in life.

Astigmatism produces two distinct focus lines with a blurred interval between them. Neither of these two focused lines provides good vision since the opposite-oriented line is highly unfocused at that plane. The region of unfocused light located between the two main focal lines is called Sturm's interval. Within Sturm's interval, there is a point in which both meridians are equally unfocused. This is called the circle of least confusion and can be used as a reference point of the clearest possible image that an astigmatic lens can produce (figure 2.3).

In ophthalmology, astigmatism is usually caused by the cornea, but it can also be originated by the lens or the retina. Clinical astigmatism can be classified between with-the-rule, against-the-rule and oblique. In with-the-rule astigmatism, the corneal vertical meridian is steeper and a horizontal line is focused at a shorter distance than the vertical line. The opposite happens in against-the-rule astigmatism, with the vertical line being focused closer than the horizontal one due to a steeper horizontal meridian. In oblique astigmatism, the steeper meridian is located either at or around 45° or 135°. Astigmatism can also be classified based on the location of the focused lines in relation to the retina. Simple myopic astigmatism occurs when a focused line falls before the retina while the other is on it. In compound myopic astigmatism, the two focused lines are produced before the retina. In simple hyperopic astigmatism, the first focused line falls on the retina while the other is located beyond it. In compound hyperopic astigmatism, both lines are located farther than the retina. And, finally, when the first line is located before the retina and the second after it, this is called mixed compound astigmatism. Being a cylindrical refractive error, astigmatism may be present together with any of the two former spherical error (myopia and hyperopia). In fact, sphero-cylindrical refractive errors are quite common.

The impact in vision caused by astigmatism varies according to the magnitude of the error as well as to the orientation of its meridian. Generally, with-the-rule astigmatism tends to affect less the uncorrected distance vision, thus, it tends to be better tolerated. However, small against-the-rule astigmatism may be beneficial for near vision in older, presbyopic patients, especially when using western alphabets

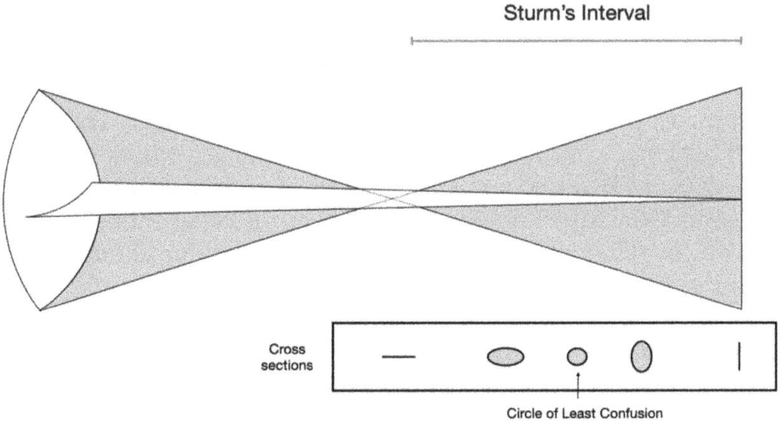

Figure 2.3. Sturm's interval.

due to the characteristics of these letters being more vertically oriented and the spaces in between them [15].

2.6 Visual corrections

2.6.1 Spectacles

The most common method of correcting refractive error is spectacles. It is a simple, non-invasive, and efficient way to correct sphero-cylindrical refractive errors, re-establishing sharp vision in the vast majority of people.

Positive and negative spherical lenses can be used to correct hyperopia and myopia, respectively, as well as cylindrical lenses for astigmatic errors. To correct the refractive error of an eye, the prescribed spectacle lens must yield the image of a distant object focused exactly on the retina.

It is important to note that for any spherical correcting lens system, the distance between the optical elements contribute to defining the refractive power of the system. In the spectacle/cornea/crystalline system, especially in the case of lenses above 5 D, the spectacle lens distance to the corneal apex (vertex distance), usually between 12 mm and 14 mm, must be taken into account when calculating the power prescription. By approaching negative lenses towards the cornea, the power of the diverging lens must be decreased in order to maintain the image on the retina. For positive lenses, the opposite occurs: their convergent power must be increased when they get closer to the cornea.

Correcting astigmatism with glasses requires a lens with different powers on its meridians. This may generate image distortion due to the different magnification of an image along distinct meridians (meridional aniseikonia). Adult individuals tend to tolerate less induced aniseikonias, while children adapt much more easily to cylindrical corrections.

2.6.2 Contact lenses

2.6.2.1 Soft

Soft contact lenses are the most used type in the world. The greatest positive aspect for these lenses is the patient comfort due to the thin edges provided by the material.

When evaluating the use of soft lenses, it is important to observe their centering and movement, which are minimal when compared to rigid lenses. Tight lenses move slightly with blinking, while loose lenses exhibit an exaggerated movement. For this type of lens, the same base curve can fit over various corneal curvatures. This way, it is rarely necessary to request different base curves for better fitting. Although they easily conform to different corneal curvatures, in irregular corneas they do not provide good vision and adaptation may not be possible.

Soft contact lenses are useful for correcting both spherical refractive errors (spherical lenses) and regular astigmatism (toric lenses). One third of contact-lens wearers have at least 1.00 D of astigmatism, requiring correction with toric lenses. Patients with astigmatism below about 1.00 D or 0.75 D may have only the spherical error corrected and still enjoy excellent vision quality.

2.6.2.2 Rigid

Rigid gas-permeable (RGP) lenses are classified as corneal, scleral, or corneal-scleral according to their support. Corneal RGP lenses have smaller chord diameters than soft lenses, allowing better circulation of tears under the lens during each blink. Parameters such as central curve, peripheral curve, and edge shape can be customized to provide comfort, tear exchange and stable vision.

As the RGP maintains its shape when placed on the cornea, it becomes the new refractive surface of the eye. The tear fills the space between the lens and the cornea constituting the tear meniscus. The difference between the curvature of the cornea and the base posterior curvature of the contact lens determines the power of the tear meniscus.

The most common fitting technique is to start with an RGP base curvature similar to the curvature of the flattest corneal meridian (K1). Using fluorescein eye drops to highlight the tear film, it is possible to study the relationship between the lens and the cornea. The fluorescein pattern is evaluated with a slit lamp (figure 2.4) that enables the observation of touching regions, in which there is no tear stained with fluorescein, or apical clearing, where there is fluorescein pooling. The upper third of the lens should ideally be under the upper eyelid, which allows the lens to move with each blink, increasing the tear change and decreasing foreign body sensation caused by the touch between the upper eyelid and lens edge.

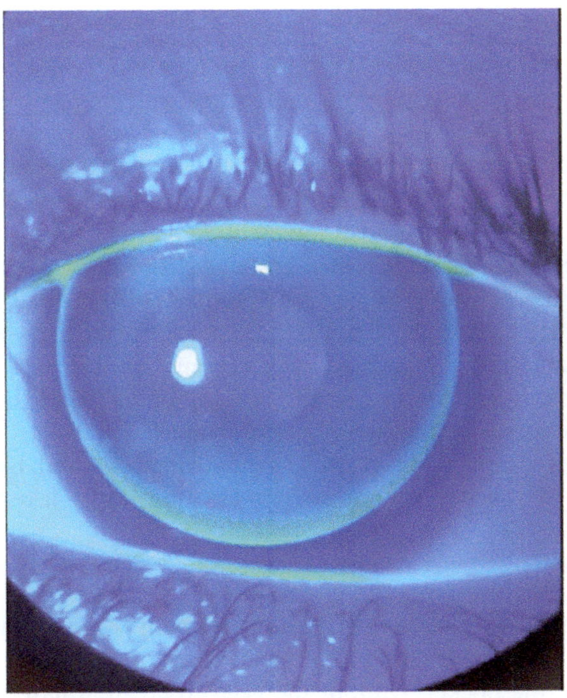

Figure 2.4. Ideal fluorescein pattern of a spherical RGP.

2.6.2.3 *Orthokeratology*

Orthokeratology refers to the overnight use of RGP contact lenses to mechanically reshape the cornea and temporarily reduce myopia. The goal is to flatten the central cornea at night with RGP so during the day, corrective lenses are unnecessary.

Temporary flattening of the cornea is a consequence of epithelial flattening. The myopia component can be reduced between −0.50 and −6.00 D, with up to 1.75 D of astigmatism.

According to the American Academy of Ophthalmology, approximately 75% experienced discomfort at some point and 30% of patients discontinued the treatment [1]. This treatment requires highly motivated patients, who are not candidates for surgical correction methods and who do not want to wear glasses or contact lenses during the day. Complications include induced astigmatism and higher-order aberrations, recurrent erosions, and infectious keratitis. The prevalence and incidence of complications is not well documented in the literature.

2.6.3 Laser vision correction—LVC

2.6.3.1 *LASIK*

LASIK means Laser *in situ* keratomileusis. The term keratomileusis comes from 'cornea' (kerato) and 'to carve' (mileusis). This technique combines two steps: creation of a stromal flap (by microkeratome or femtosecond laser) and excimer laser stromal ablation. Because the epithelial surface is preserved, this procedure promotes quick recovery of vision, and minimal patient discomfort [1].

2.6.3.2 *PRK*

Photorefractive keratectomy (PRK) is a surface ablation technique to sculpt the de-epithelialized corneal stroma altering the corneal refractive power. Compared to LASIK, PRK has a slower vision recovery and increased postoperative discomfort.

The adjunctive use of mitomycin C markedly decreased the major risk of PRK, the corneal haze. As a result, the indication of surface ablation has grown significantly. PRK has specific advantages in some indications, such as irregular or thin corneas, epithelial basement membrane dystrophy, previous corneal surgeries (such as penetrating keratoplasty or radial keratotomies), treatment of complications of LASIK flaps (like buttonholes). The treatment of high-level myopia, that could be contraindicated for LASIK, may become feasible with surface ablation. The reason for that is because in LASIK, in addition to the ablated tissue, the flap creation adds alteration in the corneal biomechanics.

PRK eliminates complications related to the stromal flap and, because the corneal nerves are not sectioned during the flap creation, it is less prone to induce postoperative dry eye.

2.6.3.3 *SMILE*

Small incision lenticle extraction (SMILE) is a refractive surgical technique that uses the femtosecond laser to create a stromal lenticle and a pocket. The lenticle is extracted through the pocket, without the need to create a flap. It is indicated for −1.00 D to −8.00 D of myopia with $\leqslant -0.50$ D of astigmatism.

SMILE results are nearly identical to those of femtosecond laser assisted LASIK. The disadvantage of SMILE is its slightly slower visual recovery, but at 6 months, no difference in vision is observed as compared to LASIK.

Low incidence of complications related to SMILE have been reported and most of them occurred early in the learning curve. Perioperative complications included epithelial abrasions (6.0% of eyes), difficult lenticule extraction (1.9%) and small tears at the incision (1.8%).

Due to the absence of a flap and the maintenance of anterior stroma lamellae, SMILE offers the relative preservation of biomechanical stability and is appropriate for patients involved in contact sports or high-risk professions. Other potential advantages are reduced aberrations and less induced dry eye.

2.6.4 Phakic lenses

Phakic intraocular lenses (pIOL) are lenses implanted into the eye for refractive error correction, without removing the eye natural lens. The phakic lens could be placed in the anterior chamber (angle or iris support) or in the posterior chamber.

Ideal candidates are those not suitable for laser vision correction such as ectatic corneas or ectasia susceptibility, as well as severe degrees of refractive error. Implantation of pIOLs withdraws the risks of corneal ectasia by excimer laser surgery and retinal detachment associated with refractive lens exchange.

Intraoperatively, the main concerns are inflammation, infection and bleeding secondary to surgical trauma. If pIOL disrupts the flow of the aqueous humor in the anterior segment of the eye, pupillary block may occur with consequent acute elevated intraocular pressure. This severe complication could be prevented with peripheral iridotomies. Chronic intraocular pressure elevation can also occur and produce permanent vision loss due to glaucoma. Another concern about pIOL is endothelial cell damage. If the pIOL is close to the cornea, a chronic endothelial cell loss can lead to corneal decompensation. Premature cataract formation is another potential complication.

2.6.5 Refractive lens exchange

Refractive lens exchange (RLE) is an exchange of the crystalline lens with an IOL for the primary purpose of refractive error correction. Myopia, hyperopia, astigmatism and presbyopia can be addressed with this procedure when other refractive methods are not adequate. Since the crystalline is replaced by the implanted lens, it is termed a pseudophakic intraocular lens.

Presbyopia is the main indication of RLE. It is considered that the crystalline lens, although without opacities, is a dysfunctional structure because it no longer has the accommodation mechanism preserved. As presbyopic patients are reaching the age when cataracts will start to appear, the exchange of the lens for an IOL is anticipated to correct the refractive error and reduce the dependence on other external correction methods, like glasses and contact lenses.

A great advantage of RLE is the expansion of the range of refractive surgery possibilities beyond what can be achieved with other available methods, without cornea alteration. Maintaining cornea structure may enhance the quality of vision, mainly for the treatment of high ametropias.

RLE carries risks and complications similar to those of routine cataract extraction with IOL implantation, although the younger the patient, the greater the odds for retinal detachment. The vitreous consistency and its greater adherence to the retina at a younger age justify the higher chances of retinal detachment. It is important to highlight that the current multifocal IOLs available may not meet the patients' high expectations in terms of quality of vision. These IOLs usually reduce contrast sensitivity compared to the monofocal versions. Advances have been made in enhanced monofocal and in Extended-Depth-of-Focus (EDoF) IOLs; the latter also sometimes referred to as extended-range-of-vision lenses.

Special attention should be paid to the RLE for myopic patients since the risk for retinal detachment rises with increased axial length. High myopia (axial length longer than 26 mm) is an independent risk factor for subsequent retinal detachment after lens extraction. A careful peripheral retinal evaluation is mandatory when considering this refractive procedure.

Candidates with hyperopia with a shallow anterior chamber due to a thickened crystalline lens or small anterior segment may have a secondary benefit from RLE. The surgery may reduce the risk of angle-closure glaucoma creating space in the anterior segment with the lens removal and smaller anteroposterior diameter IOL implantation. In highly hyperopic eyes (axial length shorter than 18 mm), nanophthalmos should be considered and a higher risk of uveal effusion syndrome and postoperative choroidal detachment exists.

Correction of astigmatism can also be targeted during RLE. There are toric IOLs available as well as multifocal toric IOLs. If a non-toric multifocal IOL is implanted and there is significant corneal astigmatism, the effect of multifocality can be disturbed and bring undesirable optical aberrations. Much of the dissatisfaction with multifocal lenses occurs when emmetropia is not achieved with the inserted lens. Refractive adjustments either with LVC or even an IOL exchange may be necessary depending on the patient complaints.

Presbyopia correction can be discussed together with refractive error correction (myopia, hyperopia and astigmatism) as well as being the unique objective of RLE. When planning to correct myopia, hyperopia and/or astigmatism with RLE, it is mandatory to preoperatively discuss how the correction for near vision will be done after surgery since the accommodative capacity will be lost after the lens removal. If a patient chooses to have both eyes implanted with pseudophakic monofocal IOLs designed for distant vision, they should be informed about the consequent dependency on reading glasses for functional near vision.

The identification of irregular astigmatism is essential for the realization of RLE. The procedure can still be performed in such cases, as in keratoconus patients, but they must understand that irregular astigmatisms and other optical aberrations may remain, biometric calculations are more inaccurate and that LVC enhancement may not be possible due to the altered corneal biomechanics.

Corneal and not the manifest astigmatism (total eye's astigmatism) should be taken into account to predict the patient's postoperative astigmatism, since lenticular astigmatism ends with lens extraction. Generally, for corneal astigmatism above 1.00 D it is interesting to choose a toric IOL, while minor astigmatism can be treated with corneal incision location or with limbal relaxing incisions and arcuate keratotomies creation by blades or femtosecond laser. LVC is also interesting to adjust remaining refractive errors. Although glasses and contact lenses can be prescribed for this purpose, the RLE patient profile generally rejects these correction methods.

Preoperative retinal OCT is suggested to identify macular pathologies that may not be easily seen in a biomicroscopy exam. This routine is especially useful when a premium IOL is indicated, since even minimal eye alterations can compromise the proper lens functioning. The study of the retinal periphery is also essential, especially in myopic patients for identification and treatment of retinal detachment predisposing lesions.

Given the high expectations of the patient, the biometry and IOL choice procedure are even more crucial than they are in a conventional cataract surgery to guarantee the postoperative high level of satisfaction. It should be noted that in extreme ametropias, both in myopia and hyperopia, as well as in irregular corneas, the biometric measurements are less accurate.

The use of the 'piggyback' IOL system, in which two posterior chamber IOLs are inserted, is an alternative when the necessary IOL power is not commercially available. Another use of 'piggyback' IOL configuration are those with important refractive surprises, in which high post-surgical refractive error exists. The implantation of an ancillary IOL in the ciliary sulcus can solve the problem, especially when LVC is not an option.

2.6.6 Other (drops, pinholes, etc)

The size of the blur circle on the retina generally increases according to the increase in pupillary size, mainly due to the increasing contribution of aberrations as light passes through wider sections of the optics. The use of a pinhole is useful to block the peripheral rays entering the eye—a phenomenon known as stenopeic effect[2]. As the central rays undergo less refraction, the pinhole effect works by decreasing the refractive error of the eye and improving visual acuity. The patient performs well in visual acuity tests even without refractive lenses use. In eyes with irregular astigmatisms, as in keratoconus cases, the function of the pinhole is even greater, as it filters more aberrant peripheral corneal regions, which contribute to visual distortion. Besides, a smaller aperture naturally yields an extended depth of focus, consequently, a pinhole implant allows sharp vision in a wider range of distances, even in situations where the accommodation has been lost, such as after lens extraction.

[2] The term pinhole in clinical ophthalmology means an aperture size of roughly 1 mm. This is a totally different scale than the pinholes used in optics and industry. In these settings, pinholes are usually required to filter a beam of light and/or cause light diffraction. In these cases, the aperture size is usually in the nanometer–micrometer range.

Pinholes can be used externally, as in spectacles, as well as surgically implanted. Intracorneal inlays with small apertures have been described for this purpose, as well as intraocular devices, with or without associated dioptric power. The disadvantage of pinholes is the lower amount of light entering the eye, which can bring a limitation in scotopic conditions.

The stenopeic effect can also be achieved pharmacologically. Eye drops of pilocarpine, a parasympathomimetic drug with direct cholinergic action on the muscarinic receptor, or even brimonidine tartrate, alpha-adrenergic agonist, promote pupillary contraction, reducing the light admittance to the eye.

Pilocarpine, although efficient in inducing miosis, has side effects that may make its long-term use impractical. Headache, visual impairment (myopia induction due to the anteriorization of the iridolenticular diaphragm), increased chance of retinal detachment (by vitreous base traction), increased salivation, reduced visual acuity under insufficient lighting, and the appearance of cataracts and irian cysts are some of its undesirable side effects.

Also, to decrease mesopic pupillary size, brimonidine can be prescribed to improve night vision or nocturnal discomforts induced by multifocal IOLs, such as halos and glares. These disturbances tend to ameliorate with neuroadaptation. In clinical practice, a very common side effect of brimonidine is allergy induction, being the main cause of drug induced follicular conjunctivitis. When this occurs, the use of the drug should be discontinued.

Chapter highlights

- The anatomy of the entire visual system was discussed in the context of vision formation.
- Refractive errors such as myopia, hyperopia and astigmatism are commonly found in the population.
- Many clinical treatments are available such as spectacles, contact lenses and eye drops.
- Intraocular lenses can be used to correct refractive errors in phakic eyes.

References

[1] American Academy of Ophthalmology 2018 *2018–2019 Basic and Clinical Science Course (BCSC)* (San Francisco, CA: American Academy of Ophthalmology)

[2] Dubbelman M, Van der Heijde G L and Weeber H A 2005 Change in shape of the aging human crystalline lens with accommodation *Vision Res.* **45** 117–32

[3] Hartridge H 1925 Helmholtz's theory of accommodation *Br. J. Ophthalmol.* **9** 521–3

[4] Schachar R A, Huang T and Huang X 1993 Mathematic proof of Schachar's hypothesis of accommodation *Ann. Ophthalmol.* **25** 5–9

[5] Glasser A and Kaufman P L 1999 The mechanism of accommodation in primates *Ophthalmology* **106** 863–72

[6] Campbell F W and Westheimer G 1960 Dynamics of accommodation responses of the human eye *J. Physiol.* **151** 285–95

[7] Suryakumar R and Allison R 2016 Accommodation and pupil responses to random-dot stereograms *J. Optom.* **9** 40–6
[8] Smelser G K 1965 Embryology and morphology of the lens *Invest. Ophthalmol.* **4** 398–410
[9] Fricke T R, Tahhan N and Resnikoff S *et al* 2018 Global prevalence of presbyopia and vision impairment from uncorrected presbyopia: systematic review, meta-analysis, and modelling *Ophthalmology* **125** 1492–9
[10] Bullimore M A and Richdale K 2020 Myopia control 2020: where are we and where are we heading? *Ophthalm. Physiol. Opt.* **40** 254–70
[11] Cooper J and Tkatchenko A V 2018 A review of current concepts of the etiology and treatment of myopia *Eye Contact Lens* **44** 231–47
[12] Zadnik K 2020 Myopia prevention: here comes the Sun *Ophthalmology* **127** 1470–1
[13] Wu P C, Chuang M N and Choi J *et al* 2019 Update in myopia and treatment strategy of atropine use in myopia control *Eye (Lond.)* **33** 3–13
[14] Torii H, Kurihara T and Seko Y *et al* 2017 Violet light exposure can be a preventive strategy against myopia progression *EBioMedicine* **15** 210–9
[15] Daiber H F and Gnugnoli D M 2020 *Visual Acuity* (Treasure Island, FL: StatPearls)
[16] von Noorden G K and Avilla C W 1990 Accommodative convergence in hypermetropia *Am. J. Ophthalmol.* **110** 287–92

Chapter 3

Optical parameters and charts

Bruno Lovaglio Cançado Trindade, Davies William de Lima Monteiro, Luiz Melk de Carvalho and Rodrigo de Abreu

To understand the optical features of an ophthalmic lens, there are a few concepts that need to be clear. In the context of this book, a lens is a transparent object in the visible spectrum that can modify the way light travels either by refraction, or diffraction, or by a combination of both. Many details are involved in this light-bending process, some of which are related to the nature of the incoming light while others relate to the lens itself. For vision, the overall intended result is the formation of an appropriate image projected on the retina. The quality and details of this image relate to how refraction or diffraction take place.

This chapter focuses on refractive lenses and will explain the mathematics behind them. It starts by highlighting different lens topologies and the variables related to each type of construction, how to calculate its refractive power, its magnification capability, and the proper positioning of the lens in an optical system. It then introduces methods to analyze the resulting image. Contrast analysis is explained using the modulation transfer function (MTF) and the point spread function (PSF). These are variables that will be referred to later in many chapters of this book, whose complete understanding is essential to the reader.

The ability of a lens to provide an extended range of focused light on the retina is important in ophthalmology because this can improve vision across different distances. This chapter will present how to measure the quality of the image at different focusing distances using the MTF through focus. It will also discuss depth of field of a lens and its importance in image formation.

Deviations of the light-ray paths from their intended trajectories may be independently or jointly caused by the topology of the lens surfaces; interfaces between media; and non-uniformity of the refractive index of the lens material, or of the propagation media. How light travels through a system can be seen as the propagation of a wavefront to which light rays are always perpendicular. The wavefront changes as it traverses a lens and any deviation of the rays can be depicted

doi:10.1088/978-0-7503-3263-7ch3

as a deformation of the transmitted wavefront. Although some of these deformations are intentionally introduced and beneficial, like when focusing light or extending the depth of focus, they are generally termed optical aberrations. The most commonly occurring aberrations will be discussed in this chapter.

Finally, simulations of image quality as well as psycho-physical clinical assessments can be performed using charts. The various types, differences and use of these charts will be introduced.

3.1 Lens formula and image formation

In geometrical optics, a light ray travels in a straight line with a constant speed in a homogeneous medium. The propagation speed depends on the refractive index n of the medium it is passing through, and is defined by equation (3.1) [1].

$$n = \frac{c}{v} \tag{3.1}$$

where c is the speed of the light in the vacuum (about 300,000 km s^{-1}) and v is the speed of the light in the medium, which depends on the light wavelength. Since v is always smaller or equal to c, for all practical materials, the refractive index is a real number equal to or bigger than 1.

When the light reaches an interface between materials with different refractive indices, it is refracted, as shown in figure 3.1. The angle θ_2 at which light leaves the interface depends on the incident angle θ_1 between the ray and the normal vector on the interception point and the difference of the refractive index between media. The light bending can be calculated using Snell's law:

$$\frac{\sin(\theta_1)}{\sin(\theta_2)} = \frac{v_1}{v_2} = \frac{n_2}{n_1} \tag{3.2}$$

where v_1 and v_2 are the speeds of light in the media 1 and 2, with refractive indices n_1 and n_2, respectively.

The simplest refractive lens is based on spherical surfaces, as shown in figure 3.2, for the case of a biconvex lens. Following the light ray direction, the first surface of the lens is referred to as anterior surface which has a radius of curvature R_1 and the second one, referred to as posterior surface, has the radius of curvature R_2. The line that passes through the center of both surfaces of the lens is the optical axis. In the

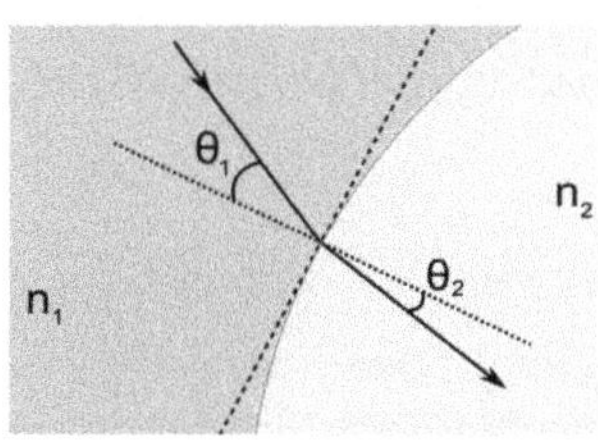

Figure 3.1. Optical refraction.

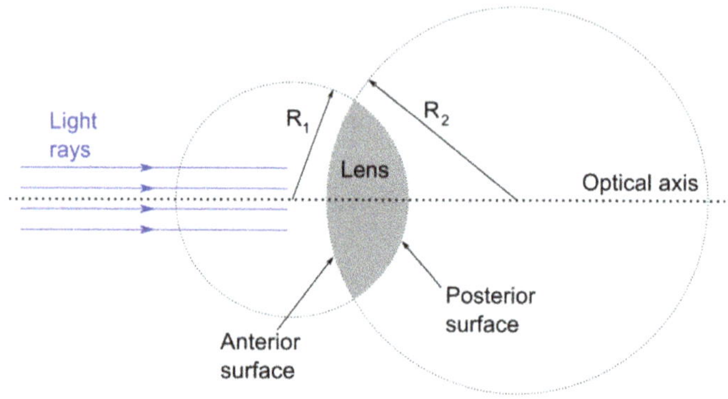

Figure 3.2. Spherical lens.

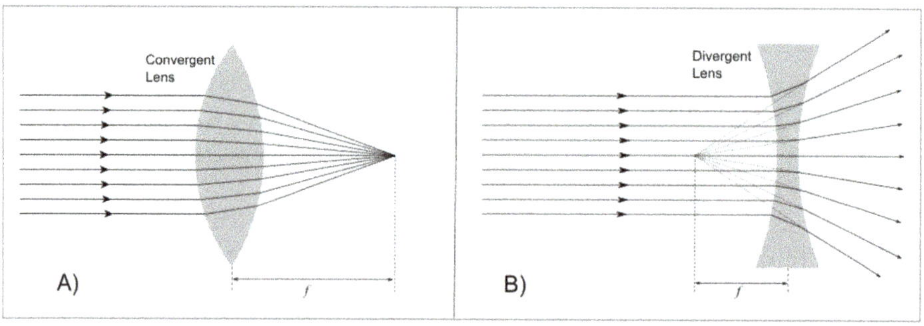

Figure 3.3. Illustrative example of: (a) a convergent and (b) a divergent lens.

lens shown in figure 3.2, by convention, the radius of curvature of the anterior surface is considered positive while that of the posterior surface is taken as negative.

A lens can be classified as convergent or divergent. In a convergent lens, a bunch of parallel rays tends to converge to a single focal point, that is located after the lens, at a focal length f, as depicted in figure 3.3(a), whereas, in contrast, divergent lenses tend to diverge the rays and the virtual focal point is located before the lens, as illustrated in figure 3.3(b).

Usually, the refractive index of the lens is higher than the refractive index of the medium where the lens is inserted. In this sense, the most common lens shapes can be classified as convergent (figure 3.4) or divergent (figure 3.5).

The extent to which light converges to a point (or diverges from a point) after traversing a lens is known as optical power, refractive power, or dioptric power, given in diopters and calculated by the inverse of the focal length, in meters.

$$P = \frac{1}{f} \tag{3.3}$$

For a convergent lens, the optical power is positive, since the focal length, formed after the lens, is also said to be positive while the optical power of a divergent lens is said to be negative. Equation (3.3) is useful if the focal length is known. The optical

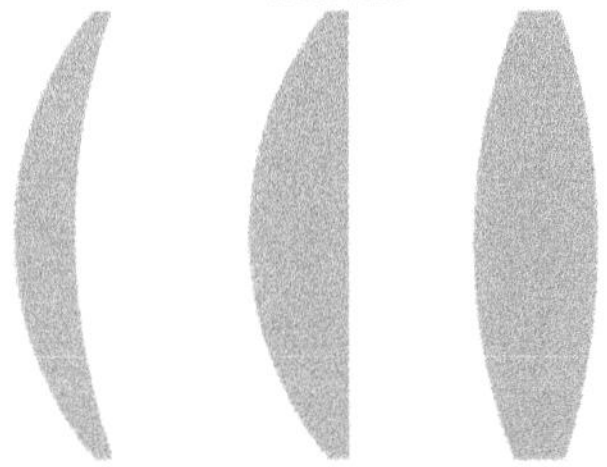

Figure 3.4. Convergent lens shapes.

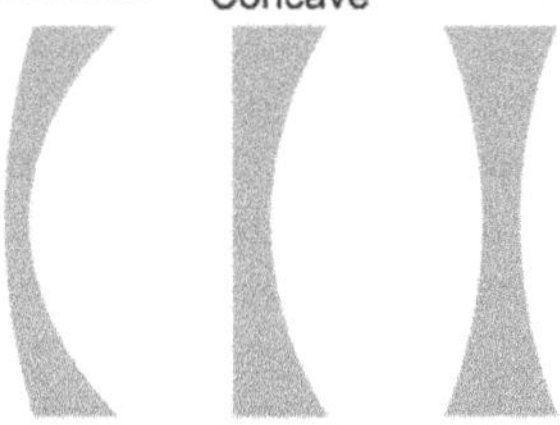

Figure 3.5. Divergent lens shapes.

power of the lens can also be calculated considering the curvature of the anterior and posterior surfaces of the lens and the refractive index of the lens n_L and of the medium n_M around it. If we consider a convergent biconvex lens with refractive index n_L, anterior radius R_1 and posterior radius R_2, immersed in a medium with refractive index n_M, the optical power of the lens P is given by the Lens Maker's formula [2]:

$$P = \frac{(n_L - n_M)}{R_1} + \frac{(n_L - n_M)}{R_2} - \frac{(n_L - n_M)(n_L - n_M). \quad t_L}{R_1 R_2 n_L} \tag{3.4}$$

where t_L is the lens center thickness. If the focal length is large enough such that $t_L \ll f$, equation (3.3) can be reduced to:

$$P = (n_L - n_M)\left(\frac{1}{R_1} + \frac{1}{R_2}\right) \tag{3.5}$$

that is known as the Lens Maker's formula version for thin lenses.

The focal length f, the object distance S_o and the image distance S_i are correlated using equation (3.6).

$$\frac{1}{f} = \frac{1}{S_i} + \frac{1}{S_o} \tag{3.6}$$

Figure 3.6 presents a lens of focal distance f, an object of height h_o at a distance S_o from the lens and the image of this object with height h_i at a distance S_i from the lens. Each point of the object is represented by a point on the image plane, as

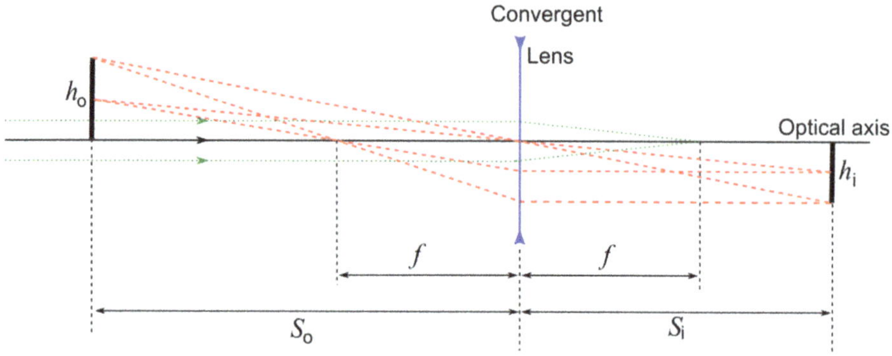

Figure 3.6. Image formation of an object with height h_o through a lens with focal length f.

projected by the dashed red lines. The larger the distance S_o, the smaller the image height h_i. Also, the distance S_i approaches the focal distance f as S_o increases. If the object distance S_o is considered as infinity, the image distance S_i is equal to the focal length f, as expected by equation (3.6).

Figure 3.6 shows that the image is formed upside down in respect to the object. In this example, the object is above the optical axis while the image is below the optical axis. The image height h_i increases as the object distance S_o gets closer to the focal distance f. The relation between the image and object heights is the transversal magnification M_T and is represented by:

$$| M_T | = \frac{h_i}{h_o} = \frac{S_i}{S_o} \tag{3.7}$$

3.2 Modulation transfer function—MTF

3.2.1 Definition

While the PSF, as will be explained in section 3.3, assesses the optical quality of a system in the spatial domain, the MTF is a metric to assess the contrast transfer quality of an optical system in the frequency domain [3, 4]. To interpret the MTF curve information it is necessary to firstly understand what the spatial frequency components of an object are. In figure 3.7(a) is presented a two-dimensional object in x and y coordinates and in figure 3.7(b) the corresponding normalized illuminance intensity of the cross-section (dashed line in figure 3.7(a)). From figure 3.7(b) it is possible to notice that the illumination cross-section presents an oscillating behavior (like a pulse train function) as a function of position (x-axis). The pulse cycles of the illuminance get narrower from left to right direction, which corresponds to a reduction in the period of each cycle. Taking the analogy with time-domain signals, it is possible to analyze the illuminance behavior in the frequency domain [3]. The oscillations in this case present an increase in frequency from left to right. Based on that, the object has spatial frequency components that are given by the inverse of the spatial-domain period [3].

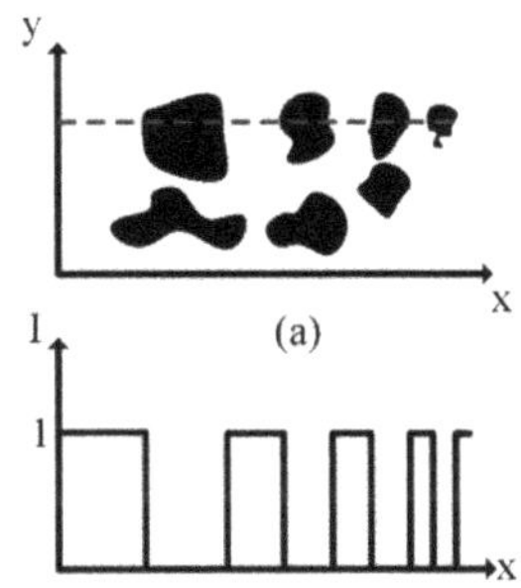

Figure 3.7. Spatial domain of an object: (a) two-dimensional object with a cross-section represented by the dashed line and (b) normalized illuminance distribution of the object cross-section as function of x position.

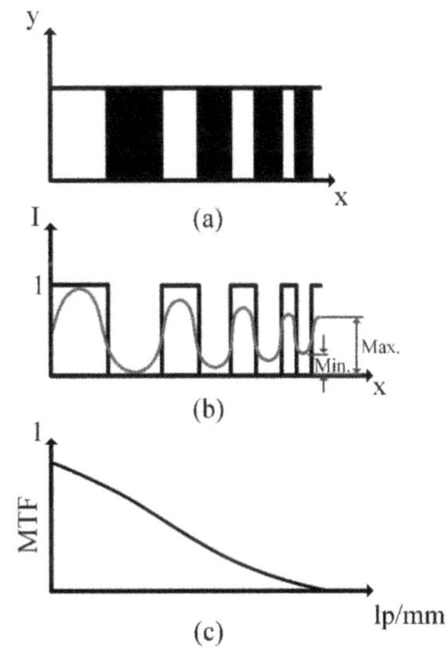

Figure 3.8. MTF representation: (a) an illuminated two-dimensional object formed by clear and opaque stripes with increasing spatial frequency; (b) light intensity right after the object, represented by a solid black curve, and the light intensity at the image plane, as represented by a gray curve; and (c) pictorial curve indicating the decreasing MTF values (modulation depth) for increasing spatial frequencies.

Consider an illuminated object formed by clear and opaque stripes with increasing spatial frequency from left to right (figure 3.8(a)) and imaged with a lens. When the sharp transition edges are projected on the image plane, they lose sharpness and contrast, as illustrated by the grey line representing the light intensity on the image plane (figure 3.8(b)) [4]. The contrast of the image expresses the illumination range between a brightest and an adjacent darkest part of the image. It can be described mathematically by a modulation depth, later called MTF, as in equation (3.8): [3, 4].

$$\mathrm{MTF} = \frac{I_{\mathrm{Max}} - I_{\mathrm{Min}}}{I_{\mathrm{Max}} + I_{\mathrm{Min}}} \tag{3.8}$$

where I_{Max} and I_{Min} are the brightest and darkest light intensities corresponding to each clear and opaque stripe, respectively.

Figure 3.8(b) shows that the image features a reduction of its modulation depth, i.e. of its contrast, when compared to that of the object, as the spatial frequency of its patterns increases. This occurs for any finite optical system and its effect depends on the aperture shape, size and on refractive or diffractive features across the aperture. More rigorously and general, the MTF is the modulus of the optical transfer function (OTF), which also contains information on phase change besides amplitude. Phase changes represent a shift in the patterns on the image plane with respect to those in the object. For now, only the change in intensity will be considered. The MTF curve represents the change in modulation depth (y-axis) on the image, as a function of the spatial frequency (x-axis), as caused by an optical system along one meridian (figure 3.8(c)) [3, 4]. When the lens is radially symmetric, the MTF at all meridians are the same. The spatial frequency is in general represented in terms of line pairs per millimeter or cycles per millimeter. When an optical system is diffraction-limited, the MTF curve is called diffraction-limited MTF and represents the best possible achievable performance of that optical system. The diffraction-limited MTF curve of an optical system is given by: [3, 4]

$$\text{MTF}(\xi) = \frac{2}{\pi}(\phi - \cos\phi \bullet \sin\phi) \tag{3.9}$$

where

$$\phi = (\lambda \bullet \xi \bullet f/\#) \tag{3.10}$$

and ξ is the spatial frequency in lp mm^{-1}, λ is the wavelength in units of millimeters and $f/\#$ is a symbol denoting the f-number of the system, which is inversely proportional to the pupil size and directly proportional to the focal length. The diffraction-limited MTF curve is affected by the wavelength of the light, the size of its clear aperture and its focal length. Other surface attributes, as asphericity or aberrations are not considered there.

The MTF curve of an optical system that is affected by aberrations, which means a non-diffraction-limited system, can be determined by the modulus of the Fourier transform applied to the bidimensional PSF of the system according to equation (3.11): [1]

$$\text{MTF}(\xi_x, \xi_y) = \left| \iint_\sigma h(x, y) e^{i(\xi_x \bullet x + \xi_y \bullet y)} dx dy \right| \tag{3.11}$$

where σ is the aperture area, ξ_x is the spatial frequency calculated from the x-axis, ξ_y is the spatial frequency calculated from the y-axis and $h(x,y)$ is the PSF of the system.

3.2.2 Importance and uses

The MTF curve is widely used to evaluate and compare the optical quality of optical systems in general because it is possible to evaluate the system performance at a range of spatial frequencies, and any detrimental factor affecting the optical performance such as ill positioning, aberrations and shape errors will impact the shape and amplitude of the MTF curve [3, 5, 6].

In ophthalmology, the MTF curve is commonly evaluated at spatial frequencies up to 100 lp mm^{-1}, which corresponds to approximately 30 cycles per degree (C/°), which is correlated to the visual acuity angle unit (1 min of arc) [7, 8]. In studies comparing different lens designs, the MTF curves are employed to identify any optical quality difference among them [9–11]. However, depending on the difference between curves, the disparity might not be significant enough to result in a noticeable impact on the visual acuity of the patient [7, 12]. This is the case for the human vision system because there is a variety of ocular and neural variables other than the individual performance of the lens that combined impact the visual acuity [12]. An alternative metric determined from the MTF curve, which has proven to be an effective parameter to correlate the optical quality of a lens with the visual acuity of the patient, is the area under the MTF curve [13, 14]. The area under the MTF (AMTF) curve is calculated within a range of spatial frequencies from 0 to 100 lp mm^{-1} and it makes the comparison process between the optical performance of different lenses simpler because the performance index is reduced to a single numeric value [7, 13, 14]. Comparative studies have shown that an optical quality difference of 15% or lower in the AMTF of two lenses is not sufficient to guarantee a noticeable difference in the visual acuity of the patient [7, 12]. On the other hand, a study comparing the performance of different multifocal intraocular lenses (IOLs) has shown that if the difference between the AMTF of the different designs reaches 25% or more, a significant distinction in visual acuity is observed [7].

Another benefit of analyzing the performance of an optical system in the frequency domain, i.e. through the MTF curve, relies on the possibility to find the overall MTF of an optical system composed by two or more optical elements by multiplying the MTF curves of each individual element of the system [3, 4]. In ophthalmology, this property can be used to determine the total neural contrast sensitivity function (CSF) of the eye by multiplying the MTF of the optical system of the eye by the neural CSF response, that considers the retina and brain perception (see equation (3.12)) [15]

$$CSF_{TOTAL} = MTF_{eye} \bullet CSF_{neural} \tag{3.12}$$

where CSF_{TOTAL} corresponds to the contrast sensitivity perception curve considering the optics of the eye, the retina and brain perception; the MTF_{eye} corresponds to the MTF curve of the optical system of the eye (cornea, crystalline lens or IOL and surrounding media) and CSF_{neural} corresponds to the retina and brain contrast sensitivity curve.

3.3 Point spread function—PSF

3.3.1 Definition

The PSF is the image of a point object as formed by an optical system [3]. Figure 3.9(a) depicts the image of a point source at infinity through a diffraction-limited lens [16]. To assess the PSF of this lens [1], as the light intensity of a point

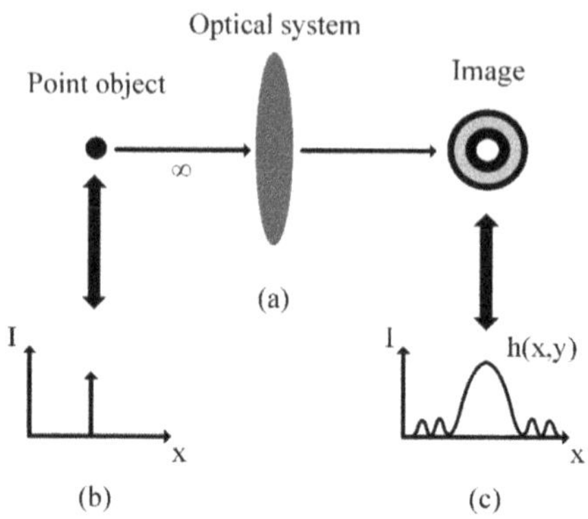

Figure 3.9. Top view of an optical system: (a) optical system forming the image (PSF) of a point object at infinity, (b) graphical distribution of the cross-section of the light intensity (y-axis) of the object as a function of position (x-axis) and (c) cross-section of the PSF amplitude (h(x,y)) as a function of image-plane position.

object has an intensity profile represented by an impulse function (figure 3.9(b)), the PSF corresponds to the impulse response ($h(x,y)$) of the optical system (figure 3.9(c)) and its dimensional interpretation is given by the intensity distribution as a function of the image-plane coordinates [3].

The PSF of a diffraction-limited optical system consists of a pattern that has a central bright disk (known as Airy disk) surrounded by concentric rings with decreasing intensity [3]. The PSF of a real system presents the pattern shown in figure 3.9(a) due to diffraction effects from the finite aperture of the system [3]. An optical system that has its performance limited only by its finite aperture is called a diffraction-limited system [4].

3.3.2 Importance and uses

The PSF is an important metric to evaluate the image quality of an optical system. As a point source can be considered as the smallest detailed information of an object, a PSF with a narrow Airy disk and high amplitude peak corresponds to a good representation of the point source, which means that the optical system yields an image with good resolution [3]. The resolution of an optical system is related to the smallest distance between two points that guarantees its distinction at the image plane [16]. Based on that, depending on how close two points are on the object plane, their images might not allow them to be resolved, i.e. to be identified as two separate points (figures 3.10(a) and (b)). This is due to a critical overlap of their PSFs [4]. Therefore, a limiting separation distance d (see figure 3.10(c)) between the centers of the Airy disks was proposed by Lord Rayleigh to indicate the minimum distance at which two points can still be distinguished. It is a parameter broadly employed to establish the resolution of an optical system, and is called the Rayleigh's criterion, given by:

$$d = 1.22 \bullet \lambda \bullet f/\# \tag{3.13}$$

where λ is the light wavelength and $f/\#$ is a symbol denoting the f-number of the optical system, which corresponds to the ratio between the focal length and the clear

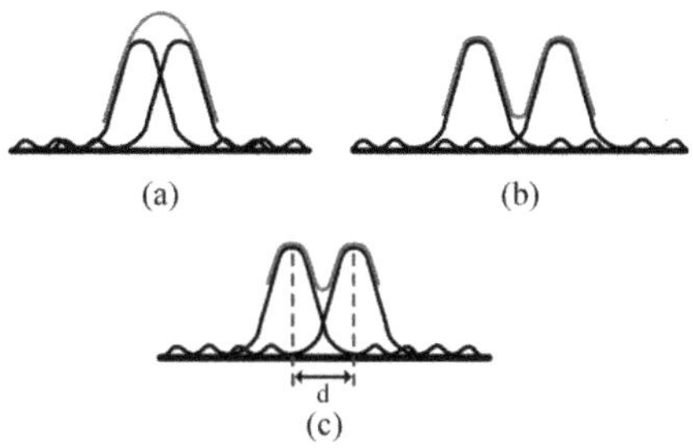

Figure 3.10. PSF of two points at different distances and the resultant intensity distribution: (a) the two points are unresolved, (b) the two points are clearly resolved and (c) the two points are at the resolution limit according to Rayleigh's criterion.

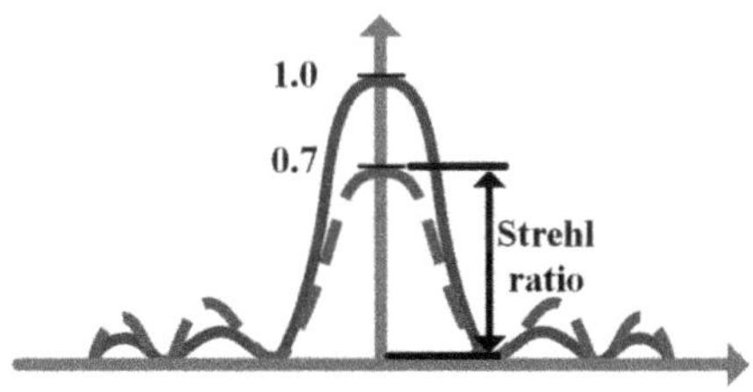

Figure 3.11. Strehl ratio of an optical system: normalized PSF of the system without aberrations (solid line) and the PSF of the system under the effect of arbitrary aberrations (dashed line).

aperture of the system. Other criteria such as Abbe and Sparrow admit an even further overlap and are also used in microscopy, photolithography and astronomy.

Another optical quality metric derived from the PSF information is the Strehl ratio, a single number given by the ratio of the peak of the PSF of an aberrated optical system to the peak of the PSF of the same system limited only by the diffraction (figure 3.11) [4, 16]. For small amounts of aberrations impacting the optical system it is possible to establish a mathematical correlation between the range of aberration amplitudes and the Strehl ratio [4, 17]. For instance, in an optical system with circular aperture, the range of the root mean square (RMS) aberration present in the system should be less than or equal to one seventh of the wavelength to guarantee a Strehl ratio of 0.8 [17].

By the PSF intensity and shape it is also possible to infer the presence of prominent primary aberrations in the optical system such as defocus, spherical aberration and coma [4]. Sometimes, especially for rotationally symmetric lenses, the line spread function (LSF) is used as a metric to qualify performance. There a thin line object is imaged through an optical system and the resulting line image is affected by the system and represents its fingerprint.

3.4 Depth of focus and depth of field

The image of an object through a lens is formed by the convolution of a perfect image of the object with the PSF of the optical system. In the human visual system, the PSF is affected by eventual spectacles or contact lens, the cornea, the crystalline lens and/or the intraocular lens, the aqueous and vitreous humors, and the pupil shape and size. The convolution process can be seen as every point of the object being replaced by the system PSF, whose size and intensity profile have a global impact on the projected image. Because the PSF is not infinitesimally thin, even when diffraction-limited, it always leads to some degree of blur in the image, even when in perfect focus. The PSF changes, usually becoming wider, for planes out of focus.

The depth of focus (DoF) is twice the maximum distance the image plane can be from the best-focus plane, along the lens optical axis, while still maintaining an acceptable image sharpness. It is twice the distance because both sides of focus are considered. At that distance the spot reaches its largest tolerated size for a given optical system [4], as highlighted in figure 3.12, where *EFFL* stands for the effective focal length of the lens in air. DoF is important in all sorts of optical imaging equipment where focus is to be preserved either when the longitudinal distance of the target object varies greatly or when the position of the focal plane is prone to misalignment or position inaccuracy. Examples are microscopes, cameras, endoscopes, telescopes and photolithographic projection systems in steppers and scanners. In a human eye with a healthy crystalline lens, the depth of focus is not a major concern, as accommodation adjusts the focal plane to the retina depending on the distance of the intended object to be viewed (chapters 4 and 13). However, a pseudophakic eye implanted with a monofocal lens can only benefit from sharply projected images on the retina of objects at different distances if their respective image planes lie within half the depth of focus to either side of the retina.

For an IOL designed for distant focus, a far-away object is projected on the retina. As the object moves closer to the eye, its best-image plane moves beyond the retina, causing a defocused image, which is only perceived sharp until a certain closer position to the eye, termed nearest field. The range of distances the object can be that correspond to the depth of focus is herein called depth of field (DOF, with a

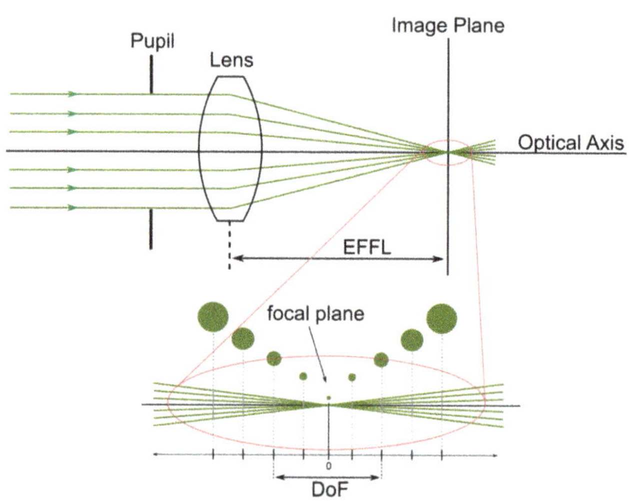

Figure 3.12. Variation on the spot size with the axial shift of the image plane. The focal spot increases as the system is defocused. The DoF is the range within which the spot size still yields to a sharp image.

capital ‘O’). If the DOF extends from the infinity to the nearest field, it is defined as the hyperfocal distance, which can be very useful in ophthalmic optics since it provides the highest DOF without compromising the sharpness of far objects.

Although DOF, and consequently DoF, can be subjective, depending on the individual’s perception of sharpness, for an ophthalmic lens designer it is convenient to have a mathematical definition. It is not a consensus in the literature which formula to use, and its description depends on the application. For optical projection printing equipment, it is common to define DoF as $k.\ NA^2$, where k is a process-dependent constant, λ is the wavelength and NA is the numerical aperture of the imaging lens [18]. In the design of ophthalmic instruments, on the other hand, the concept of circle of confusion is often used, which is the largest blur circle on the retina still perceived as a point rather than a circle and has a proposed diameter of 25.4 μm [19]. For a point object at infinity the depth of focus can be written as in equation (3.14)

$$\mathrm{DoF} = 2.\ c.\ \frac{f}{\#}, \tag{3.14}$$

where c is the diameter of the circle of confusion and $f/\#$ is the F-number of the eye compound lens system, defined as the ratio of its effective focal length in the vitreous humor and the lens aperture diameter. Equation (3.14) is a simplified version of that presented by Lambrecht *et al* [20], for object distances far greater than image distances. It shows that the smaller the aperture diameter, the larger the DoF. For completion, if close object distances are considered, then equation (3.14) must be multiplied by $\left(1 + \frac{s_i}{s_o}\right)$, where s_i/s_o is the transverse image magnification, where s_i is the position of the best image and s_o the object position. For most practical visual conditions $s_i \ll s_o$. Regarding more specifically IOL design, the DoF is often regarded as the largest span of the image plane shift for which the MTF remains above a certain minimum threshold value at a spatial frequency of $\xi = 50$ lp mm^{-1} with respect to the best-focus position. The minimum threshold can be as low as 0.05 to still allow one to distinguish object features with that spatial frequency on the retina. The distance from the image focal plane to the best-focus plane, i.e. the retina, is called focus shift (d_i'), which is further explored in the next section and in chapter 11.

It is worth noting that in most cases a smaller pupil leads to a more extended depth of focus. Well illuminated environments cause the pupil to constrict and naturally offer sharper object images throughout a wider range. When driving during a rather bright rainy day, although the crystalline lens is intentionally set to view road elements at far and intermediate distances, the smaller pupil size will cause the water dotted windshield to be simultaneously in focus. Wearing sunglasses in those conditions causes the pupil to be larger, reducing the depth of focus, therefore avoiding that the sharp image of droplets on the windshield is so clear in the observed scene.

3.5 MTF through focus

A useful graph to evaluate the DoF of an individual lens or a lens system is the MTF through focus (MTF-TF or TFM). It plots the MTF values at a specific spatial frequency for a range of focus shifts. Figure 3.13 illustrates this. The focal position axis can be represented in millimeters or diopters. In ophthalmic application, the TFM is frequently reported at the frequency of 50 lp mm^{-1}, even though 25 lp mm^{-1} and 100 lp mm^{-1} are also commonly used.

Considering the graph of figure 3.13 corresponds to the performance of a pseudophakic eye model with the IOL designed for best focus of an object at infinity and that the eye is corrected to emmetropia, the highest MTF value at 50 lp mm^{-1}, indicating the best image contrast and indicated by the black dot, occurs at the best-focus plane, at 0 mm. If the object moves closer to the eye, the whole curve shifts to the right (blue dashed line), meaning that the plane of best focus moves beyond the retina, leaving an image with lower MTF at the retina, as indicated by the red dot. We notice, therefore, that it is only necessary to know the left side of the blue curve to predict what happens to the MTF as the object moves closer. This, in practice, suffices since the object cannot move further than infinity. If the MTF threshold is that indicated by the horizontal green dashed line, the depth of focus of the IOL in this eye corresponds to the focus shift at which the curve first crosses the threshold (green dot), coming from 0 mm. The presence of two peaks on the TFM separated by a region below the threshold usually means an additional focus rather than an extended focus region.

The MTF-TF graph can result either from simulations in optical design software or from experimental test benches and dedicated equipment. It is important to report under which conditions the plot was obtained: spatial frequency, wavelength, pupil diameter and

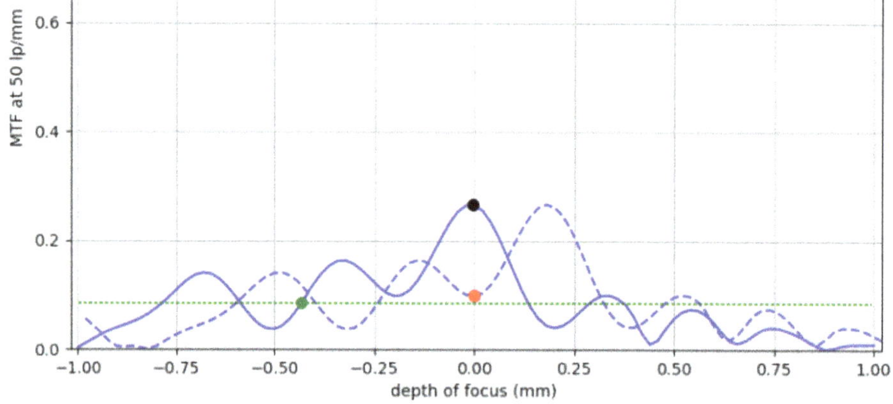

Figure 3.13. Example of an MTF-TF curve of the pseudophackic eye optics at a spatial frequency of 50 lp mm^{-1} for an object at infinity (continuous blue line) and closer to the eye (dashed blue line).

object distance. In simulations it is necessary to choose data sampling high enough to avoid aliasing, which could lead to incorrectly broader peaks, erroneously suggesting a higher DoF than the actual one. The presence of significantly high secondary peaks suggests intermediate or near foci. The farther the focus shift at which the secondary peak is, the closer its respective foci. This analysis can, however, be misleading as it represents the lens behavior at a single spatial frequency, and the curve shape can change significantly at different spatial frequencies.

3.6 Aberrations

3.6.1 Definition

Optical systems like the human eye have imperfections and limitations that cause images to be formed with aberrations. These are disturbances to the wavefront of the propagated light that can be produced by different parts of a system it traverses, such as the lenses and the media along the path. A wavefront can be interpreted as a surface of the light beam propagating through an optical element, in a similar fashion to that of a wave propagating outwards on the surface of lake when a pebble is cast. For a single plano-convex lens, considering that the amount of aberration introduced by the propagating media is negligible, either inside or outside the lens, the wavefront shape emerging from the posterior surface mostly mimics the topology of curved lens surface, with its shape shrunk or expanded depending on the differences in the refractive indices of the lens and the surrounding media. If there are topological artefacts on either surface, as a bump, this will also be imprinted on the wavefront. Because the refractive index changes for different wavelengths (colors), aberrations also change. Monochromatic aberrations will be those described for a specific color, whereas polychromatic aberrations will be those that result for a combination of colors, as for white light. In particular, longitudinal and transverse chromatic aberrations refer to the effects of different focal distances along the optical axis; and to the shift and difference in magnification at the focal plane, respectively.

Again, bridging to the analogy of a pebble hitting the water in a lake, a wavefront farther from the point of contact exhibits a longer radius of curvature, therefore a less curved shape. At a very long distance, a given section of the wavefront is basically flat. Likewise, the wavefront from a distant point light source, e.g. star, travels through the vacuum of space and becomes actually flat when it reaches the outer atmosphere of our planet. When it passes through a clear patch of the atmosphere, because of the differences of the refractive index in the atmosphere, caused by thermal fluctuations, the wavefront is modified. If instead of going through the atmosphere it hits a perfect aspheric lens, it would be modified from flat to concave, converging towards the lens focal point, resulting in a focal spot whose profile is the lens PSF.

Wavefronts can be mathematically described in multiple ways. The Zernike polynomials are one of the most used representations in ophthalmology and optical engineering. They are a set of infinite orthogonal mathematical functions (polynomial terms Z_n^m) that describe surface topologies defined within a unit circle, which is convenient in ophthalmic optics. Each function is unique and can be added together with others in a weighted sum to describe the most complex and rotationally asymmetric aberrations. Likewise, a complex aberration can be decomposed into various Zernike terms, so one knows the contribution of each of them separately. Zernike polynomials are classically divided into two different groups: low-order aberrations (zeroth, first and second orders), LOA, which include defocus ($Z^0{}_2$) and astigmatism ($Z^{-2}{}_2$ and $Z^2{}_2$) and the high-order aberrations (from third order onwards), HOA, in which the most important terms in ophthalmology are spherical aberration ($Z^0{}_4$), coma ($Z^{-1}{}_3$ and $Z^1{}_3$) and trefoil ($Z^{-3}{}_3$ and $Z^3{}_3$) (figure 3.14).

As regards to polychromatic aberrations, a beam will refract differently across the spectrum. This is called longitudinal chromatic aberration, as previously indicated. The amount of variation in the refractive index across different wavelengths of visible light is quantified by the Abbe number (V) with the following equation

$$V_D = \frac{n_D - 1}{n_f - n_c} \tag{3.15}$$

where V_D is the Abbe number of a given material, n_C, n_D and n_F are the refractive indices of the material at the wavelengths of 656.3 nm, 589.3 nm, and 486.1 nm, respectively.

The larger the variation ($n_f - n_c$), the more dispersive the material is to visible light, therefore, the lower its Abbe number. Conversely, a high Abbe number material is capable of refracting polychromatic visible light into a less dispersed image point yielding fewer polychromatic aberrations. For the same difference

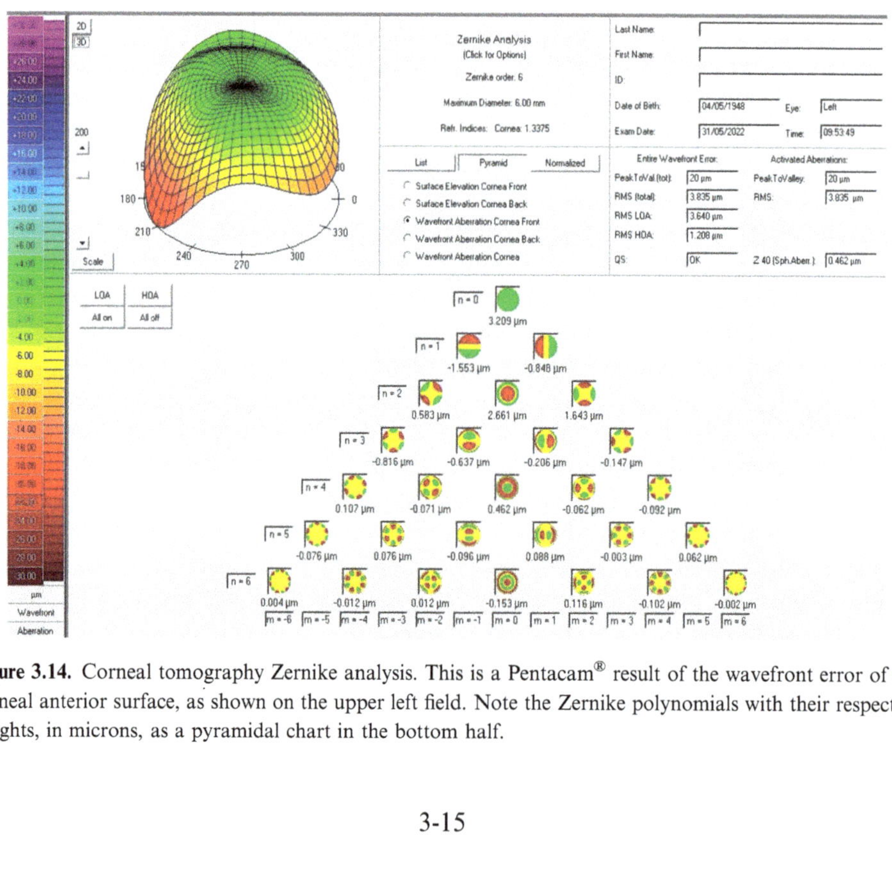

Figure 3.14. Corneal tomography Zernike analysis. This is a Pentacam® result of the wavefront error of the corneal anterior surface, as shown on the upper left field. Note the Zernike polynomials with their respective weights, in microns, as a pyramidal chart in the bottom half.

between indices for edge wavelengths, low refractive index materials feature lower Abbe numbers.

In refractive lenses blue light converges more than red light, which results in different focal distances for different colors. Lens materials with high Abbe numbers minimize this effect.

In flat diffractive lenses, due to diffraction effects the opposite occurs, i.e. red light converges more than blue light. In ophthalmic diffractive lenses, a biconvex refractive lens is usually used as the basis for a diffractive topology on one or both surfaces. Therefore, if properly designed, chromatic aberration can be minimized by using the diffractive chromatic effect to counter the refractive chromatic one. Controlling chromatic aberrations is crucial to allow a better quality image to be formed and achromatization of the diffractive IOL has been shown to provide improvement in image quality [21].

3.6.2 Importance

Aberrations degrade image quality. To produce the clearest and higher quality image, monochromatic (and polychromatic) aberrations must be overall decreased. In ophthalmology, the lower the order of the Zernike polynomial, the higher the impact on image quality an aberration produces. For instance, defocus (second order) tends to degrade the image much more importantly than secondary spherical aberration (sixth order). This is why aberrometry, a test in which high-order aberrations are quantified, requires a patient to be first fully corrected for his/her defocus/astigmatism.

3.6.2.1 Spherical aberration

A spherical lens has a constant radius of curvature across its surface. This causes light to be refracted more at its periphery than at its center. This way, light is not focused into a single point, but rather into a region of space located at the lens primary focal point. The distance through which light is being focused at is a measure of the longitudinal spherical aberration and it increases the more aberrated the lens/optical system is (figure 3.15). The higher the refractive power of the lens and the larger the pupil size, the more spherical aberration it introduces.

Spherical aberration is one of the main causes of image quality degradation in ophthalmology. IOLs are manufactured within a range of refractive powers, usually

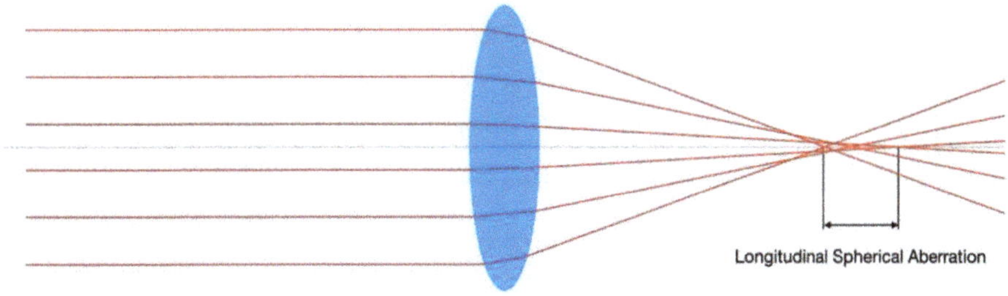

Figure 3.15. Longitudinal spherical aberration. Note the length of the region considered within acceptable focus. Aberrations are exaggerated to better illustrate.

between 10 and +30D. These lenses may have enough spherical aberration to limit patients' ability to see well, especially in mesopic conditions in which the pupil size tends to increase. Spherical IOLs are known to decrease contrast sensitivity in patients and may limit their ability to drive at night. See more detail in spherical and aspheric IOL in chapter 4.

3.6.2.2 Coma

Coma is a rotationally asymmetric aberration that causes light from a point source to be imaged as a comet-like spot, hence its name. The impact of this aberration is also proportional to the total refractive power of the optical system as well as to its aperture. It is mainly caused by misaligned or tilted lenses. In a human eye, the cornea, the pupil, the lens and the retina are not perfectly aligned. These structures have a variable degree of decentration and tilt relative to one another. The visual axis (line that connects an infinitely distant object to the fovea) and the pupillary axis (the line that passes through the center of the pupil perpendicular to the corneal surface) form an important angle between them (angle kappa) (figure 3.16). This angle is usually measured in some preoperative test and it is commonly more significant in short hyperopic eyes.

These 'imperfections' in the human eye limit image resolution to different degrees depending on the amount of coma it produces. The values of existing coma in an eye might influence the choice of IOL for a specific person. The more aberrated the eye, the less likely it is to perform well with advanced technology lenses such as the diffractive optics multifocal IOLs. This way, measuring anterior corneal surface as well as the total internal coma may be an important step in the preoperative evaluation of cataract surgery.

3.7 Visual acuity

Visual acuity is a subjective parameter that depends on neural processes that dictate how an individual perceives an object pattern or feature projected on their retina. Optical effects due to illumination level, light spectrum, diffraction, reflection, absorption, filtering, aberrations and scattering play a significant role in rendering an acceptable image of an object. These effects relate to the resolution the eye optics can have. On top of that, there is also the sensitivity, size, density, and dynamic range of the photoreceptors on the retina. An assessment of the visual acuity often boils down to evaluating contrast, more specifically, how a pattern contrast, or modulation, is transferred through the optics to its image, and

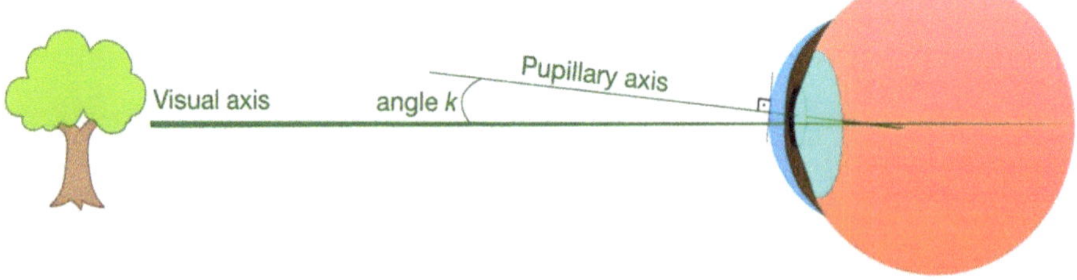

Figure 3.16. Angle kappa between the visual and pupillary axes.

then to its perceived image through the mosaic of retinal photoreceptors and the effects of neuroadaptation process.

Several patterns have been designed as targets for the visual acuity clinical test [22]. The Snellen chart is among the most used and presents several lines with capital letters with the height of the letters in subsequent lines reduced by a geometric ratio of 1.2589 [23]. The standard chart is attached to a wall 6 m (20 feet) away from the observer, who is asked to read letters on a given line. The average emmetrope will be able to read clearly a certain line labeled 20/20. Those who, at the same distance, can only read lines above the 20/20 line, i.e. with larger letters, are considered to have a lower visual acuity, whose second number indicates at which distance, in feet, an emmetrope would be able to read the same line. For instance, if one scores 20/32, it means that an emmetrope would be able to read that same line at 32 ft. The letter sizes progress linearly, according to the geometrical relation to the distance, both for lines above and below the 20/20 line. Also, an individual with 20/16 vision, can read at 20 ft what an average emmetrope would only succeed in doing at 16 ft. Many other patterns have been used to assess visual acuity, including pictures, geometrical shapes, sinusoidal grayscale images and sequences of alternating bars in different directions. The Pelli–Robson chart is a special alternative to Snellen with capital letters of decreasing contrast against a black background. It is designed to assess the contrast sensitivity for different object contrasts.

While Pelli–Robson and Snellen charts are used to assess the visual acuity for far objects, the Jaeger chart is a method commonly used to evaluate near vision. The distance the Jaeger chart should be placed is standardized as 14 inches (~36 cm) from the eye, with the J1 line of text corresponding to the Snellen visual acuity of 20/20 [22]. Even though there is a standard for distance from the eye and the letter sizes for the Jaeger chart, there are different formats of this diagram. Different versions choose to employ capital letters or upper and lower letter types, with different sizes for each acuity, and recommend at which distance it should be viewed, usually varying between 30 cm and 40 cm from the eye. Also, the scores run from J1 to JX, where X is usually 6 to 10, with increasing font sizes. Some lens manufacturers provide their own diagrams and sometimes label as J1+ the finest print.

The visual acuity can be assessed using periodic patters with different spatial frequencies. Figure 3.17 presents an optotype frequently used in clinical tests. It is similar to the capital letter ‘E’, but it represents an object composed of three black horizontal stripes and two horizontal white stripes, connected by a vertical black bar. The spatial frequency of this optotype depends on the angle of vision and it is

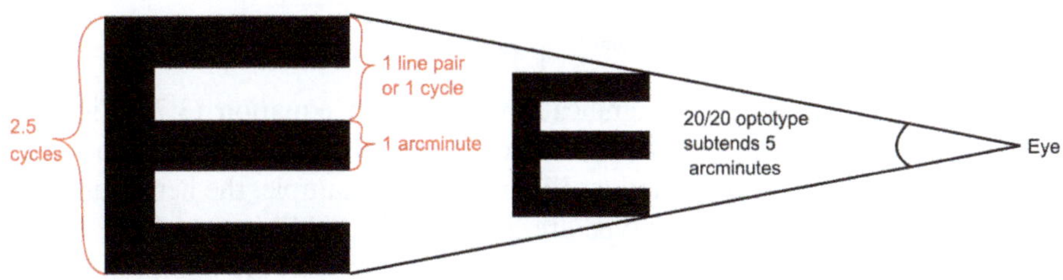

Figure 3.17. Optotype defined with three black and two white horizontal stripes connected by a vertical black stripe for a Snellen visual acuity of 20/20.

usually expressed in cycles per degrees (C/°) or 'black-and-white' line pairs per degree (lp/°). The size of the optotype depends on the visual angle it comprises and the distance from the eye.

For a Snellen visual acuity of 20/20, each stripe of the optotype, shown in figure 3.17, represents 1 arcmin of visual angle and, therefore, the optotype comprises 5 arcmin since it possesses 5 stripes or 2.5 cycles. Then, the visual acuity of the object in cycles/degree, $\xi^{o}_{c/\circ}$, to the respective Snellen visual acuity of 20/X is given by the equation (3.16).

$$\xi^{o}_{c/\circ} = \frac{600}{X} \tag{3.16}$$

where X is the denominator of the Snellen fraction. For example, an optotype for the Snellen visual acuity of 20/20 has a spatial frequency of 600/20 = 30 cycles/degree, while one for 20/40 has 15 cycles/degree. It means that the optotype with 2.5 cycles, as presented in figure 3.17, to represent a 20/40 must have twice the size of one that represents a 20/20 Snellen visual acuity.

The spatial frequency of the object can also be represented in cycles per millimeter (C mm^{-1}) or line pairs per millimeter (lp mm^{-1}). The conversion of the visual acuity of the object from C/° to C mm^{-1} is given by the equation (3.17).

$$\xi^{o}_{lp/mm} = \xi^{o}_{c/\circ} \bullet \frac{180}{\pi \bullet S_o} \tag{3.17}$$

where S_o is the distance from the object to the eye. For example, the spatial frequency of a Snellen visual acuity of 20/20 is 30 C/°, or 0.286 lp mm^{-1} for an object at 6000 mm of distance from the eye. The height of the optotype can be calculated using equation (3.18).

$$h_{obj} = \frac{2.5}{\xi^{o}_{lp/mm}} \tag{3.18}$$

For example, the height of the optotype for a Snellen 20/20 at 6000 mm of distance is 8.727 mm. If the visual acuity test is performed at 4 mm of distance, the size of the optotype of the Snellen 20/20 must be adjusted to 5.818 mm of height.

The spatial frequency can also be defined on the image plane and it is commonly reported in lp mm^{-1}, as in the MTF graph presented in figure 3.8. Assuming that for small angles, their tangent is approximately equal to the angle, in radians, the spatial frequency in the image plane, $\xi^{I}_{lp/mm}$ can be calculated using equation (3.19).

$$\xi^{I}_{lp/mm} = \frac{\xi^{o}_{lp/mm}}{|M_T|} \tag{3.19}$$

where (M_T) is the transversal magnification defined in equation (3.7). Figure 3.18 depicts an example of a 20/20 Snellen visual-acuity optotype placed at 6 m from the eye. The *EFFL* of the eye was considered as 17 mm. In this example, the height and spatial frequency of the image of the optotype are, respectively, 0.247 mm and 101 lp mm^{-1}.

Another scale to report the visual acuity is through the minimal angle of resolution (MAR), which corresponds to the minimal visual angle (α), in arcminute, from which two points can be distinguished from each other, meaning that the

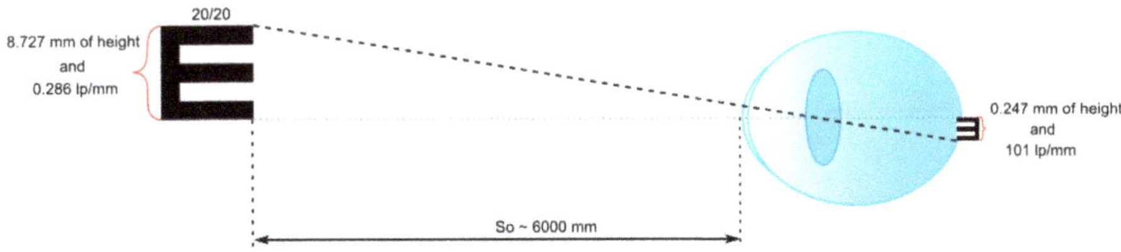

Figure 3.18. Illustrative example of the size and spatial frequency (lp mm^{-1}) of a 20/20 Snellen visual acuity optotype at 6 m of distance to the eye.

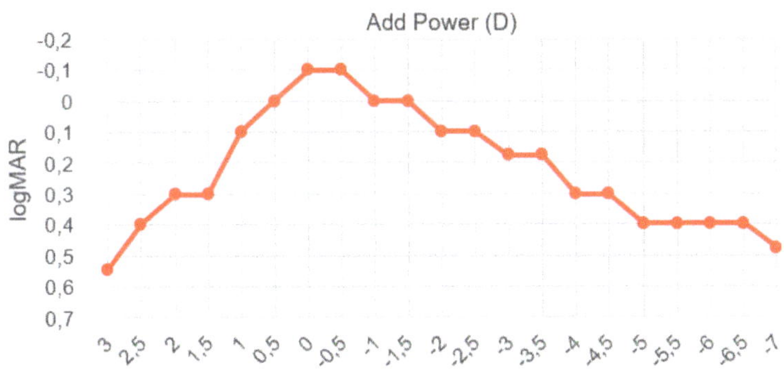

Figure 3.19. Example of a BCDVA defocus curve.

smaller the MAR, the better is the visual acuity. The visual acuity can be reported using a decimal scale, that is defined as the inverse of the MAR. The Snellen fraction is directly associated to the decimal scale. For example, a Snellen fraction of 20/40 is 0.5, which is this visual acuity in the decimal scale, which matches with 2′ of MAR. Even though the MAR is a very import visual acuity scale, its logarithm logMAR (logarithm of the minimal angle of resolution) is more commonly used.

The logMAR scale is frequently used in the clinical defocus curve, or blur curve, which is obtained in a clinical setting using a series of lenses with positive and negative optical power placed, one by one, in front of the eye, changing the focal length of the optical system (eye + lens) to simulate a variation of the object position, while the target optotype is kept at constant size and distance on a distant wall. For each added power, the visual acuity is recorded and plotted in a graph according to the added optical power of the lens used, as shown in figure 3.19. A more negative added power represents an object closer to the eye. This test can be done for both monocular and binocular vision. The logMAR graph can be reported as best corrected distance visual acuity (BCDVA), which means that the 0 D on the graph is associated with the lens power which leads to the best visual acuity. The UCDVA (uncorrected distance visual acuity) graph, on the other hand, evaluates the visual capacity without an offset correcting lens, and the occurrence of best visual acuity can be shifted from 0 D.

Alarcon *et al* [24] showed that a pre-clinical monocular defocus curve can be calculated using the area under the MTF curves up to the spatial frequency of 50 lp mm^{-1}, for different object distances, as shown in figure 3.20. It is an important

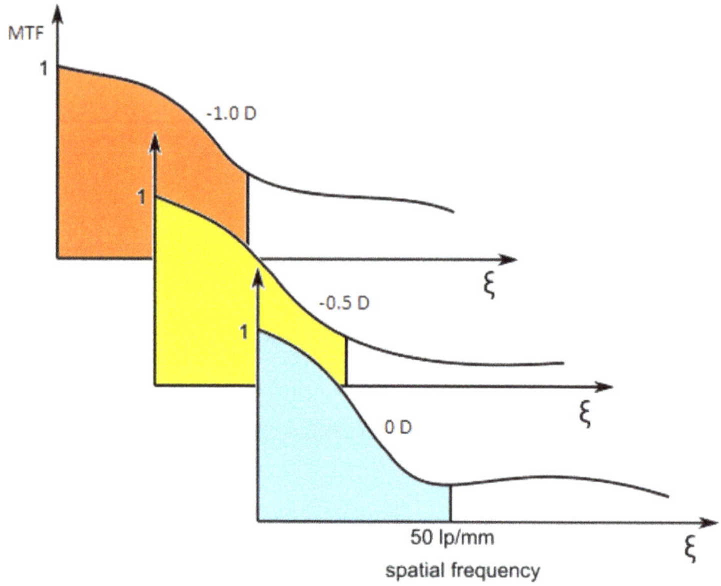

Figure 3.20. The area under the MTF curve related to the visual acuity for different additional powers.

Table 3.1. Relations between different visual acuity scales.

LogMAR	MAR (arcmin)	Decimal	Snellen (feet 20/X)	Spatial Frequency (C/°)
−0.3	0.50	2.00	20/10	60
−0.2	0.63	1.58	20/13	48
−0.1	0.79	1.26	20/16	38
0	1.00	1.00	20/20	30
0.1	1.26	0.79	20/25	24
0.2	1.58	0.63	20/32	19
0.3	2.00	0.50	20/40	15
0.4	2.51	0.40	20/50	12
0.5	3.16	0.32	20/63	9
0.6	3.98	0.25	20/80	8
0.7	5.01	0.20	20/100	6
0.8	6.31	0.16	20/126	5
0.9	7.94	0.13	20/159	4
1	10.00	0.10	20/200	3

finding, considering the defocus curve is a much more robust element to evaluate focal performance, therefore multifocality and extended depth of focus, than MTF-TF curves, which are limited to the performance at specific spatial frequencies and do not consider the eye contrast sensitivity or psycho-physical aspects of vision.

The relations between the visual acuity scales is presented in table 3.1.

Chapter highlights

- The modulation transfer function (MTF) is a very important figure of merit to assess the optical quality of a system, but each curve only provides image contrast information for objects at a single distance from the eye. It is intimately related to the point spread function (PSF) of the optical system.
- MTF through focus (MTF-TF) gives information about the expected image contrast throughout a range of distances around the main focal plane (e.g. retina), but is limited to a single spatial frequency and must be used carefully.
- Minimizing the lower order aberrations of the eye can improve the visual acuity since they tend to have higher impact on image degradation, but a certain amount of some aberrations, as spherical aberration, can benefit an extended depth of focus (DoF).
- The diversity of available visual-acuity charts allows the assessment of quality of vision for different distances, pupil sizes and contrasts.
- The psycho-physical defocus curve is a very important element in clinical analysis since it evaluates visual acuity for objects at different distances, considering both the overall image quality on the retina and neural processes. Pre-clinical defocus curves can be computed from a set of MTF curves.

References

[1] Malacara D and Malacara Z 2013 *Handbook of Optical Design* 3rd edn (Boca Raton, FL: CRC Press, Taylor and Francis)

[2] Trindade B L, Amaral F T and de Lima Monteiro D W 2016 Analysis of the optical quality of spherical and aspheric intraocular lenses in simulated nanophthalmic eyes *J. Refract. Surg.* **32** 193–200

[3] Boreman G D 2021 *Modulation Transfer Function in Optical and Electro-Optical Systems* Second edition (Bellingham, WA: SPIE Press)

[4] Smith W J 2008 *Modern Optical Engineering: The Design of Optical Systems* 4th edn (New York: McGraw-Hill)

[5] Goodman J W 2005 *Introduction to Fourier optics* 3rd edn (Englewood, CO: Roberts & Co.)

[6] Eppig T, Scholz K and Loffler A *et al* 2009 Effect of decentration and tilt on the image quality of aspheric intraocular lens designs in a model eye *J. Cataract Refract. Surg.* **35** 1091–100

[7] Felipe A, Pastor F and Artigas J M *et al* 2010 Correlation between optics quality of multifocal intraocular lenses and visual acuity: tolerance to modulation transfer function decay *J. Cataract Refract. Surg.* **36** 557–62

[8] Holladay J T 1997 Proper method for calculating average visual acuity *J. Refract. Surg.* **13** 388–91

[9] Dominguez-Vicent A, Esteve-Taboada J J and Del Aguila-Carrasco A J *et al* 2016 *In vitro* optical quality comparison of 2 trifocal intraocular lenses and 1 progressive multifocal intraocular lens *J. Cataract Refract. Surg.* **42** 138–47

[10] Son H S, Tandogan T and Liebing S *et al* 2017 *In vitro* optical quality measurements of three intraocular lens models having identical platform *BMC Ophthalmol.* **17** 108

[11] Tandogan T, Auffarth G U and Choi C Y *et al* 2017 *In vitro* comparative optical bench analysis of a spherical and aspheric optic design of the same IOL model *BMC Ophthalmol.* **17** 9

[12] Peris-Martinez C, Artigas J M and Sanchez-Cortina I *et al* 2009 Influence of optic quality on contrast sensitivity and visual acuity in eyes with a rigid or flexible phakic intraocular lens *J. Cataract Refract. Surg.* **35** 1911–7

[13] Altemir-Gomez I, Millan M S and Vega F *et al* 2020 Comparison of visual and optical quality of monofocal versus multifocal intraocular lenses *Eur. J. Ophthalmol.* **30** 299–306

[14] Cardona G, Vega F and Gil M A *et al* 2018 Visual acuity and image quality in 5 diffractive intraocular lenses *Eur. J. Ophthalmol.* **28** 36–41

[15] Michael R, Guevara O and de la Paz M *et al* 2011 Neural contrast sensitivity calculated from measured total contrast sensitivity and modulation transfer function *Acta Ophthalmol.* **89** 278–83

[16] Fischer R E, Tadic-Galeb B and Yoder P R 2008 *Optical System Design* 2nd edn (New York: McGraw-Hill)

[17] van den Bos A 2000 Aberration and the Strehl ratio *J. Opt. Soc. Am. A Opt. Image Sci. Vis.* **17** 356–8

[18] Burn J L 2010 *Optical Lithography: Here Is Why* (Bellingham, WA: SPIE Press)

[19] Belbachir A N 2010 *Smart Cameras* (Berlin: Springer)

[20] Lambrech R W and Woodhouse C 2011 *Way Beyond Monochrome* 2nd edn (Amsterdam: Elsevier)

[21] Ravikumar S, Bradley A and Thibos L N 2014 Chromatic aberration and polychromatic image quality with diffractive multifocal intraocular lenses *J. Cataract Refract. Surg.* **40** 1192–204

[22] Patel H, Congdon N, Strauss G and Lansingh C 2017 A need for standardization in visual acuity measurement *Arq. Bras. Oftalmol* **80** 332–7

[23] Holladay J T 2004 Visual acuity measurements *J. Cataract Refract. Surg.* **30** 287–90

[24] Alarcon A, Canovas C, Rosen R, Weeber H, Tsai L, Hileman K and Piers P 2016 Preclinical metrics to predict through-focus visual acuity for pseudophakic patients *Biomed. Opt. Express* **7** 1877–88

Chapter 4

Intraocular lenses

Bruno Lovaglio Cançado Trindade, Otávio Gomes de Oliveira, Luiz Melk de Carvalho, Jacqueline Beltz Franzco and Davies William de Lima Monteiro

The introduction of the intraocular lens (IOL) was, without any doubt, one of the greatest advances of all times in ophthalmology. The correction of aphakia at the time of cataract surgery improved its results remarkably and allowed much better visual performance for patients. The removal of the natural crystalline lens during cataract surgery induces a significant hyperopia due to the reduction in the total convergency power of the eye. From the original surgery in 1949, much has evolved in this field. The ability to safely implant a lens in the eye required some surgical modifications and improvements in the instruments and the technique.

Many changes were made to the intraocular lenses since their first appearance. The introduction of a foldable material allowed its implantation through a much smaller incision improving the safety and the refractive outcomes. The original intraocular lens was a rigid monofocal spherical IOL. These lenses have a constant radius of curvature in their surfaces. This feature has some manufacturing advantages but also introduces some concerns regarding the optical quality. The evolution of IOL manufacturing processes allowed companies to design improved refractive surfaces that introduced correction of high order aberrations and astigmatism (cylinder) to these implants with aspheric and toric designs, respectively.

The refinements of the surgical instruments and techniques together with the improvement in pre-operative testing allowed for a natural quest for the independence of spectacles after cataract surgery. With this, the introduction of multifocal lenses brought the possibility to partly mitigate pseudophakic presbyopia.

More recently, many new lenses were developed. The use of extended depth of focus and small-aperture optics have contributed to extend the possibility of surgical correction of presbyopia. The pursuit for a true accommodative IOL is still on. The full restoration of accommodation is still an unmet need in ophthalmology despite many current alternatives in the IOL field.

This chapter describes the various types of IOLs available. We detail their concept, indication and limitations.

doi:10.1088/978-0-7503-3263-7ch4

These concepts are required to be fully understood so the surgeon can better choose the proper IOL for each specific patient, pondering over optical performance and lifestyle.

4.1 Introduction to IOLs

The first surgeon to ever implant an intraocular lens after cataract extraction was a British doctor named Harold Ridley. This feat happened in 1949 in an operating room inside Saint Thomas Hospital in London and it changed the history of cataract surgery forever. After examining some of the British Royal Air Force pilots that had their airplanes shot during World War II, Dr Ridley noticed that shreds of the cockpit canopy found inside some pilots' eyes did not cause any sort of immune response. This led him to use this same acrylic material (poly methyl methacrylate—PMMA) to fabricate the world's first IOL. It took him four years to develop the design of this lens that was based on the same shape of the natural lens. Weighing over 100 mg, it was a heavy prosthesis, and it was first implanted in 1949 after removal of the cataract of a patient. Not happy with the stability of the implant, Ridley removed the lens and left the patient aphakic at the end of surgery. A few months later, in February 1950, the same patient was brought back to the operating room to have the IOL re-implanted and, this time, it was left inside the eye. This was the first time that aphakia was being corrected inside the eye. The refractive result of this first surgery was suboptimal and the patient ended up with −10 D of residual myopic error but that did not stop Dr Ridley improving his innovation [1]. This paved the way to the development of an incredible evolution in the materials, design and optical characteristics of these implants we see today.

However, such a revolution did not come without its condemnations. At the beginning, Dr Ridley was severely criticised by his peers and especially by academia. Professor Duke-Elder, who was a very respected and famous academic leader at the time, did not spare the young Ridley from being questioned and disapproval. Dr Ridley was unable to convince his own father, who was also an ophthalmologist at the time, of the beneficial aspects of his invention and did not even get family approval by him [2].

Years later, justice was done, and Dr Ridley received several recognitions and awards worldwide. In 1990 he underwent successful bilateral cataract surgery with implantation of an IOL at the same hospital where he first performed his surgeries in the fifties and the inventor benefitted by his own invention. In 2000, he was knighted by the British Crown at Buckingham Palace and history was able to correct its course.

Through the years, the IOL has undergone several modifications. As mentioned above, the first IOL was designed to be implanted in the posterior chamber, behind the iris. However, using the technique to extract the cataract of that time, proper lens support could not always be achieved. Extracapsular and intracapsular cataract extraction were the main techniques used to remove the opaque lens. This way, decentrations and lens migration to the vitreous cavity were somehow common. This led some researchers to develop new kinds of lenses to be implanted in front of the iris, in the anterior chamber [3–5]. Multiple different designs were conceived with clips to be fixated in the iris or haptics to secure the implant in the chamber angle the most common ones [4, 6–8]. Little was known at the time about the corneal endothelium and its importance for keeping the cornea transparent and the use of these new anterior-chamber IOLs led to the development of iritis, glaucoma and corneal decompensation in many patients [9, 10]. Later on, the use of posterior chamber IOLs was found to be the best approach and this is the main design to date [11]. Nevertheless, anterior chamber IOLs (both iris and angle fixated) can still be found, but their use is mainly limited to cases where proper IOL support is compromised.

IOL materials have also changed tremendously in the last three decades. PMMA has been used since the original Ridley's lens. It has many important properties that explain its use in the manufacturing process of IOLs. It is affordable, easy to manipulate, lightweight, very transparent, biologically inert and resistant to degradations [12]. However, it is a rigid polymer and requires a large incision for the lens to be implanted. With the gradual adoption of phacoemulsification in the late 1980s early 1990s, cataracts could be removed using a much smaller incision. However, the incision had still to be enlarged for IOL implantation which was counterintuitive. This pushed industry to develop IOLs that could be implanted into small incisions. The introduction of silicone made possible the first foldable IOLs [13]. This material was compressible and resistant yet it maintained its structural stability after unfolding inside the eye [14]. It was also inert and could be easily manipulated. This way, it was possible to implant a large lens through a small incision and the introduction of foldable IOLs was the main boost for a wider adoption of phacoemulsification as the standard procedure for cataract extraction. Since then, other polymers have been introduced in the IOL manufacturing. Hydrophilic and hydrophobic acrylics were instituted and allowed a much more controlled unfolding of the IOL once inside the eye [15]. Silicone lenses were rubbery and after implantation would unfold quickly, with no control, exposing the intraocular content to an increased risk of trauma. On the other hand, hydrophobic acrylic has a much more controlled unfolding process allowing the surgeon to correct the position and orientation of the lens during implantation, minimising risks. This material is to date the most commonly used worldwide [16, 17]. Hydrophobic acrylics are very biologically inert, and can be used even in cases of uveitic cataract. Their unfolding friendliness makes these IOLs easy to be implanted and the material sticks to the capsule allowing a more stable implant. On the other hand, hydrophilic acrylic lenses are very clear and can be compressed to be implanted into even smaller incisions. They are also less prone to scuffs and marks than the hydrophobic acrylic lenses making them surgeon-friendly for loading and implanting.

Acrylic IOLs usually have a higher index of refraction allowing thinner lenses to be produced. Nevertheless, the flatter anterior curvature of these lenses produces a mirror-like reflex when directly illuminated, usually referred to as a 'cat-eye reflex' that may be bothersome for some patients, especially those subject to public appearances. This reflex is commonly not perceived in silicone IOLs due to the lower index of refraction and the need for more curved anterior lens surface.

Another downside of high-index-of-refraction IOLs is the occurrence of negative dysphotopsia. This is a perceived dark arc in the temporal field of view and its causative factors are still under debate [18–20]. It is thought that the higher index of refraction causes total internal reflection from the oblique light rays that reach the IOL. This, in combination with the truncated optic periphery, produces a shadow over the peripheral retina that is more anteriorly located in the nasal hemisphere.

Degradation, discolouration and opacification of IOLs have been reported over the years with all the available materials [21]. PMMA lenses have been shown to degrade into a snow-flake opacification after several years of implantation. This is possibly related to UV-exposure of the material since it is usually found in the center of the lens sparing the iris-covered periphery [22]. Silicone IOLs are particular prone to adhere silicone oil used in vitreo-retinal surgery [23]. This adhesion decreases vision and it cannot be removed once it happens. Therefore, these lenses are to be avoided in patients that may develop a retinal complication that can require surgery such as high myopes and diabetics. Hydrophobic acrylics have also been shown to be particularly prone to degradations such as the occurrence of glistenings and subsurface nano-glistenings (SSNGs) [16]. Glistenings are micro-water vacuoles within the polymer structure. They are more common in lenses that have a higher index of refraction and their importance for visual performance is debatable. Mixed reports have been published showing visual compromise of glistenings varying from severe impact requiring IOL exchange to absolutely no clinical detectable difference. SSNGs are even smaller vacuoles located just under the anterior and posterior surfaces of the lenses [24]. These are found especially in the AcrySof® material and produce a white-chalk-dust appearance and can decrease contrast sensitivity in more advanced forms. However, the material that has the worst reputation regarding opacification is, by far, hydrophilic acrylic. Several of these lenses have been shown to opacify, requiring lens exchange [25, 26]. The water content of this material varies between, 30%–40% and this allows absorption of any pigment that can be found in the anterior chamber such as vital dyes used during surgery. Calcification of these lenses has also been reported, especially after vitreo-retinal surgery or endothelium keratoplasty that require insertion of an air or gas bubble in the anterior chamber. The clinical appearance of these calcifications may vary, and they are often progressive, leading to decreased vision and requiring replacement of the faulty lens. This potential complication may also limit the use of hydrophilic IOLs in patients with retinal pathologies or in those with endothelium diseases who may require a corneal transplantation in the future.

The original Ridley's implant attempted to simulate the shape of the natural human lens. This led to a bulky implant that not only required a large incision to be inserted in the eye but also had a significant weight. The design of the lens haptics, fixating structures, has also changed quite significantly since the first IOL. Various shapes have been proposed initially for the iris-fixed and the anterior-chamber-supported IOLs. However, nowadays, with posterior chamber lenses being the most popular lenses used, the designs are basically limited to one of these three: single-piece, three-piece and plate design, as depicted in figure 4.1. These lenses, unlike Ridley's IOL, comprise a central optic portion sustained in place by two (or more) haptic elements.

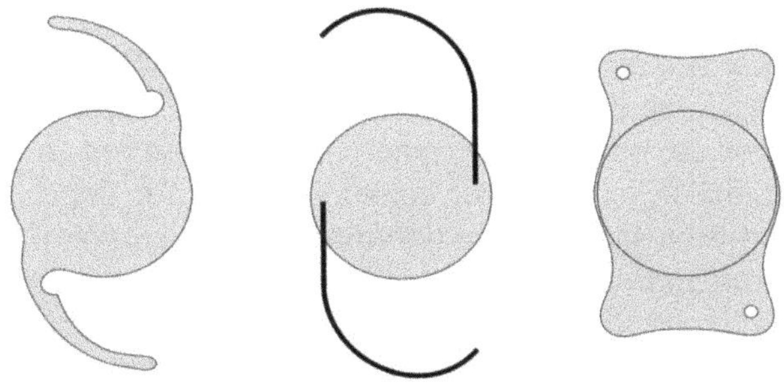

Figure 4.1. Illustration of different IOL haptic designs.

Single-piece lenses are the most commonly used to date [17]. Their manufacturing process is easier using a computer numerical control (CNC) router or an injection moulding process. They are characterised by being made of a single piece of material with the haptics being a thinned prolongation of the optic portion. These lenses allow easy implantation and safe intraocular manipulation. Three-piece IOLs are characterized by being made of a central optic portion that has the haptic elements drilled and fixated into it. The optics can be made with PMMA, silicone or acrylic and can be manufactured also using a CNC router or by moulding. The haptics are filament-like projections that can be made using different materials. PMMA, polypropylene and polyvinylidene fluoride (PVDF) are the most used. With the optic portion ready, the haptics are then drilled and fixated onto their side. This makes the manufacturing of these lenses more laborious and the delicate adhesion between the haptics and the optic may break during implantation. Plate-design IOLs are characterized by having a broader area of contact between the haptics and the optic which makes twisting of the IOL virtually impossible. They are usually made of hydrophilic acrylic and can be normally injected through a very small incision. The benefit of such design is the fact that it can be very easily manipulated once inside the eye allowing rotation to either direction (clockwise or counter-clockwise). Their manufacturing process is usually limited to moulding only.

Today, it is a consensus that the ideal site for IOL implantation is inside the capsular bag. However, there can be situations during cataract surgery in which the capsular bag is compromised and the IOL cannot be securely fixated. In these cases, the ciliary sulcus is usually a good alternative for positioning the lens. For the IOL to be safely positioned in the ciliary sulcus, there are some design characteristics that need to be respected [27]. First, the overall diameter has to be slightly larger with a 13 mm diameter from haptic tips. As shown in chapter 2, the capsular bag equatorial diameter measures around 10.5 mm whereas the ciliary sulcus is around 11.5–12.0 mm in diameter. This way, a smaller IOL may become loose in the ciliary sulcus and decentration post-operatively. The ciliary sulcus is a metabolic active site with uveal tissue around it. This way, for the IOL to be implanted in the ciliary sulcus, its haptics need to be thin and well-polished to avoid precipitation of a local inflammatory response. Finally, with the IOL in the ciliary sulcus located more anteriorly, iris movement may trigger contact between lens and the posterior iris pigment epithelium. The rubbing of these two may cause iris pigment dispersion that can lead to glaucoma and uveitis. This way, the IOL haptics should be angulated anteriorly to the optic to minimize contact of the optic edges to the posterior aspect of the iris. Single-piece acrylic lenses, despite being the most commonly used for implantation inside the capsular bag, have been shown to be harmful for implantation if the ciliary sulcus because of their thick and sharp-edged haptics, smaller diameter and uniplanar design [28]. A three-piece design should always be favoured in this scenario.

Posterior capsule opacification may happen after cataract surgery. This is the proliferation of the lens epithelial cells (LECs) that grow from the equator of the bag and tend to migrate posteriorly, where the capsule is spared during surgery. This causes moderate to severe visual degradation because of the loss of transparency and can happen any time after surgery [29]. The current treatment for this condition relies on posterior capsule opening using a photo-disruptive laser (Nd:YAG laser) [30]. Round posterior edges of the optic of the lens have been shown to facilitate cell migration and increase the incidence of posterior capsule opacification. The introduction of a square-edged design has decreased the need for posterior capsulotomy with the YAG laser [31]. The sharper posterior edges decrease the space through which the LECs can migrate centrally over the posterior capsule.

The use of filters in the optics of the IOLs has also been incorporated into the more modern lens designs. Dangerous UV light can be blocked by most of the currently available IOLs. Some lenses also block blue light in an attempt to spare the macula from the high-energy photons of this end of the visible spectrum [32]. There is no consensus in the literature about the benefit of filtering the blue light and criticism has also been published regarding these filters. There have been reported some subjective changes in color perception seen after implantation, and the circadian rhythm has also been shown to be disturbed in some patients after blocking blue light from reaching the retina [33].

Injectors are the now accepted standard method for foldable IOLs implantation in the eye. They carefully fold and protect the lens when being pushed through a small incision. This is important for not damaging the lens and not carrying any microorganism that may be found over the eye surface [34]. The behaviour of single-piece, three-piece and plate-haptic-design IOLs is different when being loaded and injected, with single-piece IOLs probably being the easiest ones to be injected and three-piece the hardest ones.

Optics of the IOLs have tremendously improved over the years and new designs are constantly being launched. The use of aspheric profiles to control spherical aberration has been seen in IOL designs for years now and toric implants have also been developed to treat corneal astigmatism. The pursuit of a true accommodative IOL still continues and not many successful mechanisms for this groundbreaking approach have been shown. Nevertheless, multiple strategies have been proposed to overcome the challenges of pseudophakic presbyopia. In the following, we will discuss each of the IOL's optics designs currently available.

4.2 Monofocal spherical IOLs

The introduction of the intraocular lens by Ridley in 1949 was a major event in the modern treatment of the cataract. The first lenses to be manufactured were standard monofocal lenses. These implants focus light to a single point giving the patient good vision at a specific distance. Glasses have to be used to allow good vision for other distances than at which the IOL was set to focus the light. These implants are still used today in many patients worldwide. The benefit of using the entire surface of the IOL to focus light into a single point is that it increases image quality and enables the best possible vision in a specific distance that can be set for near, intermediate, or distant depending on the patient's preferences. This can be crucial in specific scenarios such as for pilots, surgeons, dentists among others [35]. These IOLs are also the most appropriate for eyes that have (or may develop) any kind of diseases that may affect retinal sensitivity. Many pathologies can limit the resolution of the eye. Diseases of the tear film, cornea, vitreous, retina or the optic nerve are especially important in limiting the maximum image quality that an eye can produce. Many diseases fall into these categories such as dry eyes, keratoconus, floaters, age-related macular degeneration (AMD) and glaucoma are the most important ones. In any case of impairment of image formation due to any of these conditions, a monofocal IOL is considered the ideal implant because it can form over the macula the best possible image that an IOL can produce [36]. Limitations in retinal sensitivity can and will compromise the quality of vision of any IOL that divides light to produce multiple foci. Even though the monofocal IOL has a single point of focus to which it converges all the light, these lenses can also be used applying a technique known as monovision in which one eye (usually the dominant eye) is set to distance and the other (usually the non-dominant eye) is set to near [37]. This way, with both eyes opened, patients can see well independently of spectacle correction in multiple distances. Binocularity is the price to be paid to use this strategy however, it has been shown that, applying a minimal anisometropia of −1.25 D in the near eye allows for acceptable binocular fusion with preserved stereopsis as well as adequate near vision [38]. These lenses are commonly referred to as standard IOLs (as opposed to the advanced technology lenses) and they are usually the implant that public health authorities pay for use in the majority of the countries where cataract surgery is financed by the government.

A constant radius of curvature of the anterior and posterior surfaces of the spherical lens allowed for a simpler manufacturing process that could be fulfilled by the technology of the initial IOL industries [39]. However, the constant radius of curvature allows for more vergence to be added to the light rays at the lens periphery than at its center. This allows the light rays that reach the peripheral region of the lens to be focused in front of the ones that hit the lens more centrally. This is called longitudinal spherical aberration and it is more important the more dioptric power the lens has [40, 41]. Image quality depends on the precision on which the light is being focused. Longitudinal spherical aberration is a limitation of any spherical lens, and it compromises image quality by decreasing the contrast on image formation. The impact of spherical aberration on image quality formation depends on pupil size as well as on the IOL power. The greater these variables are, the more important the effect of spherical aberration will be on the resulting image [40]. There is a large individual variability in pupil dimension with light-coloured eyes usually having larger pupils [42]. Also, pupil diameter obviously depends on environment luminosity. However, there is a general trend of pupil size to decrease with age [42]. Therefore, the older population that have cataracts will generally exhibit a smaller pupil to lessen the effect of spherical aberration in eyes implanted with regular spherical IOLs.

4.3 Monofocal aspheric IOLs

4.3.1 Concept

A monofocal aspheric IOL is a lens that has on one or both of its surfaces a non-spheric shape, which is commonly known as aspheric surface. A generic aspheric surface with high order terms up to the 8th order is mathematically represented by [43, 44]:

$$Y = \frac{C{\bullet}X^2}{1 + \sqrt{1 - X^2{\bullet}C^2(1 + k)}} + A_4{\bullet}X^4 + A_6{\bullet}X^6 + A_8{\bullet}X^8 \tag{4.1}$$

where Y is the sag coordinate of the surface, X is the aperture coordinate, C is the curvature (the reciprocal of the radius of the curvature) of the surface, k is the conic constant and A represents the coefficients of the high order terms of a more general aspheric surface. By ignoring the high order terms of equation (4.1), different curvature shapes can be represented depending on the value of the conic constant [43]:

- If $k = 0$, sphere;
- If $k < -1$, hyperbola;
- If $k = -1$, parabola;
- If $-1 < k < 0$ and $k > 0$, ellipse.

Figure 4.2 shows the surface profile for different values of conic constant. It is possible to notice that depending on the value of the conic constant, the surface becomes more curved or flatter to the edges when compared to the spheric surface.

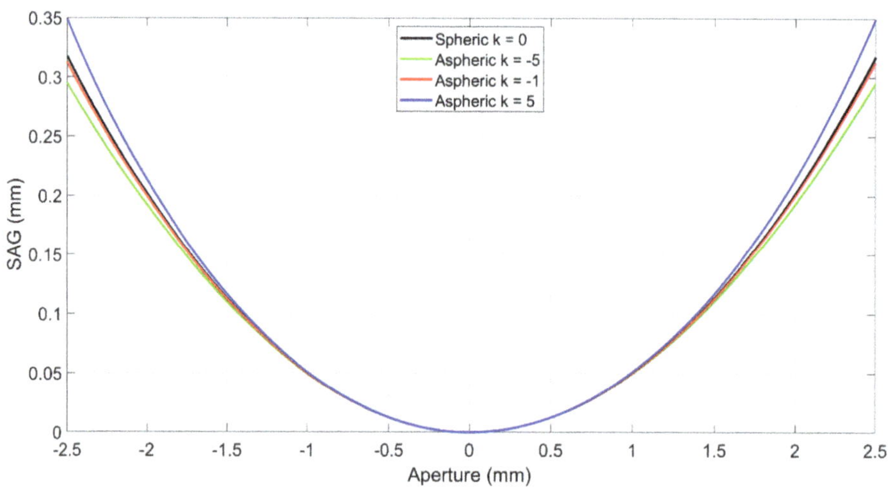

Figure 4.2. Different surface shapes due to conic constant variation considering a radius of curvature of 10 mm.

4.3.2 Indications

During the aging process, the human eye undergoes structural changes that affect the crystalline lens such as changes in its refractive index, shape and mechanical properties [45, 46]. Due to the changes involved in the aging process an increase in the spherical aberration of the human lens is observed and because of that its capability to compensate the positive spherical aberration of the cornea deteriorates [47]. As a result, the total eye aberration increases, causing a reduction in the optical quality and visual acuity with aging [48, 49].

The monofocal aspheric IOLs are designed to cancel, reduce or maintain the spherical aberration of the eye with the purpose to enhance the visual performance of the patient when compared to monofocal spheric lenses [50, 51] which contributes with positive spherical aberration (see figure 4.3(a)). The human cornea has generally a positive spherical aberration [52]. Some of the aspheric IOLs present on the market have a negative amount of spherical aberration to correct or reduce the total aberration of the eye (see figure 4.3(b)). For instance, the Tecnis Z9000 model presents −0.27 μm of spherical aberration at 6 mm aperture and the Acrysof IQ presents −0.20 μm of spherical aberration at 6 mm aperture [50]. However, it is also possible to find on the market aspheric IOLs such as the SofPort Advanced Optics IOL model that has zero spherical aberration (see figure 4.3(c)) [50].

Although aspheric IOLs present an optical performance that is theoretically superior in relation to spheric IOLs [53]. there are clinical studies in literature that reports partial or no improvement in the visual acuity of patients with aspheric IOLs in comparison to spheric IOLs [50] For instance, some studies have shown a visual acuity improvement yielded by the Tecnis aspheric IOL when compared to spheric lenses in terms of best corrected visual acuity (BCVA) and contrast sensitivity at photopic and mesopic conditions [54, 55]. However, there are studies that did not find any significant difference between the BCVA of aspheric lenses (Tecnis and Acrysof IQ models) compared to spheric lenses, but only a difference in the contrast sensitivity at some apertures [56, 57].

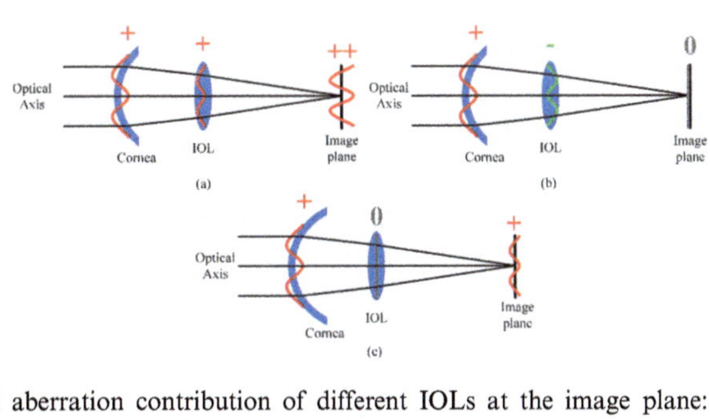

Figure 4.3. Spherical aberration contribution of different IOLs at the image plane: (a) spheric IOL with positive spherical aberration added to the spherical aberration of the cornea at the image plane, (b) aspheric IOL with negative spherical aberration correcting the spherical aberration of the cornea at the image plane and (c) aspheric IOL with null spherical aberration and keeping the cornea positive spherical aberration unchanged at the image plane.

Among other factors, one of the main reasons for the controversies observed in the clinical studies might be the wide range variation of the corneal aberration within a population [53]. As the manufacturers of aspheric IOLs offer only one aspheric design to correct a specific amount of spherical aberration, different patients, who present different amounts of spherical aberration on the cornea, will have different outcomes from the same aspheric IOL. From this perspective, theoretical studies have shown the benefits of the customization of aspheric IOLs according to the corneal biometry [53, 58, 59]. Clinical studies have also reported the benefits from selecting the aspheric IOL that best compensates the corneal spherical aberration of the patient, especially under mesopic conditions [60, 61].

4.3.3 limitation

The selection of the aspheric lens that best compensates the corneal spherical aberration of the patient is critically dependent on the precision of the pre-operative assessment of the cornea topography [62], among other factors [63]. As the spherical aberration becomes more prominent with the increase of aperture, it is important to have a high accuracy on the corneal topography, especially at the periphery [62]. For instance, a study comparing the repeatability and reproducibility of the corneal tomographer Pentacam HR (by Oculus company in Germany) has shown a reduction on the repeatability of the corneal assessment at 10 mm aperture diameter when compared to smaller apertures [64]. Based on that, the accuracy of the equipment can limit the prediction of the corneal spherical aberration.

Once the IOL is implanted into the eye some other factors can also impact its optical performance. Eppig *et al* verified that the performance of aspheric IOLs with aberration correction are highly sensitive to tilt and decenter [65]. However, the sensitivity varies with the amount of aberration correction present in the different models of aspheric IOLs [65, 66] and the degradation in optical performance caused by decentration is more relevant than tilt in aspheric lenses [67]. Studies have shown that aspheric IOLs with neutral correction (i.e. zero spherical aberration) are less sensitive to tilt and decentration when compared to aspheric IOLs that has some degree of negative spherical aberration [65, 66]. As several clinical studies have reported some amount of tilt and decentration in IOLs after cataract surgery [65], the perfect alignment and positioning of the aspheric IOL into the eye to achieve its best optical performance is a limiting factor. For instance, studies have shown that for decentration greater than 0.5 mm the optical performance of aspheric IOLs is significantly reduced [51]. Besides the presence of tilt and decentration that contributes to the increase of the amount of asymmetric aberrations, there are other high order aberrations (HOAs) into the eye represented by asymmetric terms that cannot be compensated by the asphericity of the IOLs designs that present a rotationally symmetric profile [51].

Another limiting factor of aspheric IOLs is related to the trade-off between the optical performance at the image plane (i.e. sharpness of the focal point) and the depth of focus. The sharp focus created by aspheric IOLs due to the reduction of the spherical aberration guarantees good optical performance at the image plane, but it causes a reduction in the depth of focus of such lenses. Based on that, aspheric IOLs have a reduced depth of focus when compared to spheric lenses, which compromises the intermediate and also the distance-corrected near visual acuity [68]. Although monofocal lenses are not expected to have good near and or intermediate visual acuity, the reduced tolerance of aspheric IOLs to defocus requires an accurate pre-operative assessment of the eye dimensions such as axial length and also the post-operative IOL positioning (given by the post-operative anterior chamber depth) to guarantee a good optical performance [63].

Although it is possible to verify from theoretical studies the benefits of correcting the total spherical aberration of the eye to achieve good optical performance at the focal plane, the studies were all performed considering monochromatic aberrations [53, 58, 59]. Cancelling the total spherical aberration of the eye may not correspond to the best acuity for human vision [66]. Polychromatic analysis shows that a certain amount of spherical aberration in the eye system can improve the polychromatic PSF depending on the residual defocus [69].

4.4 Toric IOLs

4.4.1 Concept

Astigmatism is a very common refractive error that can reduce the quality of vision of patients. In optical terms, astigmatism, which is often termed cylinder, describes an optical aberration that occurs when there is a maximum difference of radii of curvature at orthogonal meridians across the surface. As a result, light rays do not focus at a single point on the retina and cause the images to appear blurry. Figure 4.4 illustrates an astigmatic condition in which an optical surface presents a smaller radius of curvature in the vertical direction (higher curvature), which focuses at the focal point F_1, and a larger radius of curvature in the horizontal direction (lower curvature), which focuses at F_2.

In the eye, any of the optical surfaces can contribute with the eye's total astigmatism. The astigmatism caused by the anterior cornea is normally referred to as corneal astigmatism. The internal astigmatism accounts for the astigmatism introduced by all other surfaces within the eye. The corneal and internal astigmatism form the total astigmatism of the eye [70].

Patients with astigmatism may experience blurred vision, which has the potential to affect their quality of life [71]. Spectacles can be used to correct astigmatism.

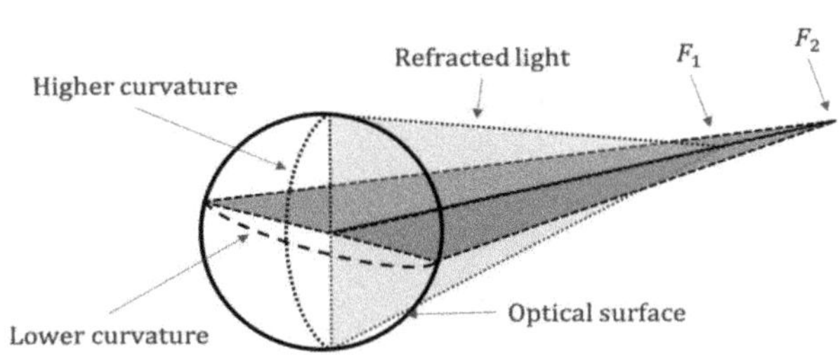

Figure 4.4. Illustration of astigmatism caused by different radii of curvature in an optical surface.

However, surgical options are also available. In particular, it can be corrected during cataract surgery, which helps patients who seek spectacle independence to improve the outcome of the cataract surgery and achieve high quality of vision [72].

Several different surgical techniques can be employed to correct astigmatism during a cataract surgery, which include corneal or limbal relaxing incisions and positioning of the phacoemulsification incision [73, 74]. In general, these techniques can provide good outcomes for small amounts of astigmatism but are very limited if the patient has 1.5 D or more of astigmatism [72, 74, 75]. Toric intraocular lenses, on the other hand, are a predictable method to correct astigmatism during the cataract surgery [71–73, 75, 76].

Toric intraocular lenses (TIOLs) are designed to compensate for corneal astigmatism. These lenses feature in one of its optical surfaces the shape of a torus, which is characterized by different radius of curvature in two meridians 90° apart from each other. Figure 4.5 illustrates a TIOL.

In TIOLs, the meridian with the smaller radius of curvature is called the steep axis and the one with the higher radius of curvature is known as the flat axis, which is 90° apart from the steep axis. The direction of the flat axis can be recognized through physical marks present in one of the optical surfaces of the lens, such as illustrated in figure 4.5.

Due to the difference in radii of curvature, the TIOL will be characterized by two optical powers, one for each axis. The lowest one is associated with the flat axis and is known as the sphere power (*Sph*). The difference between the power of the steep axis and the sphere power is known as cylinder (*C*). A TIOL is specified based on two optical parameters: its spherical equivalent power (*SE*) and its cylinder. The *SE* can be calculated from *Sph* and *C* as follows.

$$SE = Sph + \frac{C}{2} \tag{4.4.1}$$

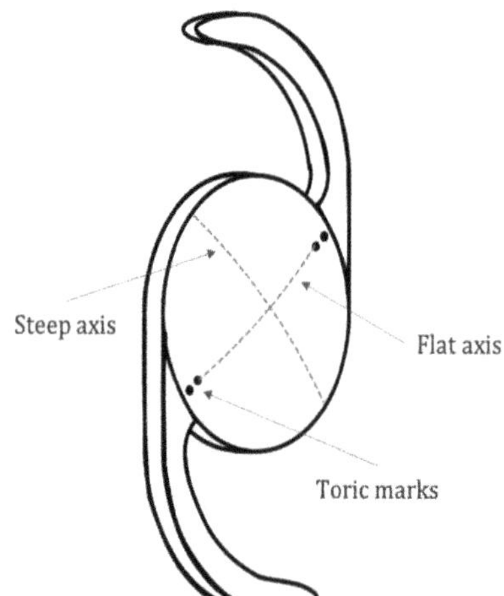

Figure 4.5. Schematic representation of a toric intraocular lens.

TIOLs are offered in a range of *SE* powers, normally in steps of 0.5 D. And a range of discrete cylinder options are associated with each *SE* power. The selection of the exact combination of *SE* and *C* to be prescribed for a specific patient's eye depends on a proper calculation, which is normally performed through online calculators made available by the TIOL manufacturer.

In general terms, the calculation of a TIOL requires the input of biometric data from the eye and the corneal keratometry (namely the flat and steep powers and axis). It also requires information about the position of the corneal incision and the surgeon's surgically induced astigmatism (*SIA*), which is dependent on the surgical procedure [77–79]. The calculators then combine such input data to estimate the total post-op corneal astigmatism after surgery and then calculate the equivalent astigmatism in a plane that corresponds to the estimated axial position of the TIOL. The TIOL parameters, as well as its orientation inside the eye can then be calculated to compensate the total post-op corneal astigmatism [80]. Since the TIOL is offered in discrete options, such compensation may not be exact. The potential portion of astigmatism estimated to remain after surgery is then calculated backwards up to the corneal plane again and is reported as the estimated post-op residual astigmatism [78]. So, in summary, the TIOL calculators show as output of the calculation the *SE* power, *C* and the orientation axis of the lens, which corresponds to the direction to which the lens marks have to be aligned during the surgery. And the calculator also shows the estimated post-op residual astigmatism to help the surgeon to make decisions on which lens to prescribe to the patient's eye.

Accurate planning of the cataract surgery using TIOLs is critical to produce good outcomes in terms of residual astigmatism, visual acuity, and patient satisfaction. The several elements that require experienced planning and execution include diagnosis, to correctly measure the pre-op keratometry and biometry, correct calculation of the TIOL to be used and carefully conducting the surgical procedure with precise corneal marking and lens positioning [78, 81]. In each step, there are variables that can impact the residual astigmatism and compromise the surgical outcomes.

4.4.2 Importance (prevalence)

Astigmatism is very common across the population of all ages and particularly in patients with cataracts [71, 74]. Several studies have already investigated the prevalence of cataracts in populations of different regions of the world.

Hoffmann and Hütz [82], for instance, analyzed data from 15 448 patients in Germany, with a median age of 74 years, and reported on the prevalence of corneal astigmatism. They found a mean corneal astigmatism 0.98 ± 0.78 D. The authors reported 63.96% of patients with astigmatism lower than 1.0 D, 27.95% of patients with astigmatism between 1.0 D and 2.0 D and a further 8% with astigmatism higher than 2.0 D [82].

Chen *et al* [83] reported on the distribution of corneal astigmatism on 4831 eyes of 2849 Chinese cataract patients. The authors reported a mean corneal astigmatism of 1.01 D, 67.7% of patients with astigmatism in the range of 0.25 D to 1.25 D and 27.5% of patients with astigmatism higher than 1.25 D.

Ferrer-Blasco *et al* [73] analyzed the prevalence of astigmatism in a cataract population in Spain. The authors collected data from 4540 eyes of 2415 patients and reported 64.6% of the patients with corneal astigmatism between 0.25 D and 1.25 D and 34.8% of patients with astigmatism of at least 1.0 D. The mean corneal astigmatism was 1.02 D.

Curragh and Hassett [72] studied the prevalence of astigmatism in a cataract population in Northern Ireland. They analyzed data from 2080 eyes of 1788 patients and reported a mean astigmatism of 1.09 ± 0.83 D. The authors found 58.7% of patients with astigmatism of 1.0 D or less, 29% of patients ranging from 1.0 D to 2.0 D and 11.6% of patients with astigmatism higher than 2.0 D.

Anderson *et al* [71] conducted a literature review that covered publications from 1996 to 2015 on the prevalence of astigmatism in cataract patients and found mean values for astigmatism reported in the range 0.79 D to 1.09 D. The authors found an overall prevalence of astigmatism of any magnitude in patients with cataracts as high as 94% in the UK and 91,8% in South Korea. The portion of patients with astigmatism higher than or equal to 1.0 D was reported to range from 23% in South Korea to 47% in China [71].

More recently, Day *et al* [74] also assessed the prevalence of pre-operative astigmatism in a large population of cataract patients in the UK. In a retrospective study, the authors reported a mean astigmatism of 1.06 D ± 0.81 D among 110 468 eyes. The authors found 78% of patients with corneal astigmatism higher than or equal to 0.5 D and 11% of patients with astigmatism higher than or equal to 2.0 D.

Corneal astigmatism is normally analyzed not only in terms of its magnitude, but also according to the direction of the steepest meridian. The corneal astigmatism is defined as with-the-rule (WTR) if the corneal steepest meridian is in the range of 90° ± 30°. On the other hand, it is defined as against-the-rule (ATR) if the same meridian is found within 180°±30°. If the steepest meridian is found at the mentioned ranges, it is then defined as oblique astigmatism (OA) [71, 74].

Although WTR astigmatism is normally the most common one across the population, it is well known that the prevalence of ATR astigmatism increases with age [71, 73, 74, 82–84], especially after the age of 60 years [84], which makes this type of astigmatism very common among cataract patients. Hoffmann and Hütz [82] reported a prevalence of ATR astigmatism of 28.8%, out of 23 239 eyes in Germany, while Hayashi *et al* [84], in a small study with 142 patients in Japan, reported 27% of cataract patients with ATR astigmatism.

Uncorrected astigmatism is known to be detrimental to the vision quality and may impact the quality of life of patients, especially those who also have cataracts [85]. Wolffshon *et al* [85] evaluated the impact of uncorrected astigmatism in patients that underwent cataract surgery on their visual function and visual performance. The authors found that even a low astigmatism may significantly reduce visual acuity and hence the ability to conduct low-contrast activities. They also report that even 1.0 D of astigmatism can affect simple routine reading tasks. The authors concluded that uncorrected astigmatism significantly reduces the patient's quality of vision with potential long-term effect over restricted independence, falls and lower quality of life and suggest that TIOL implantations should be considered as standard care [85].

The detrimental impact of uncorrected astigmatism over visual performance and functional visual activities is also reported by Read *et al* [86]. The authors highlight that even relatively low levels of uncorrected astigmatism can produce a range of visual deficits, particularly when activities require high visual acuity.

Anderson *et al* [71] analyzed the economic and humanistic burden of astigmatism in cataract patients. The authors highlighted the quality of life of cataract patients is negatively impacted by astigmatism and that this leads to increased spectacle dependence. They also point out a significant improvement in patient satisfaction after cataract surgery with implantation of TIOLs. The authors conclude that uncorrected astigmatism of 1.0 D or higher negatively influences the ability to drive at night, reduces contrast visual acuity, leads to decreased vision-related quality of life and well-being [71]. Moreover, Anderson *et al* [71] concluded that astigmatic cataract patients that undergo cataract surgery and implant TIOLs can save on costs of less expensive vision correction spectacles when compared to those patients that had standard monofocal IOLs implanted.

4.4.3 Indication

As mentioned previously, even relatively low levels of uncorrected astigmatism may produce detrimental effects on visual performance [86]. In this sense, a natural discussion is around what is the minimum clinically relevant astigmatism that should be corrected during cataract surgery. However, there is no universal consensus on this topic [71].

Villegas *et al* [87] analyzed the effect of different degrees of astigmatism on visual acuity and demonstrated an improvement in high-contrast acuity for correction of astigmatism levels as low as 0.3 D to 0.5 D, although with the caveat of a high inter-subject variability.

Sigireddi and Weikert [88] state that astigmatism of 0.25 D can be perceived by the patient for 20/25 Snellen letters, and that about 0.8 D of astigmatism becomes already objectionable to the patient. The authors add that residual astigmatism following cataract surgery of no more than 0.5 D is associated with higher patient satisfaction and hence they suggest that 0.5 D of residual astigmatism should be the target to produce optimum visual function and patient satisfaction [88].

Grunstein and Miller [20] propose an algorithm based on the magnitude of pre-op astigmatism to help the surgeon to select the best correction method for the astigmatism. The authors suggest that corneal astigmatism of no more than 1.0 D should be addressed by on-axis phacoemulsification incisions. For corneal astigmatism between 1.0 D and 3.0 D, peripheral corneal relaxing incisions can be used. Alternatively, TIOLs should be considered for corneal astigmatism higher than 1.0 D.

Visser *et al* [78] suggest 1.25 D as the minimum level of corneal astigmatism for which the implantation of a TIOL should be considered.

On the other hand, given the changes in corneal astigmatism with age, Hayashi *et al* [89] suggest that a certain degree of astigmatism should be maintained or even overcorrected.

4.4.4 Limitation

Despite the evolution of both surgical techniques and TIOLs, patients still can present significant residual astigmatism after cataract surgery. Residual astigmatism higher than 0.5 D in more than 50% of patients, 6 months after cataract surgery with TIOL implantation, has been reported by different clinical studies [90, 91]. Visser *et al* [91] reported a prevalence as high as 26% of patients with residual astigmatism higher than 1.0 D after TIOL implantation.

The main sources for residual astigmatism are normally associated with accurate placement of the TIOL during the surgery, rotation stability of the lens after the surgery and the accuracy of pre-operative measurements of the eye [78, 86, 92].

The misalignment of the TIOL during surgery may occur as a result of pre-operative markings that have been done either incorrectly or inaccurately. Studies show that manual marking procedures can induce an alignment error of 3° to 7° from the intended location [75, 93], which may reduce significantly the effectiveness of the astigmatism treatment. The effect of TIOL in correcting corneal astigmatism may be null if the TIOL is 30° out of the intended axis [94]. A misalignment higher than 10° is commonly considered a complication that requires surgical intervention for reorientation of the lens [78].

The misalignment of the TIOL can also occur due to rotation of the lens after the surgery, which may depend on the lens materials and haptics design. With early TIOLs, rotation of more than 10° could be observed in 20% to 30% of patients [86]. However, more recently, significantly lower rates have been reported with modern TIOL, with lens rotation of 10° or more being observed in about 2% of the patients [90].

The accurate planning of the surgery is another critical factor that determines the success of a TIOL implantation. A good planning must take into account not only the accuracy of the eye measurements, but also the influence of surgically induced astigmatism, posterior corneal astigmatism and estimated effective lens position within the eye's capsular bag [92]. Kock *et al* [95] analyzed 715 corneas of 435 patients and found 9% of eyes with posterior corneal astigmatism of at least 0.50 D. Studies demonstrate that the posterior corneal astigmatism contributes favorably to the total against-the-rule astigmatism of the cornea [92, 95, 96].

Misalignments or rotation of TIOLs can have an even higher detrimental effect in patients with corneal astigmatism of 3.0 D or more, since small deviations can produce higher residual astigmatism [86]. Hence, treating patients with high corneal astigmatism can be challenging sometimes.

4.5 Multifocal IOLs

The pursuit of spectacle independence after cataract surgery continues to be present for both surgeons and patients alike. The benefit of uncorrected vision across multiple distances is an object of desire for many patients that undergo cataract surgery these days. With the general increase of life expectancy across the world, cataract surgery is being performed in an increasingly demanding and active patient age group. What was once considered an elderly and limited-to-household-activities population, patients that are being submitted to cataract surgery nowadays are commonly in their peak professional years. The ability to see distance and near objects with no additional visual aide allows a much more active and young lifestyle. Not only are reading glasses considered by many as an impairment of their daily activities, they are also considered to be an age-denouncing object. This way, the quest for a truly accommodative IOL is a major goal of the industry that, so far, has not yet been fulfilled [97]. Meanwhile, many alternative strategies have been developed to overcome the challenges of pseudophakic presbyopia. The term multifocal IOL has been broadly used to describe many different implant technologies that share some properties with the intention of producing more than one foci of light to mitigate post-operative spectacle needs [98]. Multifocal IOLs behave very differently from multifocal spectacles. Spectacles produce alternate vision while multifocal IOLs will give simultaneous vision. In multifocal spectacles, patient's gaze is what selects the ideal optical power in order to see at a specific distance. By lowering gaze, additional power is added to the prescription yielding good near vision. This is why some patients report difficulties with these lenses especially when going up or down stairs. Multifocal IOLs, on the other hand, produce multiple simultaneous foci of light and it is up to subjective cortical processing to select which one of the foci is the one required to see whatever is of interest. Light is always divided for the formation of the simultaneous foci [99]. In a bifocal IOL, part of the light is directed to the distance focus, whereas another part is responsible for the near focus. In a trifocal lens, there is another intermediate focus. In all of these lenses, there is also some lost light due to scattering and dispersion [100]. Having this in mind, it is clear that these lenses will always have a lower contrast transmission than a similar monofocal IOL [101]. This way, for the spectacle-independence objective to be conquered, one will have to pay in contrast currency to achieve this. This is especially critical for patients who may have another eye condition that will cause contrast impairment such as glaucoma, AMD or even a residual refractive error [102]. Contrast lost may become problematic for some patients and they may refer to their vision as a ground glass image or a film in front of their eyes. Sometimes, when visual acuity is measured using high-contrast Snellen charts, a somehow 'normal' result may be present and the patient may still be complaining of not being able to see well. This may be an uncomfortable situation for surgeons and patients alike. The patient's complaint will only be shown if a contrast sensitivity test is applied. This way decreased perceived contrast may be present and be responsible for the patient's dissatisfaction. There are many reports of explantations of multifocal lenses due to inability of the patient to adapt to this new situation [103, 104]. Moreover, because of the simultaneous image formation, defocused light sources will appear as a halo or even as a glare. This may be confusing in certain low-light situations especially at night when driving with headlights coming from the opposite way. Even though these lenses have greatly improved over time with improved contrast transmission and less halos and glare, the inherent limitations of the technology limit their use to a more widespread population [105]. Different countries have different reimbursement systems, however, a multifocal IOL is almost always considered a premium implant with an out-of-pocket payment required by the patient. This gives surgeons a higher payment for multifocal IOL implantation than standard monofocal lenses and introduces an obvious financial bias towards indication of these premium lenses [106]. Considering this scenario, the growing number of multifocal IOL implantations has to be seen with proper caution and the economics of cataract surgery may explain why some countries/regions/surgeons have such a high rate of multifocal IOL implantation and others have a more conservative percentage.

Currently, there are two different approaches used to produce these distinct foci: refractive and diffractive optics [107]. These concepts are fundamentally different and the optical properties will differ from each of the implants with both sides having their own set of unique limitations and benefits. Refractive IOLs have alternating regions, usually designed in concentric rings, of the optics responsible for creating the different foci. By modifying the anterior and/or posterior curvature of the lens over the optical zone, light rays that reach one of the regions are focused for distance and the ones that reach another region are focused for near vision. It is a combination of two (or more) monofocal lenses built into one single optic. These lenses are very pupil-dependant because the amount of light reaching different regions is basically controlled by the pupil size. This way, low-light situations may be a problem for reading small print. The alternation of different zones of distinct dioptric powers may be easier to be manufactured [108], however, these lenses are known to cause highly noticeable halos and glare and their adoption is being gradually substituted by the less symptomatic diffractive IOLs [109]. Diffractive optics are based on diffraction of light to produce two or more foci. Diffraction occurs when light reaches an obstacle whose aperture or height is on the same order of magnitude of the incident-light wavelength. From this point onwards, the light behaves as if these obstacles/apertures are a new point source. By combining different diffractive obstacles, it is possible to produce resulting wavefronts that will interfere with each other in a constructive or destructive pattern that will enable light to be directed to a specific point or points. In IOL design, this concept is applied using concentric rings of various heights that divide light into the specific distances. By modifying the radius into which these concentric obstacle-rings are positioned, it is possible to control the amount of additional vergence inserted into the incident wavefront. And, by controlling the height of these obstacles, it is possible to modify the amount of light that is directed to each of the foci. The lower the height, the more light is directed to the main principle focus point. As the obstacle height increases, light is gradually directed more towards the additional focus points. In some IOLs the height of the concentric rings decrease as their radius increases. This is called apodization and, it allows a decrease of the amount of light that is directed to the additional focus points as light reaches more peripheral regions of the IOL optic. This way, when the pupil is more dilated, especially important when driving at night or looking into distant objects in a low-light environment, more light is directed towards the distant focus point favouring distance vision [110]. This makes these lenses less prone to cause negative optic symptoms such as halos and glare under low-light conditions.

4.6 EDoFs—extended-depth-of-focus IOL

IOLs with extended depth of focus (EDoFs) consist of a class of IOLs that enable good and relatively constant visual acuity over a range of object distances, instead of at specific focal distances as in multifocal lenses. They are also referred to as extended-range-of-vision (ERV) or wide-depth-of-focus IOLs. According to the ANSI standard Z80.35-2018, the requirements for a lens to be considered EDoF are that for a 3 mm pupil: (i) it must have a range extension of at least 0.5 D at 0.2 logMAR, when compared to the reference monofocal aspheric IOL (photopic condition, monocular, distance-corrected visual acuity); (ii) it must feature no less than 0.2 logMAR for an object placed at 66 cm from the eye; (iii) the acuity at 0 D must not be lower than that of a reference monofocal lens by 0.1 logMAR; (iv) the defocus curve must be monotonically descreasing, i.e. there should not be an increasing trend larger or equal to 0.04 logMAR; (v) for a 4.5-mm pupil, it ought to demonstrate a depth of focus of at least that of the reference monofocal aspheric lens at 0.2 logMAR [111].

On the object side an EDoF offers an extended longitudinal field range, whereas on the image plane it is translated as an extended focus range about the retina. From imaging optics, different object longitudinal positions result in different longitudinal image positions. An object located at infinity results in an image exactly on the focal plane of the lens, and any object closer than infinity results in an image beyond the focal plane. For ophthalmic optics, large distances on the object plane, from infinity to around 5 m result in an image displaced beyond the focal plane by a few micrometers. Intermediate object distances up to around 60 cm result in their images formed at around tens of micrometers from the focal plane, and closer objects from about 50 cm to 20 cm generate an image that can be a few millimeters away from the focal plane.

If a monofocal IOL is designed such that the combined focal plane of the imaging system comprising the cornea and the IOL coincide with the retina, then far objects will be imaged practically on the retina, whether intermediate objects will be imaged beyond the retina, causing a defocused (blurred) image on the retina plane. The closer the object, the more blurred the image is. How much blur is tolerated to still render an image to be perceived as sharp directly hints at the corresponding depth of field. The issue with the approach of having a lens focal length coincide with the retina is that any object located closer than infinity renders images beyond the retina, and objects can become unacceptably blurred already for intermediate distances. A common strategy to mitigate this involves the design of an IOL that, combined with the cornea, yields a focal plane before the retina. Its design can be such that the image of an object at 6 m from the eye will fall on the retina. Hence, any object closer than that will result in an image beyond the retina, and objects farther than that will result in an image closer than the retina. The object at 6 m will then be the sharpest, but a range of objects towards infinity will still be sharp enough and a range wider than the previous case will favor sufficiently sharp images of closer objects. Thus, this strategy enables a wider use of the stretch of the depth of focus for both closer and farther objects. Although this design choice helps increasing the functional depth of field, it is not enough to alone guarantee spectacle independence for near vision. There are IOLs on the market, monofocal, multifocal and EDoF, whose far-vision design is set to either

6 m or 4 m. Not surprisingly, this also favors performance evaluation in the doctor's office, where the reference visual targets for far vision (e.g. Snellen charts) are designed to be observed at 6 m in the uncorrected visual-acuity tests.

Monofocal lenses are set to far vision and the use of glasses is expected for near (and intermediate) vision. Enhanced monofocals, also referred to as monofocal plus, as the Tecnis Eyehence ICB00 lens (Johnson & Johnson Vision) or the RayOne EMV (Rayner), play with the amount of spherical aberration in the predominantly aspheric design and improve the distance range by some extent towards far intermediate, but not enough to qualify it as EDoF.

Multifocal lenses feature specific focal ranges with dips in the visual acuity in between, which introduce some degree of dysphotopic effects due to the simultaneous presence of a few clear coexisting and overlapping retinal images corresponding to these different privileged object distances.

EDoFs are designed to offer a more continuous and good enough visual acuity distributed throughout a wider range, reaching closer distances from the eye, while featuring no moving parts, as opposed to accommodative lenses.

Pseudophakic patients implanted with EDoF lenses that deliver good visual acuity from long to short distances are expected to experience reduced spectacle dependency. The same applies to phakic patients with an advanced state of presbyopia. The performance of the current models, however, often still suggest the adoption of the micro-monovision technique, in which some degree of myopia (δ 0.75 D) is deliberately introduced in one of the eyes to enable a wider range of vision when using both eyes [112].

The extended depth of focus can be achieved by tweaking the lens design in either diffractive or refractive IOLs. Diffractive models that attend to some EDoF characteristics include the Symfony XZR00 (Johnson & Johnson Vision) and AT LARA 829MP (Carl Zeiss Meditec), where the part of the diffractive features aiming at near vision are partly shifted towards intermediate vision with some power addition [113]. Refractive models with zonal modifications have also been launched, as the MiniWell lens (Sifi) with three concentric optical zones, the two inner ones with tuned positive and negative amounts of spherical aberration added. The Suprahob Infocus lens (Appasamy) also features three concentric refractive zones, with the central one for near vision, the middle one for intermediate vision and the peripheral one for far vision [112]. There are also some of the hydrophobic acrylic inverted meniscus artIOL lenses (Voptica) with EDoF characteristics that promote improved peripheral vision [112]. Other examples of candidates for refractive EDoFs are the Lentis Mplus X (Oculentis) and the Lucidis (Swiss Advanced Vision). Rayone EMV (Rayner) is an enhanced monofocal lens that plays with spherical aberration on its refractive profile and offers binocular EDoF experience when combined with some degree of monovision. A newer EDoF refractive model is the Vivity lens (Alcon) that features some concentric annular transition regions on the anterior surface, which locally redirect the refracted wavefront propagation, leading to an extension of the focal depth, reaching beyond −2 D on the corneal plane, for pupils smaller than 3 mm, with an acuity better than 0.2 logMAR for binocular vision [114]. And a similar new release, ELON (Medicontur), inferring elongated depth of focus, also with concentric refractive patterns.

4.7 Small-aperture technology

It can be difficult to optimize refractive outcomes and meet patient expectations for eyes with higher-than-average irregular corneal astigmatism or increased HOAs. These patients might not be suitable for certain categories of IOL [115]. A relatively new class of device available to reduce the burden of presbyopia and/or irregular astigmatism and HOA at the time of cataract surgery is small aperture.

4.7.1 Small-aperture optics

Focus and depth of field are enhanced by the presence of a small aperture that blocks peripheral and unfocused light rays. Depth of field refers to the distance between the closest and furthest objects that remain in focus. As well as improving depth of field, a small aperture also reduces the effect of corneal aberrations on vision. There is a known association between the size of the pupil (our natural aperture) and HOA. Paraxial light rays that are less susceptible to aberrations will pass through a small pupil to reach the retina, whilst light rays from the more peripheral and aberrated cornea will be blocked by the iris. Small pupils consequently reduce the effect of corneal aberrations but transmit less total light to the retina.

Small aperture optics (pinholes) are frequently utilized at the time of visual assessment to reduce the effects of refractive error and HOA. Small aperture optics have been applied to the field of ophthalmology in several other ways to reduce the effects of refractive error, presbyopia, irregular astigmatism and HOA. Strategies have included topical miotic agents, corneal inlays, and intraocular implants designed for the ciliary sulcus or capsular bag.

4.7.2 Small-aperture IOL

The IC-8 small aperture IOL (AcuFocus, Inc. Irvine, CA, USA) is a single-piece hydrophobic acrylic IOL that includes a central embedded opaque mask with a central aperture of 1.36 mm (figure 4.6). The aperture is similar to that of the Kamra corneal inlay (CorneaGen, Seattle, Washington, USA), however, the mask is included in the optic of an IOL that is intended for implantation into the capsular bag of the non-dominant eye at the time of cataract surgery in order to increase the depth of focus and/or decrease the impact of an irregular or aberrated cornea (figure 4.7). The IC-8 is available in powers from +10.0 D to +30.0 D in 0.5 D increments. A myopic target (−0.75 to −1.0 D) is usually employed to enhance the usability of the increased DoF and therefore to reduce the effects of presbyopia.

105 patients undergoing cataract surgery as part of one multicentre study received the IC-8 IOL in one eye with a refractive target of −0.75 and an aspheric monofocal IOL in the fellow eye with an emmetropic target [112]. At 6 months for the IC-8 eye, mean uncorrected distance, intermediate and near visual acuities were 6/7.5, 6/7.5 and 6/9, respectively. Binocular results were good with 84.8% using spectacles only occasionally or never and 95.9% of patients reporting they would undergo the same procedure again. In that study, eyes with the monofocal IOL had better contrast

Figure 4.6. The IC-8 IOL is a single-piece hydrophobic acrylic IOL that includes a central embedded opaque mask with a central aperture of 1.36 mm (image supplied by AcuFocus).

AcuFocus)

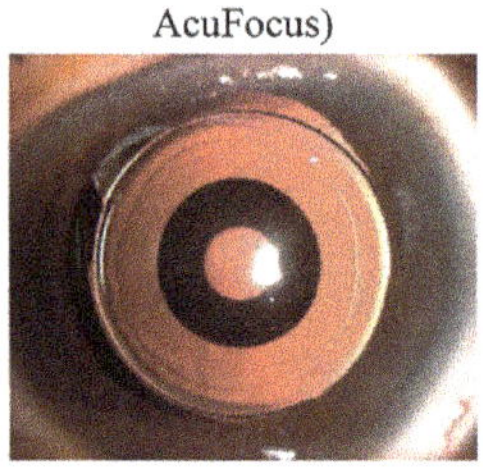

Figure 4.7. Clinical photograph showing the IC-8 IOL well-centred within the capsular bag of an eye following cataract surgery.

sensitivity than eyes with the IC-8 IOL, however binocular contrast sensitivity matched the monocular contrast sensitivity of the monofocal eye [116].

Another multi-surgeon series of 126 patients receiving mini monovision (−0.75 D target) with IC-8 found 98% of eyes without comorbidity achieved uncorrected distance visual acuity of 6/9 or better, 94% achieved uncorrected intermediate of 6/12 or better and 91% achieved 6/12 or better uncorrected for near [117]. That study also reported excellent patient satisfaction.

The above studies and others confirm that the IC-8 IOL provides an excellent option for presbyopia correction at the time of cataract surgery [116–118]. To achieve this, the IC-8 IOL is usually combined with some degree of monovision. Monovision is a well-established and successful means of reducing effects of presbyopia for patients undergoing cataract or refractive surgery, and has been described in chapter 2 [119, 120]. Monovision at the time of cataract surgery generally uses monofocal IOLs to correct the dominant eye for emmetropia and the non-dominant eye for varying degrees of myopia (common target range −0.75 to −2.00 D) [120]. This intended myopia improves the patients' intermediate and near vision depending on the degree of myopia targeted with the intention of reducing the individual patients' dependency on spectacles for some tasks. Moderate monovision is a popular approach today. The target refraction for the near eye in moderate monovision is −1.25 D (range −1.0 to −1.5 D). The aim of moderate monovision is to enhance intermediate vision which is particularly important today with our use of computers and devices and to reduce the negative effects of traditional monovision such as reduced distance visual acuity, anisometropia, poor stereopsis, reduced contrast sensitivity, and poor patient satisfaction [119, 120]. Patients with moderate monovision do often still require spectacles for some tasks, in fact one study found that only 27% of patients were totally independent from distance and reading glasses 3–4 months after cataract surgery with moderate monovision target [38]. The proportion of spectacle free patients is smaller again if mini monovision (0.75 D) is the target [121].

More patients are spectacle free with multifocal IOLs than with monovision, however patients report more symptoms (glare, haloes) with multifocal IOLs than they generally do with monovision [122]. These side effects are greatly enhanced for patients with irregular astigmatism or higher HOA.

The small aperture IOL also provides improved 'forgivability' in terms of refractive target. 1 or 2 lines of vision would be lost with a deviation of 0.5 D from refractive target with a regular monofocal IOL. Patients receiving the IC-8 IOL generally achieve approximately 6/7.5–6/9 visual acuity for distance despite refractive error ranging from 0.50 to −1.50 D [116–118]. The IC-8 IOL has been shown to have a DoF of approximately 3.0 D [117]. Aiming slightly myopic (generally −0.75 to −1.0 D) shifts the defocus curve to the left (myopic) direction, bringing this depth of focus into the usable range for intermediate and near vision. Figure 4.8 shows the binocular defocus curve for the IC-8 IOL.

Results with the IC-8 have been shown to be very good, even in the presence of up to 1.5 D of corneal astigmatism. No clinical or statistical difference

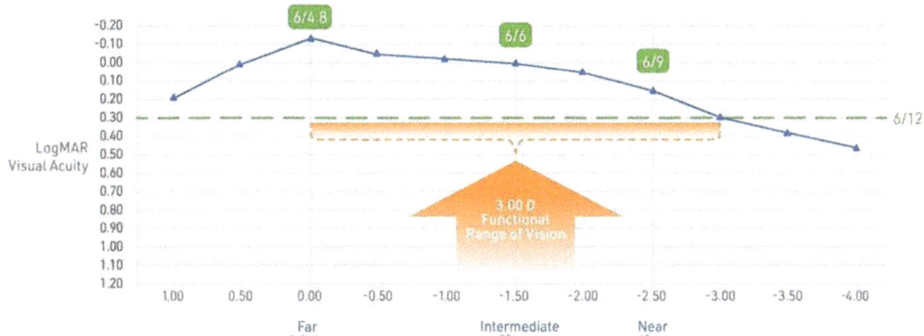

Figure 4.8. Binocular defocus curve for IC-8 IOL shows a 3.0 D range of functional vision [123].

in outcomes for patients with up to 0.75 D of corneal astigmatism versus 0.76–1.5 D of corneal astigmatism has been found [117, 124]. Trifocal and monofocal IOLs do not exhibit this same tolerance to residual refractive astigmatism [125]. This tolerance with the IC-8 reduces the need for as accurate pre-operative measurements and calculations, removes the need for accurate and sustained IOL alignment and minimises the effect of other factors such as unpredictable surgically induced astigmatism, rotation of toric IOLs, unstable and unpredictable limbal relaxing incisions and minimises the impact of pupil size on vision. For most patients with less than 1.5 D of regular or irregular corneal astigmatism the IC-8 IOL is a good option.

The IC-8 IOL may also be a promising option for the treatment of patients with increased corneal aberrations such as those that have undergone previous corneal refractive surgery [126] and patients with previous radial keratotomy (RK) [127]. In some of these cases, especially for post-refractive surgery patients who would usually have high motivation for spectacle independence but less predictable results, the IC-8 IOL might be provided as a means of reducing spectacle independence with more forgivability and fewer post-operative symptoms than multifocal IOLs. For other cases, such as the very highly aberrated post RK eyes, the IC-8 might be employed to improve the refractive forgivability and outcome, even in spectacles, especially as the cornea becomes more hyperopic over time [127].

The IC-8 IOL is contraindicated for patients with central corneal opacities, for patients with mesopic pupil size greater than 6.0 mm and for patients with known ocular pathology such as macular degeneration or advanced glaucoma. Patients with larger pupil size may experience glare or haloes, symptoms that are otherwise rarely associated with this IOL. Retinal examination remains possible at the slit lamp, via indirect ophthalmoscopy and with devices such as ocular coherence tomography, retinal photography, and angiography [128]. Although it is not possible to see through the black mask, examination can be performed both through the central aperture as well as around the edge of the mask. Vitrectomies have been performed without the need to remove the IOL [129]. Most types of imaging have now been tested with this IOL and shown to be relatively easy to perform [128]. Posterior capsular opacification can be more symptomatic given the small size of the aperture and importance of clear media within that zone. As a result, YAG laser capsulotomy might be required earlier than it would be with other IOLs and can be performed either through the central aperture or around the mask.

4.7.3 XtraFocus intraocular pinhole

The XtraFocus intraocular pinhole (IOPH) device (Morcher GmbH, Germany) is a black, foldable hydrophobic acrylic implant that is intended for implantation into the ciliary sulcus of pseudophakic patients either at the time of or as a secondary procedure after cataract surgery (figure 4.9). This device is not a lens as it includes a central aperture but does not include any dioptric power. The XtraFocus IOPH is implanted in the sulcus as an add-on to the IOL that has been placed in the capsular bag. The implant is designed to reduce the effect of irregular

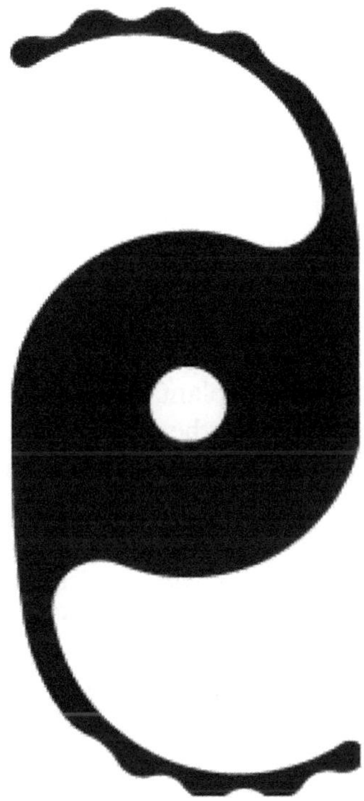

Figure 4.9. The XtraFocus intraocular pinhole device is a black, foldable implant with a central aperture intended for implantation into the ciliary sulcus. (Image supplied for publication by Morcher).

astigmatism or other corneal aberrations in eyes with complex corneal pathologies [128, 129].

The XtraFocus IOPH 'optic' has a diameter of 6 mm and contains a central aperture of 1.3 mm diameter. The primary indication for implantation is for the treatment of irregular corneal astigmatism, for example eyes with corneal ectasia, post-keratoplasty, post-trauma, or post-radial keratotomy, although it can also be used to increase depth of field and therefore to enhance near vision. The XtraFocus IOPH can be used in combination with any IOL, including toric versions. The small aperture of the implant can be used to reduce the effects of some degree of regular corneal astigmatism, but eyes with moderate to high degree regular corneal astigmatism in combination with irregular astigmatism (e.g. corneal ectasia) may benefit from the combination of a toric IOL in the capsular bag and an XtraFocus IOPH in the ciliary sulcus [130]. Similar to the IC-8, it is important that patients receiving the XtraFocus implant have a clear central cornea without opacity in the vicinity of the aperture. They should have visually significant corneal aberrations and a large pupillary diameter.

As with all surgeries, it is important to communicate expectations clearly with patients pre-operatively. Placing pinhole occlusion on top of the best refraction pre-operatively may provide adequate simulation but is not a direct predictor of the result or the success. Generally, the refractive target should be myopic (approximately −2.00 for cases with highly aberrated corneas). Depending on corneal status, patients should understand that they will likely be expected to require spectacles for all tasks after surgery.

As with other small apertures, the XtraFocus IOPH reduces light entry into the eye. Patients may notice a sensation of dark vision, especially under low-light conditions. This perception is generally well tolerated, especially in patients with highly aberrated corneas given the dramatic improvement in visual acuity and other advantages that they may have achieved, although it may lead to poor tolerance and explantation of the device [131, 133, 134]. In addition, it may be possible for patients' brightness perception to improve with time, most likely through neuro-adaptation [132]. It is extremely important to discuss brightness perception with patients pre-operatively, especially if bilateral implantation is to be considered. It should also be highlighted that it is not possible to view the peripheral retina through this device with regular equipment. The black acrylic material is transparent to infrared light only, as shown in figure 4.10, therefore allowing examination of structures posterior to the implant with infrared based equipment (such as OCT, or an infrared slit lamp filter). Patients requiring vitrectomy for retinal detachment repair or similar may require removal of the implant at the time of surgery for intra-operative visualisation and post-operative care.

Although further investigation and longer follow up is required, the XtraFocus IOPH is a promising device for pseudophakic patients with highly aberrated corneas (figure 4.11). A prospective case series looking at 21 eyes found statistically significant improvement in uncorrected and corrected distance visual acuities in cases of irregular corneal astigmatism with significant visual impairment [131]. In that study, median corrected distance visual acuity improved from 6/60 pre-operatively to 6/15 post-operatively (range 6/60–6/6). Patient satisfaction was reportedly high in that study.

Figure 4.10. The XtraFocus intraocular pinhole device is transparent to infrared light (image supplied by Morcher).

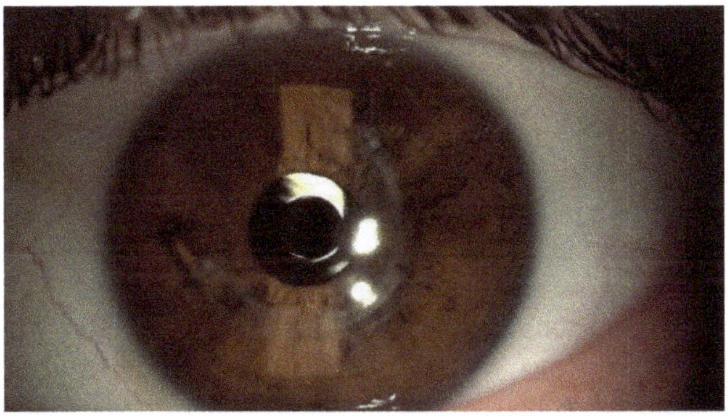

Figure 4.11. This is the left eye of a 47 year-old patient with keratoconus and an intracorneal ring segment *in situ*, 1 year after refractive lens extraction and insertion of a toric IOL in the bag and an XtraFocus intraocular pinhole device in the ciliary sulcus. This patient achieves and has maintained uncorrected visual acuity of 6/7.5 and reports excellent satisfaction (image courtesy of Drs Claudio and Bruno Trindade).

4.7.4 Refracting patients with small-aperture implants

It can be difficult to determine the final refraction of a patient with any type of small aperture device *in situ*. This relates to the extended depth of focus and lack of hard end point to refraction. The process of refraction can be slow and frustrating for patients, especially when they are functioning well independently from spectacles. Good communication with patients is essential at this time.

The most common method for refraction with a small aperture *in situ* is a midpoint refraction. During a midpoint refraction, a manifest refraction is performed and then the maximum plus lens to blur is established followed by the maximum minus lens to blur. The midpoint of those two measurements can then be calculated. A duochrome test can be used as an alternative to determine the maximum plus and minus.

4.7.5 Wrap-up on small-aperture devices

Small-aperture devices utilize straightforward optics that are simple to understand. This group of devices provides another option for presbyopia correction and also benefits patients with highly aberrated corneas and/or irregular astigmatism.

The IC-8 IOL and the XtraFocus IOPH are both effective and useful options. Whilst these two products are fundamentally different, they have been demonstrated to share some of the same properties [131, 135, 136]. Both devices employ the principles of small aperture optics to reduce the effects of irregular corneal astigmatism and corneal aberrations as well as to enhance DoF.

Intraocular implants containing small aperture optics are gaining popularity because of their easy application and versatility. Hopefully we will see more innovation in this area, leading to better options for patients seeking monovision and for patients with irregular corneal astigmatism and higher order corneal aberrations.

4.8 Accommodative

4.8.1 Concept

The accommodating intraocular lenses (AIOLs) have been developed with the intent to restore the accommodation capability of the human lens. Accommodation is the process in which the human lens dynamically changes its shape to focus objects at different distances on the retina [137].

There is a variety of possible mechanisms that have been explored in the past decades to convert the mechanical stimuli from the ciliary muscles into a dynamic change on the dioptric power of the lens [138–141]. Basically, the AIOLs can be separated in two main groups: active or passive designs depending whether the design is composed or not by an electronic actuator. Within the two main groups, the change in diopter employed in the AIOLs described in the literature are, in general, based on one or a combination of the following mechanisms: position change of one or more optical parts (the change in position can be established by axial or radial displacement depending on the lens design), change in optical shape or change in refractive index [138–141]. In the following subsection a more detailed analysis of the working principle of the existing technology is discussed.

4.8.2 Existing technology

In this subsection are presented some of the AIOLs that are available on the market and under development. The first AIOLs launched on the market were based on a single-optic design [138–141]. The BioComFold 43A model (by Morcher GmbH, Germany 1996), the 1CU model (by HumanOptics AG, Germany, 2001) and the Crystalens AT-45 model (by the former C&C Vision and now Eyeonics, California, 2002) are all examples of single-optic AIOLs with different haptic structures to convert the ciliary muscles contraction into axial anteriorization displacement of the optics to increase the lens power (figure 4.12) [138]. For instance, the performance evaluation of the Crystalens AT-45 model based on the clinical trial results of the U.S. Food and Drug Administration (FDA) organization showed that the Crystalens group needed a mean add of +1.24 D to achieve monocular best correct for near acuity which, in

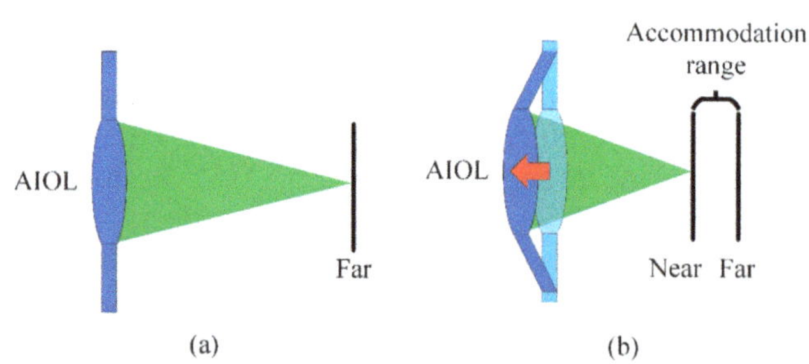

Figure 4.12. Single-optic accommodative lens working principle: (a) lens in its non-accommodating state providing far vision and (b) the lens moves to anterior direction due to the accommodation effort to provide near vision.

comparison with the monofocal reference group that required +2.32 D, corresponds to a difference of +1.08 D in accommodation power [142].

The amount of power accommodation given by an axial displacement of a single-optic AIOL depends on several factors such as lens power, axial length (AL) of the eye and corneal power [143]. Considering an average eye with AL of 24 mm, a cornea with anterior radius of curvature of 7.7 mm and an IOL power of 20 D, the amount of power change due to a 1 mm of axial displacement of the lens corresponds to 1.3 D [143]. However, studies analyzing the axial displacement of the Crystalens AT-45 and BioComFold AIOLs showed a mean change of 0.33 mm and 0.22 mm, respectively [144, 145], which is not significant enough to yield a 1 D of accommodation [140]. Based on that, one of the possible reasons that combined with the limited axial displacement of the single-optic AIOLs would explain the accommodative power of 1 D and larger observed in some clinical studies relies on the shape and alignment changes that occur on the optics of the lens during the accommodating effort [139]. For instance, a clinical study has shown changes in aberrations such as astigmatism, coma and spherical aberration in patients with the Crystalens AIOL when compared to the monofocal group of patients [146].

With the purpose to overcome the limited accommodating amplitude observed in single-optic AIOLs, new lens designs formed by two optical elements (also known as dual-optic design) were developed. Among the dual-optic IOL designs that were patented and/or have clinical trial results, it is possible to mention the Synchrony (from Abbott Medical Optics Inc., USA), Lumina lens (from AkkoLens International BV, The Netherlands) and the Turtle lens proposed by Hermans *et al* [141, 147]. Although all three previous-mentioned IOLs are dual-optic designs, they have different working principles.

The Synchrony AIOL is formed by two lenses interconnected by the haptics that work as a spring between the two components [138, 139, 148]. The anterior part of the AIOL has constant power of 32 D and the posterior part has a negative varying power to generate a dioptric variation from +15 D to +32 D [139, 148]. The posterior part of AIOL is designed to stay stationary, while the anterior part was designed to move axially due to the accommodative effort [138, 148]. For 1 mm of displacement between both parts and considering *a posterior* part with −12 D, the diopter of accommodation achieved by the lens through simulation results was of about 2.2 D [138, 148].

The Lumina AIOL has its working principle based on the centripetal movement of two lenses and the AIOL is implanted at the sulcus (anterior to the capsular bag) [140, 141]. The anterior part of the AIOL consists of a spherical lens (to achieve the refractive power) combined with a cubic surface and the posterior part is formed by a cubic surface only that combined with the sliding movement of the anterior part is responsible for creating the focal variation effect of the AIOL [149]. Simulation results based in ray-tracing have shown an accommodation amplitude of +4 D for the Lumina AIOL when the two parts are shifted by a distance of 0.75 mm [149] and a randomized clinical trial comparing the Lumina lens with a monofocal control group has reported a higher near visual acuity for the AIOL in the defocus range from −4.50 to −0.50 D [150].

The Turtle AIOL is formed by two lenses and its working principle is based on the relative rotation of the parts due to the accommodation effort of the ciliary muscles

[148]. Ray-tracing simulation results and *ex vivo* measurements of the Turtle AIOL in porcine capsule using a stretcher have shown an accommodation range of 8 D for a lens with 21 D base power [148].

All previous-mentioned techniques to achieve accommodation are based on optics with fixed shape and the movement of the lens or the lens pair provides the diopter change. As the crystalline lens changes its shape to provide dynamic variation of the optical power of the eye, a range of accommodating IOLs have been under development and study to address this feature in its optical design [139–141]. Among the AIOLs that present this working principle it is worth mentioning the NuLens (from Herzliya Pituach, Israel), the WIOL-CF (from Medicem, Czech Republic) and the FluidVision (from PowerVision Inc., USA) [139–141].

The NuLens was designed to be implanted in the ciliary sulcus with PMMA haptics system to guarantee the fixing of the lens [151]. In a constructed prototype a PMMA cylinder chamber was filled with silicone gel that was contained posteriorly by a flat PMMA piston, that fits in the cylinder chamber [151]. The silicone gel is contained anteriorly by a flexible membrane that is attached to a PMMA ring formed by the haptics system of the lens [151]. Experimental results of the implantation of the NuLens prototype in monkey eyes reported an accommodation range of about 44 D between cyclospasm and cycloplegia and a clinical experiment in humans verified an accommodation amplitude up to 10 D after 12 months period [148] (figure 4.13).

The WIOL-CF design was developed to mimic some of the key features of the crystalline lens [153]. With a total diameter ranging between 8.6 mm and 9.0 mm and a center thickness ranging from 1.1 mm to 1.4 mm, the AIOL has no haptics and it is intended to fill the entire capsule without changing its shape [152]. Basically the lens consists in a single body formed by foldable material with high water content and its shape consists of a meniscoid anterior surface and *a posterior* polyfocal hyperboloid surface [152]. The main mechanisms that hypothetically explain the accommodative feature of the AIOL are related to changes in the anterior and posterior position of the lens due to the ciliary muscles contraction and vitreous pressure that occur during the accommodation effort combined with the surface

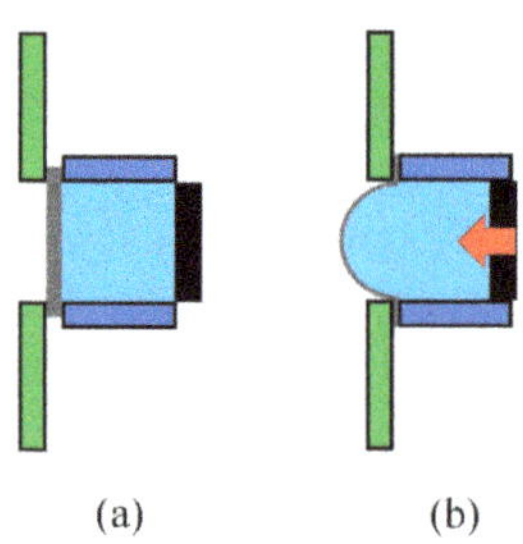

Figure 4.13. Cross-section schematic prototype of the NuLens [42]: (a) lens in non-accommodative state with PMMA cylinder (in dark blue) containing the silicone gel (in light blue), PMMA piston (in black), flexible membrane (in gray) and PMMA ring formed by the haptics system (in green) and (b) lens in accommodative state: the anteriorization of the piston compresses the silicone gel through the flexible aperture formed by the haptics creating a curved surface.

shape change [153, 154]. A clinical study trial with 40 eyes of 20 patients reported a mean objective accommodation of 2.52 ± 1.56 D for the WIOL-cf [154].

The FluidVision AIOL is formed by a hollow hydrophobic acrylic optic and haptics both filled with fluid (silicone oil) [139, 155]. The large size of the haptics is responsible for maintaining the separation of the anterior and posterior capsule with the purpose to prevent capsule opacification [155, 156]. The accommodation effort applied on the haptics is responsible for the fluid flow between the haptics and optics and therefore providing the curvature change of the optics that address the dioptric accommodation of the lens [139, 155]. A clinical trial with the lens prototype reported an accommodation power change of more than 5 D [155].

4.8.3 Promising future?

Based on the lens designs present in the state-of-the-art of accommodative IOLs, it is possible to notice that this is a technology that still needs further developments to achieve fully restoration of the natural accommodation of the eye.

Passive AIOLs with working principle based on the lens shift to achieve accommodation have proved to be unsatisfactory to provide reliable and significant refractive power change [157–159]. However, optical designs that rely on surface changes and/or refractive index change mechanisms are capable of providing high levels of dioptric power change with no need for or under minimal axial displacement of the lens [141].

In recent years, AIOLs based in electroactive mechanisms have been under development [160–162]. The working principle of these electroactive designs is based on microsensors to detect the accommodation effort from the ciliary muscles and generate a controlled signal to address the necessary dioptric change. Thus, the electroactive mechanisms are capable of providing an accommodation behavior more independent of the variations that might occur in the structures of the eye of the patient during the cataract surgery such as fibrosis and capsular shrinkage [139, 141].

In summary, a broad range of strategies have been explored to address the dynamic dioptric change of the crystalline lens and several challenges need to be overcome to guarantee a reliable and functional accommodation restoration for the patient in the long term.

Chapter highlights

- Intraocular lenses have transformed the way modern cataract surgery is performed and multiple models are available to be used in different clinical scenarios.
- Modern IOL designs are capable of correcting astigmatism and providing a wide range of uncorrected vision.
- Extended depth of focus is now being used as a strategy to elongate the useful focused distance in vision.
- Special implants can be used to correct irregular corneal astigmatism.

References

[1] Apple D J and Sims J 1996 Harold Ridley and the invention of the intraocular lens *Surv. Ophthalmol.* **40** 279–92

[2] Apple D J 2006 *Sir Harold Ridley and His Fight for Sight: He Changed the World so that We May Better See It* (Thorofare, NJ: Slack)

[3] Lieb W A and Guerry D 3rd 1957 Anterior chamber lenses; for refractive correction of unilateral aphakia *Am. J. Ophthalmol.* **44** 579–98

[4] Parry T G 1958 New design in anterior chamber lenses *Br. J. Ophthalmol.* **42** 376–9

[5] Bending G C 1960 Current status of intra-ocular acrylic lenses *Can. Med. Assoc. J.* **83** 174–7

[6] Boberg-Ans J 1961 Experience with twelve cases of intra-ocular anterior chamber implants for aphakia: two new models of lenses are described *Br. J. Ophthalmol.* **45** 37–43

[7] Troutman R C 1962 Correction of unilateral aphakia. the use of intraocular lens implants *Arch. Ophthalmol.* **68** 861–9

[8] Kalvin N H 1979 IOL design *J. Am. Intraocul. Implant. Soc.* **5** 41–2

[9] Guerry D 3rd and Geeraets W J 1962 Complications of anterior chamber lenses *Am. J. Ophthalmol.* **54** 229–32

[10] Ellingson F T 1978 The uveitis-glaucoma-hyphema syndrome associated with the mark VIII anterior chamber lens implant *J. Am. Intraocul. Implant Soc.* **4** 50–3

[11] Pearce J L 1977 Sixteen months' experience with 140 posterior chamber intraocular lens implants *Br. J. Ophthalmol.* **61** 310–5

[12] Refojo M F 1982 Current status of biomaterials in ophthalmology *Surv. Ophthalmol.* **26** 257–65

[13] Brint S F and Ostrick M 1991 The evolution of small-incision cataract surgery with foldable IOLs *J. Am. Optom. Assoc.* **62** 365–71

[14] Kulnig W and Skorpik C 1990 Optical resolution of foldable intraocular lenses *J. Cataract Refract. Surg.* **16** 211–6

[15] Pearce J L 1992 Intraocular lenses *Curr. Opin. Ophthalmol.* **3** 29–38

[16] Tetz M and Jorgensen M 2015 New hydrophobic IOL materials and understanding the science of glistenings *Curr. Eye Res.* **40** 969–81

[17] Werner L 2021 Intraocular lens evolution in the past 25 years as told by the journal of cataract & refractive surgery *J. Cataract Refract. Surg.* **47** 147–9

[18] Masket S and Fram N R 2011 Pseudophakic negative dysphotopsia: surgical management and new theory of etiology *J. Cataract Refract. Surg.* **37** 1199–207

[19] Holladay J T, Zhao H and Reisin C R 2012 Negative dysphotopsia: the enigmatic penumbra *J. Cataract Refract. Surg.* **38** 1251–65

[20] Henderson B A and Geneva I I 2015 Negative dysphotopsia: a perfect storm *J. Cataract Refract. Surg.* **41** 2291–312

[21] Werner L 2007 Causes of intraocular lens opacification or discoloration *J. Cataract Refract. Surg.* **33** 713–26

[22] Apple D J, Peng Q and Arthur S N *et al* 2002 Snowflake degeneration of polymethyl methacrylate posterior chamber intraocular lens optic material: a newly described clinical condition caused by unexpected late opacification of polymethyl methacrylate *Ophthalmology* **109** 1666–75

[23] Apple D J, Federman J L and Krolicki T J *et al* 1996 Irreversible silicone oil adhesion to silicone intraocular lenses. a clinicopathologic analysis *Ophthalmology* **103** 1555–61 discussion 61–2

[24] Rullo J and CL J 2014 Clinically significant deterioration in the quality of vision as a result of subsurface nanoglistenings in a hydrophobic acrylic intraocular lens *J. Cataract Refract. Surg.* **40** 336–7

[25] Werner L, Apple D J and Escobar-Gomez M *et al* 2000 Postoperative deposition of calcium on the surfaces of a hydrogel intraocular lens *Ophthalmology* **107** 2179–85

[26] Balendiran V, MacLean K and Mamalis N *et al* 2019 Localized calcification of hydrophilic acrylic intraocular lenses after posterior segment procedures *J. Cataract Refract. Surg.* **45** 1801–7

[27] Mehta R and Aref A A 2019 Intraocular lens implantation in the ciliary sulcus: challenges and risks *Clin. Ophthalmol.* **13** 2317–23

[28] Chang D F, Masket S and Miller K M *et al* 2009 Complications of sulcus placement of single-piece acrylic intraocular lenses: recommendations for backup IOL implantation following posterior capsule rupture *J. Cataract Refract. Surg.* **35** 1445–58

[29] Nibourg L M, Gelens E and Kuijer R *et al* 2015 Prevention of posterior capsular opacification *Exp. Eye Res.* **136** 100–15

[30] Wasserman E L, Axt J C and Sheets J H 1985 Neodymium: YAG laser posterior capsulotomy *J. Am. Intraocul. Implant Soc.* **11** 245–8

[31] Peng Q, Visessook N and Apple D J *et al* 2000 Surgical prevention of posterior capsule opacification. Part 3: intraocular lens optic barrier effect as a second line of defense *J. Cataract Refract. Surg.* **26** 198–213

[32] Olson R J, Werner L, Mamalis N and Cionni R 2005 New intraocular lens technology *Am. J. Ophthalmol.* **140** 709–16

[33] Mainster M A 2006 Violet and blue light blocking intraocular lenses: photoprotection versus photoreception *Br. J. Ophthalmol.* **90** 784–92

[34] Kohnen T and Kasper T 2005 Incision sizes before and after implantation of 6-mm optic foldable intraocular lenses using monarch and unfolder injector systems *Ophthalmology* **112** 58–66

[35] Leyland M and Zinicola E 2003 Multifocal versus monofocal intraocular lenses in cataract surgery: a systematic review *Ophthalmology* **110** 1789–98

[36] Yeu E and Cuozzo S 2020 Matching the patient to the intraocular lens: preoperative considerations to optimize surgical outcomes *Ophthalmology* **128** E132–41

[37] Ito M, Shimizu K, Iida Y and Amano R 2012 Five-year clinical study of patients with pseudophakic monovision *J. Cataract Refract. Surg.* **38** 1440–5

[38] Finkelman Y M, Ng J Q and Barrett G D 2009 Patient satisfaction and visual function after pseudophakic monovision *J. Cataract Refract. Surg.* **35** 998–1002

[39] Trindade B L C, Amaral F T, de Lima Monteiro D W and Cronemberger S 2019 Versatile optical setup customized to verify the quality of spherical and aspheric intraocular lenses *J. Opt. Soc. Am. A Opt Image Sci. Vis.* **36** B132–B7

[40] Franchini A 2006 Comparative assessment of contrast with spherical and aspherical intraocular lenses *J. Cataract Refract. Surg.* **32** 1307–19

[41] Amaral F T and Monteiro D W 2013 Alternative methodology for intraocular lens characterisation *Ophthalmic Physiol. Opt.* **33** 427–33

[42] Linke S J, Baviera J and Munzer G *et al* 2012 Mesopic pupil size in a refractive surgery population (13,959 eyes) *Optom. Vis. Sci.* **89** 1156–64

[43] Laikin M 2007 *Lens Design* 4th edn (Boca Raton, FL: CRC Press) p 478

[44] Malacara D and Malacara Z 2013 *Handbook of Optical Design* 3rd edn (Boca Raton, FL: CRC Press, Taylor and Francis)

[45] Dubbelman M, Van and der Heijde G L 2001 The shape of the aging human lens: curvature, equivalent refractive index and the lens paradox *Vision Res.* **41** 1867–77
[46] Glasser A and Campbell M C 1999 Biometric, optical and physical changes in the isolated human crystalline lens with age in relation to presbyopia *Vision Res.* **39** 1991–2015
[47] Holladay J T, Piers P A and Koranyi G *et al* 2002 A new intraocular lens design to reduce spherical aberration of pseudophakic eyes *J. Refract. Surg.* **18** 683–91
[48] Guirao A, Gonzalez C and Redondo M *et al* 1999 Average optical performance of the human eye as a function of age in a normal population *Invest. Ophthalmol. Vis. Sci.* **40** 203–13
[49] Nio Y K, Jansonius N M and Fidler V *et al* 2000 Age-related changes of defocus-specific contrast sensitivity in healthy subjects *Ophthalmic Physiol. Opt.* **20** 323–34
[50] Packer M, Fine I H and Hoffman R S 2008 Aspheric intraocular lens selection: the evolution of refractive cataract surgery *Curr. Opin. Ophthalmol.* **19** 1–4
[51] Montes-Mico R, Ferrer-Blasco T and Cervino A 2009 Analysis of the possible benefits of aspheric intraocular lenses: review of the literature *J. Cataract Refract. Surg.* **35** 172–81
[52] Wang L, Dai E, Koch D D and Nathoo A 2003 Optical aberrations of the human anterior cornea *J. Cataract Refract. Surg.* **29** 1514–21
[53] Langenbucher A, Janunts E and Seitz B *et al* 2014 Theoretical image performance with customized aspheric and spherical IOLs – when do we get a benefit from customized aspheric design? *Z. Med. Phys.* **24** 94–103
[54] Mester U, Dillinger P and Anterist N 2003 Impact of a modified optic design on visual function: clinical comparative study *J. Cataract Refract. Surg.* **29** 652–60
[55] Bellucci R, Scialdone A and Buratto L *et al* 2005 Visual acuity and contrast sensitivity comparison between Tecnis and AcrySof SA60AT intraocular lenses: a multicenter randomized study *J. Cataract Refract. Surg.* **31** 712–7
[56] Denoyer A, Le Lez M L, Majzoub S and Pisella P J 2007 Quality of vision after cataract surgery after Tecnis Z9000 intraocular lens implantation: effect of contrast sensitivity and wavefront aberration improvements on the quality of daily vision *J. Cataract Refract. Surg.* **33** 210–6
[57] Pandita D, Raj S M and Vasavada V A *et al* 2007 Contrast sensitivity and glare disability after implantation of AcrySof IQ natural aspherical intraocular lens: prospective randomized masked clinical trial *J. Cataract Refract. Surg.* **33** 603–10
[58] Einighammer J, Oltrup T and Feudner E *et al* 2009 Customized aspheric intraocular lenses calculated with real ray tracing *J. Cataract Refract. Surg.* **35** 1984–94
[59] Piers P A, Weeber H A, Artal P and Norrby S 2007 Theoretical comparison of aberration-correcting customized and aspheric intraocular lenses *J. Refract. Surg.* **23** 374–84
[60] Nochez Y, Favard A, Majzoub S and Pisella P J 2010 Measurement of corneal aberrations for customisation of intraocular lens asphericity: impact on quality of vision after micro-incision cataract surgery *Br. J. Ophthalmol.* **94** 440–4
[61] Schrecker J, Langenbucher A, Seitz B and Eppig T 2018 First results with a new intraocular lens design for the individual correction of spherical aberration *J. Cataract Refract. Surg.* **44** 1211–9
[62] Eppig T, Schrecker J, Messner A and Langenbucher A 2020 Aberration correction with aspheric intraocular lenses *Intraocular Lens* (London: IntechOpen)
[63] Whang W J, Piao J and Yoo Y S *et al* 2017 The efficiency of aspheric intraocular lens according to biometric measurements *PLoS One* **12** e0182606

[64] McAlinden C, Khadka J and Pesudovs K 2011 A comprehensive evaluation of the precision (repeatability and reproducibility) of the oculus pentacam HR *Invest. Ophthalmol. Vis. Sci.* **52** 7731–7

[65] Eppig T, Scholz K and Loffler A *et al* 2009 Effect of decentration and tilt on the image quality of aspheric intraocular lens designs in a model eye *J. Cataract Refract. Surg.* **35** 1091–100

[66] McKelvie J, McArdle B and McGhee C 2011 The influence of tilt, decentration, and pupil size on the higher-order aberration profile of aspheric intraocular lenses *Ophthalmology* **118** 1724–31

[67] Tabernero J, Piers P and Benito A *et al* 2006 Predicting the optical performance of eyes implanted with IOLs to correct spherical aberration *Invest. Ophthalmol. Vis. Sci.* **47** 4651–8

[68] Rocha K M, Soriano E S and Chamon W *et al* 2007 Spherical aberration and depth of focus in eyes implanted with aspheric and spherical intraocular lenses: a prospective randomized study *Ophthalmology* **114** 2050–4

[69] Wang L and Koch D D 2007 Custom optimization of intraocular lens asphericity *J. Cataract Refract. Surg.* **33** 1713–20

[70] Read S A, Collins M J and Carney L G 2007 A review of astigmatism and its possible genesis *Clin. Exp. Optom.* **90** 5–19

[71] Anderson D F, Dhariwal M, Bouchet C and Keith M S 2018 Global prevalence and economic and humanistic burden of astigmatism in cataract patients: a systematic literature review *Clin. Ophthalmol.* **12** 439–52

[72] Curragh D S and Hassett P 2017 Prevalence of corneal astigmatism in an NHS cataract surgery practice in Northern Ireland *Ulster Med. J.* **86** 25–7

[73] Ferrer-Blasco T, Montes-Mico R and Peixoto-de-Matos S C *et al* 2009 Prevalence of corneal astigmatism before cataract surgery *J. Cataract Refract. Surg.* **35** 70–5

[74] Day A C, Dhariwal M and Keith M S *et al* 2019 Distribution of preoperative and postoperative astigmatism in a large population of patients undergoing cataract surgery in the UK *Br. J. Ophthalmol.* **103** 993–1000

[75] Solomon K D, Sandoval H P and Potvin R 2019 Correcting astigmatism at the time of cataract surgery: Toric IOLs and corneal relaxing incisions planned with an image-guidance system and intraoperative aberrometer versus manual planning and surgery *J. Cataract Refract. Surg.* **45** 569–75

[76] Hirnschall N, Gangwani V and Crnej A *et al* 2014 Correction of moderate corneal astigmatism during cataract surgery: toric intraocular lens versus peripheral corneal relaxing incisions *J. Cataract Refract. Surg.* **40** 354–61

[77] Hayashi K, Yoshida M, Hirata A and Yoshimura K 2018 Changes in shape and astigmatism of total, anterior, and posterior cornea after long versus short clear corneal incision cataract surgery *J. Cataract Refract. Surg.* **44** 39–49

[78] Visser N, Bauer N J and Nuijts R M 2013 Toric intraocular lenses: historical overview, patient selection, IOL calculation, surgical techniques, clinical outcomes, and complications *J. Cataract Refract. Surg.* **39** 624–37

[79] Rho C R and Joo C K 2012 Effects of steep meridian incision on corneal astigmatism in phacoemulsification cataract surgery *J. Cataract Refract. Surg.* **38** 666–71

[80] Langenbucher A, Viestenz A and Szentmary N *et al* 2009 Toric intraocular lenses – theory, matrix calculations, and clinical practice *J. Refract. Surg.* **25** 611–22

[81] Bauer N J, de Vries N E and Webers C A *et al* 2008 Astigmatism management in cataract surgery with the AcrySof toric intraocular lens *J. Cataract Refract. Surg.* **34** 1483–8
[82] Hoffmann P C and Hutz W W 2010 Analysis of biometry and prevalence data for corneal astigmatism in 23,239 eyes *J. Cataract Refract. Surg.* **36** 1479–85
[83] Chen W, Zuo C and Chen C *et al* 2013 Prevalence of corneal astigmatism before cataract surgery in Chinese patients *J. Cataract Refract. Surg.* **39** 188–92
[84] Hayashi K, Manabe S I, Hirata A and Yoshimura K 2017 Changes in corneal astigmatism during 20 years after cataract surgery *J. Cataract Refract. Surg.* **43** 615–21
[85] Wolffsohn J S, Bhogal G and Shah S 2011 Effect of uncorrected astigmatism on vision *J. Cataract Refract. Surg.* **37** 454–60
[86] Read S A, Vincent S J and Collins M J 2014 The visual and functional impacts of astigmatism and its clinical management *Ophthalmic Physiol. Opt.* **34** 267–94
[87] Villegas E A, Alcon E and Artal P 2014 Minimum amount of astigmatism that should be corrected *J. Cataract Refract. Surg.* **40** 13–9
[88] Sigireddi R R and Weikert M P 2020 How much astigmatism to treat in cataract surgery *Curr. Opin. Ophthalmol.* **31** 10–4
[89] Hayashi K, Sato T and Sasaki H *et al* 2018 Sex-related differences in corneal astigmatism and shape with age *J. Cataract Refract. Surg.* **44** 1130–9
[90] Waltz K L, Featherstone K, Tsai L and Trentacost D 2015 Clinical outcomes of TECNIS toric intraocular lens implantation after cataract removal in patients with corneal astigmatism *Ophthalmology* **122** 39–47
[91] Visser N, Beckers H J and Bauer N J *et al* 2014 Toric vs aspherical control intraocular lenses in patients with cataract and corneal astigmatism: a randomized clinical trial *JAMA Ophthalmol.* **132** 1462–8
[92] Kramer B A, Berdahl J P, Hardten D R and Potvin R 2016 Residual astigmatism after toric intraocular lens implantation: analysis of data from an online toric intraocular lens back-calculator *J. Cataract Refract. Surg.* **42** 1595–601
[93] Visser N, Berendschot T T and Bauer N J *et al* 2011 Accuracy of toric intraocular lens implantation in cataract and refractive surgery *J. Cataract Refract. Surg.* **37** 1394–402
[94] Felipe A, Artigas J M and Diez-Ajenjo A *et al* 2011 Residual astigmatism produced by toric intraocular lens rotation *J. Cataract Refract. Surg.* **37** 1895–901
[95] Koch D D, Ali S F and Weikert M P *et al* 2012 Contribution of posterior corneal astigmatism to total corneal astigmatism *J. Cataract Refract. Surg.* **38** 2080–7
[96] Reitblat O, Levy A and Kleinmann G *et al* 2016 Effect of posterior corneal astigmatism on power calculation and alignment of toric intraocular lenses: comparison of methodologies *J. Cataract Refract. Surg.* **42** 217–25
[97] Zvornicanin J and Zvornicanin E 2018 Premium intraocular lenses: the past, present and future *J. Curr. Ophthalmol.* **30** 287–96
[98] Rampat R and Gatinel D 2020 Multifocal and extended depth-of-focus intraocular lenses in 2020 *Ophthalmology* **128** E164–85
[99] Ravalico G, Parentin F, Sirotti P and Baccara F 1998 Analysis of light energy distribution by multifocal intraocular lenses through an experimental optical model *J. Cataract Refract. Surg.* **24** 647–52
[100] Liao X, Lin J and Tian J *et al* 2018 Evaluation of optical quality: ocular scattering and aberrations in eyes implanted with diffractive multifocal or monofocal intraocular lenses *Curr. Eye Res.* **43** 696–701

[101] Modi S, Lehmann R and Maxwell A *et al* 2020 Visual and patient-reported outcomes of a diffractive trifocal intraocular lens compared with those of a monofocal intraocular lens *Ophthalmology.* **128** P197–207
[102] Alio J L, Plaza-Puche A B and Fernandez-Buenaga R *et al* 2017 Multifocal intraocular lenses: an overview *Surv. Ophthalmol.* **62** 611–34
[103] Kamiya K, Hayashi K and Shimizu K *et al* 2014 Multifocal intraocular lens explantation: a case series of 50 eyes *Am. J. Ophthalmol.* **158** 215–20 e1
[104] Kim E J, Sajjad A and Montes de Oca I *et al* 2017 Refractive outcomes after multifocal intraocular lens exchange *J. Cataract Refract. Surg.* **43** 761–6
[105] Rementeria-Capelo L A, Contreras I and Garcia-Perez J L *et al* 2019 Visual quality and patient satisfaction with a trifocal intraocular lens and its new toric version *J. Cataract Refract. Surg.* **45** 1584–90
[106] Todd T and Schmitz J 2019 Toric and mulitifocal lens utilization independent of patient cost and physician remuneration at a single institution *Am. J. Ophthalmol. Case Rep.* **15** 100500
[107] Yildirim Karabag R, Gunenc U and Aydin R *et al* 2018 Visual results following implantation of a refractive multifocal intraocular lens in one eye and a diffractive in the contralateral eye *Turk. J. Ophthalmol.* **48** 6–14
[108] Wenzel M R, Imkamp E M and Apple D J 1992 Variations in manufacturing quality of diffractive multifocal lenses *J. Cataract Refract. Surg.* **18** 153–6
[109] Cochener B, Lafuma A and Khoshnood B *et al* 2011 Comparison of outcomes with multifocal intraocular lenses: a meta-analysis *Clin. Ophthalmol.* **5** 45–56
[110] Choi J and Schwiegerling J 2008 Optical performance measurement and night driving simulation of ReSTOR, ReZoom, and Tecnis multifocal intraocular lenses in a model eye *J. Refract. Surg.* **24** 218–22
[111] ANSI Z80.35-2018 2019 American Vision Standards Institute, Inc., The Vision Council, The Accredited Committee Z80 for Ophthalmic Standards
[112] Bilbao-Calabuig R, Gónzalez-López F, Llovet-Rausell A, Ortega-Usobiaga J, Tejerina Fernández V and Llovet-Osuna F 2021 Lens-based surgical correction of presbiopia. where are we in 2020? *Archiv. Soc. Esp. Oftalmol.* **96** 74–88
[113] Kanclerz P, Toto F, Grzybowski A and Alio J 2020 Extended depth-of-field intraocular lenses: an update *Asia-Pacific Academy of Ophthalmology* **9** 194–202
[114] Bala C, Poyales F, Guarro M, Mesa R R, Mearza A, Varma D K, Jasti S and Lemp-Hull J 2022 Multicountry clinical outcomes of a new nondiffractive presbyopia-correcting IOL *J. Cataract Refract. Surg.* **48** 136–43
[115] de Vries N E, Webers C A and Touwslager W R *et al* 2011 Dissatisfaction after implantation of multifocal intraocular lenses *J. Cataract Refract. Surg.* **37** 859–65
[116] Dick H B, Piovella M and Vukich J *et al* 2017 Prospective multicenter trial of a small-aperture intraocular lens in cataract surgery *J. Cataract Refract. Surg.* **43** 956–68
[117] Hooshmand J, Allen P and Huynh T *et al* 2019 Small aperture IC-8 intraocular lens in cataract patients: achieving extended depth of focus through small aperture optics *Eye (Lond.)* **33** 1096–103
[118] Grabner G, Ang R E and Vilupuru S 2015 The small-aperture IC-8 intraocular lens: a new concept for added depth of focus in cataract patients *Am. J. Ophthalmol.* **160** 1176–84 e1
[119] Wright K W, Guemes A, Kapadia M S and Wilson S E 1999 Binocular function and patient satisfaction after monovision induced by myopic photorefractive keratectomy *J. Cataract Refract. Surg.* **25** 177–82

[120] Handa T, Mukuno K and Uozato H *et al* 2004 Ocular dominance and patient satisfaction after monovision induced by intraocular lens implantation *J. Cataract Refract. Surg.* **30** 769–74
[121] Goldberg D G, Goldberg M H and Shah R *et al* 2018 Pseudophakic mini-monovision: high patient satisfaction, reduced spectacle dependence, and low cost *BMC Ophthalmol.* **18** 293
[122] Greenstein S and Pineda R 2nd 2017 The quest for spectacle independence: a comparison of multifocal intraocular lens implants and Pseudophakic monovision for patients with presbyopia *Semin. Ophthalmol.* **32** 111–5
[123] Burkhard D *et al* 2017 Prospective multicentre trial of a small-aperture intraocular lens *Journal of Cataract and Refractive Surgery* **43** 956–68
[124] Ang R E 2018 Small-aperture intraocular lens tolerance to induced astigmatism *Clin. Ophthalmol.* **12** 1659–64
[125] Ang R E 2019 Comparison of tolerance to induced astigmatism in pseudophakic eyes implanted with small aperture, trifocal, or monofocal intraocular lenses *Clin. Ophthalmol.* **13** 905–11
[126] Agarwal S and Thornell E M 2018 Cataract surgery with a small-aperture intraocular lens after previous corneal refractive surgery: visual outcomes and spectacle independence *J. Cataract Refract. Surg.* **44** 1150–4
[127] Barnett V, Barsam A, Than J and Srinivasan S 2018 Small-aperture intraocular lens combined with secondary piggyback intraocular lens during cataract surgery after previous radial keratotomy *J. Cataract Refract. Surg.* **44** 1042–5
[128] Srinivasan S, Khoo L W and Koshy Z 2019 Posterior segment visualization in eyes with small-aperture intraocular lens *J. Refract. Surg.* **35** 538–42
[129] Empey G, Kuppermann B, Bababeygy S R and Gwon A 2015 An evaluation of vitreoretinal surgery through a hydrophobic small aperture intraocular lens in new zealand black rabbit eyes *Investigat. Ophthalmol. Vis. Sci.* **56** 2988
[130] Trindade C L C and Trindade B L C 2015 Novel pinhole intraocular implant for the treatment of irregular corneal astigmatism and severe light sensitivity after penetrating keratoplasty *JCRS Online Case Rep.* **3** 4–7
[131] Trindade C C, Trindade B C and Trindade F C *et al* 2017 New pinhole sulcus implant for the correction of irregular corneal astigmatism *J. Cataract Refract. Surg.* **43** 1297–306
[132] Trindade B L C, Trindade F C and Trindade C L C 2020 Toric intraocular lens combined with a supplementary pinhole implant to treat irregular corneal astigmatism *J. Cataract Refract. Surg.* **46** e22–e6
[133] Agarwal P, Navon S E, Subudhi P and Mithal N 2019 Persistently poor vision in dim illumination after implantation of XtraFocus small-aperture IOL (Morcher) *BMJ Case Rep.* **12** e232473
[134] Manzanera S, Webb K and Artal P 2019 Adaptation to brightness perception in patients implanted with a small aperture *Am. J. Ophthalmol.* **197** 36–44
[135] Schultz T and Dick H B 2016 Small-aperture intraocular lens implantation in a patient with an irregular cornea *J. Refract. Surg.* **32** 706–8
[136] Shajari M, Mackert M J and Langer J *et al* 2020 Safety and efficacy of a small-aperture capsular bag-fixated intraocular lens in eyes with severe corneal irregularities *J. Cataract Refract. Surg.* **46** 188–92
[137] David G S and Atchison A 2000 *Optics of the Human Eye* (Amsterdam: Elsevier)

[138] Menapace R, Findl O, Kriechbaum K and Leydolt-Koeppl C 2007 Accommodating intraocular lenses: a critical review of present and future concepts *Graefes Arch. Clin. Exp. Ophthalmol.* **245** 473–89

[139] Pepose J S, Burke J and Qazi M 2017 Accommodating intraocular lenses *Asia Pac. J. Ophthalmol. (Phila)* **6** 350–7

[140] Alio J L, Alio Del Barrio J L and Vega-Estrada A 2017 Accommodative intraocular lenses: where are we and where we are going *Eye Vis. (Lond.)* **4** 16

[141] Sheppard A L, Bashir A, Wolffsohn J S and Davies L N 2010 Accommodating intraocular lenses: a review of design concepts, usage and assessment methods *Clin. Exp. Optom.* **93** 441–52

[142] Cumming J S, Colvard D M and Dell S J *et al* 2006 Clinical evaluation of the crystalens AT-45 accommodating intraocular lens: results of the U.S. food and drug administration clinical trial *J. Cataract Refract. Surg.* **32** 812–25

[143] Nawa Y, Ueda T and Nakatsuka M *et al* 2003 Accommodation obtained per 1.0 mm forward movement of *a posterior* chamber intraocular lens *J. Cataract Refract. Surg.* **29** 2069–72

[144] Findl O, Kiss B and Petternel V *et al* 2003 Intraocular lens movement caused by ciliary muscle contraction *J. Cataract Refract. Surg.* **29** 669–76

[145] Marchini G, Pedrotti E, Sartori P and Tosi R 2004 Ultrasound biomicroscopic changes during accommodation in eyes with accommodating intraocular lenses: pilot study and hypothesis for the mechanism of accommodation *J. Cataract Refract. Surg.* **30** 2476–82

[146] Perez-Merino P, Birkenfeld J and Dorronsoro C *et al* 2014 Aberrometry in patients implanted with accommodative intraocular lenses *Am. J. Ophthalmol.* **157** 1077–89

[147] Hermans E A, Terwee T T and Koopmans S A *et al* 2008 Development of a ciliary muscle-driven accommodating intraocular lens *J. Cataract Refract. Surg.* **34** 2133–8

[148] McLeod S D, Vargas L G, Portney V and Ting A 2007 Synchrony dual-optic accommodating intraocular lens. Part 1: optical and biomechanical principles and design considerations *J. Cataract Refract. Surg.* **33** 37–46

[149] Simonov A N, Vdovin G and Rombach M C 2006 Cubic optical elements for an accommodative intraocular lens *Opt. Express* **14** 7757–75

[150] Alio J L, Simonov A and Plaza-Puche A B *et al* 2016 Visual outcomes and accommodative response of the Lumina accommodative intraocular lens *Am. J. Ophthalmol.* **164** 37–48

[151] Ben-Nun J and Alio J L 2005 Feasibility and development of a high-power real accommodating intraocular lens *J. Cataract Refract. Surg.* **31** 1802–8

[152] Alio J L, Ben-nun J, Rodriguez-Prats J L and Plaza A B 2009 Visual and accommodative outcomes 1 year after implantation of an accommodating intraocular lens based on a new concept *J. Cataract Refract. Surg.* **35** 1671–8

[153] Portaliou D M, Kymionis G D and Pallikaris I G 2009 The WIOL-CF accommodative intraocular lens *Eur. Ophthalm. Rev.* **3** 54–6

[154] Siatiri H, Mohammadpour M and Gholami A *et al* 2017 Optical aberrations, accommodation, and visual acuity with a bioanalogic continuous focus intraocular lens after cataract surgery *J. Curr. Ophthalmol.* **29** 274–81

[155] Floyd A M, Werner L and Liu E *et al* 2013 Capsular bag opacification with a new accommodating intraocular lens *J. Cataract Refract. Surg.* **39** 1415–20

[156] Kohl J C, Werner L and Ford J R *et al* 2014 Long-term uveal and capsular biocompatibility of a new accommodating intraocular lens *J. Cataract Refract. Surg.* **40** 2113–9

[157] Findl O and Leydolt C 2007 Meta-analysis of accommodating intraocular lenses *J. Cataract Refract. Surg.* **33** 522–7

[158] Glasser A 2008 Restoration of accommodation: surgical options for correction of presbyopia *Clin. Exp. Optom.* **91** 279–95

[159] Wolffsohn J S, Hunt O A and Naroo S *et al* 2006 Objective accommodative amplitude and dynamics with the 1CU accommodative intraocular lens *Invest. Ophthalmol. Vis. Sci.* **47** 1230–5

[160] Li G, Mathine D L and Valley P *et al* 2006 Switchable electro-optic diffractive lens with high efficiency for ophthalmic applications *Proc. Natl Acad. Sci. USA* **103** 6100–4

[161] Campin J A and Pettit G H 2013 Sensors for triggering electro-active ophthalmic lenses. US 2014/0156000 A1

[162] Venkateswaran K 2013 *Accommodative IOL – Refractive Index Change through Change in Polarizability of a Medium* (East Hanover, NJ: Novartis AG) US 2013/0226293 A1

Chapter 5

Eye models

Luiz Melk de Carvalho, Otávio Gomes de Oliveira, Pablo Nunes Agra Belmonte and Diogo Ferraz Costa

Schematic eyes that are capable of accurately modeling the human eye anatomic and biometric properties as well as its optical performance are a central element of the design process of ophthalmic lenses. Hence, eye models are also used as the basis for some metrology methods that are commonly employed in the manufacturing processes of such lenses. Due to their importance for both vision research and industrial applications, this topic has evolved giving rise to different proposals.

This chapter presents some of the most well-known computational eye models in the literature, their main characteristics and their respective correlation with the optics of the human eye. Different eye models focus on accurately representing specific sets of optical parameters and performance of the natural human eye. In this sense, during the lens design process, more than one eye model may need to be used to make sure that the lens meets the best possible overall optical performance and therefore delivers a high visual acuity to the patient.

This chapter is structured in five sections. In the first section, it presents the ISO eye model, proposed by the ISO standard. This standard establishes the requirements for intraocular lens (IOL) inspection and there are several optical metrology stations for IOLs on the market that are based on it. In the subsequent sections, three more eye models are described. These ones include elements of the eye anatomy and represent to different extents the average optical performance of the human eye, including chromatic dispersion, corneal aberration, biometric dimensions, aging and accommodation. Although the parameters of those anatomical eye models are based on an average from a certain population of individuals, some of them allow the customization of the parameters. This customization provides the optical designer with the flexibility to develop a lens design that best fits the needs of the patient, considering the patient's biometric data. In the last section, a comparison table is presented with some of the main optical biometric characteristics of the human eye, which can be modelled by each one of the three anatomical eye models discussed in the previous three sections.

doi:10.1088/978-0-7503-3263-7ch5

5.1 ISO eye model

The ISO international standard 11979 establishes all the criteria to validate the optical and mechanical properties of IOLs, test methods, biocompatibility, clinical investigations, sterilization and packaging. The second part of this standard (ISO 11979-2) specifically addresses the requirements for the optical performance of IOLs and it describes the test methods for this purpose. The prescribed eye model for the evaluation of the refractive power and modulation transfer function (MTF) of the IOL is presented in figure 5.1 [1].

In one of the possible suggested setups of materials for the model cornea and cuvette of the eye model shown in figure 5.1, the cornea is an achromatic doublet lens made of SSK4 (surfaces 1 to 2) and SF8 (surfaces 2 to 3) material and the cuvette glass is made of BK7 material [2]. In case of absence of this type of material on the market, the standard reports that the model cornea can be substituted by any other achromatic lens commercially available and that the cuvette glass material is not critical, as long as the system satisfies the requirements described in the standard [1].

In terms of the light source, the measurement shall be performed with monochromatic light with a wavelength of 546 ± 10 nm [1]. The photopic and mesopic conditions are controlled by the pupil aperture of the system and the light source should not have any vergence before reaching the model cornea [1, 3]. Based on that, the object is always considered at infinity and this is a requirement to guarantee that the magnification of the model is similar to that of the human eye [1]. For multifocal and accommodative IOL models that need an evaluation of its performance at near and/or intermediate distances, the MTF and power verification is conducted by moving the detector at the image space and not by changing the vergence of light source at the object space.

A relevant aspect of the ISO eye model is related to its cornea, that is aberration-free, which means that the cornea does not add any aberration to the optical system. As this eye model was developed before the advent of aspheric lenses, an aberration-free model cornea is the appropriate option to evaluate the aberration impact of spherical lenses on the image quality [4]. However, it is well-known that the human cornea is not only aberration free, but its mean aberration also changes with age [5–9]. With the technological advances in the computational tools for optical design and in the manufacturing process, aspheric lenses started to be developed to correct the spherical aberration of the human cornea and enhance the image contrast

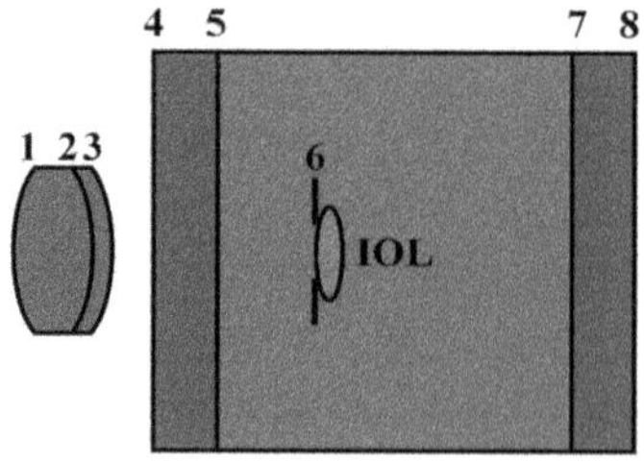

Figure 5.1. ISO eye model: 1–3 model cornea; 4–5 cuvette glass; 6 pupil aperture; 5–7 saline solution; 7–8 cuvette glass; and 9 image plane [1].

resolution on the retina [10, 11]. Based on that, an aspheric IOL that corrects the cornea spherical aberration, when evaluated in an eye model with aberration-free cornea, can result in a poor optical performance, but this does not mean that the lens will yield a poor visual quality for the patient [3]. After a review, the ISO standard also prescribed an alternative eye model (figure 5.2) with a customized aberrated model cornea made of PMMA material [1]. The customization is achieved by an equation that converts the intended spherical aberration (Zernike term $Z(4,0)$) at a pupil aperture of 5.15 mm (surface 5 in figure 5.2), which corresponds to an entrance aperture at the cornea surface of 6.00 mm, into the conic constant of the anterior surface of the cornea [1]. The equation is valid in the range from −0.2 μm to +0.5 μm of the $Z(4,0)$ term [1].

The models presented in figures 5.1 and 5.2 only differ by the model cornea. Although both models ensure that a wider range of IOL models can be assessed by the standard, it is possible that new IOL designs will arise in the market with features that cannot be assessed by any of these eye models. In this case, the performance of the new design is often first demonstrated by clinical evaluation [1, 2].

The following topics present some of the limitations of the ISO eye model:

- **Chromatic aberration:** the model is not suitable for chromatic analysis, which would allow one to evaluate the impact of the chromatic aberration on the image performance [3].
- **Flat image plane:** unlike the human retina, which presents an ellipsoidal shape and a mean radius of curvature of 11.40 mm and 11.18 mm in x- and y-axis, respectively, the ISO eye model considers a flat image plane (camera detector) [12]. Based on that, it is only possible to evaluate on-axis performance of the lenses.
- **Alpha angle:** the alpha angle is the angle between the visual axis and the optical axis [13]. In the human eye, the fovea is slightly positioned on the temporal side of the optical axis [14]. This is another aspect of the human eye that is not considered in the ISO eye model and would affect the off-axis performance of the lens.
- **Pupil displacement:** although the pupil is well-aligned with the crystalline lens, this is not the case in relation to the other elements of the eye, because a small decentration of the pupil is observed towards the nasal direction [13]. As the ISO eye model has a pupil centered with the optical axis and, therefore, all the

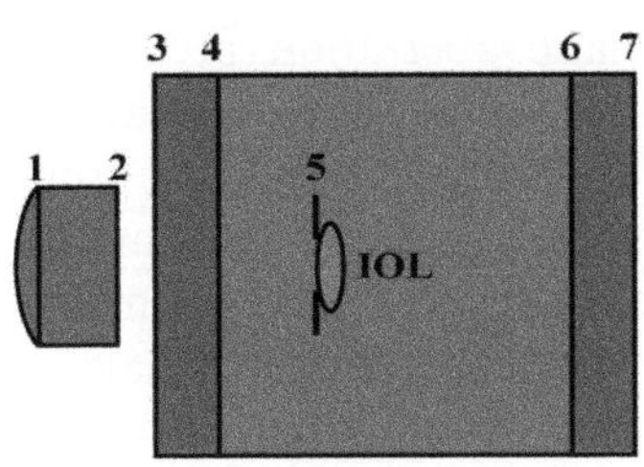

Figure 5.2. Alternative ISO eye model with custom cornea: 1–2 model cornea; 3–4 cuvette glass; 5 pupil aperture; 4–6 saline solution; 6–7 cuvette glass; and 8 image plane [1].

optical elements are well-aligned, it is not possible to evaluate the impact of the off-axis aberrations on the lens and the corresponding impact of it on the image performance.

In summary, the ISO eye model does not represent accurately the anatomy and some important optical features of the human eye. Therefore, this model should be considered only as an optical bench alternative to inspect the quality of lenses and to compare the performance of different models.

5.2 Liou and Brennan eye model

5.2.1 Background of the eye model development

In 1997, Liou and Brennan [13] proposed an eye model which became very popular due to its similarity with real eye data on both optical and anatomical parameters. In fact, the model was built using the most up-to-date empirical data of ocular parameters available at that point. Given the inaccuracy on the prediction of values for spherical aberration of the existing models, the authors intended to develop a new model that could accurately estimate not only the spherical but also the chromatic aberrations. Another motivation for the development was to build a model that could also represent the eye anatomy [13]. This last characteristic would make the model suitable to be used in simulations of conditions in which any of the eye structures could be changed. That would be the case of simulating eyes with intraocular lenses, for instance. Liou and Brennan emphasized that their model was the first to achieve both these characteristics.

Liou and Brennan's model is a finite schematic eye model that predicts retinal image quality more accurately, especially in situations when pupils are larger or objects are off-axis [15, 16]. It contains four aspheric refractive surfaces, a decentered pupil, a gradient refractive index lens, a curved retina and an off-axis fovea [16]. The equivalent optical power of the eye model is 60.35 D and its axial length is 23.95 mm [13].

The model represents closely the anatomy of an average emmetropic, non-accommodating human eye [13, 16, 17]. It has been developed using experimental data of real eyes published in the literature. The data selection followed a set of four criteria: (i) data had to be related to healthy emmetropic eyes; (ii) in case the parameter is age-dependent, the average of 45 years was used; (iii) *in vivo* studies were prioritized over *in vitro* studies; and (iv) most recently published data were preferred, in case the other criteria were the same. Parameters without support of experimental data have been considered as model variables and were estimated so that the spherical and chromatic aberration predicted by the model could be comparable to real measurement results of the human eye [13].

Liou and Brennan extracted intraocular distances from the work of Koretz *et al* [18]. In this study, the authors have measured ocular biometric parameters in 100 emmetropic patients of different ages. Anterior chamber depth, lens thickness and axial length were obtained through the average of male and female results obtained by Koretz *et al* [18]. The thickness of the central cornea was calculated through the subtraction of the anterior chamber depth, the lens thickness and the vitreous distance from the length of the globe [13].

In their eye model, Liou and Brennan represented both corneal surfaces, anterior and posterior, as rotationally symmetric conicoids, which are mathematically described in terms of their radii of curvature and asphericities [13]. For the anterior cornea, Liou and Brennan adopted the radius of curvature and asphericity values reported in the study of Guillon *et al* [19], in which 220 normal eyes of Caucasian patients have been measured using a keratometer and a keratoscope. For the posterior surface of the cornea, however, no direct measurements of its shape were available at that point. Hence, Liou and Brennan referred to a study of Dunne *et al* [20], where the authors reported a strong correlation between the shapes of the anterior and posterior corneal surfaces, with a ratio of anterior corneal radius to posterior corneal radius of 1:0.823. Thus, Liou and Brennan applied this ratio to the average anterior corneal radius reported by Guillon *et al* [19] to determine the value of the posterior corneal radius to be used in the model. The asphericity of the posterior corneal surface, however, was not well defined in the literature and hence the authors decided to consider it as a variable in the modeling process [13].

To model the lens, Liou and Brennan referred to the study conducted by Brown [21], in which the radii of curvature of the anterior and posterior surfaces of the lens were measured in 100 emmetropic eyes of patients with a wide range of ages, using slit-image photography. The radii of curvature were measured in both the center and periphery of both the anterior and the posterior lens surfaces. Liou and Brennan then assumed the lens surfaces to be rotationally symmetric conicoids and calculated the asphericities from the results reported by Brown [21]. For the gradient refractive index of the lens, Liou and Brennan adopted the results of measurements of a 16-year-old lens performed by Pierscionek and Chan [22], which they assumed to be valid near the center of the lens. Due to the lack of support from empirical data, the gradient-index distribution of the lens was considered to be a variable of the model [13].

For the refractive indices of the cornea, aqueous and vitreous humors, Liou and Brennan adopted the same values used by Gullstrand in his model eye [23] as cited in [13] and assumed these indices were valid for the wavelength of 555 nm. The authors reported lack of published experimental results on the dispersion of ocular media and decided to use in their model dispersive properties similar to those of water [13].

Liou and Brennan followed the human eye anatomy to place the pupil in front of the lens and nasally decentered by 0.5 mm from the optical axis [13]. Even though they knew from the literature available at that point that the pupil center changed with the pupil size, there were no accurate data to model the relationship between these two variables and hence they decided to consider the pupil center fixed and independent of its diameter [13].

The authors also included in their model the angle alpha, which accounts for the difference in direction between the visual and optical axis [13]. The visual axis is normally nasally deviated by 5° in relation to the optical axis in the object field.

Lastly, the shape of the retina is not defined in Liou and Brennan's eye model. When considered as a flat plane, it is not accurate to predict off-axis optical performance [17]. If a radius of curvature of −12 mm is adopted, the model represents a slight myopic condition [24].

5.2.2 Structural parameters of the Liou and Brennan eye model

Liou and Brennan built a procedure to define those parameters that have been considered as model variables due to lack of empirical data to support them. They started with the analysis of the data published by Pierscionek and Chan [22]. For simplicity, they assumed that the distribution of index of refraction followed a parabolic pattern from the center of the lens towards the surface of the lens. And they also assumed that the variation of the index of refraction was rotationally symmetric across the lens in both surfaces.

They weighted each of the data points of the study of Pierscionek and Chan [22] to account for the increasing uncertainty of the results closer to the lens surface. They investigated different weighting strategies to find out which ones would yield an overall eye power of approximately 60 D. By doing so, they found out a distribution for the gradient-index of refraction of the lens that resulted in a total optical power of 60.35 D and an axial length of 23.95 mm and decided to adopt it [13].

To define the asphericity of the corneal posterior surface, Liou and Brennan aimed at being able to predict the overall spherical aberration of the eye. They referred to their own previous study [25] in which they had reviewed and compiled published empirical data of the spherical aberration of the eye and had also established a linear relationship between the longitudinal spherical aberration and the ray height. Liou and Brennan have chosen an asphericity that generated values for the total spherical aberration of the eye at different ray heights that were within the range of empirical data that they had compiled [13].

Once all biometric data had been selected and model variables defined, the model was then built. Figure 5.3 shows a schematic representation containing all the refractive surfaces of the eye model, the retina and an imaginary plane (5), with no optical function, that divides the lens into anterior and posterior parts.

The parameters of each of the surfaces used to build the model are detailed in table 5.1, together with the reference source for the used values. The gradient-index of the lens is defined for each surface independently. For the anterior lens surface, the refractive index is defined as shown in equation (5.1) [13]:

$$n_{\mathrm{AL}}(w, z) = 1.368 + 0.049\,057z - 0.015\,427z^2 - 0.001\,978w^2 \tag{5.1}$$

and for the posterior lens surface, it is according to equation (5.2) [13]:

$$n_{\mathrm{PL}}(w, z) = 1.407 - 0.006\,605z^2 - 0.001\,978w^2, \tag{5.2}$$

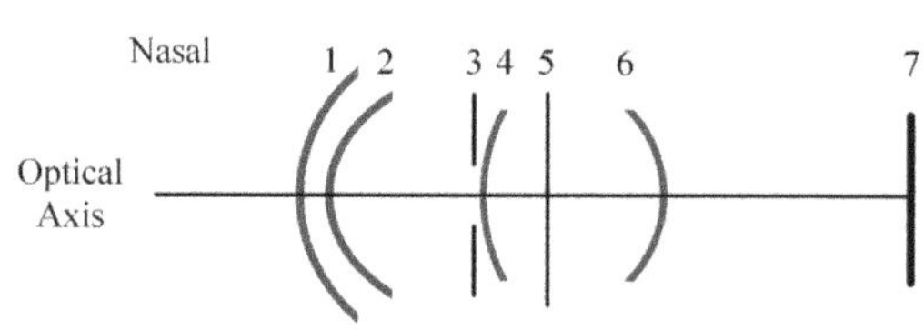

Figure 5.3. Schematic representation of Liou and Brennan eye model. Numbers on top of each structure refer to the surfaces described in table 5.1.

Table 5.1. Structural parameters of Liou and Brennan eye model.

Surface			Radius		Asphericity		Thickness		Refractive index @555 nm	
		Type		Reference		Reference		Reference		Reference
1	Anterior cornea	Aspheric surface	7.77	[19]	−0.18	[19]	0.50	Calc	1.376	[23][4]
2	Posterior cornea	Aspheric surface	6.40	Calc[1]	−0.60	Model	3.16	[18]	1.336	[23][4]
3	Pupil	Aperture							Decentered by 0.50 mm nasally	
4	Anterior lens	Aspheric surface	12.40	[21]	−0.94	Calc[3]	1.59	[18]	n_{AL}	Model[2]
5	Imaginary plane	Plane	Infinity				2.43	[18]	n_{PL}	Model[2]
6	Posterior lens	Aspheric surface	−8.10	[21]	+0.96	Calc[3]	16.27	[18]	1.336	[23][4]
7	Retina	Flat plane								

[1]Parameter calculated based on [19, 20].
[2]Parameter modeled using data from [22].
[3]Calculated using data from [21].
[4]As cited in [13].

where z is the distance along the optical axis and w is the radial distance perpendicular to the z-axis.

Although the parameters have been specified at a wavelength of 555 nm, it is possible to use Liou and Brennan eye model to simulate conditions with other wavelengths by calculating the associated ocular refractive indices through the equation (5.3) [13]:

$$n = n_{@555\,\mathrm{nm}} + 0.0512 - 0.1455\lambda + 0.0961\lambda^2 \tag{5.3}$$

5.2.3 Accuracy to the real eye

The accuracy of the eye model to the real human eye can be analyzed on two different, but interconnected perspectives: anatomy and optical performance. In an effort to guarantee the anatomical similarity with an emmetropic average human eye, Liou and Brennan built their eye model starting from the selection of biometric parameters from published empirical data and included the parameters angle alpha and decentration of the pupil. Later, however, Bakaraju *et al* [17] concluded that such decentration does not provide any significant change in the performance when compared with a centered version.

For those variables for which adequate support from the literature was not available, the authors made an effort to specify them based on the comparison of the optical performance of the eye model with empirical data. This was the case of the gradient-index of refraction of the lens, which allowed the eye model to feature an accurate axial length whilst keeping the total optical power within an experimentally acceptable level [13]. And it was also the case of the aspheric lens surfaces and posterior corneal asphericity, that together with the lens gradient-index of the lens, provided the eye model with spherical aberration values that are comparable with empirical data [13].

Bakaraju *et al* conducted a study [17] in which they evaluated comparatively several different eye models in terms of their wavefront aberration, image quality and refraction profile and also compared the results with empirical data for real eyes. The authors corroborated that the Liou and Brennan eye model predicts spherical aberration close to, although slightly lower, than the average population estimate for a 6 mm pupil. They also highlighted that, similarly to the real eye, the eye model presented relatively high levels of spherical aberration in the cornea, which was compensated by the internal optics and, as a result, the model predicted lower levels of not only total spherical aberration, but also of coma.

Although Liou and Brennan reported that the chromatic aberration of their eye model was comparable to empirical data available at that time [13], Bakaraju *et al* [17] observed that it returned the lowest estimate for on-axis chromatic aberration and highlighted that the potential reason for such behavior could be the fact that Liou and Brennan adopted the ocular media with dispersive properties similar to those of water, which, according to the authors, is not supported by actual experimental evidences.

Liou and Brennan observed that their eye model presented an MTF performance comparable to empirical results for a pupil size of 4 mm [13]. Bakaraju *et al* [17] assessed the MTF for a pupil size of 6 mm and found out that the model overestimated the MTF values in comparison with experimental data.

Bakaraju *et al* [17] also observed that the Liou and Brennan eye model predicts well myopic shifts in the periphery when compared to empirical data, although it is important to mention that the authors used a curved retina with radius of curvature of −12.40 mm, instead of a flat one. They also observed that the Liou and Brennan eye model overestimated the astigmatism by approximately 50% when compared with empirical data, which is also supported by Atchison [24].

In summary, the Liou and Brennan eye model represents reasonably well the eye anatomy and overall optical performance that, together with the possibility of using it to simulate conditions where some of the refractive surfaces are altered or even replaced, made it a very popular model.

5.3 Navarro *et al* eye model

5.3.1 Background of the eye model development

The Navarro *et al* eye model [26] describes a finite schematic eye, suitable for on-axis calculations. It predicts axial spherical aberrations with good accuracy and requires a less complex set of parameters to represent the shell structure of the lens, which is modelled with no need for a gradient index (GRIN) structure. Instead, it assumes an effective refractive index [27, 28]. This model, however, finds its limitations for off-axis predictions.

In a more recent work, Navarro *et al* also proposed a GRIN model for the lens intended to be simple enough to fit the anatomy of individual eyes, since it works with very few parameters [29]. The GRIN structure of the human lens plays a key role on the amplitude of accommodation of the eye, and is also related to detrimental aging effects [30–32]. In that new lens model, Navarro *et al* propose a parametric adaptive model, capable of describing the behavior of both the short-term accommodation, and long-term aging.

5.3.2 Structural parameters

Figure 5.4 shows the basic structure of the Navarro *et al* schematic eye model. It considers four centered quadric refracting surfaces, rotationally symmetric, where each surface has only two parameters: the radius of curvature (R) and asphericity (Q). This schematic eye is based on the Gullstrand–Le Grand model [33], which is already an improvement from the original Gullstrand *et al* model [34] as cited in [26]. However, one key difference is the use of asphericity for the Navarro *et al* model,

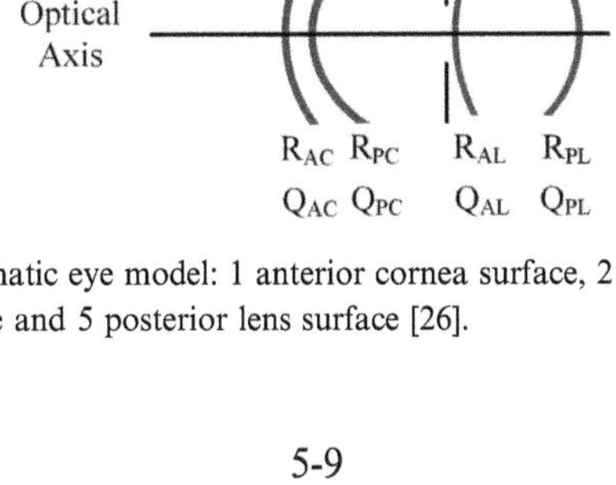

Figure 5.4. Navarro *et al* schematic eye model: 1 anterior cornea surface, 2 posterior cornea surface, 3 pupil aperture, 4 anterior lens surface and 5 posterior lens surface [26].

instead of only spherical surfaces as proposed by Gullstrand–Le Grand. Two other improvements were: (1) the inclusion of experimental data for the cornea and lens; (2) and the computation of the refractive indices for chromatic aberration adjustment.

In the Navarro *et al* model, the radius of curvature and asphericity of the cornea were both obtained from *in vivo* measurements by Kiely *et al* [35]. For the crystalline lens surfaces, the asphericities correspond to *in vitro* data from Howcroft and Parker [36], whereas the curvature radii were kept the same from Gullstrand–Le Grand model. Furthermore, the effective refractive index was also maintained from the original model, since it represents a simplification that still is a good approximation for on-axis optical performance [26]. All the parameters used in the original work are represented in table 5.2, where the resulting refractive power is 60.4 D.

Each surface of the Navarro *et al* eye model can be described as an aspheric profile, using the following equation: $x^2 + y^2 + (1 + Q)z^2 - 2Rz = 0$, where R and Q are the curvature radius and the asphericity, respectively; x, y and z are Cartesian coordinates, the optical axis being on the z-axis [26].

For the Navarro *et at* model to achieve a better fit to experimental chromatic aberration data [37, 38], the refractive indices at different wavelengths were calculated, based on the constringences of Polack [39] as cited in [26]. The Herzberger formula is employed [40], instead of the Cornu formula used by Le Grand [41] as cited in [26], which led to a better description of the optical media by Navarro *et al.* Table 5.3 shows the refractive indices that were computed to fit chromatic aberrations.

Table 5.2. Parameters for the Navarro average eye model [26].

Radius of curvature (mm)		Asphericity	
Anterior cornea (R_{AC})	7.72	Anterior cornea (Q_{AC})	−0.26
Posterior cornea (R_{PC})	6.5	Posterior cornea (Q_{PC})	0
Anterior lens (R_{AL})	10.2	Anterior lens (Q_{AL})	−3.1316
Posterior lens (R_{PL})	−6	Posterior lens (Q_{PL})	−1.00
Geometrical parameters (mm)		Optical parameters	
Corneal thickness (t_C)	0.55	Cornea refractive index (n_C)	1.367
Aqueous thickness (t_A)	3.05	Aqueous refractive index (n_A)	1.3374
Lens thickness (t_L)	4	Lens refractive index (n_L)	1.42
Vitreous thickness (t_V)	16.4	Vitreous refractive index (n_V)	1.336

Table 5.3. Computed refractive indices used on Herzberger formula to fit chromatic aberrations [40].

Wavelength (μm)	n_C	n_A	n_L	n_V
$\lambda = 0.3650$	1.3975	1.3593	1.4492	1.3565
$\lambda = 0.4861$	1.3807	1.3422	1.42625	1.3407
$\lambda = 0.6563$	1.37405	1.3354	1.4175	1.3341
$\lambda = 1.0140$	1.3668	1.3278	1.4097	1.3273

As Navarro *et al* claim that the schematic eye with four aspheric surfaces can fit the on-axis average optical performance, it is also assumed that the crystalline lens may be used to fit paraxial spherical aberrations for different accommodation states. In this way, Navarro *et al* used simple logarithmic mathematical functions to correlate the change in the cornea and lens parameters with accommodation [26]. The following equations connect both unaccommodated and accommodated anterior lens radii from the Gullstrand–Le Grand model [33] and experimental data for accommodation versus anterior radius from Ivanoff [37]. The relation between the anterior and posterior radii of the lens and the accommodation (A_{cc}) in diopters, is described, respectively, by $R_{AL}(A_{cc}) = 10.2 - 1.75 \cdot \ln(A_{cc} + 1)$ and $R_{PL}(A_{cc}) = -6 + 0.2294 \cdot \ln(A_{cc} + 1)$. The aqueous and lens thicknesses are represented, respectively, by $t_A(A_{cc}) = 3.05 + 0.05 \cdot \ln(A_{cc} + 1)$ and $t_L(A_{cc}) = 4 + 0.1 \cdot \ln(A_{cc} + 1)$.

However, only the change in the above structural parameters is not enough to predict the increment in refractive power of the lens, as the actual crystalline lens has a graded refractive index that, during accommodation, adds extra refractive power. To bypass this restriction, Navarro *et al* invoke the intracapsular mechanism of accommodation, postulated by Gullstrand [34], which considers that the effective refractive index of the lens also changes with the accommodation. This is a fictitious mathematical mechanism, used to strongly simplify the modelling. Considering that as a first approximation, a quadratic adjustment to the refractive power of the schematic eye resulted in the following formula for the effective refractive index of the lens: $n_L(A_{cc}) = 1.42 + 9 \times 10^{-5} \cdot (10 \cdot A_{cc} + A_{cc}^2)$ [26]. Additionally, Navarro *et al* consider the case where the anterior and posterior lens asphericities also change logarithmically with the accommodation, respectively, given by $Q_{AL}(A_{cc}) = -3.1316 - 0.34 \cdot \ln(A_{cc} + 1)$ and $Q_{PL}(A_{cc}) = -1 - 0.125 \cdot \ln(A_{cc} + 1)$. Finally, table 5.4 shows the effective refractive indices for two accommodation diopters (3 D and 5 D), that were calculated by keeping the chromatic aberration for each accommodation state to be equal to that of the unaccommodated state.

In the more recent work, Navarro *et al* presented a parametrized GRIN model for the lens [29]. The main goal of that model is to be customizable for individual data, to automatically adapt to the eye geometry, and also to provide age and accommodation insights. Although the model captures the main structural and functional features of aging and accommodation, it is simple and easy to understand. The gradient refractive index $n(r)$ is given by $n(r) = n_0 + \delta_n r^{2p}$, where r is the normalized radius (from the z-axis towards the crystalline-lens apex) between 0 and 1; $\delta_n = n_s - n_0$ is the difference between the edge and central index values; and p is a

Table 5.4. Computed refractive indices of the lens for two different accommodation states [26].

Wavelength (μm)	3 D	5 D
$\lambda = 0.3650$	1.4511	1.4533
$\lambda = 0.4861$	1.4298	1.43313
$\lambda = 0.6563$	1.421	1.42423
$\lambda = 1.0140$	1.4134	1.4162

parameter used to adjust the GRIN data. The parameter p is used to fit data, depending on accommodation and age.

5.3.3 Accuracy to the real eye

The Navarro *et al* model was tested for several metrics. One of them being the longitudinal spherical aberration (LSA), computed by ray tracing, where the results were in agreement with a collection of experimental data adjusted by Van Meeteren [42]. In addition, the model shows that, in accordance with El Hage and Berny [28], the lens and cornea feature opposite signs for the spherical aberrations, but the cornea has a larger absolute value.

Another metric is the chromatic aberration (CA), that was calculated through ray tracing, where the refractive indices for dispersion were obtained until experimental data was fit, as previously mentioned. The model showed a good agreement for CA with experimental data from Ivanoff [37] as cited in [26] and Wald and Griffin [38].

Furthermore, Navarro *et al* calculate the polychromatic MTF and compare it to experimental data. The polychromatic MTF was obtained by computing the monochromatic MTFs along the visible spectrum, sampled in 40 intervals. For a polychromatic MTF at a pupil aperture of 4 mm in diameter, the eye model showed a theoretical MTF that lies in between two experimental curves by Arnulf [43], as cited in [26], where the MTF was directly measured, and by Campbell and Gubisch [44], whose MTF was computed indirectly by experimental line-spread functions (LSFs).

For the accommodation-dependent model, the results from Navarro *et al* for the accommodation functions, to calculate the anterior lens radius during accommodation, are similar to the experimental average radii measured by Ivanoff [37], as cited in [26].

The lens radii and thicknesses formulas were not enough to describe the accommodation-dependent increment in the refractive power. Therefore, Navarro *et al* introduce the asphericity and the effective refractive index formulas to improve the fitting. The effective index formula for the lens is generalized for each wavelength, based on the base unaccommodated index obtained in table 5.3, and fit to have the chromatic aberrations equal to that of the unaccommodated eye. Therefore, there is a slight increase in the chromatic aberration, the larger the accommodation. This result agrees with experimental observations [45]. Furthermore, among published experimental data, there is agreement about the calculated spherical aberrations for an individual eye [46] for some levels of accommodation.

In summary, it is clear that the unaccommodated schematic eye proposed by Navarro *et al* predicts the average spherical aberration without any shape fitting, since on-axis spherical aberrations are dominated by the asphericities of the cornea and lens. However, the model cannot predict off-axis spherical aberrations, where the GRIN lens plays a major role. In addition, the refractive indices in table 5.3 are not necessarily anatomical, since they were calculated to fit the whole eye chromatic aberrations. Therefore, they should not be used individually to predicted chromatic aberrations of the different components of the eye [26].

In the accommodation-dependent model, the logarithmic relation between the radius and thickness of the lens with accommodation seems to yield a suitable fit for the average variation among individuals. In addition, although the model does not accurately predict the spherical aberration for accommodated eyes, its performance for polychromatic data fitting is reasonable.

The main issues faced by Navarro *et al* are for off-axis modelling, that can be mitigated by using an actual GRIN lens model, instead of the effective refractive index. Considering the GRIN model [29], also proposed by Navarro *et al* the data fitting was based on data published by Jones *et al* [30], that provided uniform sampling for individuals from age 7 up to 82 years old, with refractive index variations along the optical axis. A polynomial dependency with age for the parameter p was employed, as $p = 1.1 \times 10^{-7}\text{age}^4 + 2.85$, with age in years. As the formula suggests, the study found that p is for most ages close to 3, which was also found by Smith *et al* [47]. In addition, the most popular approaches to model the GRIN structure of the lens uses quadratic approximations, such as the Liou and Brennan [13] implementation, which is indeed a particular case for the Navarro model, with $p = 1$.

5.4 Atchison *et al* eye model

5.4.1 Background of the eye model development

The Atchison eye model [24] is used to construct a refractive dependent model. This is a statistical eye model based on measurements performed on 121 emmetropic and myopic patients, within an age range of 18 to 36 years. The mean age of this population is 25 years. The model is based on other unaccommodated emmetropic eyes, such as Liou and Brennan (1997) and Navarro *et al* (1985). Also, the corneal and lens shapes are based on reports that used Scheimpflug photography, *in vitro* lens refractive index measurements, MRI measurements and chromatic dispersion modelling. Some unreported measurements of anterior corneal topography and ultrasound intraocular distance were also used.

The corneas of the 121 patients were measure using videokeratography and with undilated pupils. The mean pupil diameter was measured to be 4.4 mm ± 0.8 mm, ranging from 2.7 mm to 6.0 mm.

Non cycloplegic monocular sphero-cylinder subjective refraction was performed on both eyes using a Jackson crossed cylinder in a phoropter [24].

Although two models were developed, one with centered elements and one with tilted and decentered elements, this description uses the centered model, since it shows reasonable prediction of changes with refraction (with certain limitations). It gives a good prediction of the spherical aberration in the real emmetropic eyes. This model was used to observe the effects of changing the design of ophthalmic lenses on peripheral refraction to address the hypothesis that the peripheral hypermetropia might be a stimulus in eye growth [24].

5.4.2 Structural parameters

The different eye model parameters were obtained using the phoropter lens power (PLP) as the independent variable in a linear regression. This parameter is related to the added refractive power of the lens present in a phoropter, that ranged between + 0.75 D to −12.38 D. The lens is assumed to be at a 12 mm vertex distance.

The anterior cornea radius of curvature (equation (5.4)) is highly related to the refraction. It is described by the following equation, and has a maximum regression error smaller than 0.003.

$$R_{AC}\ (\text{mm}) = 7.77 + 0.022\text{PLP} \tag{5.4}$$

Some studies have reported that the anterior radius of curvature decreases with the increase in myopia, for instance the work of Budak *et al* [52]. The equation found in this modelling is well within range of these other studies.

Since no measurements were made on the posterior cornea, the value of 6.40 mm was used [49]. There is at least one study that relates the radius of curvature of the posterior cornea to the radius of curvature of the anterior cornea [50].

For the asphericity, the value of −0.15 was used on the anterior cornea (Q_{AC}), because the linear regression did not reveal that is has a significant statistical correlation with refraction. This happens because the linear regression exhibited a constant plateau around −0.15 for all patients across the full range of the aforementioned phoropter lens power (PLP). For the posterior cornea, it is known that the shape factor (k) is related to the asphericity (Q_{PC}) and the patient's age through equation (5.5) [50]. Since the mean age of the patients measured in this study is 25 years, it leads to an asphericity of −0.275.

$$k = Q_{PC} + 1 = 0.9 - 0.007 \cdot \text{age} \tag{5.5}$$

The corneal central thickness is 0.55 mm and the refractive index is 1.376 as shown in equations (5.6) and (5.7), which is the same used in Navarro's eye model.

$$d_{C1} = 0.55\ \text{mm} \tag{5.6}$$

$$n_{C1} = 1.376 \tag{5.7}$$

The mean anterior chamber depth is 3.71 mm ± 0.29 mm. After rounding the mean value to 3.7 mm, and subtracting the cornea thickness, it leads to an anterior chamber depth of 3.15 mm. No significant statistical correlation was found between the anterior chamber depth and refraction. In a similar manner as the asphericity of the anterior surface of the cornea, the linear regression obtained from the measurements indicated a constant value across the range of the phoropter lens power (PLP).

The refractive index of the anterior chamber was based on Navarro *et al* which has a value of 1.3374. The aperture stop was placed and centered at the lens vertex.

The radius of curvature of the anterior surface of the lens was modelled based on [51] for a 3 mm zone, and can be expressed by equation (5.8).

$$R_{AL}(\text{mm}) = 12.9(\pm 0.4) - 0.057(\pm 0.009) \cdot \text{age} \tag{5.8}$$

Using the mean age of 25 years and rounding to two decimal places, it leads to 11.48 mm of radius of curvature (anterior lens). The radius of curvature of the posterior surface (equation (5.9)) was calculated in a similar manner, using equation (5.9),

$$R_{PL}\ (\text{mm}) = -6.2(\pm 0.02) + 0.012(\pm 0.006) \cdot \text{age} \tag{5.9}$$

which leads to −5.9 mm of posterior radius of curvature, for a mean age of 25 years.

The anterior surface asphericity was fitted by [51], using a 5 mm zone, which led to a mean value of −5 ± 5. Therefore, the value chosen for the anterior asphericity was −5. In a similar manner, the same authors in [51] estimated *a posterior* lens asphericity as having a value of −4 ± 5. Since the Scheimpflug technique presents errors in backward ray tracing through the eye, a value of asphericity was chosen to ensure that the model presents a Zernike spherical aberration that is consistent with the literature (which is about 0.10 μm for a 6 mm entrance pupil). Therefore, the value of asphericity used is equal to −2.

For the lens thickness (t_L), the value used was of 3.6 mm, since the linear fit did not present a significant correlation between refraction and the thickness value. Studies have shown that the lens thickness is not correlated with gender either [24]. The total lens thickness was distributed between the two lens surfaces in the same manner as done by Liou and Brennan [13], where the anterior lens surface has 40% of the total thickness. Therefore, the anterior lens thickness is of 1.44 mm and the posterior lens thickness is of 2.16 mm.

The lens refractive index was modelled using a gradient index for the anterior surface (n_{L1}) and posterior surface (n_{L2}) following equations (5.10) and (5.11):

$$n_{L1} = 1.371 + 0.065\,2778z - 0.022\,6659z^2 - 0.002\,0399(x^2 + y^2) \quad (5.10)$$

$$n_{L2} = 1.418 - 0.010\,0737z^2 - 0.002\,0399(x^2 + y^2) \quad (5.11)$$

where z is the longitudinal direction, and x and y constitute the radial directions on the perpendicular plane to the longitudinal axis.

The axial length of the eye has a strong dependence on refraction, and there is also a significant statistical difference between the length of male and female patients. The regression fit was not used. Instead, Atchison determines the lengths (d_{total}) corresponding to paraxial imagery, following equation (5.12):

$$d_{total}\,(\text{mm}) = 23.58 - 0.299\text{PLP} \quad (5.12)$$

The vitreous chamber length (t_v) is 7.3 mm shorter than the total length and is described by equation (5.13). The corresponding refractive index used was the same of Navarro *et al* and has a value of 1.336.

$$t_v(\text{mm}) = 16.28 - 0.299\text{PLP} \quad (5.13)$$

Table 5.5 summarizes all of the Atchison's eye model constructive characteristics.

5.4.3 Accuracy to the real eye

The equivalent power of the eye model is expressed by equation (5.14) and shows a small rate of increase in power as myopia increases. This happens because as the phoropter lens power decreases from +0.75 D to −12.38 D, the negative PLP contribution (meaning a myopic eye) increases the equivalent power of the eye model.

$$P_{eq} = 61.50 - 0.128\text{PLP} \quad (5.14)$$

Table 5.5. Parameters of the eye models as a function of phoropter lens power (PLP) in D [adapted from [24]].

Medium	Refractive index @555 nm	Radius of curvature (mm)	Asphericity	Distance to next surface (mm)
Air	1.0			
		7.77 + 0.022PLP	−0.15	
Cornea	1.376			0.55
		6.4	−0.275	
Aqueous	1.3374			3.15
		11.48	−5	
Anterior lens[a]	$1.371 + 0.065\,2778z - 0.022\,6659z^2 - 0.0020399(x^2 + y^2)$			1.44
		Infinity	—	
Posterior lens	$1.418 - 0.010\,0737z^2 - 0.0020399(x^2 + y^2)$			2.16
		−5.9	−2	
Vitreous	1.336			16.28 − 0.299 PLP
		R_{Rx} − 12.91 − 0.094 PLP R_{Ry} − 12.72 − 0.004 PLP	Q_{Rx} 0.27 + 0.026 PLP Q_{Ry} 0.25 + 0.017 PLP	
Retina				

[a] Stop in plane of surface vertex.

The central spherical aberration as a function of the refraction is done using ray tracing from infinity into the eye, with the rays evenly distributed across a 6.0 mm pupil diameter. The coefficient for the emmetropic eye is 0.09 μm, which is close to the estimated value of 0.10 μm [52]. The spherical aberration is highly related to the refraction and changes around 0.007 μm per diopter of myopia. This contradicts some experiments that show that there is no increase in spherical aberration with the increase in myopia [24], which is a limitation of the model.

5.5 Summary

This chapter presented some of the most well-known computational eye models employed in the analysis and optimization of IOLs. For each eye model the main characteristics of the surfaces of all the optical elements that constitute the computational model are addressed. Besides, the material properties of the elements and surrounding media and the accuracy of the optical performance of the model with real eye data are outlined. The ISO eye model was presented in section 5.1 in its two versions: the first one with a neutral model cornea, and the most recent one that allows for the configuration of the amount of spherical aberration of the cornea. The Liou and Brennan eye model, which presents a closer similarity to the anatomy and optical performance of real eyes when compared to the ISO model, was presented in section 5.2. However, the Liou and Brennan eye model cannot model some features of a real eye such as the accommodation process, which is covered in the Navarro *et al* eye model, presented in section 5.3. The Atchison eye model presented in section 5.4 was developed through statistical analysis from measurements of a group of patients, where some of its parameters were estimated by linear regression while others were inherited from well-known models, such as Liou and Brennan and Navarro *et al.* This eye model also has a benefit of considering aging effects on some of its parameters, which leads to an attempt of better representing a certain population of subjects.

In summary, each eye model presents its peculiarities in terms of the real eye properties that they can accurately model. Although the ISO eye model does not anatomically represent the human eye, and cannot reproduce some of its optical features, it is the model required by the worldwide standards to validate and compare the optical performance of different intraocular lenses. For a more accurate analysis of the optical performance of the intraocular lenses under development and its impact on human sight, it is necessary to consider other computational eye models such as Liou and Brennan, Navarro *et al* and Atchison *et al.*

Chapter highlights

- The main description of the structure of the ISO eye model is presented and the main limitations and importance of the model are discussed.
- The Liou and Brennan eye model construction is described and a comparison of its anatomical shape and optical performance with real eye data is addressed.
- The Navarro *et al* eye model parameters are described, and the optical results are compared to real eye data.
- The Atchison eye model presents a different approach by using statistical characteristics obtained through the assessment in a group of patients, and includes aging effects on some parameter variations.

References

[1] International Organization for Standardization 2014 – ISO, 11979-2—Ophthalmic implants —Intraocular lenses—part 2: optical properties and test methods

[2] Sverker Norrby N E 1995 Standardized methods for assessing the imaging quality of intraocular lenses *Appl. Opt.* **34** 7327

[3] Norrby S, Piers P, Campbell C and Van Der Mooren M 2007 Model eyes for evaluation of intraocular lenses *Appl. Opt.* **46** 6595–605

[4] Norrby S 2008 ISO eye model not valid for assessing aspherical lenses *J. Cataract Refract. Surg.* **34** 1056–057

[5] Porter J, Guirao A, Cox I G and Williams D R 2001 Monochromatic aberrations of the human eye in a large population *J. Opt. Soc. Am.* A **18** 1793

[6] Wang L, Dai E, Koch D D and Nathoo A 2003 Optical aberrations of the human anterior cornea *J. Cataract Refract. Surg.* **29** 1514–21

[7] Fujikado T *et al* 2004 Age-related changes in ocular and corneal aberrations *Am. J. Ophthalmol.* **138** 143–46

[8] Oliveira C M, Ferreira A and Franco S 2012 Wavefront analysis and Zernike polynomial decomposition for evaluation of corneal optical quality *J. Cataract Refract. Surg.* **38** 343–56

[9] Brunette I, Bueno J M, Parent M, Hamam H and Simonet P 2003 Monochromatic aberrations as a function of age, from childhood to advanced age *Investig. Ophthalmol. Vis. Sci.* **44** 5438–46

[10] Holladay J T, Piers P A, Koranyi G, Van der Mooren M and Norrby N E S 2002 A new intraocular lens design to reduce spherical aberration of pseudophakic eyes *J. Refract. Surg.* **18** 683–91

[11] Kasper T, Bühren J and Kohnen T 2006 Intraindividual comparison of higher-order aberrations after implantation of aspherical and spherical intraocular lenses as a function of pupil diameter *J. Cataract Refract. Surg.* **32** 78–84

[12] Atchison D A, Pritchard N, Schmid K L, Scott D H, Jones C E and Pope J M 2005 Shape of the retinal surface in emmetropia and myopia *Investig. Ophthalmol. Vis. Sci.* **46** 2698–707

[13] Liou H-L and Brennan N A 1997 Anatomically accurate, finite model eye for optical modeling *J. Opt. Soc. Am.* A **14** 1684

[14] Smith D A A G 2000 *Optics of the Human Eye* (Amsterdam: Elsevier)

[15] Atchison D A and Thibos L N 2016 Optical models of the human eye *Clin. Exp. Optom.* **99** 99–106

[16] Esteve-Taboada J J, Montés-Micó R and Ferrer-Blasco T 2018 Schematic eye models to mimic the behavior of the accommodating human eye *J. Cataract Refract. Surg.* **44** 627–41

[17] Bakaraju R C, Ehrmann K, Papas E and Ho A 2008 Finite schematic eye models and their accuracy to *in-vivo* data *Vision Res.* **48** 1681–94

[18] Koretz J F, Kaufman P L, Neider M W and Goeckner P A 1989 Accommodation and presbyopia in the human eye-aging of the anterior segment *Vision Res.* **29** 1685–92

[19] Guillon M, Lydon D P M and Wilson C 1986 Corneal topography: a clinical model *Ophthalmic Physiol. Opt.* **6** 47–56

[20] Dunne M C M, Royston J M and Barnes D A 1992 Normal variations of the posterior corneal surface *Acta Ophthalmol.* **70** 255–61

[21] Brown N 1974 The change in lens curvature with age *Exp. Eye Res.* **19** 175–83

[22] Pierscionek B K and Chan D Y C 1989 Refractive index gradient of human lenses *Optom. Vis. Sci.* **66** 822–29

[23] Gullstrand A 1924 Helmholtz's physiological optics *Opt. Soc. Am.* 350–58

[24] Atchison D A 2006 Optical models for human myopic eyes *Vision Res.* **46** 2236–50

[25] Liou H-L and Brennan N A 1996 The prediction of spherical aberration with schematic eyes *Ophthalmic Physiol. Opt.* **16** 348–54

[26] Navarro R, Santamaría J and Bescós J 1985 Accommodation-dependent model of the human eye with aspherics *J. Opt. Soc. Am.* A **2** 1273

[27] Lotmar W and Lotmar W 1971 Theoretical eye model with aspherics *J Opt Soc Amer* **61** 1522–529

[28] El Hage S G, Berny F and Apparatus E 1973 Contribution of the crystallyne lens to the spherical aberration of the eye *J. Opt. Soc. Am.* **63** 205–11

[29] Navarro R, Palos F and González L M 2007 Adaptive model of the gradient index of the human lens. I. formulation and model of aging *ex vivo* lenses *J. Opt. Soc. Am.* A **24** 2911

[30] Jones C E, Atchison D A, Meder R and Pope J M 2005 Refractive index distribution and optical properties of the isolated human lens measured using magnetic resonance imaging (MRI) *Vis. Res.* **45** 2352–66

[31] Smith G, Atchison D A and Pierscionek B K 1992 Modeling the power of the aging human eye *J. Opt. Soc. Am.* A **9** 2111
[32] Hemenger R P, Garner L F and Ooi C S 1995 Change with age of the refractive index gradient of the human ocular lens *Investig. Ophthalmol. Vis. Sci.* **36** 703–07
[33] El Hage S G and Grand Y L 1980 *Physiological Optics* vol 13 (Berlin: Springer)
[34] Gullstrand A 1909 Appendix in H. von Helmholtz *Physiol. Opt.* **3** 299
[35] Kiely P M, Smith G and Carney L G 1982 The mean shape of the human cornea *Opt. Acta (Lond.)* **29** 1027–40
[36] Howcroft M J and Parker J A 1977 Aspheric curvatures for the human lens *Vision Res.* **17** 1217–23
[37] Alexandre I 1953 *Les Aberrations De L'Oeil, Leur Role Dans L'Accomodation* (Sablons: La Revue d'Optique Théorique et Instrumentale)
[38] Wald G and Griffin D R 1947 The change in refractive power of the human eye in dim and bright light *J. Opt. Soc. Am.* **37** 321–36
[39] Polack A 1923 *Le chromatisme de l'oeil* (Paris: Société d'ophtalmologie de Paris)
[40] Taylor P and Herzberger M 1959 Colour correction in optical systems and a new dispersion formula *Opt. Acta Int. J. Opt.* **6** 197–215
[41] Le Grand Y 1956 *Lespace visuel, Optique physiologique* (Paris: Masson)
[42] Van Meeteren A 1974 Calculations on the optical modulation transfer function of the human eye for white light *Opt. Acta (Lond).* **21** 395–412
[43] Arnulf A 1965 Le systeme optique de l'oeil en vision photopique et mesopique *Excerpta Med. Int. Congr. Ser.* **125** 135–51
[44] Campbell F W, Gubisch R W, Campbell B Y F W and Gubisch R W 1966 Optical quality of the human eye *J. Physiol.* **186** 558–78
[45] Berny F 1969 Etude de la formation des images retiniennes et determination de l'aberration de sphericite de l'oeil humain *Vision Res.* **9** 977–90
[46] PATNAIK B 1967 A photographic study of accommodative mechanisms: changes in the lens nucleus during accommodation *Invest. Ophthalmol. Vis. Sci.* **6** 601–11
[47] Smith G, Pierscionek B K and Atchison D A 1991 The optical modelling of the human lens *Ophthalmic Physiol. Opt.* **11** 359–69
[48] Dubbelman M, Weeber H A, van der Heijde R G L and Volker-Dieben H J 2002 Radius and asphericity of the posterior corneal surface determined by corrected scheimpflug photography *Acta Ophthalmologica* **80** 379–83
[49] Lowe R F and Clark B A J 1973 Posterior corneal curvature *Br. J. Ophthalmol.* **57** 464–70
[50] Dubbelman M and Van der Heijde G L 2001 The shape of the aging human lens: curvature, equivalent refractive index and the lens paradox *Vis. Res.* **41** 1867–77
[51] Atchison D A 2005 Recent advances in measurement of monochromatic aberrations of human eyes *Clin. Experiment. Optom.* **88** 5–27
[52] Budak K, Khater T T, Friedman N J, Holladay J T and Koch D D 1999 Evaluation of relationships among refractive and topographic parameters *J. Cataract Refract. Surg.* **25** 814–20

Chapter 6

IOL power calculation

Bruno Lovaglio Cançado Trindade, Matheus Lins dos Santos and João Marcelo de Almeida Gusmão Lyra

6.1 Clinical equipment and tests

Modern cataract surgery demands precision in all levels of patient care. From the initial assessment of complaints and lifestyle of each individual to the precise implantation of a selected intraocular lens (IOL) at the end of the procedure. In order to choose the correct lens for each patient, objective precise measurement of eye dimensions is a paramount step in the process. As delineated in chapter 2, many eye structures can and have to be measured to properly calculate the total refractive power of the eye. The path that light follows inside the eye can be calculated once the refractive surfaces are known and measured.

Measuring the light path length is obviously important. Unlike cameras, that have a constant distance from the lens to the sensor plane, there is an individual variability in the eye size, and the distance from the first refractive layer of the eye (the anterior tear film) to where image formation occurs (the fovea) is the first measurement one needs to perform. As defined in chapter 2, this is called the axial length of the eye. The virtual line that connects an infinite-located object to the central part of the fovea is called the visual axis [1]. Due to an anatomical tilt of the eye, the visual axis does not coincide with the optical axis of the cornea and the lens. In fact, these axes form an important angle, referred to as alpha, and this has to be considered to measure the correct distance that light travels inside the eye [2]. To better improve these measurements, modern equipment has a fixation target to be used by the patient. The correct axial length is the distance measured from the point in which the cornea is intersected by the visual axis to the foveal retinal pigment epithelium [3].

Corneal curvature is the major contributor to the total eye refraction [4]. The difference in indices of refraction from the air to the anterior layer of the tear film is the greatest in the entire ray path towards the retina. Hence, small variations in the anterior corneal surface curvature or regularity may largely compromise image quality. The corneal curvature measurement can be obtained by various methods and each method has its own set of advantages and disadvantages [5]. The different methods to measure corneal power and main equipment that are currently available will be discussed.

During cataract surgery, the natural lens is replaced by an artificial one. Differently to the first Ridley's IOL (chapter 4) whose dimensions were similar to the original crystalline lens, modern IOLs are usually much thinner than the natural ones. Therefore, the implant sits inside a bigger envelope and its final position cannot be precisely measured preoperatively. The final IOL position is usually referred to as the effective lens position (ELP) and small variations may produce large residual refractive errors [6]. Therefore, modern formulae predict the ELP based on different measurements, which will also be discussed further in the chapter.

It is interesting to note that during the first IOL implantation by Harold Ridley in 1949, there was no biometric data and no IOL power calculation. It is said that the residual refraction of the patient was around −10.0 D which, for the standards of the time, was a big improvement over the usual aphakic high hyperopia. Times have changed and everything related to cataract surgery has improved, including a better way of calculating the IOL power to be implanted.

Axial length is one of the most important values to be computed when calculating the right IOL power to be used. There are multiple pieces of equipment available that can be used for this task. The echobiometer was first device employed to measure the ocular axial length. Using a high-frequency ultrasonic probe, this machine would measure the amount of time ultrasound waves took to reflect at the inner limiting membrane of the retina and reach back to the probe [7]. This is a linear, one-dimension scan, that is also known as A-scan (short for amplitude-scan). By knowing the speed of the ultrasonic beam in the eye, it was possible to calculate the path length [8]. When ultrasound waves travel through the eye, in regions where there is a great difference in media density there is some part of the wave that is transmitted and parts that are reflected. Because of this, there are usually four distinct peaks that are recognizable when performing an A-scan of the eye [9]. The first peak represents the cornea, the second is the anterior lens capsule while the third is the posterior lens capsule and the forth, and last, is the inner limiting retinal membrane [8]. Most ultrasound biometers yield a graph that shows the magnitude of the peaks of the ultrasound beam (plotted in the y-axis) versus the distance from the transducer (calculated by multiplying the velocity of ultrasound propagation in the eye by the time it takes to detect reflection divided by 2), plotted in the x-axis (figure 6.1(a) and (b)).

The distances between these peaks represent the dimensions of the corresponding structures. Thus, the distance between the first and fourth peak represents the entire axial length. For these measurements to be correct, the ultrasonic probe has to be perfectly aligned to the visual axis since off-axis measurements will underestimate the axial length. Moreover, it cannot indent the cornea during the measurement since minimal compressions can induce a significant error. To compensate for this, the following generation of echobiometers was equipped with the ability of performing non-contact measurements [10]. To allow this, a saline reservoir was created using a plastic shell in which the probe would be inserted. By moving away from the anterior surface of the cornea, two additional peaks are recognizable in the linear A-scan.

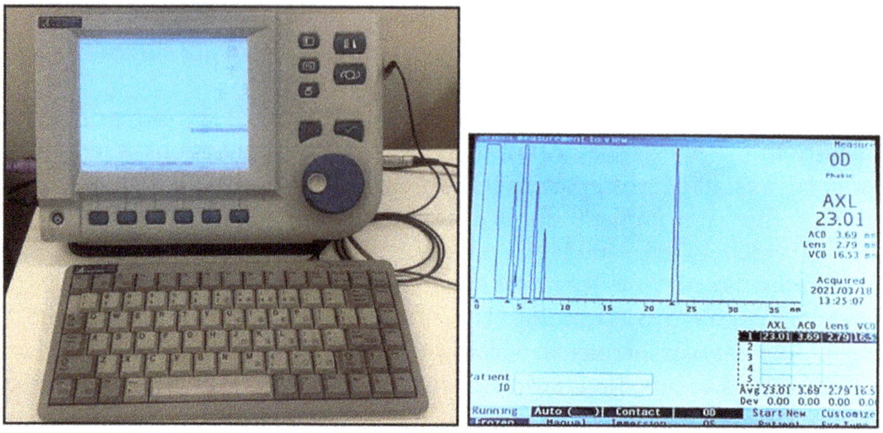

A B

Figure 6.1. (a) Picture of a modern A-scan biometer. (b) Screen showing the peaks corresponding to the cornea, anterior capsule, posterior capsule and the retina (inner limiting membrane).

The anterior and posterior surfaces of the cornea. However, the greatest advantage of the immersion technique is the fact that, since no compression of the cornea is made, there is a greater precision of the measurements [10].

Measuring the axial length using ultrasound waves, not only has limitations on positioning the beam, compressing the cornea and anesthetizing the eye, but it also has the disadvantage of using a long wavelength device. The resolution of a measurement is inversely proportional to the wavelength used to measure it. The higher the resolution, the more accurate the measurement is. The standard 10 MHz probe used in ophthalmic biometers has a resolution of about 0.03 mm, which is in the same order of magnitude of the measurement being performed [11]. Therefore, the resolution of a conventional echobiometer is not ideal to precisely measure the axial length. More recently, optical biometry was introduced [12]. By using partially coherent light with comparatively very short wavelengths, these new devices have a 9-fold increase in resolution making their measures 5–10 times more precise. To date, there are many available devices from different manufacturers that feature slight differences among them, and a direct comparison of each machine is beyond the scope of this chapter. However, we will highlight the main models and differences. The IOL Master, manufactured by Zeiss, was introduced in the 2000s delivering a faster, more precise, examiner-independent, perfectly aligned non-contact measurement of the eye [13]. By using partial coherence interferometry, it can measure the optical light path from the corneal vertex to the retinal pigment epithelium in the center of the fovea with a resolution more than 10 times better than that of ultrasound [14]. Haag Streit introduced the Lenstar a few years later. The Lenstar uses optical low-coherence reflectometry to measure not only the axial length but also central corneal thickness, anterior chamber depth, lens thickness and retinal thickness. Moreover, it measures the corneal curvature in two different regions, a 1.65 mm and 2.3 mm optical zones, enabling better calculation of the corneal power [15]. One of the main limitations of optical biometry is the inability to measure through dense cataracts. White, black or posterior subcapsular cataracts are particularly troublesome to optical biometry. In these eyes, conventional ultrasound a-scans may be performed. More recently, swept-source optical coherence tomography (OCT) has been introduced by Zeiss in the IOL Master 700 to measure axial length. This new technology has been shown to increase measurements per scans, to increase dense cataract penetration rates and to decrease standard deviation between different measurements [16, 17]. This equipment gives a true image of the entire optical path including the anatomy of the fovea. This can be of great use when screening for underlying macular pathologies before cataract surgery, such as macular holes or age-related macular degeneration [18]. Today, many other newer devices are using swept-source OCT to measure axial length including the Argos by Movu, the Anterion by Heildelberg Engineering and the Eyestar by Haag Streit.

The use of OCT allows direct measurement of the posterior corneal curvature which may improve IOL power calculation, especially in eyes that have been previously submitted to a corneal refractive surgery (as detailed below). The gradual transition from ultrasound to optical biometry has been in place since the introduction of the first IOL Master at the end of the last century. These days, optical biometry is considered the current gold standard in IOL power calculation and is not only recommended, but required especially when considering the use of premium IOLs.

Errors in axial length measurements that were common with ultrasound biometry have been greatly reduced and, to date, are not the usual culprit for eventual postoperative refractive surprises. Even in highly abnormally short or long eyes, optical biometry allows a precise measurement of the axial length. However, corneal curvature can still lead to IOL power calculation errors.

6.1.1 Corneal curvature

The relative contribution of the anterior and posterior corneal surfaces to the entire eye refractive power is of paramount importance during IOL power calculation. Measuring corneal curvature is still a major step towards the correct IOL power choice. The cornea is an aspheric refractive surface with a steeper central region and a flatter peripheral zone. For the purpose of IOL power calculation, the most important zone to be measured is the central 2.5–3.5 mm which acts as the optical zone of the cornea in most patients. Current available technology to measure corneal curvature relies on one of three different strategies. The first is based on anterior surface reflection of light. By assessing how reflected light from the corneal anterior surface diverges, it is possible to calculate its radius of curvature. Hermann Von Helmholtz proposed the use of the keratometer in the 19th century that would simulate image formation in a convex mirror to measure anterior corneal curvature [19]. The region of the cornea that is being measured varies between 2.8 mm and 4.0 mm depending on the curvature. This method has been employed since manual keratometry and more recently been used in Placido-based topographers. This is an easy and reliable strategy that can be used to calculate anterior corneal power; however, central corneal power can only be estimated based on the measurements made in the central 2.0 mm to 3.0 mm zone. Direct central measurement is not possible with this technology [20]. Moreover, posterior corneal surface is not measured using these anterior surface reflection technologies. It has been shown that, in normal eyes, the posterior corneal curvature relates to the anterior in a relatively constant ratio. This is called the Gullstrand ratio and it is calculated by dividing the radius of curvature of the posterior cornea by radius of curvature of the anterior surface [21]. The Gullstrand ratio is, on average, 0.838 (6.36 mm/7.59 mm) [22]. This ratio is profoundly affected by any corneal reshaping procedure such as refractive corneal surgery. In myopic laser vision correction surgery, the anterior central cornea is flattened, and this decreases the ratio. The opposite happens in hyperopic ablations. In these eyes, it is especially important to directly measure the posterior corneal surface to properly calculate the total corneal power [23].

The second method used to measure corneal curvature is Scheimpflug photography. These devices are gaining popularity as a way to measure not only the anterior surface but also the posterior corneal curvature. They use a rotating camera (or cameras) that captures the image of the cornea to render a 3D model of the entire anterior segment [24]. It is then possible to calculate anterior and posterior corneal curvatures as well as the corneal thickness map. By using this technology, total corneal power is measured instead of assumed based on a fixed mathematical ratio; IOL power calculation is, therefore, possible even in more complex eyes [25].

More recently, a third method of assessing corneal curvature has been used. Anterior segment optical coherence tomography is being used to measure corneal power as well [26]. This technology uses an infrared laser (1055 nm) to illuminate the tissue allowing a more precise reading of the anterior and posterior corneal surface. The IOL Master 700, by Zeiss, was the first commercially available device that used OCT-based measurements to calculate the corneal power and also lens thickness and axial length [17]. This has been shown to improve IOL power calculation accuracy, especially in challenging cases of post-refractive surgery eyes [27].

6.2 Biometric formulas

There have been ongoing developments and adaptations of mathematical calculations to estimate more precisely the dioptric power of the IOL to be implanted. Over the last few decades, great improvements occurred in the development of new designs of intraocular lenses, more reliable measuring equipment and biometric formulas in an attempt to reduce the postoperative need for spectacles [28].

The biometric formulas are instructively divided into five generations according to date of development and methodology applied [29].

6.2.1 First generation formulas

First generation formulas are typically based on theoretical models of the human eye with different ways of optimizing certain constants, both refractive ones and those related to the anterior chamber depth [30]. In 1967 Fyodorov and co-workers first estimated the optical power of an IOL using vergence formulas. In 1973, one of the first and most utilized biometric formulas was developed, the Colebrander formula [31]. In the original article, the author describes the single lens with a refractive power of 20.5 D that was in use at the time. He goes on to exemplify, step by step, how one could obtain the biometric data necessary for a tailored intraocular lens, thus minimizing refractive errors and the risk of aniseikonia, a condition detailed in chapter 7. like many other first-generation formulas, it was based on Gullstrand's schematic eye, which was developed in 1909 (see chapter 5). In 1975, Fyodorov published a formula that involved the use of biometric measures from 150 patients and his own IOL [32]. The results showed a deviation of up to 1 D in 136 of the 150 patients, i.e., 90% of the eyes within a one diopter error. Another important first-generation formula is Van der Heijde's, published in 1976 [33].

In practical terms, first generation formulas do not yield good clinical results. This may be partly justified by the lack of precision of the patient data obtained at the time. According to Binkhorst, an even simpler adjustment could have been made [34]. He proposes that adding or subtracting the manifest refractive error from an initial +18 D lens may be a better calculation method. A similar theory was presented by Kraff in 1978 [35]. It was based on the analysis of 450 eyes, but it required the use of another formula to calculate the basal value of the refractive power of the intraocular lens.

6.2.2 Second generation formulas

Second generation formulas incorporated axial length data *e* linear regression techniques. Both first and second generation formulas are hardly ever used nowadays, because their results are also quite imprecise.

The best-known second generation formulas are the Hoffer formula, the SRK formulas, and the first Olsen formula.

6.2.2.1 Hoffer formula

The Hoffer formula is derived using linear regression, thus generating a first-degree equation: [36, 37].

$$\mathrm{ELP} = (0.292 \times \mathrm{AL}) - 2.93$$

where ELP is the effective lens position (in centimeters); and AL is the axial length (in centimeters).

6.2.2.2 SRK formula

SRK formulas are also derived using linear regression methods. SRK is an acronym formed from the first letters of the authors' names: Sanders, Retzlaff, and Kraff. The first SRK was developed and published in 1980 by Retzlaff and it consisted of a linear regression method based on the evaluation of 166 patients and yielding the equation: [30].

$$P = 111.6 - 2.41 \times L - 0.87 \times K$$

where P is the refractive power of the intraocular lens (in diopters); L is the axial length (in millimeters); and K is the curvature of the cornea (in diopters).

The formula was afterwards adjusted by Sanders and Kraff, who used multiple regression techniques, still in 1980, thus giving rise to the SRK formula. Instead of using three constants for each type of intraocular lens, they left only constant A, which would vary according to the model of the lens. The simplified formula is as follows:

$$P = A - 2.5 \times L - 0.9 \times K$$

where P is the refractive power of the intraocular lens (in diopters); A is a constant; L is the axial length (in millimeters); and K is the curvature of the cornea (in diopters).

6.2.3 Third generation formulas

Third generation formulas afforded a better integration of theoretical and empirical components as well as adding greater importance to the calculation of keratometry readings, thus yielding far superior results. That is why some of them are the standard of choice to this day, such as the SRK/T and the Hoffer Q.

In 1986, Thomas Olsen published a new formula based on theoretical components and regression of pre- and postoperative data from 80 cataract patients [38]. The formula thus obtained was more complex and included in its calculation, in addition to the axial length, the anterior chamber depth, the thickness of the crystalline lens, and the shape of the cornea. Hence, the derived formula blended data science and the theoretical components of optical physics.

Not long after that, in 1988, Jack Holladay published the Holladay 1 Formula, which blended data theory and science, and brought along novelties such as a more thorough protocol to assure the quality of the data and better data predictability [39]. The Holladay 1 formula comprises a three-part system. The first part is the screening of the biometric data. Holladay attested that one of the major sources of refractive errors was not in the biometric formulas per se, but measurement errors instead. Thus, a new biometric assessment is recommended in case the patient meets any of these criteria:

1. Axial length <22 mm or >25 mm;
2. Keratometry <40 D or >47 D;
3. Difference between the calculated intraocular lens diopter to attain emmetropia and the mean diopter of the lens model superior to 3D;
4. Difference between the eyes;
5. Mean keratometry >1 D;
6. Axial length >0.3 mm;
7. Intraocular lens power to attain emmetropia >1 D.

The second part of the Holladay 1 system is the formula itself. The formula is based on theoretical concepts and was developed to show better results in view of its more accurate representation of the focal plane of the IOL in relation to foveal and corneal vertices. However, there is also a component based on data, which is part of the third step of the Holladay system 1 [39].

The third step of the three-phase system consists of surgery optimization, the so-called 'surgeon factor'. Holladay realized that, for a variety of reasons, be it the device used for biometry, the surgical technique of choice, the surgical equipment, or even the nuanced approach of the surgeon, the results could be biased one way or the other. Yet, this deviation tended to be the same. Thus, the surgeon could adjust the results using an optimized constant for his or her own practice and calculated based on a reverse solution of the Holladay formula [39].

In 1988, the trio who came up with the SRK formula made an important adaptation that led to the SRK-II. The modification was quite simple and involved just altering the value for constant A in −0.5, 1, 2, or 3 diopters in accordance with the axial length of the eye, with good results at the time [40].

Later, the SRK/T formula was published, in 1990, and it represented an upgrade to SRK-II. Unlike its predecessors, the formula had its structure made upon a theoretical foundation, but the prediction of the anterior chamber depth, the correction of retinal thickness, and the refractive index of the cornea were all improved by means of empirical methods based on data from 1677 cataract surgeries. Given its good results, this formula is still widely used to this day [41].

In 2010 the formula was revisited by Sheard and colleagues, who found out that the calculation of the corrected axial length and of the corneal height in the SRK/T did not follow physiological patterns and was a source of error in the formula [42]. Hence, the authors developed the T2 formula by altering the way to calculate corneal height to a linear regression formula that drew upon a database with 11 189 eyes and attained results superior to the SRK/T.

The Hoffer Q formula was published in 1993 by Kenneth Hoffer and its development followed a pattern on a par with the one used for the Holladay and SRK/T formulas. Not unlike the latter ones, the author's approach involved carrying out empirical optimization for theory-based formulas. Good results have validated the ample use of the formula to this day and its favorable comparison to other fourth and fifth generation formulas. One of its major differences is the use of a 'personalized anterior chamber depth (ACD)', which must be optimized for each surgeon, and a model of lens through a set of formulas also provided by the author, thus securing the best results for the patient cohort of each surgeon after some 'practice time' [43].

The year 1993 also saw the publication of the Barrett Universal formula. The first version of the Barrett Universal formula had come out in 1987 [44]. but unlike its successor, it had not received much acclaim. The Barrett Universal formula is a theoretical formula that employs empirically determined values, just as with many others of this generation. However, what makes this formula unique is the fact that it is universal. Besides determining the ACD based on both the axial length and keratometry, it also analyzes the main reflective planes of the IOL and utilizes a 'lens factor' that allows the formula to be applied to various models of lenses without the need for the surgeon to know which material it is made of and what its constant is. The formula attained a better performance than the others at the time [45]. The formula was subsequently adapted to the fourth generation in 2010 and coined the Barrett Universal II [46].

6.2.4 Fourth generation formulas

With the advent of optical biometry in the last 20 years and the concurrent availability of more biometric measurements, new formulas that use these parameters have been introduced and classified as fourth generation formulas. Most of the formulas in use today, such as the Haigis, Olsen, Holladay 2, and Barrett Universal II make part of this group, which comprises most of the current high-performance formulas empirically validated by the medical and scientific community [46, 47].

Using a novel principle for the time, Naeser published the Naeser 1 formula in 1997 [48]. Most of the other formulas employed a 'thin lens' model, in which all the IOL power was calculated as if it were on an effective lens plane, which could be streamlined for more precision without eliminating all errors due to its being an estimate or approximation. Naeser used a thick-lens model, in which the convergence power of the IOL is distributed across the anterior and posterior surfaces of the lens, thus rendering the proportions more realistic. Hence, the ACD was the distance to the first surface of the lens. The position of the posterior surface was also calculated based on the axial length and ACD. However, information regarding the architecture of the lens was necessary, something hard to obtain in view of the proprietary rights of the lens manufacturers.

The formula was revamped in 2019 and became the Naeser 2 Formula [49]. Instead of using secret and proprietary parameters, the new formula utilizes parameters calculated from public domain information and estimates as to some of the measurements of the IOL. The study compared the Naeser 1 formula, the Naeser 2, and other 5 formulas (Barrett Universal II, Haigis, Hoffer Q, Holladay 1, and SRK/T). The results were quite similar across all formulas, thus proving that secret data on the architecture of the lens is not needed to achieve reliable results.

The Haigis formula was published on a website in 1998. An article published in 2000 compared the efficacy of the formula for ultrasound biometers and a new partial coherence interferometry biometer. The original article compared only the results obtained by doing the calculation using the Haigis formula, which showed equivalence in the accuracy of the two methods. The Haigis formula is a theoretical formula based on the thin-lens model, which uses three constants to represent the IOL model: a0, a1, and a2, in which by default a1 can be replaced by 0.4 and a2 by 0.1. The inputs required by the formula are axial length, keratometry, and the postoperative measure of the ACD [50].

The Olsen formula was initially published in 1995. There are four variables for the calculation of the postoperative position of the lens: a lens constant, axial length, preoperative ACD, corneal height (based on keratometry and corneal diameter), and crystalline lens thickness [51]. A new study was published in 2006 in which it empirically optimized the formula through multivariate regression, using preoperative refraction and age of the patient were included in the equation [52]. In his study, Olsen found in descending order of relevance correlations between the ELP and axial length, preoperative ACD, crystalline lens thickness, radius of the cornea, refraction, and age. According to the same study, there was no correlation between the white-to-white distance and the ELP. The method yielded a 10% reduction in the refractive error of the previous formula. Therefore, the following linear formula was reconstructed:

$$\begin{aligned} ACD_{calc} = {} & ACD_{constant} - 4.03 + 0.19 \times A + 0.49 \times C + 0.28 \times L - 0.41 \times R \\ & + 0.0028 \times Rx \end{aligned}$$

where ACD_{calc} is the postoperative ACD; $ACD_{constant}$ is the lens constant; A is the axial length; C is the preoperative ACD; L is the crystalline lens thickness; R is the corneal radius; and Rx is the preoperative spherical equivalent.

The Holladay 2 formula and the Barrett Universal II formula did not have their details published in peer-reviewed journals. The Holladay 2 formula was distributed by means of a downloadable package on a website, whereas the Barrett Universal II formula is freely available on the APACRS website.

The variables used in the Barrett Universal II are axial length, keratometry, ACD, and preoperative refraction. Crystalline lens thickness and white-to-white distance are optional parameters.

The variables used in the calculation for the Holladay 2 formula are axial length, keratometry, ACD, crystalline lens thickness, age, white-to-white distance, and preoperative refractive data. But, according to Hoffer [29], most non-conventional parameters (age, refraction, crystalline lens thickness, and white-to-white distance) are optional, and age would probably be used to predict crystalline thickness with the following formula:

$$\mathrm{LT} = 4 + (\mathrm{age} \div 100)$$

where LT is the crystalline-lens thickness.

The Wang–Koch adjustment was specifically developed to improve the accuracy of five formulas (Haigis, Hoffer Q, Holladay 1, Holladay 2, and SRK/T) in eyes whose axial length exceeds 25 mm. The original studies showed a significant reduction in the number and intensity of hyperopic error in patients who underwent surgery with this adjustment in these formulas [53, 54].

Voytsekhivskyy published the VRF-IOL formula in 2018 [55]. It was empirically devised based on 823 eyes and the parameters are axial length, convergence power of the cornea, ACD, and horizontal diameter of the cornea. The author deemed the results comparable to those from classic formulas.

6.2.5 Fifth generation formulas

The existence of a fifth generation of formulas is a controversial topic in the medical literature. It could be explained in two ways: first, according to Hoffer, it would enable the creation of customized formulas for gender, ethnicity, and other population variables; and second, it would consider the use of newer strategies, such as artificial intelligence, for the development of formulas with complex mathematical patterns hardly achievable by humans alone [29].

The latest trend in the development of new biometric formulas is the use of strategies based on mathematical and computational models, especially artificial intelligence. Although these formulas employ the latest technologies, there is controversy regarding the superiority of their results, with mixed results as to their effectiveness for normal eyes, for which nearly all formulas work without a hitch, and atypical eyes, for which all formulas seem to incur in more errors. It is also important to point out that the methodology and data employed in the development of most of these formulas were not published in peer-reviewed journals, which makes it difficult to better understand them. The main examples of just such formulas are Hill-RBF, Fullmonte, Kane, LADAS Superformula, PEARL-DGS, and EVO Formula 2.0 [56, 57].

According to Hoffer, there is significant variability between patient eyes depending on gender, ethnicity, and other population characteristics and the Holladay 2 formula was the first one developed using mean estimates of the general population [29]. The Hoffer H5 formula was created by making some alteration to the Holladay 2 formula. The former was derived from preoperative refraction as the correction of the cataract led to great refractive alterations utterly unrelated to the earlier anatomy. It ensued from the change in the estimated biometric values used by Holladay for real measurements obtained in 7500 biometric assessments for axial length and ACD, and in other studies for crystalline lens thickness and corneal diameter. These measurements were made using an *n* ultrasound in 1980. Only the original keratometry of the formula remained unchanged. This formula was developed in 2004 and named the Hoffer H. The only refinement when compared to the others was a lower percentage of eyes with less than a 0.25 diopter error. The Hoffer H5 formula replaced the mean constants of the Hoffer H for optimized constants for gender and ethnicity. These two parameters, according to Hoffer, are the defining features of the fifth generation. The formula was tested in 2707 eyes and the results surpassed those of the Hoffer Q, Holladay 1, and SRK/T formulas. There is an ongoing study with 10 000 eyes to compare the formula with more recent ones.

The LADAS Superformula brought forth interesting ideas and it was first shown in 2015, having received moderate acclaim from the scientific community [58]. Based on literature findings, the first version was a graphical and mathematical model in which tridimensional planes representing the results of the Hoffer Q, Holladay 1, Holladay 1 with Koch adjustment, Haigis, and SRK/T formulas overlapped. Thereafter, a single tridimensional plane that mixed four of the five formulas was devised, drawing the ideal results from each one. The prospective analysis carried out in 100 eyes allowed the method to pick the best formula for the 100 eyes [57]. There is currently a new version available for tests. It is still being developed and it incorporates artificial intelligence to enhance the results.

The development of the Hill-RBF formula, published under the leadership of Warren Hill, was based on artificial intelligence algorithms of the radial basis function type, thus the RBF acronym. It is probably the best-known and most studied AI-based formula, but it does present some functional peculiarities. It is currently in its second version, but version 3.0 is already under development. According to the author, the algorithms were trained with 3445 eyes for version 1.0 and this database was expanded to 12 419 eyes, out of which over one thousand eyes were extremely short. One of the most peculiar aspects of this formula is the fact that it warns the user when the calculation is deemed 'out of bounds', that is, the system was not trained with enough data for that type of eye, thus yielding a less reliable prediction. This characteristic makes it hard to compare the formula with the others available in the market [47].

Up to the date of publication of this chapter, we saw that the Fullmonte formula had been cited in an article by Kane [56]. but the site on which it was hosted is no longer accessible. According to Kane, the Fullmonte formula is based on the Monte Carlo method, in which Markov chains are used to predict postoperative outcomes.

The Kane formula was developed in 2017 with a database containing approximately 30 000 eyes. It is a theoretical formula that incorporates regression models and artificial intelligence. The author has also created a toric formula with a similar development approach and another one for patients with keratoconus. The latter is a theoretical variation of the basic formula. The requisite parameters for the basic formula are the *A*-constant, gender, axial length, corneal curvature (in diopters), and ACD. Lens thickness and central corneal thickness may be added to enhance the predictive power of the formula. Surgically induced astigmatism (Kane recommends 0 in temporal incisions of less than 2.75 mm) and surgical incision location are necessary inputs for the toric formula.

The PEARL-DGS formula was developed in France. PEARL is an acronym for 'Prediction Enhanced by Artificial Intelligence and output Linearization', whereas DGS is an acronym for the initials of the three authors. The formula was devised from machine learning techniques in a study with 400 eyes and yielded results that surpassed those obtained with other formulas.

The EVO (Emmetropia Verifying Optical Formula) 2.0 is based on the theory of emmetropization and it denominates an 'emmetropia factor' for each eye. Based on the thick lens model, it is a primarily theoretical formula that can work for different shapes and refractive powers of IOLs. There is also an adaptation of the formula for toric lenses that is based on the simple EVO formula. It uses requisite parameters for axial length, keratometry, ACD, and *A*-constant. Lens thickness and central corneal thickness can be added for greater precision.

6.3 Comparison among the formulas

6.3.1 Establishing universal definitions

In 2015, several of the most respected authors in the field published an editorial suggesting a protocol with a series of recommendations for carrying out research into IOLs [59].

The first step is to describe the demographic data (gender, age, and ethnicity) of the study population. The second step is to calculate the mean error (ME) of the study group for each individual formula by adjusting lens factors (constants), thus reducing the bias associated with a not-optimized lens A-constant. The third step is to compare the median absolute error (MedAE) of each formula and not the ME. An ME different from zero just means that the lens *A*-constant used is too high or too low for the population. Besides, the comparison of the mean absolute error (MAE), the most often used measure in the literature, is incomplete, rendering the statistical analysis more complicated. It is also important to report standard deviation, maximum, minimum, and 95% confidence interval around the mean, as well as the ratio of eyes with a predictive error $\leqslant \pm 0.5$ D, $\leqslant \pm 1$ D *e* $> \pm 2$ D.

The ideal is to always utilize one eye per patient. In case this protocol is not followed, there is the risk that data may be altered by the correlation between the person's own eyes. To correct this error, adjustments can be made using the GEE ('generalized estimating equations') method or Bootstrap. Nevertheless, the fewer statistical adjustments the better. With all these requirements met, standardized statistical tests can be used.

The authors no longer recommend using old formulas, such as SRK I and II. The implementation of a formula needs confrontation and testing with like samples and outside the commercial setting in which it was developed, thus assuring objectivity of the comparative results. All the formulas used in the studies must be cited in alphabetical order and, in the case of the Hoffer Q and SRK/T formulas, there must be a reference to the errata [41, 43, 59].

The optimization of the Haigis formula is more complex because it relies on three different constants. The recommended method is optimization through double linear regression, but in case the data are insufficient for individual optimization, the use of the constants from the User Group for Laser Interference Biometry (ULIB) is acceptable [59]. For the biometric measurements, optical biometers are recommended. The software version of the instruments used must also be acknowledged to assure the reproducibility of the study. However, an ultrasound scan of the eye may be necessary to measure the axial length in eyes with a very dense cataract. This is preferably done through the immersion technique to minimize the risk of corneal applanation and shortening of the axial length or of the ACD (see above).

When measuring the convergence power of the cornea, one needs to disclose the instrument and methods used for the measurements, such as the area under analysis, the number of points assessed, the formulas employed, the keratometric index, among others. In the case of the ACD, the definition must be clear as to the distance between the corneal epithelium and the crystalline lens. To facilitate differentiation, the distance between the endothelium and the crystalline lens is called aqueous depth. The predictive error must not consider the surgical target since there is no relation between the two. The refractive error will be the same whether the surgeon aims for a target of 0 D or −5 D.

The postoperative subjective refraction should be preferably measured three months after the operation. If that is not possible, the measurement must be made at least one month after the surgery, when the refraction is already considered stable. As assessment of refractive error depends on visual acuity, patients whose postoperative visual acuity is below 20/40 should be excluded. The authors recommend the use of a single model of IOL for the study because it will be necessary to optimize the constants for each one of the models of lens used [59].

6.3.2 Limitations of the formulas

Despite the great advances in the field over the last few decades, biometric formulas still lag when it comes to precision in unconventional situations. The results for eyes with a long or short axial length, with atypical corneal curvature, with previous trauma, previous surgeries, or other comorbidities tend to be much worse than in virgin, average-sized eyes. Therefore, further research in the field is required, encompassing both the development of new formulas and the better understanding of the behavior of the formulas currently available and their application in each specific case.

In addition to the limitations inherent to the formula as to the predictability of results, we should pinpoint the other sources of error that may yield wrong predictions. Every surgical procedure, from the gathering of clinical and biometric data, surgical planning, surgical technique, lens choice, and all the way to pre-, peri-, and postoperative complications may negatively impact the refractive outcome of the cataract surgery. It is up to the surgeon to use his/her experience and to act in conformity with strict protocols when dealing with his/her patients. Greater homogeneity and predictability of surgical outcomes will guarantee higher rate of success.

6.4 Artificial intelligence

6.4.1 Definition of artificial intelligence

Artificial Intelligence (AI) is a branch of computer science that aims to develop machines capable of simulating the human ability to think and act. Machine learning (ML) and deep learning (DL) are currently the two most relevant divisions of AI in view of the success attained in several medical specialties, such as ophthalmology, cardiology, radiology, and oncology. The growing use of this technology in healthcare improves real time and assertive support in tasks such as patient data collection and analysis, disease prevention, early diagnosis, and customized treatment guidance. The use of big data and improvement in computer processing power and storage have enabled the expanding use of AI [60, 61].

6.4.2 The use of artificial intelligence in calculating IOL refractive power

The use of the first AI-based biometric formulas was seen in 1997, when Clarke and Burmeister published an article comparing the results of the Holladay 1 formula with those of a neural network developed by the authors showing superior results [62]. However, the formula was not incorporated by clinicians.

AI-based formulas have gradually become more accepted throughout the years, with the breakthroughs leading to clinical results that were clearly superior to those obtained with the conventional biometric formulas. As indicated in the previous section, there are five main AI-based formulas: Hill-RBF, which uses RBF as its activation function; Fullmonte, based on Markov chain Monte Carlo methods; Kane, which applies theory-elements, regression, and AI to cloud computing; LADAS, in which a mathematical method superimposes results from classic formulas and incorporates elements of AI to achieve optimization; PEARL-DGS, which is based on ML techniques.

6.4.3 Advantages of using AI

There are clear advantages of AI-based biometric formulas over traditional formulas. AI formulas can be constantly upgraded to include more cases and their algorithms can be adjusted, thus leading to progressively better results. On its third version, the Hill-RBF is a prime example of a formula that has become more precise as more cases were included. Another possibility is that results can be tailored to the user. AI formulas can provide a more thorough optimization by including more elements with data comparison beyond human understanding. This enhanced perception of patterns enables AI-based formulas to come up with continuously more precise results [63–65].

6.4.4 Limitations of AI

The patterns of answers provided by AI formulas are often incomprehensible to humans, especially when more complex technologies, such as neural networks, are employed. The difficulty in understanding the method limits the ability of clinicians to improve these formulas. There is a pressing need for interdisciplinary development teams to include engineers, physicists and physicians so that when the algorithm is being developed, biological and theoretical expertise can be used in a detailed analysis of the sample. Most vergence formulas are published in peer-reviewed scientific journals and all the calculations are explained. They may be replicated and tested by other surgeons. The AI formulas available in the market today do not have their source codes or rationale for development published, and then it is left to the scientific community to trust the work of the authors and use these formulas on the websites in which their calculators are hosted.

There are international standards for testing these biometric formulas [59], and they must be rigorously followed before studies can be published in respected scientific journals. Nevertheless, no scientific paper for an AI formula currently available was ever published, thus leaving clinicians in the dark as to how patient data were collected, how exams were carried out, and what the inclusion and exclusion criteria were, all of which are essential to the critical analysis of the scientific method.

Despite the promising proposals and excellent results, it is necessary to take some precautions when using AI formulas. They should be considered an auxiliary tool only. Decisions should be made considering the ophthalmic evaluation, the surgeon's experience and, most importantly, the other scientifically validated tools, thus providing greater safety to both patient and surgeon.

6.4.5 The future of AI

The evolution of AI-based biometric formulas will certainly occur. The development of technology directly applied to optical physics, as in the use of ray tracing, may result in huge benefits to the understanding of how the human eye works. The creation of bigger and more robust databases is having an impact on the generation of algorithms and new biometric formulas focused on more challenging clinical settings such as previously operated eyes or eyes with unusual dimensions.

There are other trends shaping the future of ophthalmology. Studies with biometric measurements taken during the intraoperative setting [66] and the development of more sophisticated and efficient equipment will enable the collection of better quality and higher precision data in the preoperative, intraoperative, and postoperative settings (see below in section 6.6.3).

The full use of AI in medicine is at its early stage. There are great challenges to be overcome before we can fully rely on algorithms that help in medical decision making. Despite the myriad of different software applications (the list of FDA approved programs keeps growing), there have not been many prospective studies to validate the tasks that could be performed by machines to predict clinical results and assist physicians. Another important aspect is the rigorous criteria for data mining to be implemented in ML systems. These criteria should give emphasis to a clinical correlation. Consequently, the approval of decision-making assistant algorithms demands rigorous studies, the publication of results in scientific journals, and clinical validation in the real world before using them in patient care.

6.5 Astigmatism

Astigmatism is an imperfection in the asphericity of the cornea in such a way that convergence of light rays occurs in two different focal points. Depending on the magnitude of the astigmatism, it may result in complaints such as low visual acuity and dysphotopsia.

6.5.1 Measuring astigmatism

There are several ways to measure astigmatism: manual keratometry, automated keratometry, corneal topography, slit-scanning technology, optical coherence tomography, and Scheimplug imaging (see above) [67]. The first three methods only evaluate the anterior surface of the cornea and do not consider the power of posterior astigmatism. The rationale behind the use of these devices is that there is a somewhat constant ratio between anterior and posterior corneal curvature. However, studies have shown that this ratio is variable [68]. Therefore, it is essential to measure the posterior corneal surface to calculate the posterior astigmatism using an appropriate method. A study has demonstrated that devices that measure only the anterior curvature have an observational error ranging from 0.2 to 0.6 diopters depending on the type of astigmatism (with or against the rule) [69].

6.5.2 Magnitude and axis

Astigmatism can be measured in two dimensions: magnitude and axis. The magnitude is the intensity of the deviation and it is measured in cylindrical diopters. It is calculated measuring the difference in curvature in diopters between two perpendicular corneal meridians, the flattest (K1) and the steepest (K2). The axis, in turn, is the relation of the two meridians with the horizontal meridian, measured in degrees. 0° is the rightmost point on the horizontal meridian, 90° the uppermost point of the eye on the vertical meridian, and 180° the leftmost point on the horizontal meridian, regardless of eye laterality [70].

It is also necessary to classify astigmatism according to its regularity and the relationship between the axes. Astigmatism is classified as regular when its flattest and steepest meridians are perpendicular, that is, they are 90° apart. Regular astigmatism can be further classified into three types. With-the-rule astigmatism (WTR) is when the steepest is within 15° of the vertical meridian (90°). Against-the-rule astigmatism (ATR) is when the steepest meridian is within 15° of the horizontal one (0°–180°). Oblique astigmatisms are the ones that do not fall into these previous categories or when the steepest meridian is between 30°–60° or 120°–150° [71].

Irregular astigmatism happens when there is no relationship between the axes. In the case of non-orthogonal irregular astigmatism, the steepest hemimeridians are not 180° apart. Asymmetrical irregular astigmatism is when the hemimeridians have different magnitudes. These types of astigmatism are more common after corneal diseases or ocular trauma and may be quantified by means of a vector analysis of the corneal topography [72].

6.5.3 Surgically induced astigmatism

Surgically induced astigmatism (SIA) is caused by the change in corneal shape resulting from surgically placed incisions. The relationship between incision type and magnitude of SIA is not well established, but there is a significant correlation with the meridian in which the incision was placed due to the oval shape of the cornea. The effects are progressively less the farther away the incision is from the optical axis and also when it is located on the temporal side of the eye [67, 73].

When making corneal and limbal incisions, especially those that affect refraction, such as in cataract surgery, surgeons must consider the effect of the incision on the final refractive result for the patient. Multiple factors affect the SIA such as the shape, size, and orientation of the incision, the orientation and magnitude of the patient's previous astigmatism and the experience of the surgeon [74, 75].

It is possible to use SIA to compensate a pre-existing astigmatism of the patient. Calculation can be made using freely available software [76].

6.5.4 Incorporating astigmatism when calculating IOLs

Cataract surgery has been shown to be an opportunity to treat a pre-existing refractive error with the use of an astigmatism-correcting IOL. Cataract surgery, whose sole purpose was to cure the blindness caused by an opacified lens, often resulted in gross refractive errors. Over the years, this surgery is now being performed in patients who can still see well and with high expectations of a good refractive result postoperatively. Surgeons have aimed at results ever closer to emmetropia.

In addition to the spherical results discussed earlier in this chapter, it is necessary to take into consideration the treatment of astigmatism. Approximately 30% of patients who undergo cataract surgery have astigmatism above −0.75 D, which requires correction. The use of toric IOLs can correct this astigmatism with an excellent refractive result. However, it is imperative that the surgeon adequately chooses the correct cylindrical correction power and adjusts the implant in the proper axis once it is inside the patient's eye. Moreover, surgeons should carefully plan postoperative results also considering some difficult-to-predict variables such as posterior corneal astigmatism, biometric variables, ELP, age-related corneal changes (tendency for slow progression to ATR astigmatism), and SIA [67, 70].

The first challenge is overcome by using nomograms, formulas, or calculators made available by authors or lens manufacturers. The second challenge is resolved by using tools that aid in the correct positioning of the IOL during surgery. The most basic technique involves the manual use of marking pens on the eye. TrueVision 3D visualization and guidance system, Callisto eye system, and VERION image guided system are amongst the most common intraoperative tools that make use of pre- and intraoperative images to project the correct axis into the visual field of the surgeon, thus enabling a more precise alignment of the intraocular lens (see below) [70].

6.5.5 Calculating toricity of the intraocular lens

The most utilized method for calculating toricity of the IOL is the Baylor nomogram This is a chart created through regression analysis that predicts posterior corneal astigmatism based on anterior corneal astigmatism magnitude and orientation [67, 70].

The Abulafia–Koch formula, which is based on a regression model and uses postoperative results of patients who received toric lenses, gives a more refined outcome. The formula was built into the Hill-RBF calculator and is available in the Lenstar biometer [70].

The Holladay toric calculator uses asphericity and the predictive ELP to define the effective toricity of the lens after implantation. It is possible to utilize the Baylor nomogram to compensate posterior astigmatism in the calculator [70].

The Barrett toric calculator was also developed based on regression analysis in patients who had received toric lens implants.

Lens manufacturers have also published private formulas. Alcon uses the Barrett toric calculator. Johnson & Johnson makes available a calculator based on the postoperative results of its lenses when associated with the Baylor nomogram. The PhysIOL calculator incorporates the Abulafia–Koch formula to calculate toricity [70].

6.5.6 Calculating toricity in patients with a history of refractive surgery

One of the greatest challenges of IOL power calculation is to perform it in patients who had previously undergone refractive surgery. These procedures consist in remodeling corneal curvature and refractive index using an excimer laser. These modifications make measurements of the corneal curvature more imprecise and the prediction of the effective postoperative IOL position even harder [70].

Aramberri developed the double-K method to increase predictability in cases after refractive surgery. It consists of using the refractive power of the cornea before refractive surgery to calculate the ELP and of using the refractive power of the cornea after refractive surgery in order to choose the refractive power of the IOL [70].

Other methods were later developed. One version of the Barrett True-K formula does not need ancillary data and the other is based on the Barrett Universal II formula. The latter calculates keratometry differently and yields a double-K solution like the one proposed by Aramberri. Barrett True-K results have been promising, equal to or superior to those obtained with the calculator at the ASCRS website (Holladay 1 with double-K correction) [70, 77].

Another promising formula was the one developed by Tang *et al* [78] which is based on precise measurements of the anterior and posterior curvatures of the cornea using OCT. The formula consists of six variables: axial length, ACD, lens thickness, central corneal thickness, anterior and posterior keratometry. The results obtained using this formula were similar to the ones obtained using Barrett True-K and the formula available at the ASCRS website. However, another study showed better results when comparing it to the average result from the ASCRS calculator or the Barrett True-K formula [79].

6.6 Assistive methods

6.6.1 Marking

Centration and orientation of the IOL is paramount to achieve success in the final visual result. This is especially true in toric implants in which the flattest and steepest meridians have to be precisely oriented perpendicular to their corneal counterpart. It is known that each 1 degree of error in the meridional alignment causes the loss of roughly 3.3% of the astigmatism-correcting power of the IOL [80]. To allow this precision in orientating the implant, the surgeon has to mark the patient's eye in order to have a reference to which align the IOL after implantation. It is important to note that the eye tends to rotate when the patient is lying down on the operating table in a phenomenon known as supine cyclotorsion [81]. Therefore, any reference meridian with the patient lying down may be off when the patient is standing up. This way, preoperative marking with the patient upright fixating on a distance object is ideal. There are many different strategies to accomplish this task [82]. The simplest method is to mark the corneal horizontal meridian with the patient looking straight ahead assisted either by the slit lamp beam positioned horizontally or a marking device with a pendular weight that ensures horizontal marking. With the horizontal meridian marked, the surgeon can use a compass (aligning 0–180 degree mark to the marked horizontal meridian) to mark the desired implantation axis. There are many different compasses that can be used for this task. The use of violet gentian ink has been proposed to mark the cornea in the correct meridian since this is an inexpensive and common tool found in most operating suites. However, ink can blur and lose precision. For that reason, a thermal marking of the cornea at the limbus has been proposed using a wet-field cautery to avoid blurring marks allowing a more standardised alignment [83].

6.6.2 Overlays

More recently, modern surgical microscopes have introduced the use of digital video overlays in the eyepiece that allow even more precision aligning the IOL [84, 85]. This technology usually consists of two separate hardware units. The first one is usually located in the clinic and can take a preoperative scan of the eye, normally during biometry. With the help of image processing technology, individual landmarks are identified such as the limbus, pupil, conjunctival vessels or nevi. Then, this image is transferred to the second equipment that is connected to the surgical microscope. In this stage, the processed image is compared to the live feed from the camera connected to the microscope and it can center the image and compensate cyclotorsion in real time. This allows projection of reference axis and marks to help surgeons to position the incisions, to better position and size the capsulorhexis and to perfectly place the implant in the x- and y-axis as well as to align the toric marks to the precise meridian. This is a modern technological approach that can give limitless information during surgery and can guide the surgeon to a more standardised procedure. However, it is a costly and time consuming technology due to required additional preoperative measurements and intraoperative calibrations. Moreover, the overlays can be extremely distractive during important parts of the procedure creating a real danger to an otherwise routine step for a competent surgeon.

6.6.3 Intraoperative aberrometry

With the intent of further improving refractive outcomes after cataract surgery, the use of intraoperative aberrometry has been recently incorporated by some surgeons to aid IOL power calculation [86]. It has been shown that around 70% of normal eyes end up within ±0.50 D of the targeted refractive error after cataract surgery. This percentage falls to below 50% when considering post-refractive surgery eyes [87]. Intraoperative aberrometry was initially proposed to be used in these eyes in which routine biometry is usually challenging. There are two commercially available pieces of equipment from different manufacturers. The first and most commonly used is called ORA and it is made by Alcon. This equipment is based on Talbot moiré interferometry and was initially designed as a means to assess the correct IOL power to be used during surgery [88]. It has two gratings that produce a fringe pattern to measure sphere, cylinder and axis. The second is called Holos, produced by Clarity, and it is based on a micro electro-mechanical system (MEMS) mirror and quad detector to measure the magnitude of wavefront displacement. Both can make aphakic and pseudophakic readings of the eye giving quantitative and qualitative data depending on the objective. The aphakic measurements can be used to calculate the correct power of the IOL to be implanted. The pseudophakic scans can further confirm the post-implantation refraction to assess the outcome. Ideally, the information collected from the aphakic refraction would guide the surgeon to the perfect IOL power choice eliminating or minimizing the need of preoperative biometric data. However, many details compromise such an approach [89]. For the scans to be performed, incisions have to be perfectly sealed with no corneal edema, intraocular pressure has to be set to a physiological level, lid speculum cannot distort the eye nor can the drapings and/or bags and, lastly, the corneal surface has to be pristine with no epithelial edema or punctate epitheliopathy. These conditions are hard to be controlled during surgery and they can add up to produce significant errors of measurements. Thus, intraoperative aberrometry has been mainly used to confirm the findings of preoperative measurements helping to decide between different implant options during surgery. This can be especially useful when implanting advanced technology IOLs such as trifocal or extended depth of focus implants in which an ideal performance is only achieved in cases where the refractive error is limited to a minimum [90]. As stated before, intraoperative aberrometry can also be used to confirm toric IOL orientation after implantation ensuring that the maximum correction of the astigmatism is obtained. After implantation of the IOL, a pseudophakic scan can be performed and repeated as many times as needed until an optimum orientation has been achieved [91]. This is especially interesting when implanting a toric IOL in an eye that has an irregular astigmatism component in which the orientation of the IOL might not be easily calculated preoperatively. Cases like post-radial keratotomy, post-corneal transplant, keratoconus, among others, are the ones that may benefit the most with this alignment. Although promising, there is no consensus in the literature whether this technology brings up a significant benefit for choosing the correct IOL power, compared to routine preoperative measurements with the use of modern formulae [89, 90, 92]. It does add up cost to the procedure and the current versions of the equipment considerably reduces the working space between the operating microscope and the patient's eye.

6.6.4 Light-adjustable lens

This is a type of IOL that can have its dioptric power modified after implantation. It was designed to allow correction of residual refractive error after cataract surgery has been finalized [93]. This is a very interesting approach that can be used to provide the ideal vision to patients when preoperative determination of the desired refraction is difficult or when there are confounding factors affecting IOL power calculations. The IOL is made of a special photo-reactive UV-absorbing silicone which can have its molecular bondings changed to allow modification of the curvature of the lens. Surgery is carried out in the same fashion as that of a routine monofocal IOL implantation. Three to four weeks after surgery, a final refraction is performed and titrated to the patient's lifestyle desire. Then, an UV-irradiating machine (light delivery device) using a 365 nm laser is used in office to irradiate specific areas of the lens depending on the desired correction [94]. This allows a precise modification of the surface curvature permitting corrections from −2 to +2 D sphere and from −0.50 to −3.00 cylinder [93, 95]. This correction seems to remain stable in the long term [96]. After the desired correction has been achieved, the lens is 'locked in' by a final irradiation of the UV laser in the entire IOL surface and the procedure is concluded. This is a paradigm shift in improving refractive outcomes after cataract surgery. The attention is shifted from preoperative data gathering to postoperative adjustments. This also allows the trial of a specific distance-adjusted vision to ascertain the suitability for the patient's requirements. This lens has recently received FDA approval and experience is growing with this implant. Because of the required irradiation of certain parts of the lens, a 7 mm pupil dilation is required for this IOL to be used and this may exclude some poorly dilating candidates from this technology. Moreover, to avoid undesired UV exposure after surgery, patients are required to wear UV-protecting glasses during all waking hours until the final lock-in procedure.

6.6.5 Refractive index shaping (RIS)

In another pursuit of postoperative adjustment of the IOL power, this exciting technology is currently being developed to promote postoperative refractive index change of any acrylic IOL [97]. This is an office-based femtosecond laser application into the IOL material that changes the index of refraction of the lens polymer. It uses a green laser (520 nm) to produce heat in the material altering the polarity of certain molecules thus changing the hydrophilicity of the lens. There is a photo-induced hydrolysis that creates new chemical groups (amines and carboxylic acids), which are strongly hydrophilic. This change induces a large, controllable and precise modification of the refractive power of the lens and can be used to adjust the dioptric power of an already implanted IOL [98]. It has been demonstrated that this adjustment can be performed in any kind of currently available IOL material. Therefore, this can be applied to virtually any patient who ends up with an undesirable refractive surprise after surgery. This, by itself, is an already revolutionary procedure that can benefit many patients providing better postoperative acuity and increasing spectacle independence. However, this treatment allows for not only the adjustment of the spherical dioptric power but also for toric magnitude and orientation correction, multifocality inducement or reversal and spherical aberration treatment [99]. Spherical and cylindrical dioptric power can be added or subtracted from any lens. A diffractive pattern can be created, altered or removed from an already implanted IOL. Moreover, the entire procedure can be repeated or even removed by creating an opposite pattern in a subjacent layer. This procedure is entirely performed without inducing any inflammatory reaction to the eye or damaging the IOL. In fact, studies have demonstrated a stable modulation transfer function (MTF) before and after IOL RIS treatment. In other scenarios, changing from a multifocal diffractive optics to a standard monofocal IOL increased the MTF after treatment [97]. This is truly a promising technology that could dramatically modify the treatment of cataract and refractive lens exchange. It could be the first major step towards an individualization and customization of the intraocular implant.

Chapter highlights

- IOL power calculation relies in solid measurements of the eye, especially the axial length and corneal curvature.
- Many different formulas can be used to calculate IOL power.
- Artificial intelligence is being used to better calculate the correct power of the IOL in complex eyes.
- Assistive methods can be used to fine tune IOL power calculation as well as its positioning during surgery.

References

[1] Zhang F, Zhang J and Li W *et al* 2020 Correlative comparison of three ocular axes to tilt and decentration of intraocular lens and their effects on visual acuity *Ophthalmic Res.* **63** 165–73

[2] Arba Mosquera S, Verma S and McAlinden C 2015 Centration axis in refractive surgery *Eye. Vis. (Lond)* **2** 4

[3] Findl O 2005 Biometry and intraocular lens power calculation *Curr. Opin. Ophthalmol* **16** 61–4

[4] Kim M, Eom Y and Lee H *et al* 2018 Use of the posterior/anterior corneal curvature Radii Ratio to improve the accuracy of intraocular lens power calculation: Eom's adjustment method *Invest. Ophthalmol. Vis. Sci.* **59** 1016–24

[5] Schultz M, Oberheide U and Kermani O 2016 Comparability of an image-guided system with other instruments in measuring corneal keratometry and astigmatism *J. Cataract Refract. Surg.* **42** 904–12

[6] Dooley I, Charalampidou S and Nolan J *et al* 2011 Estimation of effective lens position using a method independent of preoperative keratometry readings *J. Cataract Refract. Surg.* **37** 506–12

[7] Butcher J M and O'Brien C 1991 The reproducibility of biometry and keratometry measurements *Eye. (Lond)* **5** 708–11

[8] van der Heijde G L, Meinema A J and Vlaming M S 1977 Digital A-scan ultrasonography used to measure ocular distances *Am. J. Ophthalmol.* **83** 276–7

[9] Bronson N R 2nd A new A-scan ultrasonoscope *Ann. Ophthalmol.* 1984**16** 434–5

[10] Olsen T and Nielsen P J 1989 Immersion versus contact technique in the measurement of axial length by ultrasound *Acta Ophthalmol (Copenh)* **67** 101–2

[11] Hill W 2021 A-Scan Biometry2021; v. 2021, https://doctor-hill.com/iol-power-calculations/ascan-biometry/ [accessed 19/11/2022]

[12] Santodomingo-Rubido J, Mallen E A, Gilmartin B and Wolffsohn J S 2002 A new non-contact optical device for ocular biometry *Br. J. Ophthalmol* **86** 458–62

[13] Rose L T and Moshegov C N 2003 Comparison of the Zeiss IOLMaster and applanation A-scan ultrasound: biometry for intraocular lens calculation *Clin. Exp. Ophthalmol* **31** 121–4

[14] Drexler W, Findl O and Menapace R *et al* 1998 Partial coherence interferometry: a novel approach to biometry in cataract surgery *Am. J. Ophthalmol.* **126** 524–34

[15] Hill W, Osher R and Cooke D *et al* 2011 Simulation of toric intraocular lens results: manual keratometry versus dual-zone automated keratometry from an integrated biometer *J. Cataract Refract. Surg.* **37** 2181–7

[16] Akman A, Asena L and Gungor S G 2016 Evaluation and comparison of the new swept source OCT-based IOLMaster 700 with the IOLMaster 500 *Br. J. Ophthalmol.* **100** 1201–5

[17] Shajari M, Cremonese C and Petermann K *et al* 2017 Comparison of axial length, corneal curvature, and anterior chamber depth measurements of 2 recently introduced devices to a known biometer *Am. J. Ophthalmol.* **178** 58–64

[18] Hirnschall N, Leisser C and Radda S *et al* 2016 Macular disease detection with a swept-source optical coherence tomography-based biometry device in patients scheduled for cataract surgery *J. Cataract Refract. Surg.* **42** 530–6

[19] Godefrooij D A, Galvis V and Tello A 2018 Von Helmholtz's ophthalmometer: historical review and experience with one of the last surviving original devices *Acta Ophthalmol.* **96** 314–20

[20] Dave T, Ruston D and Fowler C 1998 Evaluation of the EyeSys model II computerized videokeratoscope. Part II: the repeatability and accuracy in measuring convex aspheric surfaces *Optom. Vis. Sci.* **75** 656–62

[21] Lowe R F and Clark B A 1973 Posterior corneal curvature. correlations in normal eyes and in eyes involved with primary angle-closure glaucoma *Br. J. Ophthalmol.* **57** 464–70

[22] Hasegawa A, Kojima T and Yamamoto M *et al* 2018 Impact of the anterior-posterior corneal radius ratio on intraocular lens power calculation errors *Clin. Ophthalmol.* **12** 1549–58

[23] Wang L, Mahmoud A M and Anderson B L *et al* 2011 Total corneal power estimation: ray tracing method versus gaussian optics formula *Invest. Ophthalmol. Vis. Sci.* **52** 1716–22

[24] Oliveira C M, Ribeiro C and Franco S 2011 Corneal imaging with slit-scanning and scheimpflug imaging techniques *Clin. Exp. Optom.* **94** 33–42

[25] Savini G, Hoffer K J, Schiano-Lomoriello D and Barboni P 2018 Intraocular lens power calculation using a placido disk-scheimpflug tomographer in eyes that had previous myopic corneal excimer laser surgery *J. Cataract Refract. Surg.* **44** 935–41

[26] Savini G, Schiano-Lomoriello D and Hoffer K J 2018 Repeatability of automatic measurements by a new anterior segment optical coherence tomographer combined with Placido topography and agreement with 2 Scheimpflug cameras *J. Cataract Refract. Surg.* **44** 471–8
[27] Gjerdrum B, Gundersen K G, Lundmark P O and Aakre B M 2021 Refractive precision of ray tracing IOL calculations based on OCT data versus traditional IOL calculation formulas based on reflectometry in patients with a history of laser vision correction for myopia *Clin. Ophthalmol.* **15** 845–57
[28] de Silva S R, Evans J R and Kirthi V *et al* 2016 Multifocal versus monofocal intraocular lenses after cataract extraction *Cochrane Database Syst. Rev.* **12** CD003169
[29] Hoffer K J and Savini G 2017 Clinical results of the hoffer H-5 formula in 2707 eyes: first 5th-generation formula based on gender and race *Int. Ophthalmol. Clin.* **57** 213–9
[30] Retzlaff J 1980 A new intraocular lens calculation formula *J. Am. Intraocul Implant. Soc.* **6** 148–52
[31] Colenbrander M C 1973 Calculation of the power of an iris clip lens for distant vision *Br. J. Ophthalmol.* **57** 735–40
[32] Fyodorov S N, Galin M A and Linksz A 1975 Calculation of the optical power of intraocular lenses *Invest Ophthalmol.* **14** 625–8
[33] van der Heijde G L 1976 The optical correction of unilateral aphakia *Trans. Sect. Ophthalmol. Am. Acad. Ophthalmol. Otolaryngol.* **81** OP80–8
[34] Binkhorst R D 1976 Pitfalls in the determination of intraocular lens power without ultrasound *Ophthalmic Surg.* **7** 69–82
[35] Kraff M C, Sanders D R and Lieberman H L 1978 Biometric analysis of intraocular lens power required to produce emmetropia: results of 450 implants *J. Am. Intraocul Implant Soc.* **4** 45–7
[36] Hoffer K 1984 The effect of axial length on posterior chamber lens and posterior capsule position *Current concepts Ophthalmic Surg.* **1** 20–2
[37] Hughes W F 1982 Current concepts in cataract surgery *Archives of Ophthalmology* **100** 1355–6
[38] Olsen T 1986 Prediction of intraocular lens position after cataract extraction *J. Cataract Refract. Surg.* **12** 376–9
[39] Holladay J T, Prager T C and Chandler T Y *et al* 1988 A three-part system for refining intraocular lens power calculations *J. Cataract Refract. Surg.* **14** 17–24
[40] Sanders D R, Retzlaff J and Kraff M C 1988 Comparison of the SRK II formula and other second generation formulas *J. Cataract Refract. Surg.* **14** 136–41
[41] Retzlaff J A, Sanders D R and Kraff M C 1990 Development of the SRK/T intraocular lens implant power calculation formula *J. Cataract Refract. Surg.* **16** 333–40
[42] Sheard R M, Smith G T and Cooke D L 2010 Improving the prediction accuracy of the SRK/T formula: the T2 formula *J. Cataract Refract. Surg.* **36** 1829–34
[43] Hoffer K J 1993 The Hoffer Q formula: a comparison of theoretic and regression formulas *J. Cataract Refract. Surg.* **19** 700–12
[44] Barrett G D 1987 Intraocular lens calculation formulas for new intraocular lens implants *J. Cataract Refract. Surg.* **13** 389–96
[45] Barrett G D 1993 An improved universal theoretical formula for intraocular lens power prediction *J. Cataract Refract. Surg.* **19** 713–20
[46] Kuthirummal N, Vanathi M and Mukhija R *et al* 2020 Evaluation of barrett universal II formula for intraocular lens power calculation in Asian Indian population *Indian J. Ophthalmol.* **68** 59–64

[47] Melles R B, Holladay J T and Chang W J 2018 Accuracy of intraocular lens calculation formulas *Ophthalmology* **125** 169–78
[48] Naeser K 1997 Intraocular lens power formula based on vergence calculation and lens design *J. Cataract Refract. Surg.* **23** 1200–7
[49] Naeser K and Savini G 2019 Accuracy of thick-lens intraocular lens power calculation based on cutting-card or calculated data for lens architecture *J. Cataract Refract. Surg.* **45** 1422–9
[50] Haigis W, Lege B, Miller N and Schneider B 2000 Comparison of immersion ultrasound biometry and partial coherence interferometry for intraocular lens calculation according to Haigis *Graefes Arch. Clin. Exp. Ophthalmol.* **238** 765–73
[51] Olsen T, Corydon L and Gimbel H 1995 Intraocular lens power calculation with an improved anterior chamber depth prediction algorithm *J. Cataract Refract. Surg.* **21** 313–9
[52] Olsen T 2006 Prediction of the effective postoperative (intraocular lens) anterior chamber depth *J. Cataract Refract. Surg.* **32** 419–24
[53] Wang L, Holladay J T and Koch D D 2018 Wang–Koch axial length adjustment for the Holladay 2 formula in long eyes *J. Cataract Refract. Surg.* **44** 1291–2
[54] Wang L, Shirayama M and Ma X J *et al* 2011 Optimizing intraocular lens power calculations in eyes with axial lengths above 25.0 mm *J. Cataract Refract. Surg.* **37** 2018–27
[55] Voytsekhivskyy O V 2018 Development and clinical accuracy of a new intraocular lens power formula (VRF) compared to other formulas *Am. J. Ophthalmol.* **185** 56–67
[56] Kane J X, Van Heerden A, Atik A and Petsoglou C 2017 Accuracy of 3 new methods for intraocular lens power selection *J. Cataract Refract. Surg.* **43** 333–9
[57] Ladas J G, Siddiqui A A, Devgan U and Jun A S 2015 A 3-D 'Super Surface' combining modern intraocular lens formulas to generate a 'Super Formula' and maximize accuracy *JAMA Ophthalmol.* **133** 1431–6
[58] Hoffer K J 2016 Origin of multiple formula use to calculate intraocular lens power *JAMA Ophthalmol.* **134** 847–8
[59] Hoffer K J, Aramberri J and Haigis W *et al* 2015 Protocols for studies of intraocular lens formula accuracy *Am. J. Ophthalmol.* **160** 403–5e1
[60] Ting D S J, Foo V H and Yang L W Y *et al* 2021 Artificial intelligence for anterior segment diseases: emerging applications in ophthalmology *Br. J. Ophthalmol.* **105** 158–68
[61] Turing A M 1950 I.—Computing machinery and intelligence *Mind.* **LIX** 433–60
[62] Clarke G P and Burmeister J 1997 Comparison of intraocular lens computations using a neural network versus the holladay formula *J. Cataract Refract. Surg.* **23** 1585–9
[63] Connell B J and Kane J X 2019 Comparison of the Kane formula with existing formulas for intraocular lens power selection *BMJ Open Ophthalmol.* **4** e000251
[64] Darcy K, Gunn D and Tavassoli S *et al* 2020 Assessment of the accuracy of new and updated intraocular lens power calculation formulas in 10 930 eyes from the UK National Health Service *J. Cataract Refract. Surg.* **46** 2–7
[65] Savini G, Di Maita M and Hoffer K J *et al* 2021 Comparison of 13 formulas for IOL power calculation with measurements from partial coherence interferometry *Br. J. Ophthalmol.* **105** 484–9
[66] Yoo Y S, Whang W J and Hwang K Y *et al* 2019 Use of the crystalline lens equatorial plane as a new parameter for predicting postoperative intraocular lens position *Am. J. Ophthalmol.* **198** 17–24
[67] Ventura B V, Wang L and Weikert M P *et al* 2014 Surgical management of astigmatism with toric intraocular lenses *Arq. Bras. Oftalmol.* **77** 125–31

[68] Koch D D, Ali S F and Weikert M P *et al* 2012 Contribution of posterior corneal astigmatism to total corneal astigmatism *J. Cataract Refract. Surg.* **38** 2080–7

[69] Koch D D, Jenkins R B and Weikert M P *et al* 2013 Correcting astigmatism with toric intraocular lenses: effect of posterior corneal astigmatism *J. Cataract Refract. Surg.* **39** 1803–9

[70] Yanoff M, Duker J S and Augsburger J J 2004 *Ophthalmology* 2nd edn (St. Louis, MO: Mosby)

[71] Smolin G, Foster C S, Azar D T and Dohlman C H 2005 *Smolin and Thoft's The Cornea: Scientific Foundations and Clinical Practice* 4th edn (Philadelphia, PA: Lippincott Williams & Wilkins)

[72] Alpins N A 1998 Treatment of irregular astigmatism *J. Cataract Refract. Surg.* **24** 634–46

[73] Nunez M X, Henriquez M A and Escaf L J *et al* 2019 Consensus on the management of astigmatism in cataract surgery *Clin. Ophthalmol.* **13** 311–24

[74] Hasegawa Y, Okamoto F and Nakano S *et al* 2013 Effect of preoperative corneal astigmatism orientation on results with a toric intraocular lens *J. Cataract Refract. Surg.* **39** 1846–51

[75] Hill W 2008 Expected effects of surgically induced astigmatism on acrysof toric intraocular lens results *J. Cataract Refract. Surg.* **34** 364–7

[76] Alio J L, Pinero D P, Tomas J and Plaza A B 2011 Vector analysis of astigmatic changes after cataract surgery with implantation of a new toric multifocal intraocular lens *J. Cataract Refract. Surg.* **37** 1217–29

[77] Abulafia A, Hill W E and Koch D D *et al* 2016 Accuracy of the Barrett True-K formula for intraocular lens power prediction after laser *in situ* keratomileusis or photorefractive keratectomy for myopia *J. Cataract Refract. Surg.* **42** 363–9

[78] Tang M, Li Y and Huang D 2010 An intraocular lens power calculation formula based on optical coherence tomography: a pilot study *J. Refract. Surg.* **26** 430–7

[79] Ma J X, Tang M and Wang L *et al* 2016 Comparison of newer IOL power calculation methods for eyes with previous radial keratotomy *Invest. Ophthalmol. Vis. Sci.* **57** OCT162–8

[80] Hirnschall N, Findl O and Bayer N *et al* 2020 Sources of error in toric intraocular lens power calculation *J. Refract. Surg.* **36** 646–52

[81] Terauchi R, Horiguchi H and Ogawa T *et al* 2020 Posture-related ocular cyclotorsion during cataract surgery with an ocular registration system *Sci. Rep.* **10** 2136

[82] Popp N, Hirnschall N, Maedel S and Findl O 2012 Evaluation of 4 corneal astigmatic marking methods *J. Cataract Refract. Surg.* **38** 2094–9

[83] Bhandari S and Nath M 2016 Anterior stromal puncture with staining: a modified technique for preoperative reference corneal marking for toric lenses and its retrospective analyses *Indian J. Ophthalmol.* **64** 559–62

[84] Varsits R M, Hirnschall N, Doller B and Findl O 2019 Evaluation of an intraoperative toric intraocular lens alignment system using an image-guided system *J. Cataract Refract. Surg.* **45** 1234–8

[85] Ohashi T and Kojima T 2020 Comparison of the orientation of the corneal steep meridian determined by image-guided system and manual method in the same eye *Clin. Ophthalmol.* **14** 4135–44

[86] Hemmati H D, Gologorsky D and Pineda R 2nd 2012 Intraoperative wavefront aberrometry in cataract surgery *Semin. Ophthalmol.* **27** 100–6

[87] Ianchulev T, Hoffer K J and Yoo S H *et al* 2014 Intraoperative refractive biometry for predicting intraocular lens power calculation after prior myopic refractive surgery *Ophthalmology* **121** 56–60

[88] Davison J A and Potvin R 2017 Preoperative measurement vs intraoperative aberrometry for the selection of intraocular lens sphere power in normal eyes *Clin. Ophthalmol.* **11** 923–9

[89] Raufi N, James C, Kuo A and Vann R 2020 Intraoperative aberrometry vs modern preoperative formulas in predicting intraocular lens power *J. Cataract Refract. Surg.* **46** 857–61

[90] Davison J A, Makari S and Potvin R 2019 Clinically relevant differences in the selection of toric intraocular lens power in normal eyes: preoperative measurement vs intraoperative aberrometry *Clin. Ophthalmol.* **13** 913–20

[91] Modi S S 2020 Clinical outcomes after aphakic versus aphakic/pseudophakic intraoperative aberrometry in cataract surgery with toric IOL implantation *Int. Ophthalmol.* **40** 3251–7

[92] Blaylock J F and Hall B 2020 Astigmatic results of a diffractive trifocal toric IOL following intraoperative aberrometry guidance *Clin. Ophthalmol.* **14** 4373–8

[93] Chang D F 2019 Disruptive innovation and refractive IOLs: how the game will change with adjustable IOLs *Asia Pac. J. Ophthalmol. (Phila)* **8** 432–5

[94] Villegas E A, Alcon E and Rubio E *et al* 2014 Refractive accuracy with light-adjustable intraocular lenses *J. Cataract Refract. Surg.* **40** 1075–84

[95] Chayet A, Sandstedt C A and Chang S H *et al* 2009 Correction of residual hyperopia after cataract surgery using the light adjustable intraocular lens technology *Am. J. Ophthalmol.* **147** 392–7

[96] Schojai M, Schultz T and Schulze K *et al* 2020 Long-term follow-up and clinical evaluation of the light-adjustable intraocular lens implanted after cataract removal: 7-year results *J. Cataract Refract. Surg.* **46** 8–13

[97] Nguyen J, Werner L and Ludlow J *et al* 2018 Intraocular lens power adjustment by a femtosecond laser: *in vitro* evaluation of power change, modulation transfer function, light transmission, and light scattering in a blue light-filtering lens *J. Cataract Refract. Surg.* **44** 226–30

[98] Sahler R, Bille J F and Enright S *et al* 2016 Creation of a refractive lens within an existing intraocular lens using a femtosecond laser *J. Cataract Refract. Surg.* **42** 1207–15

[99] Sahler R and Bille J F 2019 Refractive index shaping: *in vivo* optimization of an implanted intraocular lens (IOL) ed J F Bille *High Resolution Imaging in Microscopy and Ophthalmology: New Frontiers in Biomedical Optics* (Cham: Springer)

Chapter 7

Aniseikonia

Andrew J Toole, Thomas W Raasch and Marjean Taylor Kulp

Normal human vision involves fusing the images arising from each individual eye into a single binocular percept. Many binocular vision disorders exist that can interfere with this process. Here we focus on one such disorder: aniseikonia.

7.1 Background

Aniseikonia is a binocular vision condition where the images perceived by each eye differ in size and/or shape. Although aniseikonia has been described since the nineteenth century [1], it is still a condition that can plague patients and seasoned eye care professionals alike. In fact, aniseikonia may be more prominent today due to our aging population and advances in eye care, such as cataract surgery, that often maintain clarity of vision throughout life. The prevalence of clinically significant aniseikonia has been reported to range from 3% to 33% [2–5] with the lower end of this range likely reflecting the prevalence in the general population.

Symptoms of those afflicted with aniseikonia can range from minimal/none to significantly affecting quality of life. The most common symptoms include eyestrain and headaches while some experience photophobia (light sensitivity), difficulty reading, diplopia (double vision), distortions in space perception, loss of binocular depth perception (stereopsis), nausea, and dizziness [6–10]. Aniseikonia affecting stereopsis and leading to alterations in space perception deserves further elaboration.

7.1.1 Normal stereopsis (binocular depth perception)

To understand the distortions of space perception that aniseikonia can cause, we first must briefly review the normal physiological process of stereopsis. The lateral displacement of our two eyes provides a slightly different vantage point for each eye of the visual scene. Figure 7.1(a) shows a representation of this with a top-down view of an observer viewing two pencils where the yellow pencil 1 is at a further distance than the blue pencil 2. This results in the angular separation of the pencils to be smaller for the right eye than for the left eye. Figure 7.1(b) represents the views of the pencils for the left and right eyes individually

doi:10.1088/978-0-7503-3263-7ch7

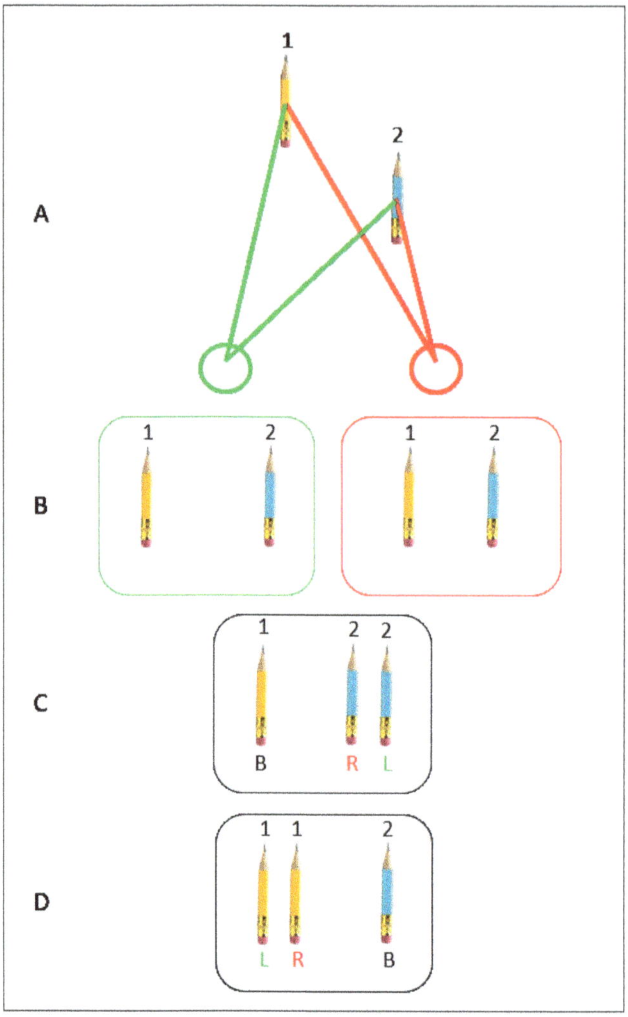

Figure 7.1. Depiction of normal stereoscopic vision. (a) Top-down view of an observer viewing two pencils where the yellow pencil 1 is at a further distance than the blue pencil 2, resulting in a smaller angular separation of the pencils for the right eye than for the left eye. (b) Views of the pencils for the left and right eyes individually with the separation of pencils smaller in the right eye's view due to the smaller angular separation. (c) The right and left eyes images of the yellow pencil 1 are superimposed or 'fused', resulting in a single image of pencil 1 and two separate images of the blue pencil 2. The images of the blue pencil have a 'crossed' relative disparity compared to the yellow pencil (the right eye's image of the blue pencil is on the left, the left eye's image is on the right). This relative disparity is the basis for stereoscopic depth perception. This 'crossed disparity' leads to the perception that the blue pencil 2 is closer than the yellow pencil 1. (d) Images of the blue pencil are superimposed or fused, resulting in an 'uncrossed' relative disparity for the yellow pencil relative to the blue pencil (right eye's image of the yellow pencil is on the right and the left eye's image is on the left). This 'uncrossed' disparity of the yellow pencil 1 compared to the blue pencil 2 leads to the perception that the yellow pencil 1 is further away than the blue pencil 2.

with the separation of pencils smaller in the right eye's view due to the smaller angular separation. Figure 7.1(c) represents a 'fused' image created by overlaying the right and left eye's images such that the images of the yellow pencil 1 are superimposed. As can be seen, this results in the images of the blue pencil 2 not coinciding and this separation of the two eyes images of the blue pencil is described as a relative disparity of the blue pencil 2 compared to the yellow pencil 1. This relative disparity is the basis for stereoscopic depth perception. In this example, the right eye's image of the blue pencil is to the left of the left eye's image of the blue pencil. This may be termed 'crossed disparity' and leads to the perception that the blue pencil 2 is closer than the yellow pencil 1. Figure 7.1(d) illustrates overlaying the two eye's views such that the images of the blue pencil 2 are superimposed. The relative disparity shown here for yellow pencil 1 relative to blue pencil 2 can be described as 'uncrossed disparity' as the right eye's image of the yellow pencil is to the right of the left eye's image. The uncrossed disparity of the yellow pencil compared to the blue pencil leads to the perception that the yellow pencil 1 is further away than blue pencil 2. This process can also be used to create the illusion of depth by presenting the images in figure 7.1(b) dichoptically to the corresponding eyes and is the basis for 3-D movies and more recently the basis for the perceived depth in virtual reality headsets. It is worth noting that monocular cues to depth, such as relative distance magnification, have been ignored in this description.

7.1.2 Stereoscopic distortions of depth due to aniseikonia

Aniseikonia can alter how we detect the relative disparities of objects in the visual scene and thus create distortions of perceived relative distances between objects. The following example illustrates this effect. Figure 7.2(a) shows an observer looking at two pencils that are the same distance from the observer (i.e. the pencils lie in a frontal parallel plane). In normal binocular vision the angular separation of the two pencils would be the same for each eye and thus no relative disparity would exist between them. Therefore, stereoscopically the pencils would be accurately perceived to be at the same distance. However, now consider the scenario where the left eye perceives a larger image of the visual scene in the horizontal meridian compared to the right eye. Figure 7.2(b) illustrates the effect this would have on each eye's image with the separation of the pencils being larger in the left eye's image compared to the right eye's image. Figure 7.2(c) represents the fused image created by superimposing the images for the yellow pencil 1. Here it can be seen that a crossed disparity exists for blue pencil 2 relative to yellow pencil 1 and thus the observer would erroneously perceive that the blue pencil was closer to them than the yellow pencil (that is, they would perceive them to be arranged as in figure 7.1(a)).

Stereoscopic distortions of depth due to aniseikonia can cause walls and/or floors to appear tilted. People with aniseikonia may complain of computer monitors appearing closer on one end than the other. These distortions of perceived depth are common following obtaining a new glasses prescription and are usually temporary (at least in familiar settings). However, for some these perceived distortions can persist. As will be discussed in a later section 7.3.1, these stereoscopic distortions of depth can be used to measure the amount of aniseikonia present.

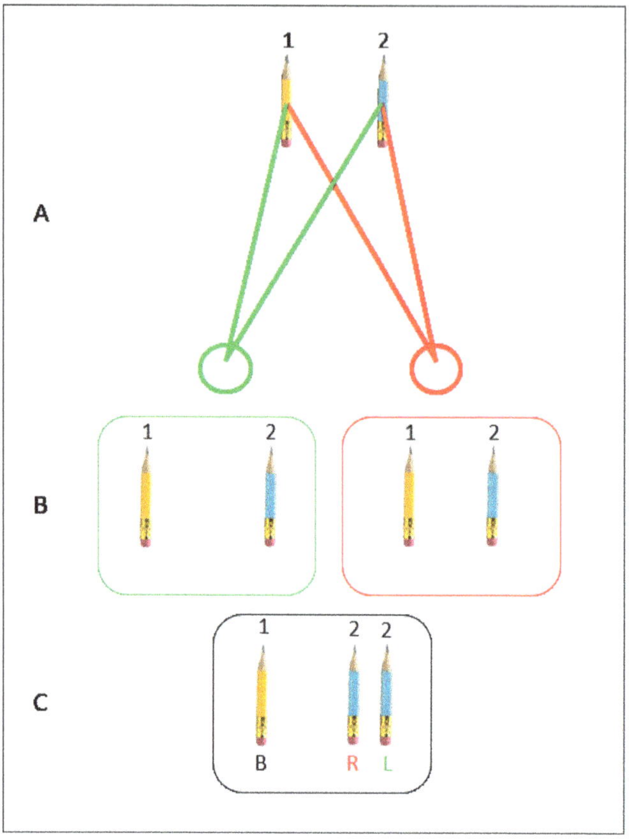

Figure 7.2. Example of stereoscopic distortions resulting from aniseikonia causing a larger perceived image in the horizontal meridian of the left eye. (a) A top-down view of an observer looking at two pencils that are the same distance from the observer (i.e. the pencils lie in a frontal parallel plane). In normal binocular vision the angular separation of the two pencils would be the same for each eye and thus no relative disparity would exist between them. Therefore, stereoscopically the pencils would be accurately perceived to be at the same distance. (b) The left eye perceives a larger image of the visual scene in the horizontal meridian compared to the right eye, resulting in the separation of the pencils being larger in the left eye's image compared to the right eye's image. (c) Images of the yellow pencil 1 are fused, resulting in a crossed disparity for the blue pencil 2 relative to the yellow pencil 1. Thus, the observer would erroneously perceive that the blue pencil was closer than the yellow pencil.

7.2 Causes of aniseikonia

The etiology of perceived image size differences can typically be traced back to one or more of three possible origins: optically induced unequal image sizes formed on the retinas, differences in photoreceptor spacing between the retinas, and/or scaling differences along the visual pathway and visual cortex.

7.2.1 Optically induced aniseikonia

The first step in the perception of vision is the formation of an image of the visual scene on the retina. This image formed on the retina is commonly referred to as the retinal image. The retinal image is formed by the combined effects of the optical elements of the eye, most notably the positive or converging power of the air–cornea interface and the positive-powered intraocular crystalline lens, and any external lenses used to correct refractive error, such as nearsightedness or farsightedness. Unequal retinal image sizes most commonly occur when the two eyes have unequal refractive errors, a condition known as anisometropia. Not surprisingly, unequal retinal image sizes is a common cause of aniseikonia. The origin of the anisometropia affects the resulting retinal image size discrepancies.

Axial anisometropia is a condition where the refractive error differences between the eyes is due to differences in the anterior–posterior axial length of the eye. Here the longer eye would carry the higher myopic (nearsighted) refractive error (if the secondary focal point of the eye was in front of the retina) or the lower hyperopic (farsighted) refractive error (if the secondary focal point was behind the retina). In an uncorrected state, the longer eye would have the larger retinal image due to the retinal being further from the secondary nodal point of the eye's optical system [11]. Correcting axial anisometropia with spectacle lenses may in fact decrease the retinal image size differences due to the minimizing effects of the more negative or diverging powered lens required to correct the longer eye.

Alternatively, refractive anisometropia exists when the secondary focal points of the eyes fall in different planes. Here the higher-powered eye (most commonly due to a higher-powered cornea) has the shorter secondary focal length and therefore is the more myopic (or less hyperopic eye). Since the axial lengths are similar in refractive anisometropia, the retinal images are similar sizes when the refractive error is not corrected (although one image is more blurred than the other). However, when correcting refractive anisometropia with spectacle lenses dissimilar retinal images can be generated due to the asymmetric lens powers required to correct the anisometropia. Further, meridional retinal image size differences can be generated due to any asymmetries in astigmatic refractive error corrections required. Additional information regarding axial versus refractive anisometropia can be found in section 7.4.1 under Knapp's Law.

7.2.2 Retinally induced aniseikonia

Retinally induced aniseikonia is due to the differences in photoreceptor spacing between the two eyes. Therefore, the next step in the perception of vision is the detection of the retinal image by the retinal photoreceptors. Photoreceptor spacing plays a key role in the perceived angular size of an object in visual space. This role can be illustrated with a digital camera to display monitor analogy. For simplicity we can consider a camera's image detection array to be comprised of individual detection units, camera pixels, and that these units are connected to a display monitor in a one-to-one fashion such that one detection unit drives one display pixel. If an object is imaged on the detection array, the size of the object on the display will be directly related to the density of the detection units on the camera. If the detection units are packed in tighter then the object will be imaged on more detection units and the displayed image will be larger. Conversely more sparsely arranged detector units would result in the image falling on fewer detector units and a smaller

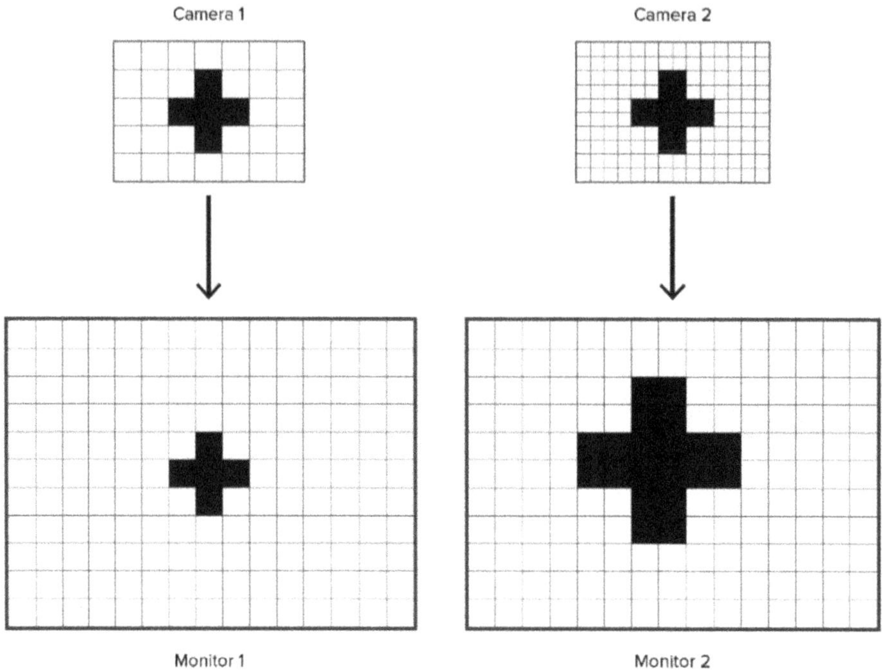

Figure 7.3. Representation of an equal sized image of a cross being focused on two cameras' sensor arrays where the camera pixel density of camera 2 is double the pixel density of camera 1. When these images are displayed on a monitor in a 1-camera pixel-to-1-display pixel fashion, the displayed image originating from camera 2 is twice the size of the image from camera 1.

displayed size on the monitor. This concept is depicted in figure 7.3 where two equal sized images of a cross are focused on the camera pixel arrays of two cameras. Here the camera pixel density for camera 2 is double the pixel density of camera 1 (represented by the size and number of squares in the grid). When these images are projected onto two monitors with equal resolution, the image from camera 2 is displayed as twice the size of the image originating from camera 1. Considering the detection units here to represent the retinal photoreceptors and the displayed image to represent the perceived image, we can see that more tightly packed photoreceptors should lead to a larger perceived image. Thus if the right and left eyes are both receiving the exact same size retinal image, a perceived size difference can exist if the photoreceptor packing density is different between the two eyes.

Retinally induced aniseikonia generally occurs due to retinal disease processes or mechanical insults in one or both eyes such as epiretinal membranes (macular pucker), retinal detachments, macular edema, or macular holes [9, 12–16]. As such, the retinal disruption from these conditions typically varies across the retina and results in varying amounts of aniseikonia across the visual field. This field dependency poses extra challenges in the treatment of the condition. Further, these conditions can result in more random variances in photoreceptor distribution in a given retinal location resulting in image distortions beyond simple scaling differences known as metamorphopsia.

7.2.3 Higher order neural image processing

The final (large) step in vision is to transmit the image signal from the retina along the visual pathway to visual cortex and beyond to reach perception. Thus, any unequal scaling that occurs along the two eye's visual pathways could result in aniseikonia. This potential origin of perceived size differences has been less studied and is often not considered in the clinical evaluation of aniseikonia. However, it has been suggested that this scaling is likely adaptive and may account for the significant number of individuals that would be predicted to have unequal retinal image sizes, due to anisometropia, that do not experience aniseikonia symptoms even when switching between spectacle and contact lens refractive error corrections [11].

7.3 Subjective evaluation of aniseikonia

As was just discussed, aniseikonia can result from multiple causes. Of these potential causes, only retinal image size differences can reasonably be predicted based on objective measurement of the eyes structures and will be discussed in a later section. Thus, subjective measurements of perceived size difference are of clinical importance. Tools available for subjective assessment of aniseikonia can be broken into two categories: stereoscopic assessment and direct comparison techniques.

7.3.1 Stereoscopic assessment of aniseikonia

As was described in section 7.2, aniseikonia can cause stereoscopic distortions of space. An example was presented where an aniseikonic individual, who perceived a magnified image of the horizontal meridian of the left eye, was viewing two pencils standing in a frontal parallel plane. Here the disparities created by the aniseikonia caused the pencil on the right to appear closer to the observer than the pencil on the left. One method of measuring the present aniseikonia could be to physically move the right-side pencil back away from the observer until the two pencils are perceived to be equally distant. One significant downside to this method (aside from the calculations required to determine the image size adjustments necessary to obtain this effect being cumbersome) is that by physically sliding the pencil further from the observer, the angular size of the pencil shrinks and this size change can in itself alter distance perception. A more accurate method would be to introduce a meridional afocal magnifying lens (or lens system) before the right eye that specifically magnified in the horizontal meridian. Different magnifications could be tried until the observer perceived the pencils to be equally distant. This method would thus measure the correction necessary to eliminate the aniseikonia present. The pencil target described here has a substantial limitation though. The target can only be used to measure image size differences in the horizontal meridian. Thus, the space eikonometer target was developed.

The space eikonometer target is a three-dimensional setup that consists of five vertical lines and two diagonal lines. Four of the vertical lines are positioned as the vertical corners of a cube; thus, two vertical lines lie in a near frontal parallel plane and two lie in a more distant frontal parallel plan. The fifth vertical line is positioned centered within the outer four lines. The two diagonal lines form an 'X' and intersect

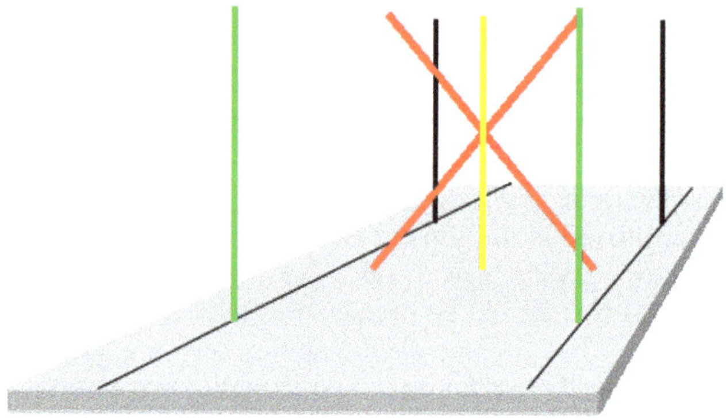

Figure 7.4. Representation of the space eikonometer target (the base plate has been added to aid in displaying the depth of the target only).

at the midpoint of the central vertical line such that the 'X' and the central vertical line all lie in a third frontal parallel plane. Figure 7.4 shows a representation of the space eikonometer target (the base plate has been added for illustrative purposes only), which has the advantage over our pencil target in that horizontal, vertical, oblique, or overall magnification differences each create unique perceived distortions of the target. Thus, afocal magnifiers can be used to measure these effects by neutralizing these distortions.

Originally, space eikonometer targets were created by suspending rods in an observation room. Subsequently, the American Optical Company created a desktop stereoscope model that included built-in variable afocal magnifiers. This office model space eikonometer significantly simplified measurement and went on to become the gold standard for quantifying aniseikonia. This role is rapidly fading, though, as the apparatus has not been commercially available for many years and is becoming increasingly difficult to locate.

Stereoscopic assessment of aniseikonia can make very accurate measures but has two main downsides. The first is the limited availability of commercial devices. The second, as this section title implies, is that these devices require observers to have good stereoscopic depth perception, an ability that many individuals with retinally induced aniseikonia do not have [14, 17].

7.3.2 Direct comparison techniques

A simpler method for measuring aniseikonia is by directly comparing the right and left eyes' perception of the size of an object or objects. A host of techniques have been proposed for accomplishing this. Here we will focus on the simplest option and the two most popular techniques today.

7.3.2.1 Alternate occlusion

The most basic of direct comparison techniques is to fixate on an object (such as a symmetric clock face) while alternating between covering the right and left eyes. Here the observer tries to judge any perceived size difference with the right versus the left eye viewing. This technique is likely only sensitive enough to pick up large perceived differences (5% or more) and is often self-discovered by individuals suffering from retinally induced aniseikonia. Further, observers can try to estimate the percent size difference experienced. However, clinical observations of one of the authors (AJT) suggests that patients tend to substantially overestimate the perceived size difference with this technique compared to the more quantitative measures described next.

7.3.2.2 New aniseikonia tests

The new aniseikonia test (https://www.good-lite.com/products/137800), or sometimes referred to as the Awaya Test, utilizes red/green anaglyphically presented adjacent semicircles printed on the white pages of the test book (shown in figure 7.5). The observer wears a red filter in front of one eye and a green filter in front of the other eye; when the test is viewed through these filters, the eye behind the red filter only sees the green semicircles and the eye behind the green filter only sees the red semicircles. This presentation allows for simultaneous direct comparison of the size of the long dimension of the semicircles seen by each eye. The test book contains multiple pairs of targets with varying amounts of physical size disparities between the right and left eye's targets allowing the observer to choose the pair that matches in size perceptually. The orientation of the book can be changed to measure in different meridians and the test distance can be altered to change the overall angular size of the targets, thus assessing for field dependency of the aniseikonia present. The

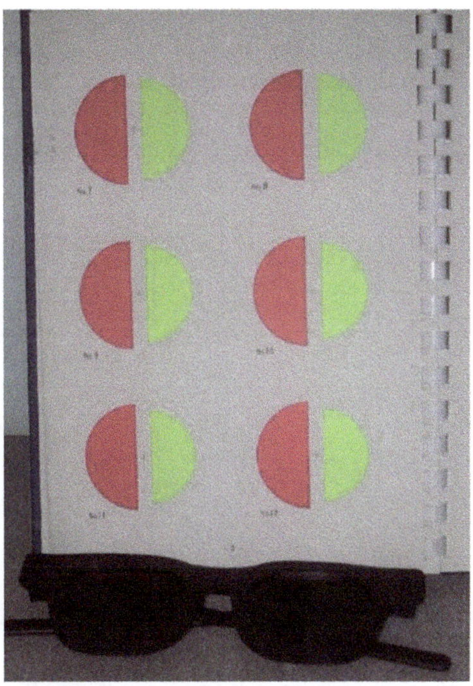

Figure 7.5. The new aniseikonia tests (Awaya).

simplicity of the test is appealing, however, users should be aware that the test has been shown to underestimate the magnitude of the aniseikonia present [18].

7.3.2.3 Aniseikonia inspector

The Aniseikonia inspector (https://www.opticaldiagnostics.com/products/ai/index.html) utilizes a similar target presentation as the new aniseikonia tests but with anaglyphic adjacent rectangles being presented on a computer monitor (figure 7.6). The software presents varying size disparities in a forced choice paradigm to arrive at the magnitude of the perceived size difference. Targets may be presented horizontally, vertically, and at both diagonal angles. Additionally, the overall size of the targets can be altered. The aniseikonia inspector has the added feature of fitting the results from the four meridians tested to a best fit single cycle sine wave. This allows the results to be described as a best predicted overall magnification combined with a meridional correction in an analogous fashion to a sphero-cylindrical spectacle prescription. Earlier versions of the software were shown to underestimate the aniseikonia present [19], however, more recent versions that utilized forced choice paradigms and limited view durations may have nulled this effect, but the accuracy still falls short of that of the space eikonometer, nevertheless [20, 21].

7.4 Optical principles of aniseikonia and treatment

Correction of aniseikonia involves manipulating the design of the spectacle lenses that correct the refractive error of the eyes. Lens properties that are adjusted to control magnification are the front surface power, thickness, and refractive index. In addition, the distance of the lens from the eye can be altered to change magnification. On occasion, aniseikonia correction can involve the use of glasses in combination with contact lenses. The optical principles involved in the control of magnification are described in the following sections.

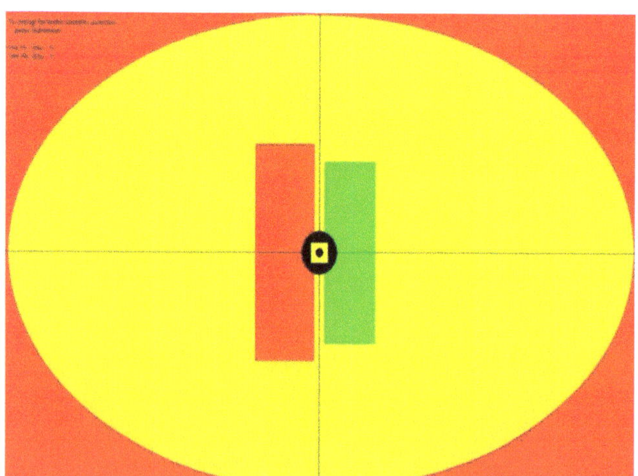

Figure 7.6. Aniseikonia inspector target presentation.

7.4.1 Spectacle magnification

Spectacle magnification (SM) is defined as the ratio of the size of the retinal image in the corrected eye to the image size in the uncorrected state. SM is calculated as the product of two factors: the shape factor (SF) and the power factor (PF) [22]. The SF is a function of the front surface power, lens thickness, and refractive index of the lens material.

$$\text{The shape factor is: SF} = \frac{1}{1 - \frac{t}{n}F_1} \tag{7.1}$$

where t is lens center thickness (meters), n is refractive index, and F_1 is the front surface power (diopters).

$$\text{The power factor is: PF} = \frac{1}{1 - hF_b} \tag{7.2}$$

where F_b is the back vertex power of the lens, and h is the distance from the back surface of the lens to the relevant reference point of the eye.

7.4.1.1 Relationship to the magnification of a telescope

Both the shape factor and power factor can be recognized as a formula for the angular magnification of an afocal telescope. The shape factor can be derived from formulas (7.3) and (7.4) for telescope magnification and length:

$$\text{Telescope magnification: } M = -\frac{F_2}{F_1} \tag{7.3}$$

$$\text{Telescope length: } \frac{t}{n} = \frac{1}{F_1} + \frac{1}{F_2} \tag{7.4}$$

Essentially, the single lens is treated as an afocal telescope of reduced length t/n, with objective lens power of F_1. In formula (7.3), F_2 is the power of the ocular lens of the afocal telescope. Here, to express this lens as an afocal telescope, this power (termed F_2') is the opposite of the vergence arriving at the back surface of the lens, i.e. $F_2' = -V_2$. It is not the power of that back surface. (This vergence arriving at the back surface of the lens, added to the power of the back surface of the lens, is the back vertex power. This back vertex power is used in the calculation of the PF, discussed next.) F_1 and F_2', separated by t/n, makes this an afocal telescope. The vergence arriving at the back surface is: $-F_2' = V_2 = \dfrac{F_1}{1 - \frac{t}{n}F_1}$.

Substituting F_2' for F_2 in equation (7.3), and simplifying yields $M = \dfrac{1}{1 - \frac{t}{n}F_1}$, which is the SF, as previously shown as equation (7.1).

In nearly all cases, the front spectacle lens surface is convex, making F_1 a positive dioptric power. For an object at infinity, the vergence arriving at the back surface is also positive, and the opposite of that power makes this an afocal Galilean telescope with magnification greater than unity. Only in very unusual cases is the front surface not convex. For example, for an extremely myopic eye, the front surface may be flat, in which case the SF magnification would be unity, or even concave, producing magnification less than unity.

Figure 7.6 illustrates the change in shape factor magnification across a range of front surface powers and lens thicknesses, for a given fixed refractive index. Figure 7.7 shows the SF magnification across a range of front surface powers and selected refractive indices for specific lens materials, with a fixed lens thickness.

The PF can also be treated as an afocal telescope. It is a function of the back vertex power of the lens and the distance from the back surface of the lens to the entrance pupil of the eye, i.e. the vertex distance. The back vertex power is the vergence leaving the back surface of the lens when the object vergence is zero, i.e. when the object is at infinity. In a clinical setting, the prescribed power of a lens is the back vertex power, which compensates for the refractive error of the eye. PF magnification can be manipulated by adjusting the back vertex power and vertex distance. If either is changed by a substantial amount, the other requires changing so that the needed vergence arriving at the eye is maintained.

As shown earlier as equation (7.1), the PF is: $\mathrm{PF} = \dfrac{1}{1 - hF_\mathrm{b}}$, where F_b is back vertex power in diopters, and h is the vertex distance in meters. Vertex distance is commonly measured clinically as the distance from the back surface of the lens to the corneal vertex, which is typically 12–14 mm. For the purpose of calculating PF magnification, this distance is more appropriately the distance from the back of the lens to the entrance pupil of the eye, which is approximately 3 mm behind the corneal vertex, resulting in typical PF vertex distances between 15 mm and 17 mm.

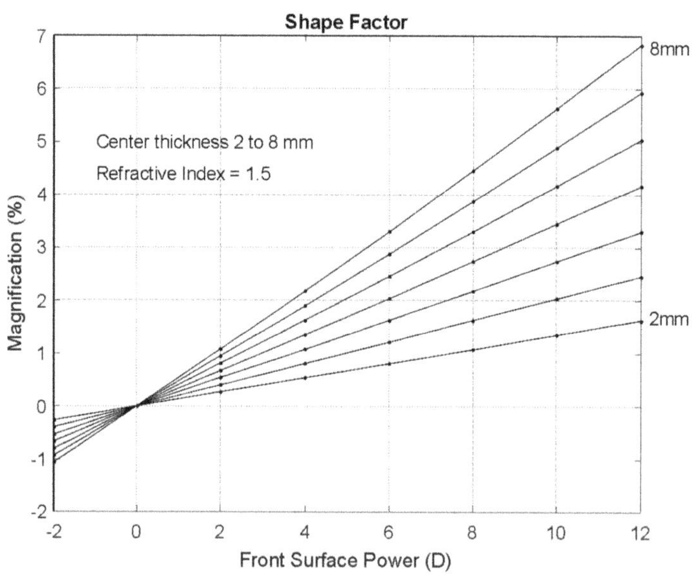

Figure 7.7. Change in shape factor magnification with varying front surface power and lens thickness.

The PF is also derived from the telescope equations. In this case, the objective lens of the telescope (F_1) is the back vertex power of the lens (F_b), and the ocular lens power of the telescope (F_2) is the opposite of the vergence arriving at the entrance pupil of the eye (termed V_2). Telescope length, t/n, is the vertex distance, h, expressed in meters. The vergence of light arriving at the eye's entrance pupil is: $V_2 = \frac{F_b}{1 - hF_b}$.

Substituting into equation (7.3): $M = \frac{-F_2}{F_1} = \frac{\frac{F_b}{1 - hF_b}}{F_b}$, and dividing top and bottom by F_b, we arrive at $M = \frac{1}{1 - hF_b}$, which is the PF (equation (7.2)).

For a myopic eye, which requires a negative lens power for proper correction, the PF will always yield magnification less than unity; for a hyperopic eye (requiring a positively powered lens), the PF will always be greater than unity.

Figure 7.8 illustrates the PF across a range of refractive error corrections and vertex distances.

The total magnification, or SM, is the product of the shape factor and power factor. Clinically significant aniseikonia is a binocular vision problem. That is, the perceived size of an object between left and right eyes is sufficiently disparate to create problems. Therefore, addressing aniseikonia typically involves adjusting one lens to increase or decrease magnification, while doing the opposite in the other.

7.4.1.2 Knapp's Law

Knapp's Law addresses the retinal image size in spectacle-corrected ametropia [22]. Two types of ametropia are considered: axial and refractive ametropia. Axial

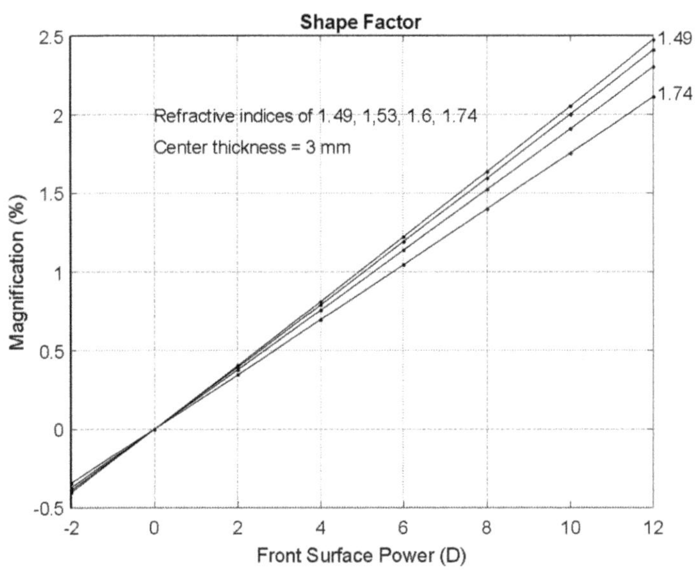

Figure 7.8. Change in SF magnification with varying front surface power and refractive index.

ametropia is when the eye differs from an emmetropic eye by being shorter or longer than that emmetropic eye. Refractive ametropia is when the eye differs from the emmetropic eye by having a different optical power. Knapp's Law states that in axial ametropia, when the correcting lens is positioned at the anterior focal plane of the eye, the corrected retinal image size is the same as in the emmetropic eye. Because the spectacle lens position typically is not very far from the anterior focal plane of the eye, a somewhat relaxed version of Knapp's Law states that in axial ametropia, spectacle lens correction will make the retinal image size equal to that of the emmetropic eye. When considering the two eyes of a patient, if the optical components are similar, but differ in axial length (and therefore differ in refractive error), correction of that refractive error with spectacles will tend to equalize the retinal image sizes. In contrast, if contact lenses are used to correct for refractive errorin axial ametropia, the retinal image sizes will differ. The same is true in cases when corneal refractive surgery is performed. If two eyes differ in refractive error due to axial length differences, and those refractive errors are corrected with refractive surgery, according to Knapp's law, differences in retinal image sizes will be created.

Conversely, in refractive ametropia and, in particular, if the two eyes are different only in their corneal powers, then contact lenses will tend to equalize retinal image sizes. Similarly, refractive surgery correction will also tend to equalize retinal image sizes. Spectacle lens correction will tend to make the retinal image sizes different.

7.4.1.3 Astigmatism and meridional aniseikonia

If one or more of the optical components of the eye are not rotationally-symmetric, an astigmatic refractive error will result. Just as astigmatism is a difference in refractive error by meridian, image size differences may also differ by meridian. The previous equations for the shape and power factors can be expanded to accommodate astigmatic and meridional magnification errors.

The SF as described above is calculated here in a slightly different form, $\mathrm{SF} = \left(1 - \frac{t}{n}F_1\right)^{-1}$, where the scalar value representing a spherical power, F_1, is instead expressed as a matrix representing an astigmatic lens. It takes the form of a 2×2 dioptric power matrix [23, 24]: $\mathbf{D} = \begin{bmatrix} d_{11} & d_{12} \\ d_{21} & d_{22} \end{bmatrix}$.

The various elements are d_{11}, the dioptric power in the horizontal meridian, d_{22} power in the vertical meridian, and $d_{12} = d_{21}$ is half the difference in power between the 45° and 135° meridians.

Substituting $\mathbf{D}$ for F_1, the astigmatic form of the shape factor becomes: $\mathbf{SF} = \left(\mathbf{1} - \frac{t}{n}\mathbf{D}\right)^{-1}$, where $\mathbf{1}$ is the 2×2 identity matrix, and $\frac{t}{n}$ is, as before, the reduced thickness of the lens. This results in another 2×2 matrix: $\mathbf{M} = \begin{bmatrix} m_{11} & m_{12} \\ m_{21} & m_{22} \end{bmatrix}$, where m_{11} is the magnification in the horizontal meridian, m_{22} is the magnification in the vertical meridian, and $m_{12} = m_{21}$ is half the difference in magnification in the two diagonal meridians [25].

The PF can be similarly expressed in matrix form: $\mathbf{PF} = (1 - h\mathbf{F_b})^{-1}$. The term h is, as before, the distance from the back surface of the lens to the entrance pupil of the eye, and $\mathbf{F_b}$ is the back vertex power expressed as a 2×2 dioptric power matrix. This calculation also results in a 2×2 matrix representing the magnification in the horizontal, vertical, and diagonal meridians.

These two 2×2 magnification matrices from the astigmatic forms of the SF and PF are multiplied to yield the total SM, which is also in the form of a 2×2 matrix. This total magnification matrix can be resolved into the orientations of the major and minor axes of magnification, and their magnitudes, by their eigenvectors and eigenvalues.

As an example, consider a lens with a 3 mm thickness, refractive index of 1.5, and a spherical front surface power of +6 D. The back vertex power, in clinical form is: + 2.50DS = −3.00DC × 20 (i.e. a +2.5 D sphere combined with a −3 D cylinder with axis at 20°). The lens is positioned so the back surface is 15 mm from the entrance pupil of the eye.

The front surface power, as a dioptric power matrix, is: $\mathbf{F}_1 = \begin{bmatrix} 6 & 0 \\ 0 & 6 \end{bmatrix}$, and the back vertex power is: $\mathbf{F}_b = \begin{bmatrix} 2.149 & 0.964 \\ 0.964 & -0.149 \end{bmatrix}$. That front surface power, lens thickness and index, entered into the matrix form of the shape factor yields: $\mathbf{SF} = \begin{bmatrix} 1.021 & 0 \\ 0 & 1.021 \end{bmatrix}$. The back vertex power and vertex distance, entered into the matrix form of the power factor yields: $\mathbf{PF} = \begin{bmatrix} 1.0335 & 0.0149 \\ 0.0149 & 0.9980 \end{bmatrix}$.

Multiplying the shape factor by the power factor results in an expression for total SM: $\mathbf{SM} = \begin{bmatrix} 1.046 & 0.015 \\ 0.015 & 1.010 \end{bmatrix}$.

The eigenvectors of this matrix are: $\begin{bmatrix} 0.342 & -0.940 \\ -0.940 & -0.342 \end{bmatrix}$.

Each column contains the x- and y-coordinates of the unit vectors at the orientations of the principal meridians of magnification. In this case, those two perpendicular meridians are:

$$\tan^{-1}\left(\frac{-0.94}{0.342}\right) = 110 \text{ and } \tan^{-1}\left(\frac{-0.342}{0.94}\right) = 20$$

The eigenvalues of the SM matrix are 1.005 and 1.052, representing the magnification at the corresponding eigenvector orientation. The magnification in the 110° meridian is 1.005x, or 0.5%, and in the 20° meridian 1.052x, or 5.2%.

Of course, if there is only one toric surface, as in the previous example, this matrix approach is not strictly necessary, as each principal meridian can be addressed separately using the scalar formulas. Even if both the front surface and back vertex powers are toric, if their principal meridians coincide, both meridians can again be treated separately. If, however, the principal meridians do not coincide, this matrix approach would be applicable. In addition, if an aniseikonic correction were being designed in which meridional magnification did not coincide with the principal meridians of the refractive error, this matrix approach would apply there as well.

7.4.2 Prismatic effects due to correction of anisometropia

A closely related issue to aniseikonia is anisometropia, i.e. disparate refractive errors between the two eyes. Eyes that are highly anisometropic are more likely to also have clinically significant aniseikonia, as well being more susceptible to developing amblyopia. Amblyopia is a condition where the vision does not develop properly in one or both eyes and can lead to permanent vision loss if not treated in youth. Thus significant anisometropia must be corrected with spectacles or contact lenses quickly when it is detected in children. Refractive surgery has been considered when compliance with these more traditional methods is inadequate [26].

Correction of anisometropia with spectacles, even with lenses designed to minimize aniseikonic differences, will be different in power. That can be the source of a different type of problem, that of differential prism effects. Each position on the lens has a prismatic effect with a magnitude and direction. At the optical center, the prismatic effect is zero, but away from the center, that prismatic effect increases with distance from the center. Consider a −5 D lens; at 1 cm away from the optical center, there is a prismatic magnitude of 5 prism diopters, or 5^{Δ}, the common clinical notation and unit of angular measurement. A prism diopter is defined as a deviation of 1 unit over a distance of 100 units, and for small angles it is equal to 1/100 of a radian. For a concave lens, the direction of the deviation is away from the optical center (as with any prism, toward the 'base' of the prism).

In cases of corrected anisometropia, differential prism effects between right and left eyes can be a source of problems, such as eyestrain and diplopia. Consider someone with an anisometropic correction of −5 D for the right eye and −2 D for the left. If they look through the optical centers of the lenses, there is no differential prism effect. If they look 1 cm below the optical centers, the right eye has 5^{Δ} base down prism, the left 2^{Δ} base down, for a differential prismatic effect of 3^{Δ} base down for the right eye. Especially in the vertical direction, this differential prismatic effect is likely to generate symptoms of eyestrain or diplopia. This type of effect occurs even in lenses that have been designed to minimize aniseikonic differences between the two eyes.

An effective way to deal with differential prism in anisometropic corrections is to correct the refractive error with contact lenses. Of course, a contact lens moves with the eye, so the line of sight will always be comparatively close to the optical center of the lens. So even with highly anisometropic corrections, differential prism effects are minimal. If those two eyes are anisometropic due to axial length differences, however, contact lens correction will tend to maintain retinal image size differences.

7.4.3 Prediction of aniseikonia based on measurement of the optical elements of the eye

In a clinical setting, several types of ocular biometry may be available, and those measurements may lend insight into the potential for aniseikonic problems and the appropriate treatment options. A nearly ubiquitous measurement is of refractive error. If the refractive error of the two eyes is equal (i.e. isometropia), aniseikonia is unlikely. With anisometropia, however, aniseikonia becomes more likely, and the question then becomes what the source is of that anisometropia. The cornea is the major source of power of the eye (approximately 2/3 of the total power of the eye), and if the two corneas are about equal in curvature (another common clinical measurement), and under the assumption that the optics of the eyes' crystalline lenses are also similar, the anisometropia is likely to be due to differences in axial length. That leads to unequal retinal image sizes, with the longer (more myopic) eye having the larger retinal image. According to Knapp's Law, correction with spectacles will tend to equalize those retinal image sizes. Conversely, correction with contact lenses (because the contact lens is relatively far away from the anterior focal plane of the eye) will tend to maintain that retinal image size difference, potentially leading to aniseikonic symptoms.

Increasingly sophisticated technologies for ocular biometry, such as ultrasound and optical coherence tomography, are becoming more widely available. Those can be used to measure more precisely the location and curvature of the posterior corneal surface, the surfaces of the lens, and of the retina. In principle, these technologies could allow a biometry-based computation of aniseikonia. However, detailed measurement of the optical components of the eye still cannot reveal the perceived image size, since higher neural levels of the visual system will affect it. While biometry can inform aniseikonic calculations, subjective measurements of image size differences are generally used as the measure of aniseikonia rather than a strictly biometric assessment of the optical components of the eye.

7.4.4 Treatment

7.4.4.1 Contact lenses

As mentioned previously, contact lenses can play an important role. From an optical standpoint, contact lenses can be thought of as simply changing the power of the cornea. The same can be said of corneal refractive surgery, which literally creates a change in power of the cornea. In cases where the two eyes of a patient differ only in their corneal powers, correction with contact lenses (or refractive surgery) will not only correct the refractive error of each eye, but also maintain equal retinal image sizes. In contrast, correction with spectacle lenses will tend to create differences in the retinal image. The more myopic (or less hyperopic) eye will require a more negatively powered lens. The PF for the more negative lens will lead to a smaller retinal image for that eye.

The effect of a contact lens or refractive surgery can also be evaluated in the same way a spectacle lens is evaluated. Assume a contact lens sits 3 mm in front of the entrance pupil of the eye. The PF can be applied using that distance, and the power of the lens (or the magnitude of power change as a result of refractive surgery). For example, consider an eye that requires a −10 D contact lens to correct refractive error. The PF produces magnification of: $M = \frac{1}{1 - (0.003)(-10)} = 0.971x$, or −2.9% magnification. That same eye would require spectacle lens correction, at

12 mm in front of the cornea, of approximately −11.25 D. The PF in that case produces magnification of:

$$M = \frac{1}{1 - (0.015)(-11.25)} = 0.863x, \text{ or } -13.7\% \text{ magnification.}$$

7.4.4.2 Spectacle lens-design options and limitations

The SF and PF manipulate magnification with changes in lens thickness, refractive index, front surface power, back vertex power, and vertex distance. There are constraints on all those properties. Minimum lens thicknesses are specified mainly for safety reasons to prevent lens breakage. Maximum lens thickness is limited by patient acceptance and comfort; lens thicknesses greater than 7 mm to 8 mm will generally not be accepted. Different lens materials with different refractive indices are available, but fall within the range of 1.49 to 1.74. Front surface powers are generally optimized to minimize aberrations with normal viewing. Altering a front surface power too far from this optimized value can degrade function. Finally, vertex distance and back vertex power can be altered to change magnification with the PF. The vertex distance depends upon the fit and adjustability of the eyeglass frame and how the lens is edged to fit within the frame. It cannot typically be made to be less than 10 mm (from the lens to the cornea), or much longer than about 20 mm. If the vertex distance is changed substantially, in combination with higher power lenses, the lens power may need to be adjusted so that the correct vergence arriving at the eye is maintained.

7.4.4.3 Contact lenses + spectacle lenses

Contact lenses can be used in combination with spectacle lenses, and in particular if larger magnification differences are to be compensated. As noted earlier, retinal image sizes in an eye will be very different between contact lens and spectacle lens correction, particularly for higher levels of refractive error. Figure 7.9 shows the difference in magnification between contact lens and spectacle lens correction of refractive error. This is dominantly driven by the PF, and shows the large difference in image sizes for these two modes of refractive error correction.

In unusual instances, a contact lens is used in combination with a spectacle lens on the same eye. That combination can easily be understood as an afocal Galilean telescope. For example, a +10 D spectacle lens in combination with a −11.75 D contact lens, separated by 15 mm, is an afocal telescope with magnification of 1.175*x*, or 17.5%. Magnification of less than 1*x* can be created by using a negatively powered spectacle lens and a compensating positive power contact lens. This combination is also a Galilean telescope, but in reverse. With either approach, to also correct refractive error, the refractive error correction would be added to the power of the contact lens.

7.4.4.4 Intraocular lenses

The optics of the pseudophakic eye are quite different than the eye with the natural crystalline lens, including the size of the retinal image. Consider a model eye, based on the Gullstrand simplified schematic eye. This eye has axial length of 24 mm, a

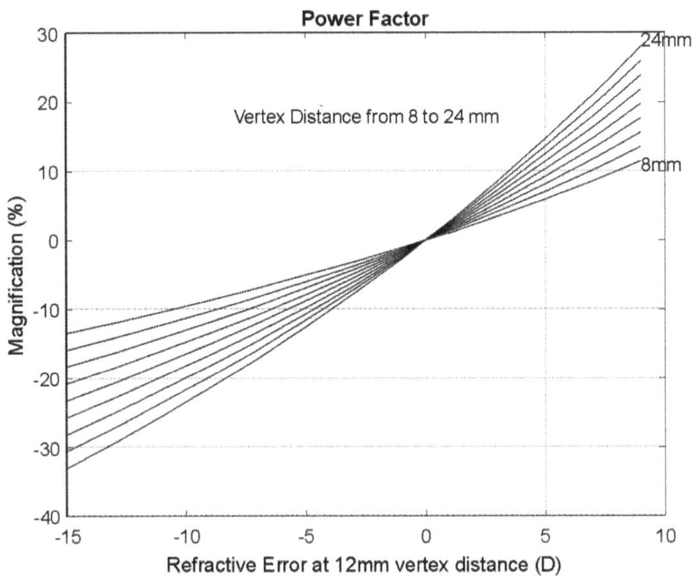

Figure 7.9. Change in PF with varying refractive error and vertex distance.

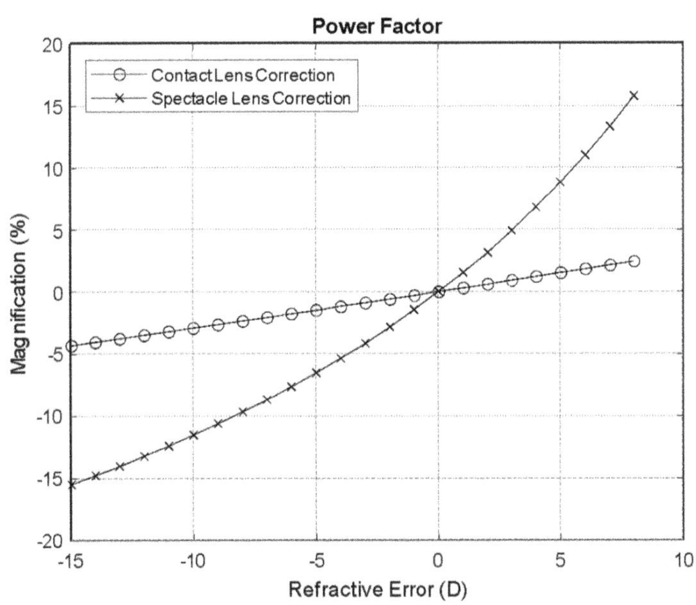

Figure 7.10. Effect of spectacle lens correction versus contact lens correction on PF.

+43.08 D cornea, and a lens that has a +8 D front surface and a +13.33 D back surface. The front surface is 3.6 mm behind the cornea, the lens is 3.6 mm thick, and the aqueous and vitreous fluids have a refractive index of 1.336. This eye is nearly emmetropic (−0.06 D myopic), and has an equivalent power of +60.23 D. That eye is made pseudophakic by replacing the natural lens with an IOL, modeled as a thin lens, positioned 3.6 mm behind the cornea. To make that eye emmetropic, the power of the IOL must be +16.76 D in the eye. That pseudophakic eye has an equivalent power of 57.89 D. The ratio of those equivalent powers is the ratio of image size differences, i.e. 60.23/57.89 = 1.04*x*. That is, the image size in the pseudophakic eye is 4% larger than in the phakic eye. If this is a case of unilateral pseudophakia, that 4% difference is a potential source of visual symptoms.

Aphakia is now very rare, although less unusual in developing regions of the world, where cataract surgery without implantation of an IOL may still occur. In cases of aphakia, options for correction include contact lens correction and spectacle lens correction. If the eye considered above is made aphakic, and corrected with a contact lens (power of +12.5 D), the equivalent power of that contact lens/eye combination is +55.67 D, and the magnification in comparison to the phakic eye is 60.23/55.67 = 1.082*x*, or 8.2%. Finally, if that aphakic eye were corrected with a spectacle lens (power of +11 D at 12 mm vertex distance), the equivalent power of the spectacle lens/eye combination is +48.36 D, and the magnification relative to the phakic eye is 60.23/48.36 = 1.25*x*, or 25%. In cases of unilateral aphakia, the differences in image size, even with contact lens correction of the aphakia, is probably too large to be adequately equalized with manipulations of spectacle lens design. In the rare instance of adult unilateral aphakia, the most practical approach is probably to choose one eye or the other for full correction, while leaving the other eye uncorrected.

7.5 Future consideration in aniseikonia

Aniseikonia has been being diagnosed and treated by manipulating spectacle lens designs for around 100 years. These designs are limited in two significant ways. First, the magnification effect feasibly obtainable with this method is limited to approximately 5%. Contact lens–spectacle telescope designs can exceed this amount but can be fraught with difficulties due to unwanted prismatic effects and are useless for patients unable or unwilling to wear contact lenses. And second, these methods provide no option for varying the magnification effect across the visual field. These are especially limiting factors for treating retinally induced aniseikonia where the central size disparities often well exceed this limit as well as varying substantially across the visual field. Retinally induced aniseikonia may be predicted to become a more frequent impairment as medical advances continue to lead to an aging population. These current limitations may pave the path for future explorations.

7.5.1 Assessment

One area that could be explored is in seeking better options for measuring variances in aniseikonia across the visual field. Could objective measurements be utilized to predict retinally induced aniseikonia? As adaptive optics techniques in the laboratory now allow us to image the photoreceptor mosaic, this could indeed become a future possibility. Could feasible methods be developed to subjectively measure these variances? A group at SUNY State College of Optometry have recently made the first attempt at such an endeavor by utilizing a dichoptic location task at locations across the visual field and fitting the data utilizing Zernike coefficients [27]. Of course, these measurements are of little benefit if improvements in treatment do not accompany them.

7.5.2 Treatment

Freeform optics have become widespread in the production of spectacle lens designs, especially in progressive addition designs. Could freeform optics be utilized to create a variable eikonic design? If so, this design would still be of limited benefit as a spectacle lens as any eye movements would alter the magnification across the field. Perhaps this variable magnification effect is more feasible to incorporate into a binocular head mounted display. Here, forward facing camera feeds could be displayed to the eyes with the appropriate image distortions incorporated. This could be combined with eye tracking technology to adjust the distortions for the current position of gaze.

7.6 Summary

Aniseikonia is a binocular vision disorder where the size and/or shape of the perceived images of the visual scene do not match between the eyes. This condition can cause headaches, eyestrain, distortions of space, and double vision. Aniseikonia can occur due to several conditions including unequal refractive errors between the eyes, unilateral pseudophakia or aphakia, or retinal disorders. Assessment of the condition includes a thorough review of symptoms, objective assessment of the ocular structures, and subjective assessment of the perceived image size differences. Treatment of aniseikonia can involve contact lenses, spectacles, or a combination of both. Often, spectacle lens designs are modified to manipulate the spectacle magnification generated in order to correct for all or a portion of the image size differences experienced while still accurately correcting the refractive error present. With these treatment options many patients experience enhanced binocular function and reduced symptoms. However, some cases of aniseikonia are not amenable with these traditional treatments and these patients must resort to occluding an eye or simply resign themselves to living with their symptoms and visual disturbances.

Chapter highlights

- Aniseikonia is a condition where there is a difference in the size and/or shape of the perceived images coming from the right and left eyes.
- Aniseikonia most commonly occurs due to correcting differing refractive errors between the two eyes (anisometropia), or due to retinal disease.
- Aniseikonia can be assessed by evaluating the stereoscopic distortions that it causes or by directly comparing the perceived image sizes.
- Treatment for aniseikonia can include contact lenses, specially designed spectacle lenses, or a combination of both with the goal of minimizing the perceived image size difference.

References

[1] Donders F C, Gregory G, James R and Duke-Elder S 1864 *University College London. Library Services. On the Anomalies on Accommodation and Refraction of the Eye with a Preliminary Essay on Physiological Dioptrics* (London: New Sydenham Society) Available from: https://archive.org/details/b21284842

[2] Bannon R E 1952 Incidence of clinically significant aniseikonia *Opt. World* **40** 32–3

[3] Burian H M, Walsh R and Bannon R E 1946 Note on the incidence of clinically significant aniseikonia *Am. J. Ophthalmol.* **29** 201–3

[4] Duke-Elder S and Stewart W 1949 *Text-Book of Opthalmology* (St. Louis, MO: CV Mosby)

[5] Hawkswell A 1974 Routine aniseikonic screening *Br. J. Physiol. Opt.* **29** 126–9

[6] Bannon R E 1954 *Clinical Manual on Aniseikonia: A Lecture Series* (Buffalo, NY: American Optical, Instrument Div)

[7] Jimenez J R, Ponce A, del Barco L J, Diaz J A and Perez-Ocon F 2002 Impact of induced aniseikonia on stereopsis with random-dot stereogram *Optom. Vis. Sci.* **79** 121–5

[8] Mitchell L and Kowal L 2010 Aniseikonia causing diplopia: retinal and optical causes (including 'successful' cataract surgery) *Clin. Exp. Ophthalmol.* **38** 17

[9] Rutstein R P and Currie D C 2019 Topical review: retinally induced aniseikonia *Optom. Vis. Sci.* **96** 780–9

[10] Rutstein R P, Fullard R J, Wilson J A and Gordon A 2015 Aniseikonia induced by cataract surgery and its effect on binocular vision *Optom. Vis. Sci.* **92** 201–7

[11] Taylor-Kulp M A, Raasch T W and Polasky M 2006 Patients with anisometropia and aniseikonia ed W J Benjamin *Borish's Clinical Refraction* 2nd edn (Oxford: Butterworth-Heinemann) pp 1479–508

[12] Chung H, Son G, Hwang D J, Lee K, Park Y and Sohn J 2015 Relationship between vertical and horizontal aniseikonia scores and vertical and horizontal OCT images in idiopathic epiretinal membrane *Invest. Ophthalmol. Vis. Sci.* **56** 6542–8

[13] Currie D 2012 Partial correction of irregular aniseikonia secondary to retinal traction *Optom. Vis. Sci.* **89** 1081–6

[14] Okamoto F, Morikawa S, Sugiura Y, Hoshi S, Hiraoka T and Oshika T 2020 Preoperative aniseikonia is a prognostic factor for postoperative stereopsis in patients with unilateral epiretinal membrane *Graefe's Arch. Clin. Exp. Ophthalmol.* **258** 743–9

[15] Rutstein R P 2012 Retinally induced aniseikonia: a case series *Optom. Vis. Sci.* **89** e50–5

[16] Ugarte M and Williamson T H 2005 Aniseikonia associated with epiretinal membranes *Br. J. Ophthalmol.* **89** 1576–80

[17] Okamoto F, Sugiura Y, Okamoto Y, Hiraoka T and Oshika T 2015 Stereopsis and optical coherence tomography findings after epiretinal membrane surgery *Retina* **35** 1415–21

[18] McCormack G, Peli E and Stone P 1992 Differences in tests of aniseikonia *Invest. Ophthalmol. Vis. Sci.* **33** 2063–7

[19] Antona B, Barra F, Barrio A, Gonzalez E and Sanchez I 2007 Validity and repeatability of a new test for aniseikonia *Invest. Ophthalmol. Vis. Sci.* **48** 58–62

[20] Fullard R J, Rutstein R P and Corliss D A 2007 The evaluation of two new computer-based tests for measurement of aniseikonia *Optom. Vis. Sci.* **84** 1093–100

[21] Kehler L A, Fraine L and Lu P 2014 Evaluation of the aniseikonia inspector version 3 in school-aged children *Optom. Vis. Sci.* **91** 528–32

[22] Keating M P 2002 *Geometric, Physical, and Visual Optics* 2nd edn (Boston, MA: Butterworth Heinemann)

[23] Harris W F 1997 Dioptric power: its nature and its representation in three- and four-dimensional space *Optom. Vision Sci.* **74** 349–66

[24] Long W F 1976 Matrix formalism for decentration problems *Am. J. Optom. Phys. Opt.* **53** 27–33

[25] Keating M P 1982 The aniseikonic matrix *Ophthalm. Physiol. Opt.* **2** 193–204

[26] Kraus C L and Culican S M 2018 New advances in amblyopia therapy II: refractive therapies *Br. J. Ophthalmol.* **102** 1611–4

[27] Willeford K T, Butera M, LeBlanc J and Sample A 2020 Field-wide quantification of aniseikonia using dichoptic localization *Optom. Vis. Sci.* **97** 616–27

Chapter 8

Intraocular and contact lens manufacturing

Luiz Melk de Carvalho, Otavio Gomes de Oliveira and Carlos Henrique Lauro

Intraocular and contact lenses are two different fields in the ophthalmology industry and are both associated with huge markets. Although the value chain is significantly different after such products have left the manufacturing facility, the processes to transform raw material into finished products share elements of technology, procedures, resources and even supply chain. In this context, this chapter aims at providing an overview of the most used manufacturing technologies and procedures to produce and inspect intraocular and contact lenses.

As in any industrial transformation process, the flow requires inputs, a manufacturing method with its technologies and procedures, and the intended outputs that eventually represent the finished products. The inputs include the raw materials and geometrical parameters that describe the intended optical and mechanical functions of the lens being manufactured. The manufacturing method comprises the preferred technology and the set of steps to transform the raw material into the actual lenses. And the outputs are the products inspected, labelled, packaged and ready to be shipped.

The present text follows the sequence of the above-mentioned flow and is hence organized in four sections. The first section presents the design parameters required to manufacture an intraocular or a contact lens. It describes the characteristics of the different raw materials that are normally used to manufacture these lenses. In the sequence, it covers the specific optical and platform parameters that are required to set up the production process.

The subsequent sections describe two different manufacturing processes: machining and molding. In the first one, the machining process is broken down into two major steps: high-precision machining and polishing and cleaning. The methods, setup parameters, typically used equipment and the most common defects are described for each of these steps, including a specific subsection to discuss the main causes of such defects. The molding process is presented in the third section. It covers the main characteristics of this process, the method, its advantages and limitations, and requirements associated with the manufacturing of intraocular and contact lenses.

doi:10.1088/978-0-7503-3263-7ch8

The last section presents the main parameters that must be verified to guarantee the quality of the manufactured lenses. It describes the typical processes for checking such parameters and addresses the verifications that must be made to comply with the requirements of international standards. It also covers the packaging step as a final procedure before the products are ready to be shipped to distributors and customers.

8.1 Design parameters

Intraocular lenses (IOLs) and contact lenses are optical devices used in a variety of applications in ophthalmology, all under the broader objective of improving or restoring people's vision. To address the different ophthalmic conditions for which such devices are required, the lenses must be capable of safely delivering optical and mechanical functions. Hence, safety of use, optical performance and mechanical behavior are characteristics of the lenses that must be assured from the concept and design to the finished product.

In the context of design and manufacturing of lenses, the safety and required optical and mechanical functions can be achieved through the combination of three main components: materials, optical surfaces, and platform geometries. In the subsequent sessions, each of these components will be described with a focus on the characteristics and parameters that are critical to the manufacturing.

8.1.1 Materials

IOLs have been first implanted in 1949 by Sir Harold Ridley in London [1]. They were made of polymethylmethacrylate (PMMA), which was observed to be relatively inert in the eye. Despite the requirement for large incision sizes, PMMA continued to be used for a long time as the material of choice for IOLs [2]. However, the evolution of the surgical techniques and the increasing demand for faster visual rehabilitation started to drive the need for the development of other materials that could allow for smaller incisions and improved clinical outcomes [3, 4].

Scientists engaged in the search for new IOL materials frequently referred to the general criteria for implantable materials that have been proposed by Scales [5]. According to such criteria, to be used in implants, materials should: (i) not be modified by tissue fluids; (ii) be inert; (iii) not stimulate tissue reactions to a foreign body; (iv) be non-carcinogenic; (v) not induce allergy; (vi) be mechanically stable; (vii) allow for industrial feasibility; and (viii) allow for sterilization.

In this context, several different options of foldable materials have been developed over time and have played an important role in making cataract surgery one of the most common and successful surgical procedures conducted worldwide.

8.1.1.1 Rigid materials—PMMA

Polymethylmethacrylate (PMMA) is a synthetic polymer that uses methylmethacrylate (MMA) as its basic molecule [6, 7]. It was the material of choice for the first ever implanted IOL [1] and has an extensive clinical record of safety [8]. It is well known for its good biocompatibility [9], although inflammatory responses have been reported among post-operative complications in early IOL implants [7, 10].

PMMA is a rigid material with glass transition temperature of approximately 110 °C [3]. It is suited to either machining or molding manufacturing processes [7] and it is inexpensive. It presents hydrophobic characteristics, with an angle of contact of 70° [8] and high transparency with index of refraction of 1.49 [11].

Since it is a non-foldable material, PMMA IOLs require large incision sizes to be implanted [8], which may be associated with high post-operative astigmatism. Hence, the need for improving clinical outcomes pushed the development of not only surgical procedures that could allow for smaller incisions, but also materials that can be folded to be inserted through such small incision sizes.

8.1.1.2 Foldable materials

Foldable materials used in the manufacturing of lenses are normally from one of two categories: silicone-based materials and methacrylate-based materials, which include hydrogels and acrylics [3, 4].

Silicone materials were first implanted in the 1970s and started to gain the market in the 1980s. Earlier lenses were quite thick due to the low refractive index of the polymers. New generations of silicone materials have since been developed with higher refractive indices, thus allowing the lenses to be thinner.

Acrylic-based lenses can also be made foldable, when the side-chain chemistry of PMMA is substituted by methacrylates [3, 4, 7, 8]. Different side-chain molecules can be used to produce materials with either hydrophilic or hydrophobic properties [8].

8.1.1.2.1 Foldable materials—silicone

Foldable silicone IOLs have been developed as an attempt to overcome the disadvantage of a large incision size required by PMMA lenses. They were first implanted in 1978 [12]. Silicone-based materials are biocompatible and present good transparency. The early generations of silicone IOLs had an index of refraction in the order of 1.41, which required these lenses to be thicker than acrylic IOLs [13]. They were also associated with visual quality lower than that for PMMA and acrylic IOLs [4], since the lower index of refraction would require these lenses to have higher curvatures which, in turn, would introduce higher levels of spherical aberrations and, hence, reduce modulation transfer function (MTF). These materials have then evolved, and the latest generations feature higher indices of refraction.

Silicone lenses are hydrophobic and normally have water contact angle higher than hydrophobic acrylic IOLs [7, 11, 13]. The water contact angle is a measure of wettability of a given surface. Higher water contact angles are associated with hydrophobic materials while lower water contact angles define hydrophilic surfaces.

Silicone lenses are malleable and elastic [2], which can be good properties for certain types of lenses. However, these lenses can be more challenging to be handled during implantation, especially because they unfold quickly when inside the eye [2, 3, 7], which makes it challenging and risky for the surgeon to deal with any mispositioning of the lens. Silicones have a low glass transition temperature, normally between −91.7 °C and −119.6 °C, and hence present rubber-like characteristics at room temperature [3, 13].

Despite their good biocompatibility, silicone lenses are associated with severe complications in patients with posterior uveitis that require vitreoretinal surgery with silicone oil [7, 14] as discussed in chapter 4.

8.1.1.2.2 Foldable materials—hydrophobic acrylic

Foldable hydrophobic acrylic IOLs were introduced in the early 1990s [7]. These materials derive from PMMA through the substitution of side-chain molecules with the purpose of making it foldable, while keeping its hydrophobic characteristic. They normally absorb water to levels as low as 1% and have typical angle of contact of 73° [2, 13, 15].

These materials normally have glass transition temperatures around 12 °C and feature indices of refraction that can be as high as 1.55 [7]. Hence, foldable hydrophobic IOLs are normally thin and light. In comparison with hydrophilic IOLs, the hybrophobic acrylic IOLs are normally slower to recover their shape after being implanted and can get deformed during the injection to an extent that their optical performance is reduced [16].

Such materials are tacky, present low tendency to move and adhere to the capsule bag [2, 11]. Although this characteristic may require precaution from the surgeon when positioning the lens inside the capsule bag, it is beneficial for the mechanical stability of the IOLs, especially for toric ones. Another aspect of such bioadhesiveness is that this characteristic has been reported to be associated with lower levels of posterior capsular opacification (PCO) [17], which is a benefit in relation to hydrophilic acrylic IOLs or silicone IOLs.

A complication typically associated with hydrophobic acrylic IOLs are the glistenings. These are fluid-filled vacuoles that appear within these IOLs after they are implanted into the eye [2, 17, 18]. Glistenings can be found in all types of IOLs but are more prevalent in hydrophobic acrylic IOLs [18]. Glistenings can degrade vision quality and even lead to the IOL explantation in extreme cases.

8.1.1.2.3 Foldable materials—hydrophilic

Similarly to hydrophobic compounds, hydrophilic acrylic materials are derived from PMMA through the same mechanism of substituting side-chain molecules by other molecules aiming at obtaining a foldable material. Hydroxyethylmethacrylate (poly-HEMA) is commonly used to produce hydrophilic acrylic materials [7, 19], which become foldable when hydrated [2]. These materials absorb and retain water to an extent that normally ranges from 18% to 38% [13, 19]. Their mechanical and optical properties may change according to the water content.

These materials present high transparency and normally have an index of refraction in the range from 1.45 to 1.47. At room temperature and when dry, these materials are rigid, which makes them easy to be machined during manufacturing [2]. When immersed in fluid, they absorb such fluid and swell. Therefore, it is critical that the lenses are manufactured scaled down to a swell factor, so that when hydrated they reach the final dimensions that will produce the designed optical function. Typical hydrophilic lenses can increase their dimensions by between 10% and 15% with hydration.

Hydrophilic acrylics are not tacky and can be more easily injected and moved within the capsular bag [2, 7, 19], which are important characteristics for implantable lenses. The hydrophilic surface and water content contribute for these materials to present good biocompatibility [7, 19]. However, the lower bioadhesiveness may prevent it from adhering to the capsule bag and this has been reported to contribute to posterior capsular opacification (PCO) [2, 7, 19], a complication commonly associated more with hydrophilic acrylic IOLs than to hydrophobic ones [14]. Hydrophilic acrylic IOLs normally present higher tolerance to Nd:YAG laser procedures, given their lower adherence to the capsule bag. Differently from silicone lenses, they have lower tendency to attract silicone droplets and then may be used in eyes that may require vitreoretinal surgery using silicone oil [7, 8, 19] (table 8.1). Table 8.1 summarises positive aspects and limitations of materials commonly used in the manufacturing of IOLs.

8.1.1.3 Materials for contact lenses

PMMA used to be employed in the manufacturing of contact lenses in early times. Although PMMA lenses can have excellent optical quality, their inability to transmit oxygen represents an important limitation that can lead to corneal hypoxia [20, 21]. Such lenses are also often uncomfortable, and difficult to get used to. PMMA is rarely used nowadays for prescribed lenses [22].

Table 8.1. Comparison among the materials commonly used in the manufacturing of IOLs.

Material	Typical indices of refraction	Positive aspects	Limitations
PMMA	1.49	Inexpensive Good biocompatibility Long clinical experience	Rigid material, requires large incisions
Silicone	1.41–1.46	Good biocompatibility Highly elastic, what can be good for certain types of lenses	Complications with vitreoretinal surgery Normally thicker lenses Cannot be manufactured through machining
Hydrophobic acrylic	Up to 1.55	Good biocompatibility Thin lenses Associated with lower PCO rates	Normally associated with glistenings Slower unfolding
Hydrophilic acrylic	1.45–1.47	Good biocompatibility Can be easily manufactured through machining Easily injected and manipulated within the capsular bag	Normally associated with higher rates of PCO

Rigid gas-permeable materials, or simply RGP, have been developed as an attempt to preserve the outstanding optical quality offered by PMMA whilst improving critical properties such as permeability to oxygen. Their composition normally combines fluoro, silicone or siloxane and hydrophilic monomers such as HEMA [20]. RGP lenses have high oxygen permeability and therefore can be fitted closer to the cornea than PMMA, which makes these lenses more comfortable to wear. The oxygen permeability of the material is normally measured through the parameter *Dk*, which is the result of the product between diffusion (*D*) and solubility (*k*). Sometimes, what is reported is the oxygen transmissibility *Dk*/*t*, which is the ratio between the oxygen permeability and the lens thickness (*t*). Recommended safe levels of oxygen transmissibility are $Dk/t = 21 \times 10^{-9}$(cm/ss)(mlO_2/ml × mmHg) for daily wear and $Dk/t = 80 \times 10^{-9}(\mathrm{cm}/(\frac{\mathrm{mlO_2}}{\mathrm{ml}} \times \mathrm{mmHg}))$ for extended wear [23]. Although RGP lenses currently account for approximately 9% of market share in the USA [22], their use has been gradually decreasing over the past decades [24].

Currently, almost 90% of the contact lens market share is concentrated on soft lenses [22], of which almost 2/3 comprise silicone hydrogels. Silicone contact lenses are flexible, present very high oxygen permeability and are often durable [20]. Due to its hydrophobic properties, these materials are reported to be associated with high levels of lipid and proteins deposits [25]. Silicone hydrogels have water content that commonly ranges from 20% to over 50%. These materials are normally associated with higher comfort and longer wear periods than other hydrogels.

Soft lenses can also be made of HEMA-based hydrogels. These materials contain hydroxyethylmethacrylate (HEMA), which provides the material with oxygen permeability and improves the comfort of the contact lenses [20]. However, the oxygen permeability is still limited, and prolonged use can result in serious corneal complications [21]. This material allowed for the introduction of disposable lenses to the market, which could be manufactured by cast molding. It also appeared in the market as an alternative to RGP lenses in an attempt to overcome its drawbacks. Since the development of silicone hydrogels, its use has been reduced, but HEMA-based hydrogels still account for almost 1/3 of the market share of soft lenses in the United States [22].

8.1.2 IOL design

As shown in chapter 4, an IOL is usually formed by two basic structures: optic body and haptics. The former provides the device with the required optical function and the latter has a mechanical function, which may include the fixation of the lens in the correct position within the eye, the control of the axial movement of the optic body as a response to post-op capsule shrinkage, and positional and rotational stability. These two major components are covered in the following subsections.

An IOL may include other features such as square edges and haptics angulation. These structures play a role in reducing the occurrence of posterior capsule opacification (PCO) [26], a common post-op surgical complication. Alignment marks may also be present. Marks on the optic body are normally used for rotational orientation of the lens, which is critical for lenses that correct astigmatism.

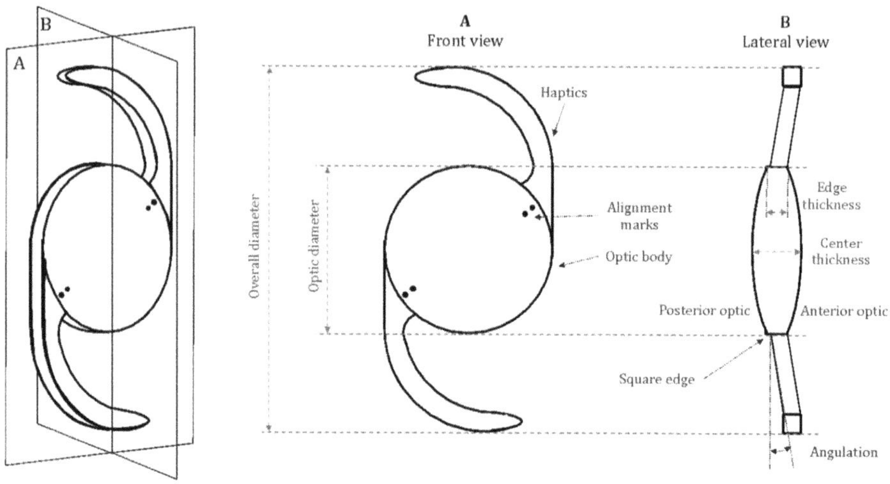

Figure 8.1. The cross-sections of a typical C-loop IOL.

Orientation structures may also be present around the optic body or in the haptics of lenses with symmetric haptics design, to allow the surgeon to differentiate between anterior (facing the cornea) and posterior (facing the retina) surfaces.

Figure 8.1 illustrates a typical C-loop IOL with two perpendicular cross-sections: the front view and the lateral view. Producing these cross-sections is a central element of the IOL design work, since the manufacturing is based on these two views. The lateral view contains information about the optical surfaces, square edges and haptics design. The front view specifies lens dimensions, haptics design, alignment marks and orientation structures.

8.1.2.1 Optical surfaces

Cataract surgeries have been performed for centuries. For over 200 years prior to the development of IOLs in 1949, the surgical procedure consisted in only removing the cataract from inside the eye so that light would not be blocked anymore. However, without the crystalline lens, the optical power of the aphakic eye is significantly lower, which required patients to use spectacles with very high optical power. These were inconvenient because they were very thick, heavy and introduced high levels of optical aberrations which were frequently intolerable to patients that were submitted to unilateral surgery [1].

In this context, IOL implantation represented huge progress in cataract surgery since patients were left with significantly less refractive error after surgery. Therefore, patients would require common glasses with much lower optical power and fewer optical aberrations.

The evolution of surgical techniques, IOL design, materials and IOL calculations allowed for the development of lenses that could correct the patients' refractive errors, thus leaving the patient with no need for glasses to see distant objects clearly.

The capability of manufacturing IOLs with rotationally asymmetric surfaces opened the possibility to produce lenses with toric surfaces, in which the optical power of the lens measured in two perpendicular transverse directions is different. With proper IOL formulas to correctly calculate the ideal lens for each individual patient, toric lenses can be employed to correct for the patients' corneal astigmatism.

The growing understanding of the optical performance of the human eye and the visual demands of patients, together with the evolution of diagnostic technology, surgical procedures and IOL manufacturing technology nurtured the development of lenses with increasing optical complexity. Lens surfaces, for instance, can be built out of aspheric shapes rather than spherical ones. Such asphericity can be used with the purpose to reduce the overall spherical aberration of the lens itself or to compensate, to some extent, the positive spherical aberration normally present in the patient cornea.

More recently, the interest for full spectacle-independence motivated the development of lenses that could provide good visual acuity not only for distance vision, but also for closer objects. A range of different technologies has been developed to address this demand, starting with the multifocal IOLs. These IOLs have as basic principle the generation of two or more focal points, hence, the simultaneous and overlapping projection over the retina of images from objects placed at different distances from the observer. These lenses have the drawback that the image of closer objects forms a hazy background for images of distant objects, and vice-versa, causing a loss of contrast at the focal planes. This inherent characteristic of multifocal IOLs requires some level of neuroadaptation (or resignation) for the patient to get used to such lenses. However, not every patient is suitable for multifocal lenses, since some patients do not tolerate the loss of contrast.

Multifocal intraocular lenses can be either refractive or diffractive, depending on which optical principle is used to split light energy into different focal points. Refractive lenses normally use concentric zones of different radii of curvature, thus, different optical powers. The relative sizes of the different zones determine the percentage of light energy that is directed to each focal point. Such light distribution is affected by the eye's pupil size, as the proportion of the exposed zones change. Therefore, the optical performance of these lenses is pupil dependent.

Diffractive IOLs, on the other hand, can be designed to be pupil independent. These lenses rely on the diffraction of light caused by microscopic patterns of concentric steps built in one of the lens surfaces. Each of these steps form a discontinuity on which light diffracts and splits into different diffraction orders. Such diffraction orders represent different directions, distributed around a central axial direction, in which portions of the light energy propagate. These step profiles can be designed so that the diffracted light interferes constructively in two or more focal points.

Interest has been growing on alternatives to multifocal IOLs, trying to avoid their drawbacks while providing good near and intermediate vision. One such concept is that of extended depth of focus, whose core idea is to guarantee the patient good visual acuity at distance vision and intermediate vision. Some different technologies have been explored to achieve this effect. One of them is also known as the low-add multifocals. These are basically derived from the conventional multifocals that provide distant and near vision, but with lower additional optical power, so that they deliver distance and intermediate vision. This would then potentially reduce the drawbacks of multifocals and yet deliver a good visual acuity at intermediate vision. Another technology that has been explored is one that controls the depth of focus by the type and level of optical aberrations deliberately introduced on the lens design, especially coma and spherical aberration. This can also be achieved through the choice of the surface asphericity.

The optical design of IOLs can be as simple as a spherical biconvex lens, or as complex as diffractive multifocal lenses. There are different optical parameters that characterize the optical performance of the lens, and these are closely associated with geometrical parameters that describe the lens geometry and the surfaces that need to be manufactured. Table 8.2 presents such parameters for the most common IOL optical platforms and figure 8.2 illustrates the main geometrical parameters.

The design work of IOLs is normally conducted through optics engineering software, such as Zemax OpticStudio [27], Breault Research ASAP [28] or Lambda Research OSLO [29], which can determine the geometric parameters that optimize the optical performance of the lens according to the desired requirements.

8.1.2.2 Haptics design

The first ever implanted intraocular lens [1] consisted of a single body of PMMA containing an anterior radius of curvature of 17.8 mm, *a posterior* radius of curvature of 10.7 mm, central thickness of 2.40 mm and diameter of 8.35 mm. The lens contained circular grooves in the periphery of both sides to allow for manipulation using forceps. Since then, the IOL design has changed significantly. The development of haptics allowed the lenses to become thinner and smaller, whilst providing positional stability, centration, and the intended mechanical behavior inside the eye [30–32].

In broader terms, IOLs can be classed as three-piece or single-piece lenses. Three-piece lenses, such as illustrated in figure 8.3(a), have haptics made of different materials, normally more rigid such as PMMA, and assembled on the optical body in a separate manufacturing step. This makes its manufacturing process more laborious since there is the need for making holes in the optical body and then fixating each haptic. Three-piece lenses normally offer more resistance to capsule compression, due to the rigid haptics [33]. The IOL CT LUCIA 602 from Zeiss [34] is an example of a three-piece lens. It is a monofocal, aspheric, foldable IOL made of hydrophobic acrylic material and polyvinylidene fluoride (PVDF) monofilament haptics with a C-loop design and 5° angulation. Its overall diameter is 13 mm and its optic diameter is 6 mm. The lens MN60AC from Alcon [35] is another example. It is also made of hydrophobic material, but its C-loop haptics are made of PMMA and they are designed with an angulation of 10°. This lens has the same overall and optic diameters as the one from Zeiss. The lens ASPIRA 3P-aVA from HumanOptics [36] is one more example of three-piece lenses but made from hydrophilic material. Its haptics are made of polyethersulfone (PES) in the C-loop design with no angulation. This lens has the same overall and optic diameter as the models from Zeiss and Alcon.

Single-piece lenses, on the other hand, comprise lenses in which the optics and the haptics form a single body and the whole lens is manufactured at once. This makes the process much more flexible and allows lenses with more complex shapes to be

Table 8.2. Optical parameters of lenses with different optical functions.

	Optical surfaces	Optical parameters	Geometrical parameters
Spherical	Refractive biconvex with both surfaces being spherical	P, MTF	d, t_{center}, t_{edge}, R_1, R_2
Aspheric	Refractive biconvex with one or both surfaces containing some degree of asphericity	P, W_{sph}, MTF	d, t_{center}, t_{edge}, R_1, R_2, A_1, A_2
Toric	Refractive surface on one side and refractive toric surface on the other, which can be combined with multifocal refractive and diffractive lenses	Seq, C, W_{sph}, MTF	d, t_{center}, t_{edge}, R_1, R_{2-flat}, $R_{2-steep}$, A_1, A_2, C
Multifocal refractive	Refractive surfaces containing one or more concentric and radially symmetric zones of different optical power	P, (P_{add}) for near and/or intermediate focal points, if applicable, W_{sph}, (MTF) for each focal point	d, t_{center}, t_{edge}, R_1, A_1, (R_{zi}, A_{zi}, r_{zi}) for each refractive zone i.
Diffractive lenses	Refractive surface on one side and a diffractive surface on the other side	P, P_{add}, W_{sph}, light loss, light energy split between each focal point, (MTF) for each focal point	d, t_{center}, t_{edge}, R_1, A_1, R_2, A_2, (r_{si}, h_{si}) for each diffractive step
Higher-order optical aberrations	Refractive biconvex with at least one surface containing a radially symmetric profile that includes higher-order optical aberrations	P, aberration profile of the lens or depth of field, MTF	d, t_{center}, t_{edge}, R_1, A_1, geometric profile of the second surface

P: optical power of the lens; W_{sph}: lens spherical aberration; d: Optics diameter; t_{center}: central thickness; t_{edge}: edge thickness; R_1: radius of curvature of one surface; R_2: radius of curvature of the second surface; A_1: asphericity of one surface; A_2: asphericity of the second surface; C: cylinder power; Seq: spherical equivalent power; R_{2-flat}: radius of curvature at the flatter direction of the toric surface; $R_{2-steep}$: radius of curvature at steeper direction of the toric surface; P_{add}: additional power; R_{zi}: radius of curvature of a given refractive zone i; A_{zi}: asphericity of a given refractive zone i; r_{zi}: radius of circle that delimits the given refractive zone; h_{si}: height of a given diffractive step; r_{si}: radius of a given diffractive step.

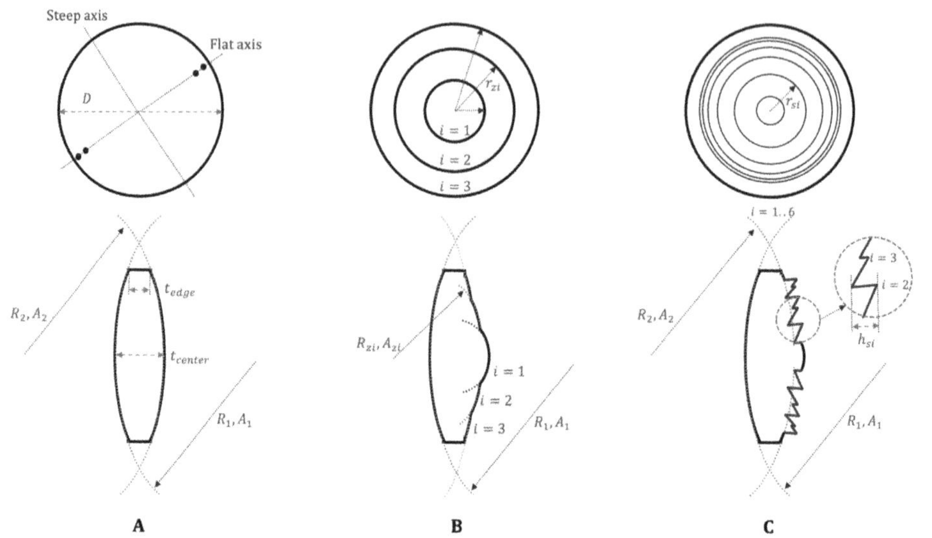

Figure 8.2. Optical platforms for (a) spherical, aspheric and toric IOLs, (b) multifocal refractive IOLs and (c) multifocal diffractive IOLs.

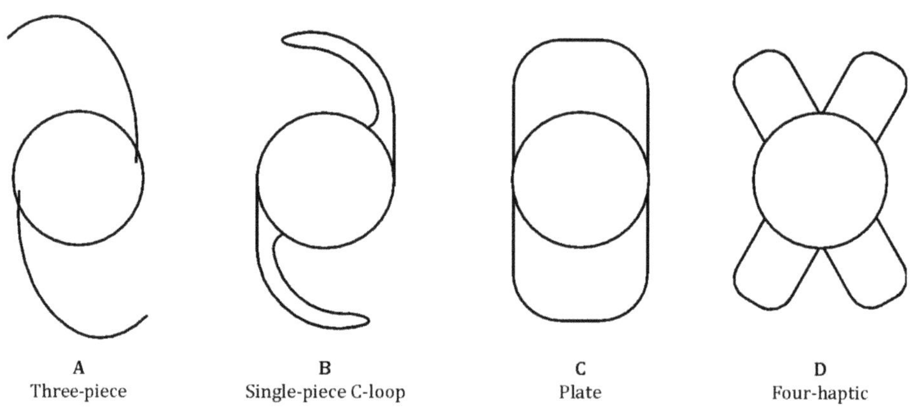

Figure 8.3. Different IOL platforms.

produced. All lenses illustrated in figure 8.3(b)–(d) are single-piece lenses, but with different haptics shapes. The C-loop design is the most commonly used by manufacturers and has a long-term clinical history. The lens SN6CWS from Alcon [37] is an example of a single-piece, C-loop, monofocal IOL, made of hydrophobic material, with an overall diameter of 13 mm, an optic diameter of 6 mm and no haptics angulation. The lens ASPIRA-aA from HumanOptics [38] is another example, but made of hydrophilic material, with an overall diameter of 12.5 mm, an optic diameter of 6 mm and no haptics angulation.

Plate-haptic lenses are reported by some studies as associated with higher rates of PCO [7], due to lack of capsule bending at the optic edge. Other researchers, however, attested low PCO rates and excellent rotational stability for these lenses [31], which makes them very suitable for toric models. The lenses CT ASPHINA 409 M/MP from Zeiss [39] and VERSARIO Multifocal MICS from Bausch+Lomb [40] are both examples of plate-haptic IOLs made of hydrophilic materials with overall diameter of 11 mm, optic diameter of 6 mm and no angulation. The former is a monofocal, aspheric lens, with neutral spherical aberration while the latter is a multifocal IOL with negative asphericity.

More recently, manufacturers have been trying to develop new haptics designs that may have the potential to improve the lens stability and centration [41] within the capsule bag whilst keeping some degree of adjustability to capsules of different sizes. Several different designs can be found on the market. Many of them make use of the concept of four-point contact, which means that the lens haptics is designed to provide contact with the capsule bag around each of four specific regions. The overall diameter of these lenses is usually between 10 mm and 11 mm, thus smaller than that of most other designs. Figure 8.3(d) illustrates the main concept of the design of a four-haptic IOL, although the specific shape may be significantly different among manufacturers. The ASPIRA-aQA from HumanOptics [42] is an example of a four-haptic IOL. It is a monofocal, hydrophilic, aberration-neutral IOL with the branded Quattro haptics. This lens has an overall diameter of 10.5 mm, an optic diameter of 6 mm and no haptic angulation. The Akreos MICS from Bausch+Lomb [43] is another monofocal, hydrophilic, aberration-neutral IOL with four-haptic design. However, this lens is offered with sizes that vary with the lens optical power. The reason is that myopic eyes, which require lenses with lower optical power, are normally larger than hyperopic eyes. Lower optical powers of this lens model, up to +15.0 D, feature 6.2 mm of optic diameter and 11.0 mm of overall diameter. Higher optical powers, between +22.5 D and +30.0 D, are offered with 5.6 mm of optic diameter and 10.5 mm of overall diameter. These lenses have a 10° angulation of the haptics. Baush+Lomb also offers in its portfolio the Quatrix Aspheric Evolutive [44], which is a monofocal, hydrophilic, negative aspheric IOL with a different four-haptic design. The optic diameter decreases from 6.15 mm for lower optical powers to 5.85 mm for higher ones, while the overall diameter changes from 10.8 mm to 10.3 mm. Its haptics present 6° of angulation.

As can be noted from the previous examples, there is a large number of different IOL designs available on the market. And this includes not only lenses intended to be implanted inside the capsule bag, but also those designed to be fixated at other places such as sulcus, anterior chamber and sclera.

When designing an IOL, it is critical that not only the requirements associated with the specific lens application are considered as inputs to the design work, but also those imposed by the international standard ISO 11979 part 3 [45]. This standard specifies design requirements and test methods for certain mechanical properties of IOLs. Such requirements include the minimum optic diameter of 4.25 mm, tolerances for optic and overall diameter, maximum tilt of 5° and maximum decentration of 10% of the clear optic when the lens haptics are compressed.

According to the standards mentioned in the previous paragraph, IOLs have to be tested using specified procedures and experimental apparatus. The tests aim at characterizing the mechanical behavior of the IOL and include compressing the lens to specific diameters, which vary according to the intended application of the lens, and measuring the compression forces applied by the haptics, the axial displacement, optic decentration, optic tilt and angle of contact between the haptics and the compressing element. The lenses have still to be assessed in terms of dynamic fatigue to ensure that no haptics breaks due to repeated compression movement.

To design an IOL engineers normally use CAD software such as SolidWorks [46], in which the optical surface geometry generated by optics engineering software can be integrated with the haptics design and any other desired features. This CAD software can then generate the required machine files to be used to manufacture the lenses.

8.1.3 Geometric parameters of contact lenses

Contact lenses are convex–concave lenses. They are called corneal lenses if their diameter is suitable for fitting over the cornea or scleral lenses, if they are larger and are fitted to the sclera. Their optics are normally spherical, aspheric, toric or refractive multifocal, just as the intraocular lenses previously mentioned. Besides the optics, contact lenses feature additional surface structures that commonly aim at providing comfort to the patient by guaranteeing proper fitting of the lens.

The design of the cross-section of a typical contact lens is shown in figure 8.4. The convex shape is the anterior surface of the lens and the concave one is the posterior surface, which covers the corneal region. The central region of the lens, comprising both surfaces, which has an optical function, is the optic zone. Out of the optic zone, the lens may have one or more secondary zones and a peripheral zone. These zones are designed to allow for proper fitting and comfort at the edge of the lens. Their sizes are defined based on the intended use of the lens as well as on the choice of material.

Within the optic zone, anterior and posterior surfaces are characterized by their radii of curvature R_a and R_p and by their respective asphericities A_a and A_p. This is applicable to both spherical and aspheric lenses.

Toric lenses add a steeper axis in one of the surfaces and orientation marks. Toric lenses can be found on the market with a range of different orientations for the steeper axis. Because toric lenses need to have their steeper axis properly aligned to provide suitable visual correction, lenses normally present some sort of rotational stabilization feature, such as prism ballast. This technique consists of adding prism to the vertical axis of the lens, which results in a thicker edge at the bottom of the lens, and hence contributes to preventing the lens from rotating, since the base of the prism will tend to be oriented inferiorly due to gravity. It is up to the ophthalmologist or optometrist to prescribe the proper orientation to correct the astigmatism of the individual patient's eye.

Refractive multifocal lenses, on the other hand, add to one of the surfaces one or more concentric rings of increased optical power. The sizes of such rings as well as their additional optical power are elements of the design work of the lens.

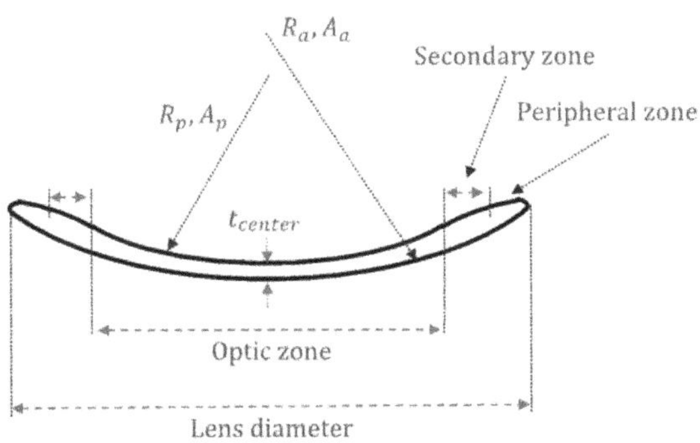

Figure 8.4. Cross-section of a typical contact lens.

8.1.4 Manufacturing processes

According to Sahler *et al* [60], the IOLs require micrometer accuracy to avoid shape and dimensions deviations during the manufacturing. That is, manufacturing of monofocal IOLs still are limited by the technologies and expertise of molding or turning processes, limits that are accentuated for the premium IOLs, increasing the costs.

Generally, because of their higher thicknesses, the machined lenses are easier to handle than the lenses made through spin casting. However, the machining is more time-consuming because each lens is made separately [57].

When the subject is the customized IOL, Chassagne and Canioni [58] highlighted that the manufacturing process may unduly increase the cost because a patient requires few units, emphasizing the importance of high precision of the machine tools. These authors suggested the additive manufacturing, 3D printing, as an interesting alternative; however, this technology is limited by the materials that are available and the accuracy of the current machines.

Debellemanière *et al* [59] analyzed the reproduction of an existing IOL using a current 3D printing technology. Although the printed IOL has presented great transparency, an average 75% transmission in the visible spectrum, they observed deviation between 1.8% and 17.3% in the IOL geometry, and the surface roughness did not attend to the requirements for implantation.

8.2 Machining process

In general, the manufacturing process of IOLs and contact lenses employs a diamond tool to machine the surfaces of each individual lens [47]. According to Olufayo and Abou-El-Hossein [48], although the precision machining appeared a few decades ago, its use in polymers is still an untapped field, containing several gaps. The ultraprecision single-point diamond (USPD) machining is the most effective process to obtain high-quality functional surfaces out of polymers for optical purposes [49].

The key steps of the manufacturing process of lenses through machining are shown in figure 8.5. The process starts with the preparation of raw material, which is

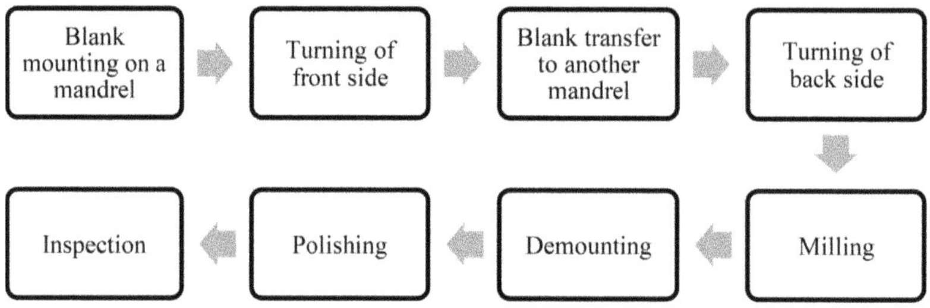

Figure 8.5. Machining process flow.

normally supplied in the form of disks, also known as blanks, with diameter in the range of 15 mm to 20 mm. A blank is mounted over a mandrel commonly using a soluble wax. The mandrel is then positioned in the lathe and a diamond tool is used to cut the front side of the lens. In fact, more than one diamond tool, with different finishing characteristics, can be used to cut a single surface. Once the front side has been finished, the blank is transferred to another mandrel, so that its back side is now exposed. The blank transferring is normally conducted with specialized equipment to guarantee accurate alignment. The mandrel is then positioned in the lathe so that the back side of the lens can now be machined. At the end of this process, the mandrel is taken to the milling machine, where a milling tool is used to mill the lens borders and haptics, if it is a single-piece model. This step gives the final shape to the lens. Once the lens is milled, it can be removed from the mandrel. The lens is then cleaned and enters the polishing step, where finishing of both borders and optical surfaces is improved. After being polished, the lens is ready for inspection and quality checks.

The manufacturing process of both IOLs and contact lenses share the basic key steps shown in figure 8.5. At the end of the machining process, IOLs go into a clean room for deep cleaning and preparation for sterilization. After sterilization, they are ready to be packaged and shipped. Contact lenses, on the other hand, do not require sterilization.

Depending on the materials used, lens design and features, the manufacturing process may include several additional steps. Aiming at increasing cost-efficiency of the process, some suppliers of raw materials offer blanks in which the front surface is already made. In this case, the lens manufacturer only has to conduct one turning operation, saving time and resources.

8.2.1 High-precision machining

In machining of lenses, either natural or synthetic diamond tools are used to obtain high dimensional accuracy, quality of surface, and also to increase tool life. Single-point diamond (SPD) turning is advanced machining technology widely used to manufacture optical components, which provides great control of the surface roughness. However, there are gaps of the knowledge about the optical surface quality [50].

During the cut of polymeric materials, the resulting quality depends not only on materials properties, such as glass transition temperature (T_g), melt temperature

(T_m), molecular weight and viscosity, and others, but also on cutting parameters, such as the cutting speed, cutting tool geometry, tribological properties, and others [7]. According to Liman *et al* [51], the machining of contact lenses requires great surface finish and dimensional accuracy.

The cutting parameters combination, like cutting speed (v_c) or spindle speed (n), feed rate (f) or feed speed (v_f) and the depth of cut (DoC), can be defined or limited by machine, such as its stiffness, power, and others. In the literature, several studies on the machining of biopolymers were developed to analyze different responses relating to quality of lens, like surface roughness [51], process consumption, like cutting force (F_c) [48], productivity, like the material remove rate (MRR) [51], and other responses. Liman and Abou-El-Hossein [51] concluded the electrostatic charges ($ESCs$) encountered in a cutting tool when turning advanced contact lenses are important, as they reflect the quality and condition of the tool, machine, fixture, and sometimes even the finished surface. Table 8.3 shows some cutting parameters and responses found in the literature.

The optimization is a hard task due to nonlinear structure of the machining process, involving the cutting parameters, experimental design and measuring techniques [51]. Liman and Abou-El-Hossein [51] studied the turning of PMMA with more than 99.5% hardness (Shore D) of 93, temperature at 23° and humidity of 50%. The optimal combination was, approximately, cutting speed of 3700 rpm, feed rate of 2.8 μm/rev, and depth of cut of 37 μm when the responses observed were surface roughness ($R_a \approx 0.0017$ μm), electrostatic charges (0.009 kV) and material removal rate (23.972 mm^3).

In the turning of PMMA for IOL, Mishra *et al* [52] analyzed the effects of the cutting parameters, spindle speed (n), feed rate (f), and depth of cut (DoC), using a diamond tool with nose radius of 1 mm and rake angle of 0°. They observed that, for

Table 8.3. Cutting parameters employed in the machining of biopolymer.

Researchers		Material	Input variables	Responses
Turning	Liman and Abou-El-Hossein [51]	PMMA	n (200, 2100, 4000 rpm) v_f (2,7,12 mm min^{-1}) DoC (10, 25, 40 μm) Temperature 23 ° CHumidity 50%	R_a (0.0017 to 0.0837 μm) ESC (0.0098 to 0.4102 kV) MRR (1.07 to 25.64 mm^3/min)
	Olufayo and Abou-El-Hossein [48]	Silicon acrylate co-polyme	v_c (9, 80, 150 m min^{-1}) f (2, 7, 12 μm/rev) DoC (10, 25, 40 μm)	F_c (0,042 to 0,141 N)
	Liman *et al* [51]	ONSI-56® Onsifocon-A	n (200 and 4000 rpm)v_f (2 and 12 mm min^{-1}) DoC (10 and 40 μm)	R_a (0.006 to 1.184 μm)
	Mishra *et al* [52]	PMMA	n (1000, 2000, 3000 rev) f (1, 5, 10 μm/rev) DoC (5, 20, 35 μm)	R_a (0.0195 to 0.0423 μm)

surface roughness, the percentage contribution of spindle speed is 9.11% and 85.65% for the feed rate. The higher level of spindle speed caused a fracture on the IOL, even though the surface roughness decreased. Aiming at lower R_a surface roughness, the optimized combination was the spindle speed of 3000 rpm, feed rate of 10 μm/rev, and depth of cut of 5 μm, providing R_a surface roughness about 0.02 μm.

However, some care must be taken in the diamond machining. In the diamond machining of biopolymers, despite low hardness, the presence of abrasive particles should be avoided because they can damage the cutting edge [49]. In addition, the high temperature during the cut causes tool wear, reducing of hardness in the cutting edge, and poor surface quality, for which a countermeasure is, generally, the use of cutting fluids, such as water and oil emulsions, when a metal is machined. However, oil-based lubricants are not recommended in precision machining because they are ineffective in cooling. In single-point diamond precision machining, for the non-ferrous alloys, generally, methylated spirit or kerosene, carbon tetrachloride, and compressed air sprayed through nozzles are applied [53].

The machining of PMMA to make lenses can happen under dry condition due to the high stiffness at room temperature. For the more flexible polymers, which are generally characterized by glass transition temperatures well below room temperature, the machining process can be satisfactorily executed if the polymer stock contains very little water [48]. Li *et al* [53] applied the nanodroplet cutting fluid (NDCF), promising cutting fluid for the optical polymers, in the machining of PMMA with a flow of 50 ml h^{-1} at 2 bar. They observed a reduction of nanosurface irregularities and an improvement in the surface finish.

According to Mamalis and Lavrynenko [49], for the stable single-point diamond machining, the establishment of the maximal possible tool wear is necessary to guarantee the surface quality. However, the wear of the diamond tool is difficult to be assessed by direct visual inspection because it is characterized by sub-micrometric destruction of the micro blades on the clearance face. The authors chose as wear criterion the cutting wedge clearance face (hcf) lower than 0.6 μm, guaranteeing the high level of finished quality for optical and bioengineering components.

8.2.2 Polishing and cleaning process

The manufacturing of intraocular and contact lenses commonly requires a polishing process, which is intended to minimize the sub-surface damage or cracks caused by the machining of hard and brittle materials [54]. In the conventional polishing process, high pressure is employed between the workpiece and the pad, however, for the polishing of the lenses the use of magnetorheological (MR) fluid is recommended [55].

During the manufacturing, several parameters influence quality of lens, like the chemical composition. The polishing of polymer materials, like hydrogel, are highly sensitive because the selected parameters can cause damage that compromises the interaction with the tear film, consequently its biocompatibility [56]. Contact lenses should be polished less than 50 s, avoiding heating and/or warpage causing changes in curvature and surface hydrophobicity. In the fine surface finishing, manual sanding and polishing can be effective, but achieving adequate transparency is challenging. Furthermore, grit paper can be used to achieve a certain level of smoother surface, following a polishing with acrylic cleaner or microfiber cloths [57].

The increasing of the rotational speed can improve the quality of the surface, however, the number of abrasive grains increased. At an excessive rotational speed, the fluid can no longer act as a polishing pad because the chain-like structure is destroyed [55]. After the polishing process, the lenses are cleared using the ultrasonic process and are dried using filtered air in a clean room [7].

8.3 Molding process

Besides the machining processes, Moreddu *et al* [61] affirmed that the contact lenses also can be manufactured by two other processes. One is spin-casting, where the material is poured into a mold that is attached on a spinning cylinder and UV light is employed to polymerize the material. The other one is the injection molding, in which the material is placed into the mold and pressed until reaching the lens-shaped and polymerized under UV light. At the end of both processes, the edges are polished, the lens is hydrated, inspected, and packaged.

The evolution of equipment has improved the precision and reliability since the 1980s, which is critical since, unlike conventional injection molding, the plastic optics molding requires strict tolerance, high accuracy, high process stability and repeatability, clean production environment, and more [62]. According to Bozukova *et al* [63], the lenses that are manufactured by injection molding can provide low-cost and high-quality, like contact lenses.

The thermoplastics are widely used in injection molding, mass-produced, due to their hydrophilicity and ability to modify the channel surfaces being crucial for biological applications [64]. Pertici [65] cited that a polymer can be molded by injection, reactive injection, compression, or rotational molding. Among these processes, the injection molding is the most common one, since it allows for complex shapes with accuracy. However, in the injection molding, the material can present greater shear mechanical and thermal stresses [66].

Numerous polymers are used for the manufacturing of biocomponents, like the poly-l-lactic acid (PLLA), polyglycolic acid (PGA) and poly-DL-lactic-co-glycolic acid (PLGA). [67]. Pandey *et al* [68] highlighted the use of poly(methylmethacrylate) (PMMA) and Poly(2-hydroxyethylmethacrylate) (PHEMA) in the manufacturing of intraocular lenses. However, some of the most used thermoplastics for the injection-molded lenses are polylactic acid (PLA), poly(glycolic acid) (PGA), and copolymers [66].

To manufacture lenses, several steps are need, like the design of lenses and respective shell molds, machining of molds, injection molding, lens casting, and hydration [69]. Thus, to improve the quality of the contact and intraocular lenses, in this section, the molding of these lenses is debated, indicating the attributes that are important to obtain high-quality lenses.

8.3.1 Mold

In the mass low-cost production of different contact lens types, the molding process requires molds of complex smooth surfaces, providing high accuracy and minimal deformation [70]. For the great quality of the part, it should be ensured that the design and machining of the mold guarantees the complete filling, flow of material, high pressure to avoid leakage, and uniform cooling, which increases the costs [66].

The mold materials should be carefully selected, ensuring great conduction, convection, and radiation to conduct the heat flux inside the mold that is resulting from the melted material. The cooling performance and cyclic time of mold can improve by the design of cooling channels [71].

Because of the requirements for rigorous control, various authors report in the literature the use of several software and technologies, like computer-aided design (CAD) and computer-aided engineering (CAE), to improve the project and process. To ensure the quality of the shell molding, Vu *et al* [69] suggested the development of numerical simulations to analyze the molding parameters before experimental tests, like mold temperature (T_W), dwell under pressure (p_N), etc.

Kanbur *et al* [71] highlighted the use of CAD and CAE software to model and develop the thermal and mechanical analysis for the molds, respectively. Thermal-fluid dynamics and mechanical analysis were developed to evaluate the cooling and stress performance of conformal cooling channels.

Each year, the use of contact lenses is continuously growing, which requires a higher number of casting molds for massive processes, since they are made of polymer and discarded after the molding [72]. In the shell molding, two parts are needed, anterior and posterior shell molds, which are fabricated by injection molding [69].

Vu *et al* [69] cited that, because the shell mold shapes for dry contact lens are complex (thin and convex), the polypropylene globalene 6331 (PP) is used due to its high thermal stability and high flowing property. The melted material, with temperature in the range of 190 °C to 270 °C, is easily filled into the mold inserts. Shrinkage errors along the axial axis (Z-shrinkage error) typically account for an incidence of 1.4%.

According to Lin *et al* [72], the shape of conformal cooling channels can reduce the contraction and warping due to the uniform temperature distribution during the cooling. Using numerical simulation, the authors observed that the cooling of sprue bush influences the cooling cycle without any positive correlation between efficiency and cooling time. Moreover, the injection molding cycle can be shortened by approximately 20% through the addition of the conformal cooling channel in the connecting between nozzle of injection machine and injection molding, the sprue bush.

The IOL molds can be made of non-ferrous metals such as electroless nickel, because they present low porosity, excellent corrosion resistance, and high hardness [7]. However, the contact lens molds can include several materials, likes alicyclic polymers or polyoxymethylene, which are non-polar and polar. The quality of contact lenses is affected by the interaction between mold surface and material [20]. According to Li *et al* [73], the coating on the mold can increase the life of the mold and reduce the interfacial reaction, which improves the surface quality of lenses.

Thus, among other aspects, on polymerization should be ensured that there will be no adhesion of material on the mold surface, which could influence the wettability of the hydrogels. The quartz (SiO_2) mold can be used because it has the non-sticking property [20]. Koronkevich *et al* [74] analyzed diffractive–refractive bifocal intraocular lenses (BIOLs) molded on quartz molds. Moreddu *et al* [61] highlighted graphene coating, using chemical vapor deposition (CVD), in the quartz mold to avoid the formation of wrinkles. Bensingh *et al* [75] suggested evaluating the quality of molded lenses through the molds geometry, likes the radii of curvatures, waviness (W_a) and surface roughness (R_a).

8.3.2 Molding

The quality of molded IOLs is related to the injection parameters, such as temperature, pressure, and velocity. They influence filling, holding, and cooling steps [62]. According to Dick *et al* [76], to guarantee the quality in the molding process, a complex combination of the material, design, and process parameters should be defined. Thus, a molding parameters combination, melt temperature (T_m), mold temperature (T_W), dwell under pressure (p_N), work time (t_N), and cooling time (t_{rk}), can largely influence the process.

The melt and mold temperatures are related to flow and cooling conditions, i.e., low melt temperature inhibits the flow material and low mold temperature provides higher level of residual stress and warpage. The dwell under pressure and the dwell time are critical parameters in compensating for shrinkage and reducing the warpage [77]. According to Badur *et al* [78], the melt temperature is considerably higher for crystalline and semi-crystalline polymers than for monomers. For the cross-linked polymers, because they do not present a defined melt temperature, the glass transition temperature (T_g) is the relevant parameter which characterizes the temperature range.

In a study of the injection-molded bi-aspheric soft contact lens using Polypropylene Globalene 6331 (PP), Chen *et al* [70] analyzed three variables, melt temperature (210 °C and 230 °C), injection velocity (40 and 60 mm s^{-1}) and dwell under pressure (500 and 700 bar). The authors observed that the Z-shrinkage error increased with the higher level of the injection velocity and with the higher melt temperature, but reduced for the higher level of dwell under pressure. In a similar study, Vu *et al* [69] analyzed also that increasing the cooling time from 15 to 45 s reduced the Z-shrinkage error. For the dwell under pressure, an increase from 700 to 900 bar, increased the Z-shrinkage error. Moreover, the Z-shrinkage error reduced when the injection velocity increased from 60 to 80 mm s^{-1}.

Analyzing the increase of the dwell under pressure, 700 to 1400 bar, Dick *et al* [76] observed that for the value of 1000 bar, the form deviation was reduced from 30 μm to 4.6 μm (rms) and the optical path difference increased from 130 to 330 nm cm^{-1}. Above this pressure, the residual stress caused stability of both quality parameters. The authors highlighted that values for both parameters should be minimal. Comparing the dwell time between 10 and 40 s, these authors observed that the increase of time up to 30 s minimized the form deviation, 15.8 μm to 4.4 μm (rms), and the optical path difference increased from 217 to 290 nm cm^{-1}.

In the molding of soft contact lenses, Cheng and Chang [79] analyzed the behavior of polypropylene homopolymer PP 6331 varying the mold temperature

(30 °C to 70 °C), melt temperature (190 °C to 230 °C), dwell under pressure (150 to 250 bar), and cooling time (6 to 10 s). Through the Taguchi analysis, the authors observed that the dwell under pressure was the main parameter for the Z-shrinkage error. The optimal condition, consisting of mold temperature (50 °C), melt temperature (230 °C), dwell under pressure (250 bar), and cooling time (6 s), provided an estimated Z-shrinkage error of 1.531%.

For the cooling time, varying between 60 and 240, the optical path difference tended to achieve stability at 290 nm cm^{-1}. The influence on the form deviation can be observed until 180 s, reducing from 18.2 μm to 4.3 μm (rms). It can be caused by longer mold constraint at longer cooling times and, thereby, better cooling conditions [76]. Li and Hsieh [73] employed the FEM to analyze the molding of Poly (methyl methacrylate) (PMMA), varying the filling time, melt temperature, mold temperature, and dwell time. The optimal condition was found when employing a filling time of 0.5 s, a melt temperature of 250 °C, a mold temperature of 60 °C, and a dwell pressure time of 2 s, which provided about 75% residual stress and birefringence reduction compared to other molding conditions.

Bozukova *et al* [63] developed a micro-molding study employing a Battenfeld Microsystem™ M50 injection molding and thermoplastic polyurethane elastomers (Elastollan 1190 A U), which is a material widely used in the manufacturing of several other medical devices, except the IOL. Although a few geometrical deviations may have been influenced by poor quantity of the polymer in the equipment, the micro-molded IOL presented a clear optic and entirely replicated haptics.

Fang *et al* [62] cited the development of a micro-adjustment system to compensate tool offset for the double convex lens in the manufacturing of biconvex lens with a diameter of 50 mm. The initial deviations were about 42 μm and 31 μm, in *X* and *Y* directions, which were reduced to approximately 1.87 μm and −0.79 μm, respectively.

8.3.3 Main defects

The manufacturing of IOLs employing the molding processes can provide benefits, such as low quantity of material, low operator intervention, compatible to mass production and easy reproduction of complex shapes. However, several factors are critical for the quality of plastic optics, such as the design (mold and product), machine, materials, parameters and postprocessing. Furthermore, the shape accuracy, residual stress, and imaging quality can be considered of main interest. It is important to emphasize that in the machining or in the molding processes, the environmental conditions influence the product and that they are not always compensated effectively [62].

Residual stress, an essential quality criterion for polarized light within transmission optics, can provide different densities per spatial directions, changing the directional property of the refractive index [76]. The residual stresses can cause dimensional deviations after the demolding and/or during the use [66]. The volumetric shrinkage during the cooling process can cause residual stress, resulting in local slight variations in the birefringence [80]. When the cooling is fast and inhomogeneous, an equilibrium state is not obtained, which can affect the dimensional stability and optical quality. This concept, free volume, can be well predicted by applying the equation below (8.1) [81].

$$\delta = \frac{V(t) - V_{\infty}(T, p)}{V_{\mathrm{ref}}} \tag{8.1}$$

where δ is the fractional free volume; V is the specific volume at current time t; V_{∞} is the equilibrium volume at current time t; V_{ref} is the equilibrium volume at reference temperature; T is the temperature and p the pressure.

The manufacturing of the aspheric lens requires tolerance in the micrometric range, which asks for great control of injection molding to avoid lens distortion due to the volumetric shrinkage and its distribution [82]. According to Bensingh *et al* [75], during the molding, the environmental conditions can be considered an important factor for the accuracy for optical devices because they can affect the shrinkage, warpage, thermal and mechanical stress, water absorption and heat deflection. The alignment error or offset/deformation in the mold can cause decentration defects due to the high shear and, after demolding, non-uniform Z-shrinkage error [62].

In injection molding, the thermal isolation of skin layer and lower thermal conductivity of polymer melts cause a high temperature in the core that relaxes the weakly oriented polymer chains [62]. In addition, because the material is submitted to high temperature over an extended period, the molding requires particular attention, since the thermal degradability and the gas release can cause small voids in the parts [66].

8.4 Quality control and packaging

The quality control process consists of the inspection, during the production phase, of each feature and parameter of a manufactured product that would affect the satisfaction and safety of the end user. The quality requirements and processes to inspect contact lenses and IOLs are described by international standards. The standards establish not only the optical and mechanical requirements of the lens, which will affect its optical performance, but also the sterilization and packaging requirements that will, respectively, be responsible for the protection of the patient against contamination and will guarantee the stable transportation of the product. The fulfilment of the standards criteria is one of the prerequisites to distribute the lenses on the market. In this context, this section describes the quality requirements and methods for the inspection of the optical and mechanical quality of the lenses, their sterilization and the packaging process, based on the main international standards applied to contact lenses and IOLs.

8.4.1 Standards and tests

In the manufacturing process of IOLs and contact lenses, it is crucial to guarantee that the final product is compliant to all the requirements defined by the official standards. In general, the standards state not only the requirements that a product must meet but also the methods to verify and how to report the analysis of each requirement.

The ISO (International Organization for Standardization) is the main organization responsible for stablishing the standards for the eye care industry and for several other businesses, and it is applied to all member countries around the globe [83]. The standard body of each country adopts and adapts the ISO standards to its national market such as ANSI (American National Standards Institute) from the United States and EN (European Standard), which is the representative body of the European Union.

The ISO 11979 standard is divided into 10 parts and defines the requirements and methods applied to intraocular lenses such as: vocabulary and terms definitions (part 1); optical requirements and test methods for optical properties evaluation (part 2); mechanical requirements and test methods for mechanical properties evaluation (part 3); labelling and information requirements that are provided within or on the packaging (part 4); lens material biocompatibility requirements and methods (part 5); methods and requirements for shelf-life and transport stability of the final product into its final packaging (part 6); clinical study requirements and recommendations for the lens performance and risk investigation (part 7); fundamental requirements that are applied to all IOL models and technologies (part 8); specific requirements and methods applied to the evaluation of multifocal IOLs (part 9) and the clinical requirements for the investigation of phakic IOLs, that are employed to correct higher amounts of refractive errors in patients that have a health crystalline lens (part 10). Some parts of ISO 11979 also refer to other ISO standards and scientific papers that present additional definitions and methods to support the fundamentals of the requirements [84].

The ISO 18369 standard is divided in four parts and it is applicable to contact lenses. It addresses the requirements and test methods for rigid (corneal and scleral lenses) and soft contact lenses such as: vocabulary and terms definitions (part 1); the tolerances applied to the optical and dimensional parameters (part 2); methods to assess the dimensional and optical properties (part 3) and the methods to evaluate the physicochemical properties (part 4) [85]. Although ISO 18369 covers the main aspects and requirements for contact lenses, it also refers to other ISO standards and scientific papers to complement and support the quality and risk assessment of the lenses.

Besides the two previous mentioned ISO standards, there are other standards that must be followed by the manufacturer such as ISO 14971 and ISO 22979 that address the requirements for risk assessment and the requirements for clinical investigations, respectively [86, 87]. However, this section is focused only on ISO 11979 and 18369, that explicitly describe the processes and requirements that should be evaluated by the manufacturer to guarantee the quality and safety of the lenses.

The next subsections discuss in detail the requirements applied to IOLs and contact lenses that relate to the geometry and optical properties as well as the sterilization and packaging processes.

8.4.2 Geometrical analysis

The dimensional parameters of IOLs and contact lenses have a relevant impact not only on the safety of the patient, but also on the optical performance of the lens.

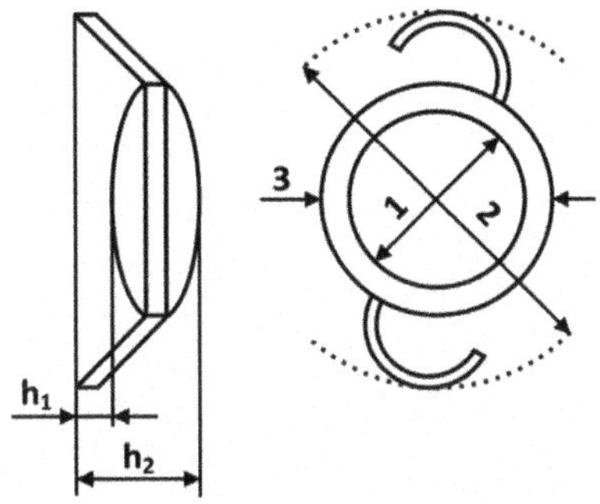

Figure 8.6. ISO dimensional parameters on side and top view: h_1 vault height; h_2 sagittal distance; 1 clear optic; 2 overall diameter and 3 body [45].

Regarding IOLs, the ISO standard 11979-3 establishes the dimensional and mechanical properties of the lens and the processes to make the inspection. According to the standard, the dimensional tolerances of an IOL are defined within three main groups of lenses: anterior chamber lenses, multipiece posterior chamber lenses and other lenses. The range of the tolerances for the geometrical parameters of the IOLs (see figure 8.6) are summarized in table 8.4 (the specific tolerances for the three groups mentioned previously can be found in [45]). Specifically for the clear optic, it is defined as minimum value for the diameter size of 4.25 mm in any meridian [45].

The most common equipment used to evaluate the parameters presented in figure 8.6 is the profile projector (also known as shadowgraph). In this equipment, the object to be measured (in this case the IOL) is placed on a platform and it is illuminated, and then the image of the illuminated object passes through a magnification optical system to be projected onto a viewing screen [88]. Through the viewing screen it is possible to assess the dimensions of the IOL. Currently, it is possible to find on the market not only metrology equipment that is based on manual measurement of the lens dimensions [89], but also automated options that allow the assessment of the dimensions through imaging detection software [90]. Although the profile projector is largely used by IOL manufacturers, there are different and even more accurate methods to assess the dimensions of the lens such as fringe projection, scanning electron microscopy and optical coherence tomography [91–93].

The geometrical requirements of contact lenses are described in ISO 18369, which is also used as reference in the American National Standard ANSI Z80.20 that applies to contact lenses. The dimensional tolerances of paralimbal[1] and corneal contact lenses depend on the material of the lens, which is separated in two main groups: rigid and soft. The rigid group comprises PMMA, RGP and any other non-hydrogel material that has a water content less than 10%, such as silicone elastomer [94]. The soft group comprises two subgroups of hydrogel material: one group of materials with water content greater than or equal to 10% and less than or equal to

[1] Paralimbal contact lenses have a total diameter greater than the visible iris diameter.

Table 8.4. Geometrical tolerances range for intraocular lenses.

Parameter	Range of tolerance for the three groups	Reference
Overall diameter [2]	From ±0.20 mm to ±0.30 mm	[7]
Vault height (h_1)	From ±0.20 mm to ±0.35 mm	[7]
Sagittal distance (h_2)	From ±0.35 mm to ±0.45 mm	[7]
Clear optic [1]	±0.15 mm	[45]
Body [3]	±0.10 mm	[7]

Table 8.5. Geometrical tolerances (mm) for rigid contact lenses [95].

Parameters	PMMA	RGP
Back optic zone radius curve (r_0)	±0.025	±0.050
Back optic zone radii of toric base curve:		
$0.00 < \Delta r \leqslant 0.20$	±0.025	±0.050
$0.20 < \Delta r \leqslant 0.40$	±0.035	±0.060
$0.40 < \Delta r \leqslant 0.60$	±0.055	±0.070
$\Delta r > 0.60$	±0.075	±0.090
Back or front peripheral radius (r_1, r_2, r_{a1}, ...)	±0.10	±0.10
Total diameter (Φ_T)	±0.10	±0.10
Back optic zone diameter (Φ_0)	±0.20	±0.20
Back peripheral zone diameter (Φ_1, Φ_2, ...)	±0.20	±0.20
Center thickness (t_c)	±0.02	±0.02

70% and a second group of material with water content greater than 70% [94]. Scleral lenses tolerances are not separated in material groups as the other models [94]. Table 8.5 summarizes the geometrical tolerances for PMMA and RGP corneal and paralimbal contact lenses. A detailed summary of the tolerances for the other contact lenses such as soft corneal (in the group of hydrogels and non-hydrogels) and scleral lenses can be found in [94]. The Δr term present in the toric base curve tolerance (see table 8.5) corresponds to the difference between the steeper and flatter radii of curvature on toric surfaces [94]. The dimensional parameters associated to the symbols present in table 8.5 are presented in figure 8.7.

The ISO standard 18369-3 stablishes the methods for the verification of not only the dimensional properties, but also of the optical properties of contact lenses [96]. Besides the methods proposed in the standard, such as optical projectors, micro-spherometer, ultrasound and v-gauges, there are other methods commercially available on the market to assess the dimensional properties of contact lenses [94, 97].

8.4.3 Optical analysis

The requirements for the optical inspection of IOLs are established in the ISO standard 11979-2 [83]. The standard establishes not only the tolerances of the

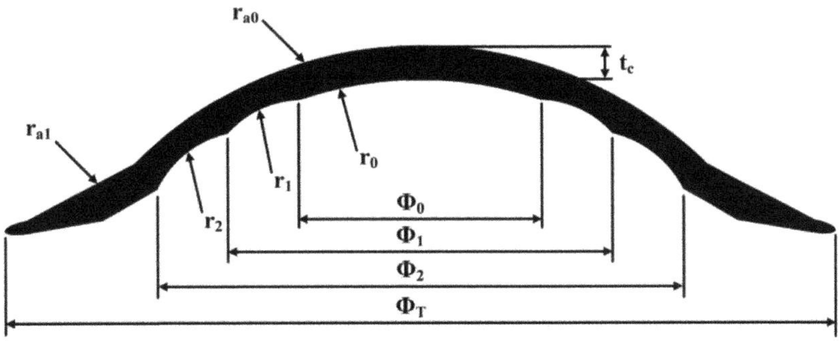

Figure 8.7. Contact lens with tricurve optical zone and the standard symbols correspondence to the lens dimensions.

parameters under inspection, but also the methods to verify each parameter. For the optical analysis of an IOL is required the inspection of three main features of the lens: the optical power, the image quality and the spectral transmittance. The optical power of an IOL is verified according to the lens design. Monofocal IOLs have four different tolerance values defined in the following ranges of the nominal power: from 0 to 15 D it is ±0.3 D; from 15 to 25 D it is ±0.4 D; from 25 to 30 D it is ±0.5 D and, finally, above 30 D it is ±1.0 D [83]. If the lens is toric, besides the tolerances applied to the base power, there are also the tolerances defined for the cylindrical power, which are defined in the following three ranges: from 0 to 2.5 D; from 2.5 to 4.5 D and above 4.5 D [83]. For multifocal IOLs, besides the tolerances applied to the base power, are defined the tolerances for every additional power. The tolerances are defined in three ranges of the additional power: from 0 to 2.5 D; from 2.5 to 4.5 D and above 4.5 D [83]. If the lens is multifocal and toric all the previous tolerances described must apply [83]. For accommodative lenses, the far power must be within the tolerances defined for the monofocal lens and the power change during accommodation can be defined in theoretical or in optical bench eye models [83]. The range of tolerances applied to the base power (P), cylinder (C) and additional (A) power for each lens category is within ±0.3 to ±1.0 D [7]. A detailed list of the optical tolerances applied to each lens type and dioptric range can be found in [83].

The image quality of an IOL is determined, according to the ISO standard, by resolution efficiency or by MTF [83]. The MTF curve is the metric commonly used in the industry and in the scientific community. To evaluate the MTF performance of monofocal IOLs, the MTF should be inspected at a spatial frequency of 100 lp mm^{-1} and the measurement must be done at a pupil aperture diameter of 3 mm [83]. One of two criteria shall apply to the MTF verification of a monofocal lens: the MTF must be higher or equal than 0.43 or it must be greater than or equal to 70% of the diffraction limit of the lens design [83]. The MTF of toric IOLs must be within the previous described limits at each one of the meridians. For multifocal IOLs the MTF curve must be reported at 50 lp mm^{-1} for each one the focus points created [83]. The far focus must be analyzed at 3 and 4.5 mm aperture and all the other additional focuses must be analyzed at 3 mm aperture. The MTF minimum limit for multifocal IOLs should be defined by the manufacturer and it must guarantee, by clinical data, that this minimum requirement corresponds to a visual acuity greater than the 0.3 logMAR threshold [83, 98]. For accommodative IOLs the MTF analysis should be done in increments not greater than 0.5 D for the entire range of accommodation of the lens and the requirements of the monofocal lens should apply to the far focus of the lens [83].

The ISO 11979-2 establishes not only the criteria and tolerances applied to optical power and MTF measurements, but also describes different methods to assess both metrics [83]. For instance, the dioptric power can be determined from the measurement of the magnification of the IOL, from the measurement of the back focal length (BFL), or effective focal length (EFL), or even from the measurement of the IOL dimensions [83]. Regarding the MTF, the measurement method is based on an eye model proposed by the standard [83]. On the market it is possible to find metrology equipment capable of assessing both metrics (dioptric power and MTF). The OptiSpheric® IOL PRO 2 is an optical bench developed by the Trioptics company and it is in compliance with the model eye proposed by ISO standard 11979-2 [99]. The OptiSpheric bench can measure the dioptric power and MTF curves (among other optical metrics) of a wide range of IOL models: monofocal (aspheric, spheric and with controlled aberration), toric (monofocal and multifocal), multifocal (refractive and diffractive) and extended depth of focus (refractive and diffractive) [99]. The Lambda-X, is another well-known company on the market that commercializes two metrology equipment applied to IOL metrology: PMTF and Nimo [100]. The PMTF is similar to the OptiSpheric® IOL PRO 2 equipment in terms of optical metrics capability and range of IOL models. The Nimo equipment has a proprietary measurement method based on phase shifting technique [101, 102] and can be only used in refractive lenses (monofocal, toric and multifocal).

The spectral transmittance of the IOL material is another requirement established by the ISO standard 11979-2. The material transmittance should be evaluated in the range from 300 nm to 1,100 nm at 3 mm aperture and in aqueous medium [83]. The resolution of the analysis should be at least 5 nm and the transmittance accuracy between ±2% [83]. Although the standard does not establish any limits for the transmittance of the material, this is an important parameter to be considered in particular situations where the UV component of ambient light spectra can harm the retina and in diagnostic and treatment procedures that use lasers [83]. The spectral transmittance can be assessed by a spectrophotometer equipment, which is designed to measure the light energy that passes through a certain medium (in this case the IOL material) at each specified wavelength.

The optical parameters required to be inspected in contact lenses are the back vertex power, the cylinder power and cylinder axis (for toric lenses), the additional power (for multifocal lenses), the prismatic power and the spectral transmittance [94]. The back vertex power (F'_v) is the contact lens power measured with a focimeter (also known as lensometer) in air medium [94]. For measurements in saline solution the standard recommends employing a moiré deflectometer or a Hartmann measurement method [94]. The focimeter can also be applied for the measurement of the cylinder power, cylinder axis, additional power and prismatic power [94]. The spectral transmittance of the material can be measured by a spectrophotometer [94]

Table 8.6. Optical tolerances for rigid and soft contact lenses.

Parameters	PMMA	RGP	Reference
$0.00 < \|F'_v\| \leqslant 5.00$ D	±0.12		[95]
$5.00 < \|F'_v\| \leqslant 10.00$ D	±0.18		[95]
$10.00 < \|F'_v\| \leqslant 15.00$ D	±0.25		[95]
$15.00 < \|F'_v\| \leqslant 20.00$ D	±0.37		[95]
$\|F'_v\| > 20.00$ D	±0.50		[95]
$0.00 < \|F'_c\| \leqslant 2.00$ D	±0.25		[95]
$2.00 < \|F'_c\| \leqslant 4.00$ D	±0.37		[95]
$\|F'_c\| > 4.00$ D	±0.50		[94]
Cylinder axis (±°)	±5		[95]
Add (±D)	±0.25		[94]
Prismatic power (±cm m^{-1})			
$\|F'_v\| \leqslant 10$ D	±0.25		[94]
$\|F'_v\| > 10$ D	±0.50		[94]

as described previously for the IOL. Table 8.6 summarizes the tolerances applied to the back vertex power, the cylinder power, cylinder axis, additional power and prismatic power for PMMA and RGP corneal and paralimbal contact lenses. The term F'_c in table 8.6 corresponds to the back vertex power difference of the two main meridians [94]. A detailed list of the tolerances applied to contact lenses can be found in [94].

8.4.4 Microscopic analysis

The surfaces of the lenses manufactured must not only be within the dimensional tolerances and guarantee the optical performance required by the standard, but also must be uniform and free from defects. Regarding IOL surface quality, the ISO standard 11979-2 establishes that the surfaces of the lens must be verified by microscopy analysis under optimum light conditions and at 10× magnification [83].

The cosmetic inspection is intended to detect any deviation on the surface or bulk defects that are not a feature of the optical design such as discoloration, scratches, protrusions, roughness, bubbles, among others [83]. Although some defects might not impact the optical performance of the IOL, because it can be a shallow and small scratch on the optics periphery or it can be located outside the IOL clear optic (haptics for example), such irregularities can impact the post-operative process and the long-term optical performance of the lens [103, 104]. The presence of sharp edges and cracking irregularities on the IOL surface or haptics for example can harm the intraocular tissues and consequently generate or aggravate an inflammatory process in the eye [103, 104]. It is expected that the packaged IOL is free of irregularities and bulk defects. However, the incidence of surface damage during the packaging and/or transportation of the lens should not be discarded and, because of that, it is a safe measure to check the IOL on the operating microscope prior to the lens implantation [104].

For contact lenses the cosmetic inspection must be done using a microscope or a shadowgraph to identify any bulk defects such as surface irregularities and/or material nonhomogeneities [94]. The surface inspection requires a magnification of 7× under controlled illumination and the edge inspection must be done with a magnification of 20× using preferably a binocular microscope [94]. The assessment of the finishing of the surface and edge of the contact lens is a qualitative process that shall meet the requirements specified by the manufacturer [94].

8.4.5 Sterilization process

The sterilization process consists in eliminating all forms of life, mainly microorganisms, that may be present on the product to be sterilized. Regardless of the sterilization process and the period it is applied to the product, it is not possible to guarantee that microorganisms have been completely eliminated. Thus, the effectiveness of the sterilization is defined by the probability of the existence of any microorganisms in the end of the process.

ISO standard 14630 establishes the general requirements for non-active surgical implants, which consequently embraces IOLs [105]. According to this standard, the surgical implants are separated in three main levels (from 1 to 3) and the IOLs is part of level 1, while the other levels are related to a more restricted type of implant designated, for example, for cardiovascular surgery, neurosurgery or joint replacement [105]. There are different processes that can be employed to sterilize an IOL such as ethylene oxide, irradiation, steam, chemical liquid agents, aseptic processing, among others, and all of them have specific standards that addresses the requirements for each method [105]. In general, the sterilization process should be verified and validated by the manufacturer and it shall guarantee a probability of microorganism existence, after the sterilization, of equal to or less than 1×10^{-6}, which corresponds to the probability of 1 in 1 000 000 of finding a non-sterile unit [105]. It is important to verify during the selection of the sterilization method if the material of the IOL has any interaction with the substances used during the manufacturing and packaging process at any step of the sterilization process [106].

ISO standard 14534 defines the fundamental requirements applied to contact lenses and contact lens care products [107]. The standard also refers to the same sterilization processes applied to IOLs as presented previously [107]. According to this standard (ISO 14534), all hydrogel contact lenses must be delivered sterile considering the same sterility assurance level (SAL) of the IOLs, which corresponds to a probability of microorganism existence of less than or equal to 1×10^{-6} [107]. The other models of contact lenses can be delivered non-sterile but the manufacturing and packaging process must guarantee through a shelf-life study that the average bioburden is less than 100 cfu (colony-forming units) per lens [107].

8.4.6 Packaging

The packaging of a final product is the last step to be verified before the product is ready to be shipped to the customers. The inspection at this stage is important to verify if the packaging can guarantee that the sterility of the product is maintained when it is in transit and that there is no damage and/or degradation under environmental circumstances specified by the manufacturer, such as temperature, humidity, and pressure during the transportation [105]. Another relevant aspect of the packaging is related to the labelling information. The information provided on and inside the package of the product should bring clear and standardized information that can help the surgeon (in the case of an IOL) or the user (in case of a contact lens) to identify relevant information about the characteristics of the product and its usability.

Part 4 of ISO standard 11979 describes the labelling and information requirements applied to IOLs [108]. The requirements are applied to the primary package, which is the sterile IOL container, and to the storage container, which is the outside package that contains the previous IOL container, IOL manual and any other item included by the manufacturer such as adhesive labels [108]. On both packages the following information must be present: the name or the trade name of the manufacturer; the trade name and/or the IOL model; the serial or batch number; the word 'STERILE' indicating that the product was sterilized; the dioptric power of the lens; the overall diameter in units of mm and body diameter (see figure 8.6) in mm [108]. Depending on the IOL model (toric or multifocal lenses for example) additional information is required such as spherical equivalent and cylinder power and additional power [108]. Besides the information that must be present on both packages, the standard also defines information requirements that must be only on the storage container [108]. A detailed list of specific symbols and metric units required for the labelling information is also presented in [108].

The labelling requirements stablished for contact lenses are specified in ISO standard 14534. If the contact lens is sterile the word 'STERILE' must be in the label of the packaging [107]. However, if the contact lens is non-sterile, the manufacturer must provide instructions regarding the contra-indications, warnings or any sort of information that can affect the safe use of the product [107]. In contrast to IOLs, contact lenses are commonly replaced from time to time and the period depends on the lens properties such as material and intent of use. Based on that, the manufacturer must also provide information about the replacement intervals [107]. Regarding the disinfecting products applied to contact lenses, the manufacturer of contact lenses shall provide information about the product use and the steps necessary to guarantee the correct cleaning of the lens to promote safety wear for the user [107].

8.5 Summary

This chapter covered elements of the manufacturing processes of IOLs and contact lenses, starting with design parameters that represent input to the manufacturing procedures. Initially, the materials normally used in the manufacturing of IOLs were presented in two groups: rigid and foldable materials. PMMA, hydrophilic, hydrophobic and silicone materials had their properties described and were compared in terms of their advantages and limitations. Materials used to make contact lenses are described as either soft or RGP types.

Besides the material to make a lens, the geometrical parameters that characterize the lens are required as input to the manufacturing. Such parameters include not only those that define the optical surface, such as radius of curvature and asphericity, but also the dimensional parameters of the lens, such as the shape of the haptics of IOLs or the shape of peripheral curves in contact lenses. The required set of parameters to characterize a given lens is dependent on the intended optical performance as well as on the mechanical behavior that the lens is supposed to deliver.

Once the design variables of a lens are defined, the lens can be manufactured through either machining or molding methods. Key elements of both methods have been presented, including typical ranges of values used in process variables, limitations and potential defects that may affect the quality of the finished product.

The key international standards used by industries and regulators in different countries are presented. The main tests and associated methods required to characterize both the mechanical and optical performance of the lenses are described together with typical reference values for some metrics that have to be monitored during the manufacturing process of both IOLs and contact lenses. Lastly, sterilization processes and packaging requirements are briefly introduced.

It is important to highlight that although this text is comprehensive in the approach of manufacturing of IOLs and contact lenses, it covers the topic from a generic perspective. In an industrial context, the manufacturing procedures are designed for a specific product or a narrow range of products and hence defined to a much deeper level of details.

Chapter highlights

- The processes to manufacture IOLs and contact lenses have many common steps and share similarities in required inputs as well as in quality assessment.
- The choice of material is a critical parameter to define the manufacturing process.
- The most commonly used methods for manufacturing IOLs and contact lenses are high-precision machining and molding.
- International standards describe not only which quality and safety parameters have to be assessed, but also what methods can be used for such assessment.

References

[1] Ridley H 1952 Intra-ocular acrylic lenses; a recent development in the surgery of cataract *Br. J. Ophthalmol.* **36** 113–22

[2] Argal S 2013 Newer intraocular lens materials and design *J. Clin. Ophthalmol. Res.* **1** 113

[3] Tehrani M, Dick H B, Wolters B, Pakula T and Wolf E 2004 Material properties of various intraocular lenses in an experimental study *Ophthalmologica* **218** 57–63

[4] Kohnen T 1996 The variety of foldable intraocular lens materials *J. Cataract Refract. Surg.* **22** 1255–8

[5] Scales J T 1953 Tissue reactions to synthetic materials *Proc. R. Soc. Med.* **46** 647–52

[6] Chehade M and Elder M J 1997 Intraocular lens materials and styles: a review *Aust. NZ J. Ophthalmol.* **25** 255–63
[7] Yu N, Fang F, Wu B, Zeng L and Cheng Y 2018 State of the art of intraocular lens manufacturing *Int. J. Adv. Manuf. Technol.* **98** 1103–30
[8] Kretz F, Scholtz S and Auffarth G 2014 A brief history of IOL materials *The Ophthalmologist* https://theophthalmologist.com/subspecialties/a-brief-history-of-iol-materials
[9] Hollick E J, Spalton D J, Ursell P G and Pande M V 1998 Biocompatibility of poly(methyl methacrylate), silicone, and AcrySof intraocular lenses: randomized comparison of the cellular reaction on the anterior lens surface *J. Cataract Refract. Surg.* **24** 361–6
[10] Wolter J 1982 Foreign body giant cells on intraocular lens implants *Graefe's Arch. Clin. Exp. Ophthalmol.* **219** 103–11
[11] Bellucci R 2013 An introduction to intraocular lenses: material, optics, haptics, design and aberration *Cataract* (Basel: S. Karger AG) pp 38–55
[12] Zhou K Y 1983 Silicon intraocular lenses in 50 cataract cases *Chin. Med. J. (Engl.)* **96** 175–6
[13] Werner L 2008 Biocompatibility of intraocular lens materials *Curr. Opin. Ophthalmol.* **19** 41–9
[14] Abela-Formanek C, Amon M, Kahraman G, Schauersberger J and Dunavoelgyi R 2011 Biocompatibility of hydrophilic acrylic, hydrophobic acrylic, and silicone intraocular lenses in eyes with uveitis having cataract surgery: long-term follow-up *J. Cataract Refract. Surg.* **37** 104–12
[15] Cunanan C M, Ghazizadeh M, Buchen S Y and Knight P M 1998 Contact-angle analysis of intraocular lenses *J. Cataract Refract. Surg.* **24** 341–51
[16] Bozukova D, Pagnoulle C and Jérôme C 2013 Biomechanical and optical properties of 2 new hydrophobic platforms for intraocular lenses *J. Cataract Refract. Surg.* **39** 1404–14
[17] Pagnoulle C, Bozukova D, Gobin L, Bertrand V and Gillet-De Pauw M-C 2012 Assessment of new-generation glistening-free hydrophobic acrylic intraocular lens material *J. Cataract Refract. Surg.* **38** 1271–7
[18] Werner L 2010 Aug 1 Glistenings and surface light scattering in intraocular lenses *J. Cataract Refract. Surg.* **36** 1398–420
[19] Panahi-Bazaz M R, Zamani M and Abazar B 2014 Hydrophilic acrylic versus PMMA intraocular lens implantation in pediatric cataract surgery *J. Ophthalm. Vis. Res.* **4** 201–7
[20] Musgrave C S A and Fang F 2019 Contact lens materials: a materials science perspective *Materials (Basel)* **12** 261
[21] Athreya P K and Bhardwaj G K 2018 Contact lens materials and modalities *Trends Ophthalmol.* **1** 1–5
[22] Nichols J J and Starcher L 2020 Contact lenses 2019 *Contact Lens Spectr.* **35** 18–25
[23] Cohen S W 1998 A history of oxygen permeability *Contact Lens Anter. Eye* **21** 35–40
[24] Efron N 2010 Obituary – rigid contact lenses *Contact Lens Anter. Eye* **33** 245–52
[25] Omali N B, Subbaraman L N, Coles-Brennan C, Fadli Z and Jones L W 2015 Biological and clinical implications of lysozyme deposition on soft contact lenses *Optom. Vis. Sci.* **92** 750–7
[26] Buehl W and Findl O 2008 Effect of intraocular lens design on posterior capsule opacification *J. Cataract Refract. Surg.* **34** 1976–85
[27] Zemax. OpticStudio 2020 [Computer software] Retrieved from https://www.zemax.com/pages/opticstudio

[28] Breault Research Org. ASAP 2020 [Computer sofware] Retrieved from https://breault.com/asap/

[29] Lambda Research Corportation 2020 [Computer Software] Retrieved from https://lambdares.com/oslo

[30] Remón L, Siedlecki D, Cabeza-Gil I and Calvo B 2018 Influence of material and haptic design on the mechanical stability of intraocular lenses by means of finite-element modeling *J. Biomed. Opt.* **23** 035003

[31] Prinz A, Neumayer T, Buehl W, Vock L, Menapace R and Findl O *et al* 2011 Rotational stability and posterior capsule opacification of a plate-haptic and an open-loop-haptic intraocular lens *J. Cataract Refract. Surg.* **37** 251–7

[32] Dewey S, Michael Colvard D, Arbisser L B, Jick S L, Gangadhar D V and Ricardo Glikin C G *et al* 2015 Understanding intraocular lenses: the basics of design and material *FocalPoints Clin. Modul. Ophthalmol.* **33** 1–21

[33] Bender L E, Nimsgern C, Jose R, Jayaram H, Spalton D J and Tetz M R *et al* 2004 Effect of 1-piece and 3-piece AcrySof intraocular lenses on the development of posterior capsule opacification after cataract surgery *J. Cataract Refract. Surg.* **30** 786–9

[34] Carl Zeiss Meditec A G CT LUCIA 602 [Internet] 2019 [cited 2020 Sep. 18] Available from https://zeiss.com/meditec/int/product-portfolio/iols/monofocal-iols/3-piece-iols.html

[35] Alcon Laboratories I Acrysof MN60AC [Internet] 2014 [cited 2020 Sep. 4] Available from http://embed.widencdn.net/pdf/plus/alcon/dxgu8ogiib/40–500-049_us_en.pdf?u=4rqn9d

[36] HumanOptics A G ASPIRA 3P-aVA [Internet] [cited 2020 Sep. 4] Available from https://humanoptics.com/en/physicians/intraocular-lenses/monofokal-3p-aspira3p/

[37] Alcon Laboratories Product Information 2013 Acrysof SN6CWS [Internet] [cited 2020 Sep. 4] Available from http://embed.widencdn.net/pdf/plus/alcon/nlmtz9qegy/40-500-096-AD_I_AcrySof_IQ_CERT.pdf?u=4rqn9d

[38] HumanOptics A G ASPIRA-aA [Internet] [cited 2020 Sep. 4] Available from https://humanoptics.com/en/physicians/intraocular-lenses/monofocal-1p-aspira/#d2

[39] Carl Zeiss Meditec A G CT ASPHINA 409M/MP [Internet] 2019 [cited 2020 Sep. 5] Available from https://zeiss.com/meditec/int/product-portfolio/iols/monofocal-iols/ct-asphina-family.html

[40] Bausch+Lomb Incorporated VERSARIO Multifocal MICS [Internet] [cited 2020 Sep. 5] Available from https://bauschsurgical.eu/products/cataract/intraocular-lenses-monofocal-and-premium/versario-multifocal-mics/

[41] Choi M, Lazo M Z, Kang M, Lee J and Joo C K 2018 Effect of number and position of intraocular lens haptics on anterior capsule contraction: a randomized, prospective trial *BMC Ophthalmol.* **18** 78–87

[42] HumanOptics A G ASPIRA-aQA [Internet] [cited 2020 Sep. 5] Available from https://humanoptics.com/en/physicians/intraocular-lenses/monofocal-1p-aspira/#d7

[43] Bausch+Lomb Incorporated Akreos MICS [Internet] [cited 2020 Sep. 5] Available from https://bauschsurgical.eu/products/cataract/intraocular-lenses-monofocal-and-premium/akreos-mics/

[44] Bausch+Lomb Incorporated Quatrix Aspheric Evolutive [Internet] [cited 2020 Sep. 5] Available from https://bauschsurgical.eu/products/cataract/intraocular-lenses-monofocal-and-premium/quatrix-aspheric-evolutive/

[45] International Organization for Standardization – ISO 2012 11979-3 - Ophthalmic implants - intraocular lenses - part 3: mechanical properties and test methods Available from https://www.iso.org/obp/ui/#iso:std:iso:11979:-3:ed-3:v1:en

[46] Systemes D SolidWorks [Internet] 2020 [cited 2020 Sep. 12] Available from https://solidworks.com
[47] Heberle J, Häfner T and Schmidt M 2018 Efficient and damage-free ultrashort pulsed laser cutting of polymer intraocular lens implants *CIRP Ann.* **67** 197–200
[48] Olufayo O A and Abou-El-Hossein K 2015 Evolution of high precision manufacturing of contact lenses *Materials Science Forum* (Wollerau: Trans Tech Publications Ltd) pp374–80
[49] Mamalis A G and Lavrynenko S N 2007 On the precision single-point diamond machining of polymeric materials *J. Mater. Process. Technol.* **181** 203–5
[50] Hatefi S and Abou-El-Hossein K 2020 Review of hybrid methods and advanced technologies for in-process metrology in ultra-high-precision single-point diamond turning *Int. J. Adv. Manuf. Technol.* **111** 427–47
[51] Liman M M and Abou-El-Hossein K 2019 Modeling and multiresponse optimization of cutting parameters in SPDT of a rigid contact lens polymer using RSM and desirability function *Int. J. Adv. Manuf. Technol.* **102** 1443–65
[52] Mishra V, Singh G, Yadav C B, Karar V, Prakash C and Singh S 2019 Precision machining of biopolymers: a brief review of the literature and case study on diamond turning *J. Thermoplast. Compos. Mater* **34** 5577–8
[53] Li L, Wong H C and Lee R B 2020 evaluation of a novel nanodroplet cutting fluid for diamond turning of optical *Polymers (Basel)* **12** 2213
[54] Chen W 2019 Effect of free abrasive on sub-surface damage in rolling friction contact of optical lens *Int. J. Adv. Manuf. Technol.* **100** 1243–51
[55] Lee J W, Hong K P, Cho M W, Kwon S H and Choi H J 2015 Polishing characteristics of optical glass using PMMA-coated carbonyl-iron-based magnetorheological fluid *Smart Mater. Struct.* **24** 065002
[56] Maldonado-Codina C and Efron N 2005 Impact of manufacturing technology and material composition on the surface characteristics of hydrogel contact lenses *Clin. Exp. Optom.* **88** 396–404
[57] Alam F, Elsherif M, AlQattan B, Ali M, Ahmed I M G and Salih A *et al* 2020 Prospects for additive manufacturing in contact lens devices *Adv. Eng. Mater.* **23** 2000941
[58] Chassagne B and Canioni L 2020 Analytical solution of a personalized intraocular lens design for the correction of spherical aberration and coma of a pseudophakic eye *Biomed. Opt. Express* **11** 850
[59] Debellemanière G, Flores M, Montard M, Delbosc B and Saleh M 2016 Three-dimensional printing of optical lenses and ophthalmic surgery: challenges and perspectives *J. Refract. Surg.* **32** 201–4
[60] Sahler R, Bille J F, Enright S, Chhoeung S and Chan K 2016 Creation of a refractive lens within an existing intraocular lens using a femtosecond laser *J. Cataract Refract. Surg.* **42** 1207–15 .
[61] Moreddu R, Vigolo D and Yetisen A K 2019 Contact lens technology: from fundamentals to applications *Adv. Healthc. Mater.* **8** 1900368
[62] Fang F, Zhang N and Zhang X 2016 Jan Precision injection molding of freeform optics *Adv. Opt. Technol.* **5** 303–24
[63] Bozukova D, Bertrand V, Pagnoulle C, De and Pauw-Gillet M-C 2015 Evaluation of a class of polyurethane materials for intraocular lens manufacturing *J. Biomed. Mater. Res. Part B Appl. Biomater.* **103** 1274–86

[64] Patabadige D E W, Jia S, Sibbitts J, Sadeghi J, Sellens K and Culbertson C T 2016 Micro total analysis systems: fundamental advances and applications *Anal. Chem.* **88** 320–38

[65] Pertici G 2017 Introduction to bioresorbable polymers for biomedical applications *Bioresorbable Polymers for Biomedical Applications* 1st edn (Amsterdam: Elsevier BV) pp 3–29

[66] Damodaran V B, Bhatnagar D and Murthy N S 2016 *Biomedical Polymers: Synthesis and Processing* 1st edn (Piscataway, NJ: Springer) p 71

[67] O'Brien F J 2011 Biomaterials and scaffolds for tissue engineering *Mater. Today* **14** 88–95

[68] Pandey S P, Shukla T, Dhote VK K, Mishra D and Maheshwari R 2019 Use of polymers in controlled release of active agents ed R K Tekade *Basic Fundamentals of Drug Delivery.* (New York: Academic) pp 113–72

[69] Vu L T, Chen C-C A, Lee C-C and Yu C-W . 2018 Compensating additional optical power in the central zone of a multifocal contact lens forminimization of the shrinkage error of the shell mold in the injection molding process *Appl. Opt.* **57** 2981

[70] Chen C-C A, Vu L T and Qiu Y-T . 2017 Study on injection molding of shell mold for aspheric contact lens fabrication *Procedia Eng.* **184** 344–9

[71] Kanbur B B, Suping S and Duan F 2020 Design and optimization of conformal cooling channels for injection molding: a review *Int. J. Adv. Manuf. Technol.* **106** 3253–71

[72] Lin Y F, Wu J R, Liu B H, Wei W C J, Wang A B and Luo R C 2017 Improved contact lens injection molding production by 3D printed conformal cooling channels *2017 IEEE/SICE Int. Symp. on System Integration (SII), Taipei* (Piscataway, NJ: IEEE) pp 89–94

[73] Li K, Xu G, Huang X, Chen Q, Xie Z and Gong F 2020 Surface evolution analysis of CrxWyNz coatings on WC mold in glass molding process *Surf. Coat. Technol.* **393** 125839

[74] Koronkevich V P, Korolkov V P, Lenkova G A, Treushnikov V M, Viktorova E A and Iskakov I A *et al* 2007 Phototechnologies for fabrication of bifocal intraocular lenses ed V Panchenko, A Larichev and G Zheltov (Bellingham, WA: International Society for Optics and Photonics) 67340Z

[75] Bensingh R J, Machavaram R, Boopathy S R and Jebaraj C 2019 Injection molding process optimization of a bi-aspheric lens using hybrid artificial neural networks (ANNs) and particle swarm optimization (PSO) *Measurement.* **134** 359–74

[76] Dick L, Risse S and Tünnermann A 2012 Injection molded high precision freeform optics for high volume applications *Adv. Opt. Technol.* **1** 39–50

[77] Lin C-M and Hsieh H-K 2017 Processing optimization of Fresnel lenses manufacturing in the injection molding considering birefringence effect *Microsyst. Technol.* **23** 5689–95

[78] Badur T, Dams C and Hampp N 2018 High refractive index polymers by design *Macromolecules* **51** 4220–8

[79] Chen C-C A and Chang S-W 2008 Shrinkage analysis on convex shell by injection molding *Int. Polym. Process.* **23** 65–71

[80] Chung C-Y 2019 Oct Integrated optimum layout of conformal cooling channels and optimal injection molding process parameters for optical lenses *Appl. Sci.* **9** 4341

[81] Lee Y B and Kwon T H 2001 Apr Modeling and numerical simulation of residual stresses and birefringence in injection molded center-gated disks *J. Mater. Process. Technol.* **111** 214–8

[82] Jialing W and Pengfei W 2005 Jun The simulation and optimization of aspheric plastic lens injection molding *J. Wuhan Univ. Technol. Sci. Ed.* **20** 86–9

[83] International Organization for Standardization – ISO 2014 11979-2 - Ophthalmic implants - intraocular lenses - part 2: optical properties and test methods Available from https://www.iso.org/obp/ui/#iso:std:iso:11979:-2:ed-2:v1:en
[84] International Organization for Standardization – ISO ISO 11979-1:2006(en), Ophthalmic implants—intraocular lenses—part 1: vocabulary [Internet] 2006 [cited 2020 Nov. 30] Available from https://iso.org/obp/ui#iso:std:iso:11979:-1:ed-2:v1:en
[85] International Organization for Standardization – ISO ISO 18369-1:2017(en), Ophthalmic optics—contact lenses—part 1: vocabulary, classification system and recommendations for labelling specifications [Internet] 2017 [cited 2020 Nov. 30] Available from https://iso.org/obp/ui/#iso:std:iso:18369:-1:ed-2:v2:en
[86] International Organization for Standardization – ISO ISO 14971:2019(en), Medical devices—application of risk management to medical devices [Internet] 2019 [cited 2020 Nov. 30] Available from https://iso.org/obp/ui/#iso:std:iso:14971:ed-3:v1:en
[87] International Organization for Standardization – ISO ISO/TR 22979:2017(en), Ophthalmic implants—intraocular lenses—guidance on assessment of the need for clinical investigation of intraocular lens design modifications [Internet] 2017 [cited 2020 Nov. 30] Available from https://iso.org/obp/ui/#iso:std:iso:tr:22979:ed-2:v1:en
[88] John R and Turner R K 1951 Profile projector and optical comparator *US Patent* 2,552,23811
[89] Mitutoyo Profile Projector PJ/PV/PH Series [Internet] 2006 [cited 2020 Dec. 1] 24 Available from https://mitutoyo.co.jp/eng/support/service/catalog/04/E14005.pdf
[90] Optimec Optimec Is830 Auto [Internet] 2020 [cited 2020 Dec. 1] Available from https://optimecsystems.com/solutions/optimec-is830-auto/
[91] Vahey D W, Mills M D, Patterson M R and Mueller E P 1982 Noncontact profiling techniques for intraocular lens edge measurements *Appl. Opt.* **21** 766
[92] Nanavaty M A, Spalton D J, Boyce J, Brain A and Marshall J 2008 Edge profile of commercially available square-edged intraocular lenses *J. Cataract Refract. Surg.* **34** 677–86
[93] Jianping Wei A B 2008 Measurement of lenses and lens moulds using optical coherence tomography *US Patent* 7,416,300 B2
[94] ANSI ANSI – ANSI Z80.20 – For Ophthalmics Contact Lenses—Standard Terminology, Tolerances, Measurements and Physicochemical Properties | Engineering360 2016 109
[95] Hough D A 1998 Contact lens standards *Contact Lens Anter. Eye* **21** S41–5
[96] International Organization for Standardization – ISO ISO – ISO 18369-3:2017—Ophthalmic optics—contact lenses—part 3: measurement methods [Internet] 2017 [cited 2020 Dec. 1] Available from https://iso.org/standard/66340.html
[97] Coldrick B J, Richards C, Sugden K, Wolffsohn J S and Drew T E 2016 Developments in contact lens measurement: a comparative study of industry standard geometric inspection and optical coherence tomography *Contact Lens Anter. Eye* **39** 270–6
[98] International Organization for Standardization – ISO 2018 11979-7 - Ophthalmic implants - intraocular lenses - part 7: clinical investigations of intraocular lenses for the correction of aphakia Available from: https://www.iso.org/obp/ui/#iso:std:iso:11979:-7:ed-4:v1:en
[99] Trioptics OptiSpheric ® – Testing of optomechanical parameters by TRIOPTICS [Internet] 2020 [cited 2020 Dec. 1] Available from https://trioptics.com/products/optispheric-focal-length-and-radius-measurement/
[100] Lambda-X Ophthalmics | Lambda-X [Internet] 2020 [cited 2020 Dec. 1] Available from https://lambda-x.com/en/ophthalmics#intraocular-lens

[101] Joannes L, Dubois F and Legros J-C 2003 Phase-shifting schlieren: high-resolution quantitative schlieren that uses the phase-shifting technique principle *Appl. Opt.* **42** 5046

[102] Joannes L 2004

[103] Kohnen T, Magdowski G and Koch D D 1996 Scanning electron microscopic analysis of foldable acrylic and hydrogel intraocular lenses *J. Cataract Refract. Surg.* **22** 1342–50

[104] Nguyen D Q, Saleh T A, Pandey S K and Bates A K 2006 Irregularities on the surface of single-piece AcrySof SA60AT intraocular lenses *J. Cataract Refract. Surg.* **32** 495–8

[105] International Organization for Standardization – ISO 2012 14630 - Non-active surgical implants - general requirements Available from https://www.iso.org/obp/ui/#iso:std:iso:14630:dis:ed-5:v1:en

[106] Ovchinnikova A V, Kagramanova A V, Gorbunov Y S, Egorova E V, Osipov V B, Solyanina I P, Tal'roze V L, Trofimov V I, Fedorov S N and Shub T A 1986 IIC. Radiation sterilization of intraocular lenses *Med. Tekh.* **5** 141–7

[107] International Organization for Standardization – ISO 2011 14534 - Ophthalmic optics - contact lenses and contact lens care products—fundamental requirements Available from: https://www.iso.org/obp/ui/#iso:std:iso:14534:ed-3:v1:en

[108] International Organization for Standardization – ISO 2008 11979-4 - Ophthalmic implants - intraocular lenses - part 4: labelling and information Available from: https://www.iso.org/obp/ui/#iso:std:iso:11979:-4:ed-2:v1:en

IOP Publishing

Advances in Ophthalmic Optics Technology

Davies William de Lima Monteiro and Bruno Lovaglio Cançado Trindade

Chapter 9

Optimization deployed to lens design

Lucas de Souza Batista and Luiz Melk de Carvalho

The optimization process is one of the most important and essential steps in an optical design. Essentially, it consists of an iterative search for the parameter values of the system, regarding some constraints and specifications to guarantee the best performance of the problem merit function. The performance of the optical system at the end of the optimization process depends not only on the optimization method, but also on the knowledge of the optical designer. The experience of the optical designer is crucial for the selection and configuration of the optimization algorithm parameters such as the starting solution of the design, the constraints parameters, the merit function and stop criteria. The optical designer also has an important task at the end of the optimization process, which is to evaluate the performance of the final design and verify if there is any margin for improvement. In that case, the optimization process starts a new iteration phase. Once the optimization process is concluded, the next phase consists in performing the tolerancing analysis to evaluate the impact of the manufacturing precision on the performance of the final design.

This chapter presents the basic concepts related to optimization and how it applies to a lens design. The chapter is structured in two sections. The first section discusses the basic concepts of optimization, regarding three main topics. The first one states an optimization problem and core definitions related to design vector, design constraints, objective function, and optimal solutions. Secondly, we present the way an optimization problem can be classified based on a set of design features. Finally, principles of deterministic and stochastic optimization strategies are addressed. The second section presents the steps for the optimization of a lens design, which is divided in three main stages. Firstly, we establish a correlation between the parameters of an optical system and the optimization concepts described in the previous section. Then, the second stage presents the construction of the mathematical statement of an optimization problem based on the example of a monofocal intraocular lens (IOL) design and two strategies to apply the optimization algorithms commonly available in ray-tracing software to optimize optical systems. Finally, the third stage shows the steps to re-optimize the system if necessary and how to make the tolerancing analysis of the design.

In summary, this chapter aims at presenting in a broad view the steps necessary for an optical engineer to develop a lens design from its initial constructive and performance specifications to a feasible design compliant to fabrication in large scale. The optimization principles and methods applied to different types of optimization problems that an optical designer might come across are outlined. General aspects related to the lens material, mechanical and optical properties of the lens and manufacturing process are presented in an interconnected manner. This is meant to guide an expert in the field of optical engineering through some elements that must be considered in IOL optimization. It also presents to a more general reader an accessible overview of the meanders of this process to achieve a lens design that can enhance the visual performance of an implanted patient.

9.1 Basics of optimization

Optimization is the act of finding the best result under given circumstances. In the design, construction, operation, and maintenance of any engineering system, practitioners must take many technological and managerial decisions at several stages. The goal of all such decisions is either to minimize the cost or to maximize the desired benefit. Since cost or benefit in any practical situation can be expressed as a function of certain decision variables, optimization can be defined as the process of finding the conditions that yield the minimum or maximum value of a function [1–6].

In this section, an optimization problem is stated, together with some related concepts. A discussion regarding the classification of optimization problems is also addressed, considering different properties. Finally, classical and modern methods for the optimization of nonlinear programming problems are outlined. Herein, a programming problem can be understood as a mathematical representation at programming or planning the best possible solution for a problem.

9.1.1 Statement of an optimization problem

In particular, considering the maximum of a mathematical function can be found by searching for the minimum of its negative counterpart, herein all the definitions will arbitrarily assume a minimization optimization problem. In this sense, a general optimization or single-objective mathematical programming problem can be stated as follows [1–6]:

$$\text{Find } x = (x_1, x_2, \ldots, x_n) \text{ which minimizes } f(x) \tag{9.1}$$

subject to the constraints

$$g_i(x) \leqslant 0, \quad i = 1, 2, \ldots, p$$

$$h_j(x) = 0, \quad j = 1, 2, \ldots, q$$

where x is an n-dimensional vector called the design vector, $f(x)$ is the objective function, and $g_i(x)$ and $h_j(x)$ are known as inequality and equality constraints, respectively. These constraint functions define the feasible search space Ω, such that $\Omega = \{x | g_i(x) \leqslant 0 \, \forall i \text{ and } h_j(x) = 0 \, \forall j\}$. Figure 9.1 illustrates some of these concepts.

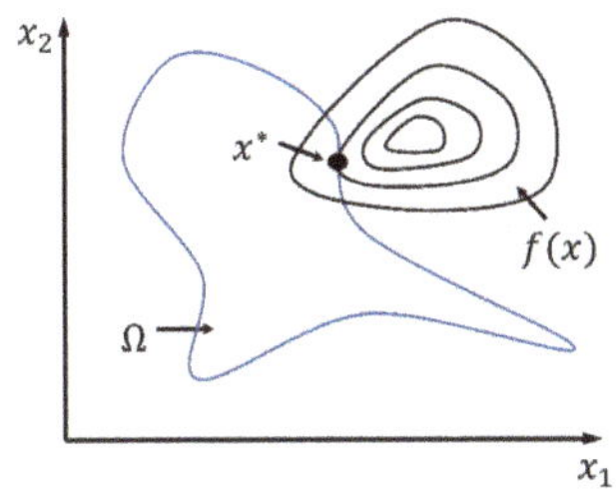

Figure 9.1. Illustration of a two-dimensional single-objective optimization problem, in which Ω indicates the feasible search space (composed of vectors inside and on the surface of Ω), $f(x)$ a minimization function represented by contour lines, and $\boldsymbol{x}^*$ the optimum solution vector, comprising optimal values for x_1 and x_2.

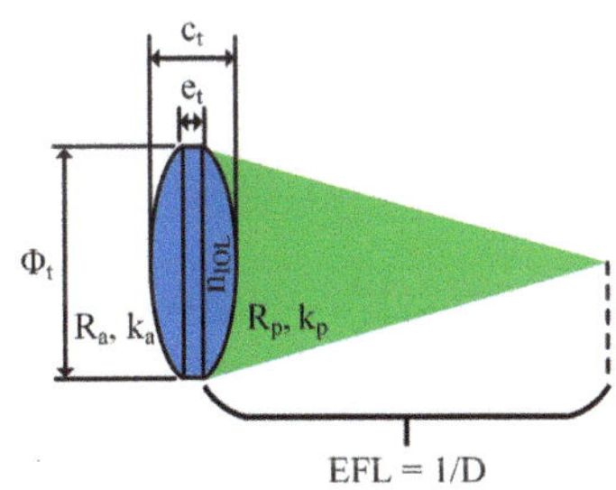

Figure 9.2. Lens design parameters of a monofocal IOL: Φ_t—clear optic diameter, e_t—edge thickness, c_t—center thickness, R_a—anterior radius of curvature, k_a—conic constant of the anterior surface, R_p—posterior radius of curvature, k_p—conic constant of the posterior surface, n_{IOL}—lens refractive index, EFL—effective focal length.

9.1.1.1 Design vector

Any engineering system is defined by a set of quantities some of which are viewed as variables during the design process [1–6]. Some of these quantities are usually fixed beforehand and are called preassigned parameters. The remaining ones are considered as variables in the optimization problem and are called design or decision variables x_i, $i = 1, 2, \ldots, n$. Each vector x in the n-dimensional design space is called a design solution and represents either a feasible (if $x \in \Omega$) or an infeasible solution (if $x \notin \Omega$) to the design problem.

As an example, suppose the lens design of a monofocal IOL intended to provide a dioptric power of 20 D (reciprocal of the focal length in meters) and the best image quality possible at the focal plane (aberration minimization or modulation transfer function (MTF) maximization). As illustrated in figure 9.2, according to the designer preferences, both the material refractive index (n_{IOL}) and the lens optical diameter (Φ_t) can be defined *a priori* as preassigned parameters. On the other hand, the edge thickness (e_t) and anterior and posterior surface parameters (radii of curvature and conic constants) can be selected as decision variables.

9.1.1.2 Design constraints

In many practical problems, the design variables cannot be defined arbitrarily. Rather, they must satisfy certain specified requirements [1–6]. The restrictions that must be satisfied to produce an acceptable design are called design constraints. In this sense, constraints that represent limitations on the behavior or performance of the system are termed behavior or functional constraints. On the other hand, constraints that represent limitations on design variables, such as material availability, manufacturability, and clinical compliance, are known as side constraints. Regarding the previous example of lens design, the dioptric power can be considered as a functional constraint and the center thickness can be considered as a side constraint.

9.1.1.3 Objective function

In general, the conventional way of designing aims at finding an acceptable or adequate design that satisfies the functional and other requirements of a problem. However, since there will probably be more than one feasible design, optimization tools can be used to find the best one of a set of acceptable designs. In this sense, a criterion must be defined in order to enable the comparison of different alternatives and the selection of the most suitable one. This criterion with respect to which the design is optimized, when expressed as a function of the design variables, is known as the merit or objective function [1–6].

Essentially, the choice of an objective function is governed by the nature of the problem. For instance, power loss would be an interesting objective function to be minimized in electrical systems. Additional examples may include the minimization of weight in aircraft and aerospace design problems; the maximization of stability in control systems; the maximization of reliability in engineering systems design; among others. Concerning the lens design example, optical aberrations can be considered an objective function to be minimized.

Thus, the choice of an objective function appears to be straightforward in most design problems. However, there may be situations where the optimization of a specific criterion may lead to results that may not be satisfactory with respect to another, sometimes concurrent, criterion [1–6]. For instance, an electrical machine transmitting the maximum power may not have the minimum energy consumption. Similarly, in a monofocal lens design, the maximization of the MTF of the system at the focal plane may not lead to the maximization of the depth of focus. In this way, the selection of the objective function can be one of the most important decisions in the whole optimization design process.

9.1.2 Statement of a multiobjective optimization problem

In some situations, multiple and often conflicting criteria should be optimized simultaneously. An optimization problem involving multiple objective functions is known as a multiobjective programming problem and can be stated as follows [7–9]:

$$\text{Find } x = (x_1, x_2, \ldots, x_n) \text{ which minimizes } F(x) = \{f_1(x), f_2(x), \ldots, f_m(x)\} \quad (9.2)$$

subject to

$$g_i(x) \leqslant 0, \quad i = 1, 2, \ldots, p$$

$$h_j(x) = 0, \quad j = 1, 2, \ldots, q$$

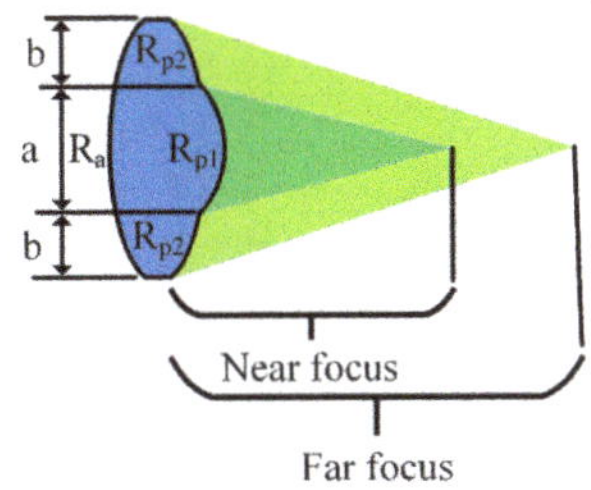

Figure 9.3. Lens design parameters of a bifocal refractive IOL.

where F represents the m-dimensional vector objective function and $f_1, f_2, \ldots, f_m$ denote the scalar objective functions to be minimized simultaneously.

For instance, suppose the multiobjective optimization of a refractive bifocal lens design as illustrated in figure 9.3. In this example, it should be interesting to maximize the MTF curve at both foci (far and near focus), considering as design variables the radius of curvature of the anterior surface (R_a), the radii of curvature of both segments of the posterior surface (R_{p1} and R_{p2}) and the aperture size of each segment (a and b). This optimization problem presents a trade-off on the optical performance of both foci. Thus, to guarantee a good MTF performance at the near focus for example, a contrast improvement is required by reducing the impact of the background light coming from the far focus (which can be achieved by increasing the aperture region a) and, hence, the far focus will have a poor MTF performance.

In general, assuming the existence of conflict among the objective functions, no solution vector x exists that minimizes all the m objective functions simultaneously [7–9]. Hence, a useful concept, known as Pareto dominance [7, 8], is used in multiobjective optimization in order to define a set with the most suitable solutions for the problem, where each alternative design represents a specific trade-off regarding all the criteria considered [7–9]. Based on the Pareto dominance concept, it is said that a feasible solution $x' = (x_1', x_2', \ldots x_n') \in \Omega$ Pareto dominates another feasible solution $x'' = (x_1'', x_2'', \ldots, x_n'') \in \Omega$ if, and only if, $f_i(x') \le f_i(x'')$ for all $i = 1, 2, \ldots, m$ and $f_i(x') < f_i(x'')$ for at least one i. For simplicity, this dominance is usually expressed as the vector relation $F(x') \preccurlyeq F(x'')$. Furthermore, a feasible solution $x^* \in \Omega$ is called Pareto optimal if $F(x^*) \preccurlyeq F(x)$ for any other feasible solution ($x \in \Omega$). In other words, a feasible vector x^* is called Pareto optimal if there is no other feasible solution x that would reduce some objective function without causing a simultaneous increase in at least one other objective function. Since different optimal trad-eoff solutions can be defined for a multiobjective problem, the Pareto optimal set is stated as $PS = \{x^* \in \Omega \mid \nexists x \in \Omega: F(x) \preccurlyeq F(x^*)\}$. The image of these solutions into the objective space represents the Pareto optimal front $PF = \{F(x^*) | x^* \in PS\}$. Notice that PS is stated in the design space, while PF is defined in the objective domain. These concepts are illustrated in figure 9.4.

An example in the ophthalmology field that can be used to describe the concept of Pareto dominance concerns the optimization of a diffractive bifocal IOL problem. For this case, it is expected to optimize the amount of income light energy that is going to be split onto the two main foci (which represent the objective functions): far

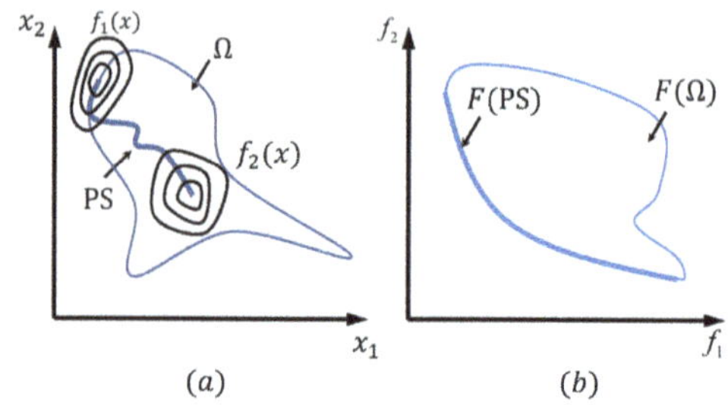

Figure 9.4. Illustration of the decision and objective domains of a multiobjective optimization problem. In (a), a two-dimensional decision space is considered, highlighting the contours of the objective functions ($f_1(x)$ and $f_2(x)$), the feasible space Ω (composed of vectors inside and on the surface of Ω), and the Pareto optimal set *PS* (composed of vectors that belong to the *PS* surface). In (b) is shown the image of the feasible space ($F(\Omega)$) and the Pareto optimal front ($PF = F(PS)$) regarding a two-objective minimization problem. Note that any solution inside the region $F(\Omega)$ is feasible, but only the solutions at the bold surface $F(PS)$ are efficient, i.e., non-dominated with respect to the entire feasible domain.

and near. The two objective functions represent a trade-off because in case more light is directed to the far focus, consequently less light will be directed to the near focus. Based on that, it is possible to have a range of efficient non-dominated solutions given by the variety of combinations of light distribution between the far and near focus, which will define the Pareto front of the optimization problem. By comparing each solution on the Pareto front against each other, it is not possible to find a solution that has a higher light energy on both foci. In this situation, an adequate trade-off solution should be selected by the decision-maker regarding some design specifications.

9.1.3 Classification of optimization problems

Optimization problems can be classified based on some different properties. For instance, as has been previously discussed, depending on the number of objective functions, optimization problems can be classified as single- and multiobjective programming problems. Additionally, any optimization problem can be classified as constrained or unconstrained, depending on whether functional constraints exist in the problem. Since the knowledge of these characteristics is extremely useful for the selection of an appropriate optimization strategy, some other features are pointed out hereafter. A comprehensive discussion regarding the classification of optimization problems can be found in [1–6].

9.1.3.1 Classification based on the nature of the design variables

Optimization problems can be classified into two broad categories based on the nature of the design variables [1–6]. In the first one, it is desired to find values to a set of design parameters that optimize some predefined function (of these parameters) subject to certain constraints. The lens design example illustrates this type of problem. In general, such problems are called *parameter* or *static optimization problems*, since once a suitable solution is implemented, it remains unchanged. In the second category of problems, it is aimed to find a set of design parameters, which are all continuous functions of some other parameter, that optimize an objective function subject to a set of constraints. For instance, there are problems in which the design variables are functions of the parameter time. As an example, the amount of energy delivered by a utility to households depends on how the demand varies throughout the day. Such problems, where each design variable is a function of one or more parameters, are known as *trajectory* or *dynamic optimization problems*, since the specifications of the optimal solution may vary over time.

9.1.3.2 Classification based on the nature of the equations

Another important classification of optimization problems is related to the nature of equations used to model the objective function and constraints. Fundamentally, optimization problems can be classified as linear, nonlinear, geometric, and quadratic programming problems [1–6]. This classification is extremely useful from the computational point of view because there are many specific methods available for the efficient solution of each class of problem. In this way, one of the main tasks of a designer would be to investigate the class of the goal problem. This may often dictate the types of solution strategies to be adopted in solving the problem.

A *nonlinear programming* (NLP) *problem* is the most general programming problem and all other problems can be regarded as its special cases. Essentially, if any of the equations among the objective and constraint functions in equation (9.1) is nonlinear, the problem is called an NLP problem. This is exactly the case of the problems addressed in this chapter. A *quadratic programming* (QP) *problem* is an NLP problem with a quadratic objective function and linear constraints. If the objective function and all the constraints in equation (9.1) are linear functions of the design variables, the mathematical programming problem is called a *linear programming* (LP) *problem*. Finally, a *geometric programming* (GMP) *problem* is one in which the objective function and constraints are expressed as posynomials in x. A posynomial is a function where all the design variables and coefficients are positive real numbers and the exponents are real numbers. A detailed description of these classes of problems can be found in [1–6].

In the case of NLP problems, it is important to highlight that some of them are stated based on multimodal functions, i.e., they have multiple optimum solutions, of which many are local optimal solutions. A multimodal optimization problem usually causes difficulty to any optimization algorithm because there are many attraction basins in which an algorithm can get stuck. Finding the global optimum solution (or even a good local optimum) in such problems becomes a challenge to any optimization algorithm, mainly for classical approaches, which are often monotonically convergent and based on deterministic procedures. The multimodality concept can be understood by means of the popular *Rastrigin* function defined in equation (9.3) and illustrated in figure 9.5, regarding a two-dimensional problem (i.e., for $n = 2$), in which $-2 \leqslant x_i \leqslant 2$, $i = 1, 2$.

$$f(x) = \sum_{i=1}^{n}(x_i^2 - 10.\cos(2\pi x_i)) \tag{9.3}$$

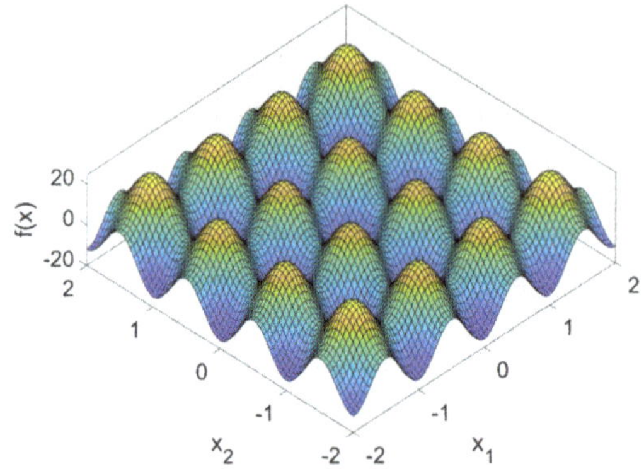

Figure 9.5. Illustration of multimodality by means of the Rastrigin function.

9.1.3.3 Classification based on the possible values of the design variables

Optimization problems can be classified as integer and real-valued programming problems depending on the values allowed for the design variables [1–6]. Essentially, if some or all the decision variables $x_1, x_2, \ldots, x_n$ of an optimization problem are restricted only to integer (or discrete) values, the problem is called an *integer programming problem*. On the other hand, if all the decision variables are allowed to assume any real value, the optimization problem is called a *real-valued programming problem*. As will be discussed afterwards in this chapter, the lens design problem represents a real-valued programming problem.

9.1.3.4 Classification based on the deterministic nature of the variables

Optimization problems can be classified as deterministic and stochastic programming problems [1–6]. Fundamentally, a *stochastic programming problem* is an optimization problem in which some or all of the parameters (preassigned parameters and, or, design variables) follow a probability distribution, i.e., they are probabilistic (nondeterministic or stochastic). Since stochastic models possess some inherent randomness, the same set of parameter values will probabilistically lead to an ensemble of different outputs. Robust optimization approaches are usually applied in this context. On the other hand, a *deterministic programming problem* is formulated with known parameters and the output is fully determined by the parameter values defined. According to this definition, the lens design problem considered in this chapter represents a *deterministic programming problem*.

9.1.4 Optimization techniques

Fundamentally, if the expressions for the objective function and constraints are fairly simple in terms of the design variables, the classical analytical methods of optimization (i.e., differential calculus-based methods) can be used to solve the problem. On the other hand, if the optimization problem involves the objective function and, or, constraints that are not stated as explicit functions of the design variables or which are too complicated to manipulate, it is not possible to solve it by using analytical approaches. Many engineering design problems hold this feature and, in such cases, numerical methods of optimization are needed [1–6]. The basic philosophy of these approaches is to produce a sequence of improved approximations to the optimum, wherein an estimate solution is sought by proceeding in an iterative way by starting from an initial solution.

Since this chapter deals with nonlinear programming problems, only nonlinear numerical approaches are addressed in the following subsections.

9.1.4.1 Classical unconstrained optimization methods

Rarely would a practical design problem be unconstrained, however, some of the powerful and robust methods of solving constrained optimization problems require the use of unconstrained optimization techniques. In general, several methods are available for solving an unconstrained optimization problem and these approaches can be classified into two broad categories as *direct search methods* and *descent methods* [1–6].

The direct search methods require only the objective function values but not the partial derivatives of the function, and hence are often called *non-gradient methods.* Since derivative information is not used, the direct search methods can be applied to many problems without a major change in the algorithm. However, for the same reason, they are usually slow and require many function evaluations for convergence. The descent techniques require, in addition to the function values, the first and in some cases the second derivatives of the objective function. Since more information about the function being optimized is used, descent methods (also known as *gradient methods*) quickly converge near an optimal solution but are not efficient in non-differentiable or discontinuous problems. Furthermore, there are two main difficulties with most classical direct and gradient-based techniques [1–6]:

1. The convergence to an optimal solution depends on the chosen initial solution.
2. Most algorithms tend to get stuck to a local optimal solution.

Since nonlinearities and complex interactions among problem variables often exist in real-world optimization problems, the search space usually contains more than one optimal solution, of which most are undesired locally optimal solutions having inferior objective function values. Unfortunately, given an initial solution, when classical methods get attracted to any of these locally optimal solutions, there is no escape [1–6]. This difficulty is illustrated in figure 9.6, considering the popular *Peaks* function stated in equation (9.4), in which $-3 \leqslant x_i \leqslant 3$, $i = 1, 2$.

$$f(x) = 3(1 - x_1)^2 \; e^{-x_1^2-(x_2+1)^2} - 10\left(\frac{x_1}{5} - x_1^3 - x_2^5\right)e^{-x_1^2-x_2^2} - \frac{1}{3} \; e^{-(x_1+1)^2 - x_2^2} \tag{9.4}$$

The general approach of these methods can be understood in the following way [1–6]. Most classical algorithms use a deterministic procedure to approach the optimum solution. At first, such algorithms start from an initial trial solution, usually suggested by the designer. Thereafter, based on a pre-specified transition rule, the algorithm suggests a search direction, which is often arrived at by considering local information. A unidirectional search is then performed along the search direction to find the local best solution (which belongs to the same attraction basin of the initial solution). This best solution becomes the new current solution, and the procedure above is repeated several times. It is

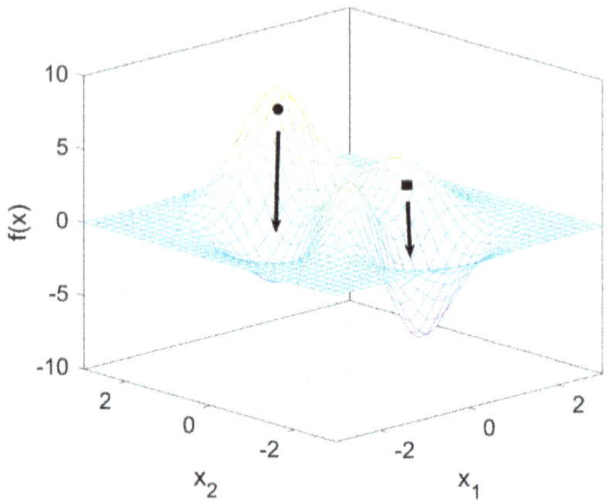

Figure 9.6. Illustration of the effect of the starting point on the convergence of a classical optimization algorithm. For instance, if the initial solution is given by the square point, the global optimum solution can be achieved; however, if it is given by the cycle point, only a local optimum may be found.

important to note that these classical algorithms differ from one another mostly in the way the search direction is defined at each intermediate solution; additionally, they require an initial point to start the iterative procedure, which strongly influences the quality of the final solution.

An interesting example of a gradient-based algorithm is the damped least-square (DLS) method [10], which represents the most widespread optimization tool available in commercial ray-tracing software [11].

9.1.4.2 Classical constrained optimization methods

There are many techniques available for the solution of a constrained nonlinear programming problem and these approaches can be classified into *direct methods* and *indirect methods* [1–6]. In the direct methods, the constraints are handled in an explicit way and some examples include the sequential linear programming (SLP) and sequential quadratic programming (SQP) approaches [1–3]. On the other hand, in most of the indirect methods, a constrained problem is solved as a sequence of unconstrained optimization problems. Undoubtedly, the indirect methods are the most commonly used due to their interesting trade-off between simplicity and performance. Additionally, algorithms for unconstrained optimization of rather arbitrary functions are well studied and generally quite reliable.

The most popular indirect methods are based on the penalty function approach [1–6], which is used to transform a constrained optimization problem into an alternative unconstrained formulation. In this approach, assuming the basic optimization problem as given in equation (9.1), this one is converted into an unconstrained optimization problem by constructing a penalty function of the form [1–6]:

$$\phi(x, r) = f(x) + r\left(\sum\nolimits_{i=1}^{p} G_i[g_i(x)] + \sum\nolimits_{j=1}^{q} H_j\left[h_j(x)\right]\right) \qquad (9.5)$$

where G_i and H_j are some functions of the constraints g_i and h_j, respectively, r is a positive constant known as the penalty parameter, and the second term on the right side of equation (9.5) is called the penalty term, which indicates the values of violations of inequality and equality constraints. Essentially, while minimizing the pseudo-objective function $\phi(x,r)$, a penalty is added whenever a constraint is violated, such that the solution of ϕ approximates that of the original problem stated in equation (9.1). The value of the constant r can be adjusted during optimization to change the contribution of the penalty term relative to the magnitude of the objective function $f(x)$. A detailed discussion of constrained optimization approaches can be found in [1–9].

9.1.4.3 Scalar approaches for multiobjective optimization

Several vector and scalar strategies have been developed to solve a multiobjective optimization problem [7–9, 12, 13]. In the former, an optimization algorithm indeed deals with the vector formulation of the problem, while in the latter a scalar approach is used to convert the original problem into a single-objective optimization problem. Basically, most of these strategies estimate a set of Pareto optimal solutions and apply some additional criterion or rule for decision making, so as to select an adequate trade-off design as the final solution of the goal problem [7–9, 12, 13]. In this chapter, only the two most popular scalar strategies are addressed, which are known as weighted sum and ϵ-constraint [7, 8].

The weighted sum method scalarizes a set of objectives into a single merit function by multiplying each objective with a user-defined weight. This approach basically converts the original problem in equation (9.2) to a single-objective optimization problem as follows [7, 8]:

$$\text{Find } x = (x_1, x_2, \dots, x_n) \text{ which minimizes } u(x) = \sum\nolimits_{k=1}^{m} w_k f_k(x) \tag{9.6}$$

subject to

$$g_i(x) \leqslant 0, \quad i = 1, 2, \dots, p$$

$$h_j(x) = 0, \quad j = 1, 2, \dots, q$$

This method is the simplest approach and is probably the most widely used among the scalar strategies. The weight w_k of the k-th objective is usually chosen in proportion to the objective's relative importance in the problem and it is the usual practice to choose weights such that their sum is one, i.e., $\sum_{k=1}^{m} w_k = 1$. Additionally, setting up an appropriate weight vector also depends on the scaling of each objective function. When such objectives are weighted to form a composite objective function, it would be important to scale them appropriately so that each has approximately the same order of magnitude.

Figure 9.7(a) illustrates how the weighted sum approach can find Pareto optimal solutions of the original problem. The composite function u can be stated once the weight vector is defined. Since u is a linear combination of both objectives f_1 and f_2, a straight line is expected as the contour line of u on the objective space, and its slope is related to the choice of the weight vector. The contour surfaces of u can be visualized in the objective space, as shown by lines $\lambda_1, \lambda_2, \dots, \lambda^*$. Since the problem

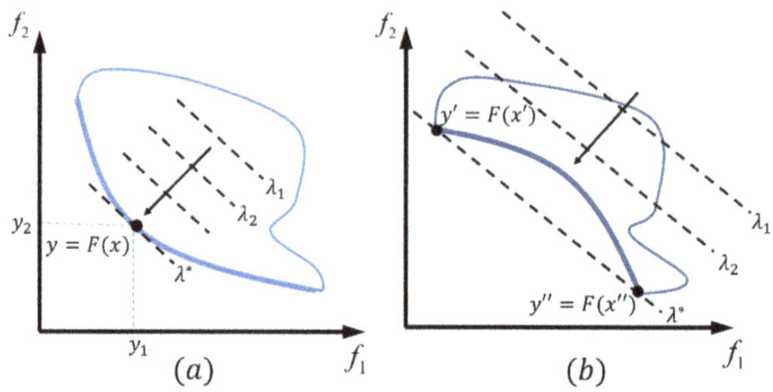

Figure 9.7. Illustration of the weighted sum approach. In (a), a sequence of contour lines λ_1, λ_2, … , λ^*, defined by a specific weight vector, are evolved towards a specific Pareto optimal solution y. Since the bold surface represents a convex Pareto front, all the Pareto solutions can be reached by the weighted sum approach. In (b), the Pareto front is nonconvex and, except for the two extreme optimal solutions (y' and y''), the trade-off surface marked with a bold line cannot be reached by this method.

stated in equation (9.6) requires minimization of u, the task is to find the contour line with the minimum u value. This happens with the contour line which is tangential to the objective space and lies in the bottom-left corner of this space. The tangent point y is the minimum solution of u and is consequently the Pareto optimal solution corresponding to the weight vector.

When a different weight vector is used, the slope of the contour line would be different and the previous procedure, in general, would result in a different optimal solution. This property is observed for multiobjective problems with a convex Pareto front, such as the one illustrated by the bold curve in figure 9.7(a). In this way, for a convex Pareto front, multiple Pareto optimal solutions can be found by solving the problem in equation (9.6) with multiple weight vectors, one at a time, each time finding a different Pareto optimal solution [7–9].

It is worth noting, however, that a uniformly distributed set of weight vectors does not ensure the achievement of a uniformly distributed set of Pareto optimal solutions. Since this mapping is not usually known, it becomes difficult to set the weight vector to obtain a Pareto optimal solution in a desired region of the objective domain. Furthermore, the weighted sum approach cannot find certain Pareto optimal solutions in the case of a nonconvex Pareto optimal front [7–9]. In this situation, there will be Pareto optimal solutions that will not be reached by any weight vector. For example, only the extreme solutions (y' and y'') of the nonconvex Pareto front (bold surface) indicated in figure 9.7(b) can be accomplished by the weighted sum strategy.

In order to overcome the difficulties faced by the weighted sum approach, the ϵ-constraint method can be used. In this strategy, the original problem in equation (9.2) is reformulated by just keeping one of the objectives and restricting the remaining ones within user-specified values. The modified problem is stated as follows [7, 8]:

$$\text{Find } x = (x_1,\ x_2,\ \ldots,\ x_n) \text{ which minimizes } f_\mu(x) \tag{9.7}$$

subject to

$$f_k(x) \leqslant \epsilon_k, \ \ k = 1, 2, \dots, m \text{ and } k \neq \mu$$

$$g_i(x) \leqslant 0, \quad i = 1, 2, \dots, p$$

$$h_j(x) = 0, \quad j = 1, 2, \dots, q$$

in which ϵ_k represents an upper bound of the value of f_k. Figure 9.8 illustrates the working principle of the ϵ-constraint approach, in which f_1 is kept as an objective and f_2 is constrained such that $f_2 \leqslant \epsilon$. The resulting problem with this constraint divides the original feasible objective space into two portions, i.e., a feasible one for $f_2 \leqslant \epsilon$ and an infeasible for $f_2 > \epsilon$. In such manner, the Pareto optimal solution that minimizes f_1 subject to $f_2 \leqslant \epsilon$ is given by y.

In this approach, a vector of ϵ values representing, in some sense, the location of the Pareto optimal solution is needed, and the ϵ vector must be chosen so that it lies within the minimum and maximum values of the individual objective function. Furthermore, this method can be used for any arbitrary problem with either convex or nonconvex Pareto front, as indicated in figure 9.8. Essentially, a solution of the ϵ-constraint problem stated in equation (9.7) is Pareto optimal for any given upper bound vector $\epsilon = (\epsilon_1, \epsilon_2, \dots, \epsilon_{\mu-1}, \epsilon_{\mu+1}, \dots, \epsilon_m)$. Additionally, different Pareto optimal solutions can be found by using different ϵ_k values [7–9].

9.1.4.4 Modern methods of optimization

In the last decades, some optimization methods that are conceptually different from the traditional mathematical programming techniques have been developed [14, 15]. These methods are termed as modern or nontraditional strategies of optimization and most of them are either metaphor-based metaheuristics or swarm intelligence algorithms. These techniques can be easily adapted for different classes of problems, and some examples include genetic algorithms (GAs), differential evolution (DE), particle swarm optimization (PSO), evolution strategies (ESs), among others [14, 15]. Since GAs are extremely popular and have been used in the last decades on the optimization of complex optical

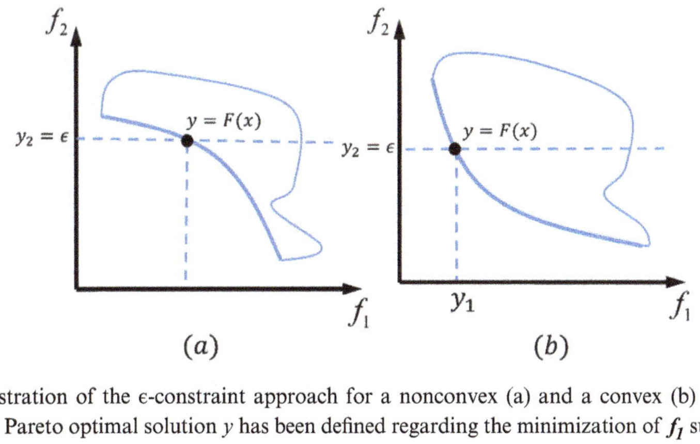

Figure 9.8. Illustration of the ϵ-constraint approach for a nonconvex (a) and a convex (b) Pareto front, in which a specific Pareto optimal solution y has been defined regarding the minimization of f_1 subject to $f_2 \leqslant \epsilon$.

systems [16], only this approach is addressed hereafter. The GAs were first presented systematically by J H Holland and the basic ideas of analysis and design, based on the concepts of biological evolution, can be found in [17].

In general, the development of these strategies has been motivated to deal with increasing challenges of real-world problems [14, 15]. For instance, many practical design problems are characterized by mixed continuous-discrete variables, and discontinuous and nonconvex design domains. It is worth noting that if standard nonlinear programming techniques are used for this type of problem, they will be inefficient, computationally expensive, and, in most cases, find a local optimum that is closest to the starting point (i.e., a local optimum inside the same attraction basin of the starting point). On the other hand, GAs (and some other modern methods) are well suited for solving such problems and, in most cases, they can find the global optimum solution with a high probability.

GAs have been extensively used as search and optimization tools in various problem domains, including the sciences, commerce, and engineering [14, 15]. Among the reasons for their success are their broad applicability, ease of use and global perspective [14, 15, 18].

GAs are based on the principles of natural genetics and their basic elements (i.e., selection, crossover, and mutation operators) are used in the evolutionary search procedure. The main features that distinguish GAs from the traditional methods of optimization are pointed out in the following [14, 15, 18]:

1. A population of candidate solutions (trial design vectors) is used to start the procedure instead of a single design point. Since several points are used as candidate solutions, GAs are less likely to get trapped at an undesired local optimum.
2. GAs use only the values of the objective function and constraints (i.e., they are non-gradient methods and the derivatives are not required in the search procedure).
3. Due to previous features, GAs can handle problems characterized by mixed continuous-discrete variables, discontinuity and nonconvexity.
4. Due to previous features, constraints can be handled in a much better way than in classical search and optimization algorithms, since there are a lot of flexible non-gradient-based operators and strategies to deal with this task.
5. The objective function value related to a design vector plays the role of the fitness, which is used within the selection operator to indicate the quality of this candidate solution.
6. In every new generation, a new set of candidate solutions is produced by using parents' selection and crossover/mutation operations from the old generation.
7. Although stochastic, GAs are not simple random search techniques, but, in turn, they efficiently explore the new combinations with the available knowledge (current candidate solutions) to find a new generation with better objective function value or fitness.

Figure 9.9 shows a flowchart of the working principle of a GA. Unlike classical search and optimization methods, a GA begins its search with a random (or

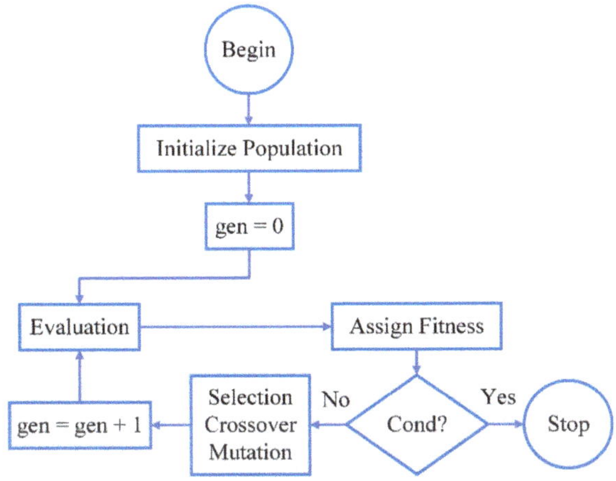

Figure 9.9. A flowchart of the working principle of a GA.

predefined) set of solutions, instead of just one candidate alternative. Once a population of solutions is created, each alternative is evaluated in the context of the underlying NLP problem and a fitness is assigned to each solution. The evaluation of a solution means calculating the objective function value and constraint violations. Thereafter, a metric must be defined by using the objective function value and constraint violations to assign a relative merit to the solution (called the fitness). A termination condition is then checked and if the termination criterion is not satisfied, the population of solutions is modified by three main operators and a new (and hopefully better) population is created. The generation counter is incremented to indicate that one generation (or iteration) of the GA is completed. As can be seen, the flowchart shows that the working principle of a GA is simple and straightforward. Some reference books about metaphor-based meta-heuristics and swarm intelligence algorithms are included in [12–15].

9.2 Optimization of lens design

The optimization of a lens design is an iterative process to guarantee that the entire optical system is within the specifications and requirements established during the early stages of a project [19, 20]. There are several methods for the optimization of a lens design, but, due to the complexity of a lens design problem, there is no singular approach that would surely converge to the best possible solution without any previous analysis of the system [19–21]. The complexity of an optical design optimization comes from its merit function, which is described by nonlinear functions [19].

Several software tools (also known as ray-tracing software) with capabilities to design, analyse, optimize and perform tolerance analysis in optical systems, are available on the market and they are very helpful to automate the optimization process [21, 22].

Basically, the lens design optimization can be outlined in three main phases: pre-optimization phase, optimization phase and post-optimization phase. In the following subsections, the details of each phase will be described. Also, a correlation between the concepts discussed in the previous section and the parameters and metrics of an optical system will be established, considering an IOL as baseline.

9.2.1 Pre-optimization phase

The pre-optimization phase is the early stage of an optimization process and consists in the gathering of all information related to the optical system to be optimized. At this stage it is very important to have the specification of all the design inputs, which will support the definition of the basic parameters, the boundary conditions and the performance criteria of the system. The basic parameters are, for instance, the number of optical elements, the radii of curvature, the separation between elements, the lens material, the clear optical radius, the magnification, the field of view, the design wavelength or spectrum, etc. The boundary conditions correspond to limits on the variation of specific parameters like center thickness, edge thickness, aperture location, among others. The performance criteria are all the metrics that will quantify how good the optical quality of the system is [19, 20].

Considering the design of an IOL and the basic parameters outlined in this subsection, an example outlines a brief list of important aspects that can be identified during this phase and that are relevant on the development of the starting point of the next optimization phase. The following main items detail which parameters should be defined as preassigned, decision variables or design constraints:

- **Material:** the material is one of the most critical parameters to be selected in an optical design, because it will have impact not only on the optical performance of the lens, but also on the mechanical properties. There are several considerations to be taken during the material selection of an IOL. The material must be biocompatible, as it will be implanted in the human eye [23], it could be hydrophobic or hydrophilic, and it must also be available on the market in industrial scale. Although the biggest IOL companies on the market often have their own material formulated in-house and protected by patent laws [24–27], there are also third-party companies specialized in the manufacturing of biocompatible materials for IOLs [28, 29]. The type of material also has an important impact on the incision size of the cataract surgery. For micro-incision (<2 mm), for example, which is a common incision in cataract surgeries nowadays, the material must be foldable [30]. Based on all previous considerations about the material, it is in general considered as a preassigned parameter in an IOL optimization process.
- **Optical size:** the optical size of an IOL should not be less than 4.24 mm (diameter size) according to ISO standard 11979-3 [31]. The IOLs on the market have an optical diameter ranging from 5 mm to 7 mm [32, 33]. It is common to consider this parameter as a preassigned parameter for a given dioptric power, but it may vary within the dioptric power range of the IOL design, because the higher the dioptric power of the lens, the higher the curvatures, hence the higher the overall IOL volume, which impacts directly on the incision size.

- **Edge thickness:** the edge thickness of the lens is the area from where the haptics are attached (for a three piece IOL) or from where they are constructed (for a single piece IOL) [33]. Although it is preferable to have a thin edge thickness to guarantee a reduced overall IOL volume, if it is too thin, it cannot provide the necessary mechanical support to the haptics [34, 35]. In general, the IOLs have an edge thickness ranging from about 0.15 mm to 0.45 mm [36–38]. Based on that, this thickness is considered a decision variable and its value depends on the mechanical properties of the IOL material.
- **Surface shape:** the surface shape of an IOL is one of the most important parameters for the optical performance of the lens. Depending on the overall optical design of the IOL, which can be monofocal, multifocal, extended depth of focus (EDoF) or accommodative, the shape of the lens surface can be spherical, aspherical, toric, diffractive, asymmetric, with zones, freeform or even a combination of them [24]. For each type of surface there are several parameters that can be considered in the optimization process such as the radius of curvature, conic constant, high order polynomial terms, diffractive step heights, diffractive step shape, diffractive step position, among others. Based on that, the surface shape parameters are, in general, considered as decision variables in the lens optimization process.
- **Center thickness:** the center thickness of an IOL depends on the three previously mentioned parameters: the edge thickness, surface shape and optical size. Besides, it is often designed to be completely accommodated inside the capsular bag. It can be considered as a design constraint to guarantee that the overall IOL volume will not exceed a limiting value, which after that would incidentally require a bigger incision.
- **Distance between elements:** there are some accommodative IOL designs that present more than one optical element [39–41]. For this type of design, the distance between the parts is another parameter to be considered in the optimization process. Depending on the working principle of the lens, the distance between elements can be considered a design constraint or a decision variable. For instance, as a design constraint, it can be used to guarantee that the parts will never get in contact and, as a decision variable, it can be optimized to achieve a desired optical power.

It is important to notice that the definitions presented previously may vary depending on the optical design that is being optimized. One of the surfaces of the lens, for example, can be fixed as a preassigned parameter while only the other surface is considered as a decision variable. In summary, there are several aspects to be considered in an optical design before even starting the optimization process. For instance, considering an IOL design, an in-depth understanding of the anatomy and optical properties of the human eye, optical properties of materials, light propagation, imaging techniques and manufacturing process is crucial for the configuration of the optimization algorithm to guarantee a feasible optimized design. In the following subsection the optimization steps of a lens design are presented, and the two examples mentioned in section 9.1 (the monofocal lens and the refractive bifocal lens) are covered in detail.

9.2.2 Optimization phase

The optimization phase starts with the mathematical statement of the problem, as presented in section 9.1 (see equations (9.1) and (9.2) for a single-objective and a multiobjective problem, respectively). Once all relevant data about the optical design is listed, these pieces of information need to be categorized as decision variables, preassigned parameters, design constraints (equality and/or inequality constraints) or as merit functions.

The merit function of an optical system can be defined in distinct manners. For instance, it can be described by the spot size on the image plane; the root mean square of the wavefront error; the MTF or a linear combination of first and high-order aberration terms [19]. Mathematically, it is preferable for it to be continuous, differentiable in all decision variables and to demand a low computational cost [22]. However, the merit function of a lens design is highly nonlinear, and a range of linearization approaches have been studied in the past to find an efficient optimization method [42]. Research in the last decades has found that the best approach to model the merit function of a lens design is based on the weighted sum of the square of the image quality parameters of the system (see equation (9.8)) [22, 42]. In general, the parameters f_i represent the imperfections of the system and thus the optimization process consists in the minimization of the merit function [19, 42]:

$$\Psi = \sum_{i=0}^{m} f_i^2 \tag{9.8}$$

where Ψ is the representation of the overall objective function of the optical system, m is the total number of objective functions that describe the quality of the system and f_i is the ith weighted quality indicator given by equation (9.9) [42]:

$$f_i = w_i(e_i - t_i) \tag{9.9}$$

where w_i is a weight factor, e_i is the current value of the quality indicator and t_i is the target value of the quality indicator. The weight balance of each term present in the merit function is crucial to achieve the preferred optical performance from the optimization process. Several examples of ray-tracing software already have different default merit functions with the weights established for a variety of possible image distortions [42]. However, if a lens design does not reach the desirable performance after the optimization process, there are different methods that can be used by the designer to define a new combination of weights in order to lead the convergence to the expected performance target [42].

The boundary conditions, which represent a kind of design constraint, limit the variation of the decision variables. To guarantee a viable optimized design, it is important to make choices considering aspects of the manufacturing process. Otherwise, the optimization process can converge to solutions that cannot be manufactured due to the machinery limitations or that cannot be afforded due to the costs of the high precision elements required [42].

In ray-tracing software the deviations from the boundary conditions and design constraints are added in the merit function as a new distortion term to be minimized [20, 42]. There are different methods to address the constraint violations such as absolute control, penalty control and variable bounds [20].

The decision variables, also known as degrees of freedom, are the parameters to be changed by the optimization process to achieve the best performance possible and, together with the preassigned parameters, they define the final constructive shape of the lens.

The two optical optimization problems mentioned in the previous section (the monofocal and bifocal IOL) are used as examples to describe the process of construction of the mathematical formulation presented in equations (9.1) and (9.2). For the monofocal IOL, the following requirements are considered:

1. The lens must be biconvex and have a dioptric power of 20 D.
2. The lens must yield null spherical aberration at an aperture of at least 5 mm diameter.
3. The lens is intended to be implanted through a micro-incision surgery.
4. The optic performance and dimensions of the lens must comply with the required standards.

Firstly, the mathematical formulation of the optimization problem is presented and then the steps and considerations taken into account are outlined. The formulation for this problem, considering the previous mentioned requirements, is given by:

Minimize:

$$Z_4^0(R_a, k_a, R_p) \tag{9.10}$$

Subject to:

$$\text{EFL} = 50\ \text{mm} \tag{9.11}$$

$$R_a > 0\ \text{mm} \tag{9.12}$$

$$R_p < 0\ \text{mm} \tag{9.13}$$

$$0.15\ \text{mm} < e_t < 0.45\ \text{mm} \tag{9.14}$$

$$c_t < 1.20\ \text{mm} \tag{9.15}$$

$$5\ \text{mm} < D_{\text{overall}} < 7\ \text{mm} \tag{9.16}$$

The merit function represented by equation (9.10) corresponds to the contribution of the spherical aberration (Zernike term Z_4^0 [43]) of the lens on the image plane (retina) that must be cancelled for an aperture of at least 5 mm diameter. The optimization process will minimize the merit function by changing three design variables: the anterior radius of curvature (R_a), the conic constant of the anterior surface (k_a) and the posterior radius of curvature of the lens (R_p). The choice for three design parameters was guided by the discussions made in section 9.2.1 and are sufficient to guarantee the minimization of the spherical aberration by varying the conic constant. Besides, the radii of curvature will guarantee the required dioptric power [44]. The effect focal length (EFL in equation (9.11)) is the equality constraint related to the dioptric power requirement of 20 D and is given, in meters, by the inverse of the dioptric power. Equations (9.12) and (9.13) guarantee that the anterior surface and posterior surface of the lens are convex because there is a convention established between the sign of each radius and its concavity [19]. Equation (9.14) represents the edge thickness of the lens and its limits were established based on the commercial IOLs already discussed in section 9.2.1. The maximum value of center thickness of the lens is restricted in equation (9.15) and its limit was based on the center thickness of foldable off-the-shelf IOLs [35]. Equation (9.16) defines the minimum and maximum limits of the overall optic diameter of the lens, which was based on the IOLs available on the market, and it also guarantees that the lens will correct the spherical aberration for at least 5 mm aperture. The center thickness, edge thickness and overall diameter are considered as inequality constraints because they control the volume of the lens, thus, they will satisfy the third requirement of the optimization problem related to the incision size. The fourth requirement is fulfilled by guaranteeing that the MTF is above 0.43 at 100 lp mm^{-1} [45], which will be achieved by the aberration minimization. The overall optic diameter is above 4.24 mm [31], which is guaranteed by equation (9.16) [31, 45]. The international standard (ISO 11979) establishes the wavelength of light and the model eye to be employed in the assessment of the IOL [45]. Regarding the material, which is a preassigned parameter, the selection should be based on one that allows the construction of a foldable lens.

As a second example, considering the optimization problem of the refractive bifocal IOL mentioned in the previous section of this chapter (see figure 9.3), the mathematical formulation is built taking into account the following, not extensive, requirements:

1. The lens must be biconvex and the far focus and near focus must have, respectively, 20 D and 23 D.
2. The lens must have the best MTF possible at 5 mm pupil diameter, for both foci, and the ratio between the MTF at far and near focus must be within the range of 0.95 and 1.05.
3. The lens is intended to be implanted through a micro-incision surgery.
4. The optic performance and dimensions of the lens must comply to the required standards.

The mathematical formulation of the optimization problem related to the previously mentioned requisites is given by:

Maximize:

$$\begin{aligned} &\mathrm{MTF}_{\mathrm{IOL}}\left(R_a, k_a, R_{p1}, k_{p1}, R_{p2}, k_{p2}, a\right) \\ &\quad = \left\{\mathrm{MTF}_{\mathrm{far}}\left(R_a, k_a, R_{p1}, k_{p1}, R_{p2}, k_{p2}, a\right), \mathrm{MTF}_{\mathrm{near}}\left(R_a, k_a, R_{p1}, k_{p1}, R_{p2}, k_{p2}, a\right)\right\} \end{aligned} \tag{9.17}$$

Subject to:

$$\mathrm{EFL}_{\mathrm{far}} = 50\ \mathrm{mm} \tag{9.18}$$

$$\mathrm{EFL}_{\mathrm{near}} = 43.5\mathrm{mm} \tag{9.19}$$

$$R_a > 0\ \mathrm{mm} \tag{9.20}$$

$$R_{p1} < 0\ \mathrm{mm} \tag{9.21}$$

$$R_{p2} < 0\ \mathrm{mm} \tag{9.22}$$

$$0\ \mathrm{mm} < a < 5\ \mathrm{mm} \tag{9.23}$$

$$0.15\ \mathrm{mm} < e_t < 0.45\ \mathrm{mm} \tag{9.24}$$

$$c_t < 1.20\ \mathrm{mm} \tag{9.25}$$

$$5\ \mathrm{mm} < D_{\mathrm{overall}} < 7\ \mathrm{mm} \tag{9.26}$$

These requirements characterize a multiobjective problem (see equation (9.17)), which is represented by the MTF at far and near focus. Seven design variables are considered (see figure 9.3 to identify the correspondence on the IOL): the anterior radius of curvature of the lens (R_{a}), the conic constant of the anterior surface (k_{a}), the posterior radius of curvature of the central refractive region (R_{p1}), the conic constant of the central posterior surface (k_{p1}), the posterior radius of curvature of the outer refractive region (R_{p2}), the conic constant of the outer posterior surface (k_{p2}) and the diameter size of the central refractive region of the posterior surface (a).

The first six design variables are responsible for guaranteeing the required dioptric power of each refractive region (20 D and 23 D), which are stated by equations (9.18) and (9.19) as a constraint. They are also responsible for maximizing the MTF at both foci, by controlling the aberration of the lens through the conic constants. The last design parameter (a) is responsible for balancing the incoming light energy that is separated onto the two main foci (far and near). Equations (9.20), (9.21) and (9.22) have the same purpose of the previous optimization problem, namely, to guarantee that the lens is biconvex (see first requirement). Equation (9.23) establishes the variation limits of the diameter of the inner region, that is responsible for the vergence of the near focus. This region cannot be equal to zero, otherwise the lens would be a monofocal lens of 20 D, and, on the other hand, it cannot be bigger than 5 mm diameter, because, in that case, the lens would also become a monofocal lens of 23 D. The three last design constraints (equations (9.24), (9.25) and (9.26)) have the same purpose as already presented in the monofocal IOL optimization problem, which is to respect the two last requirements: micro-incision surgery and standard requirements.

It is important to discuss a few important aspects of the mathematical formulation and the decisions taken that impact the optimization process and the performance of the final design. Although the requirement establishes that the MTF is the metric to be optimized, it is important to define at which spatial frequency it is supposed to be registered. The ISO standard 11979-2 recommends the assessment of the MTF of multifocal IOLs at 50 lp mm^{-1} spatial frequency [45], therefore, it will be the one used.

Another important point is related to the aperture at which the MTF is evaluated. As the example establishes a requirement on the MTF performance at a specific aperture (5 mm), the final design will present the MTF ratio between 0.95 and 1.05 exclusively for this aperture. However, if it is desired to keep the same MTF ratio requirement at different apertures, more refractive zones must be added onto the posterior and/or anterior surface of the lens to achieve that. Therefore, the optimization problem of the lens can be separated in several sub-problems that can correspond to each desired aperture. For instance, it can be desired to keep an equal MTF performance at far and near focus at 3, 4 and 5 mm aperture. In this case, the first sub-problem to be solved is at the 3 mm aperture and then, its optimized design parameters will be preassigned parameters for the next sub-problem (4 mm aperture) and so on. At the end of the process, a refractive lens with multiple zones will guarantee a bifocal performance within the requirements.

After the mathematical statement of the optimization problem, the optical designer has all the essential information to configure the optimization tool of the ray-tracing software by: defining and configuring the merit function, setting the preassigned parameters, defining the constraints and setting the design variables. Once all software configuration is done, it is necessary to define a starting point to begin the optimization process.

The starting point is the baseline design considered by the optimization software to begin the iterations towards a good design, and it can be set up in different manners. It can be based on the experience of the designer; it can come from books, papers, patents or any other scientific source describing a similar design; it can be based on a first-order calculation design from scratch; or it can even be a design defined by two flat surfaces, leaving the work of finding the best curvatures up to the optimization algorithm [19, 20, 22, 42]. The latter possibility to generate a starting point, in general, does not result in a good final design and, in that case, requires intervention from the designer to re-establish a new starting point with curved surfaces. However, with the advances in artificial intelligence (AI), it will be possible to see in the future the starting point design being defined by a machine learning process integrated into the ray-tracing software [42].

Once the starting point is defined, the algorithm selection is the next step. Ray-tracing software not only has embedded optimization algorithm tools, but can often also communicate with third-part software like MATLAB® to access third-part algorithms, that can control, analyse and optimize the ray-tracing optical design.

One of the main optimization methods present in ray-tracing software is the DLS. The DLS is a deterministic algorithm based on a modification of the least-square method and is the most widespread optimization tool available in commercial ray-tracing software [10, 22]. The least-square method was developed to determine the solution of a system of linear equations through the minimization of the sum of the square of the residuals [19]. A limitation of this method for a lens optimization process relies on the non-linearity of the aberration terms that are considered in the merit function [22]. Although it is possible to apply the Taylor expansion technique on the merit function and keep the terms up to the first order to make it linear, the aberration terms diverge so substantially from a linear behavior, that the optimization convergence is not reached [19]. The DLS method applies a damping factor onto the magnitude of the variation of the decision variables to limit the wide oscillations observed in the least-square method and thus guarantees the convergence of the optimization process to a local minimum [10, 22].

Another important algorithm that has been used in the optimization of optical designs is a GA [16]. As described in section 9.1.4, the GA is a stochastic optimization method and is part of a group of evolutionary strategies that mimic the natural selection process to solve an optimization problem [16, 22].

Considering both optimization methods previously mentioned, two optimization approaches will be described as an example for lens design. The first approach consists in applying exclusively the DLS algorithm to optimize the lens design. As it is a local search method, its success is highly dependent on the starting point [22]. To illustrate this approach and its limitation, figure 9.10 illustrates a hypothetical plot, where the x-axis corresponds to the combination of all possible values of the decision variables described for the optimization of the monofocal lens design (equation (9.10)) versus the merit function of the system (y-axis), that is represented by the spherical aberration (equation (9.10)). In case the optical designer chooses a starting point represented by point C (figure 9.10), the DLS algorithm converges to the local minimum represented by point D (considering a minimization problem), but this point does not correspond to the best possible merit function value (represented by point B). However, if the starting point chosen by the optical designer corresponds to the design parameters represented by point A, the DLS algorithm converges to point B, which is the best possible solution. In summary, to guarantee that the DLS algorithm converges to the best possible solution, the starting point must be in the convergence basin of the global minimum value of the merit function (considering a minimization problem).

The second approach of optimization consists in applying both methods (GA and DLS) in sequence. To understand the purpose of this approach, the set of solution distribution found by the GA method is shown in figure 9.11. This is obtained when

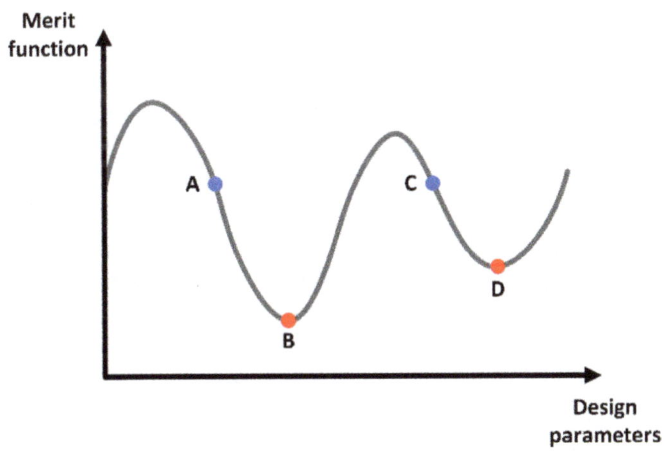

Figure 9.10. Hypothetical performance of the merit function versus the values of design parameters for a monofocal lens design and possible starting points (blue dots) and corresponding final solutions (red dots) to be found by the DLS algorithm.

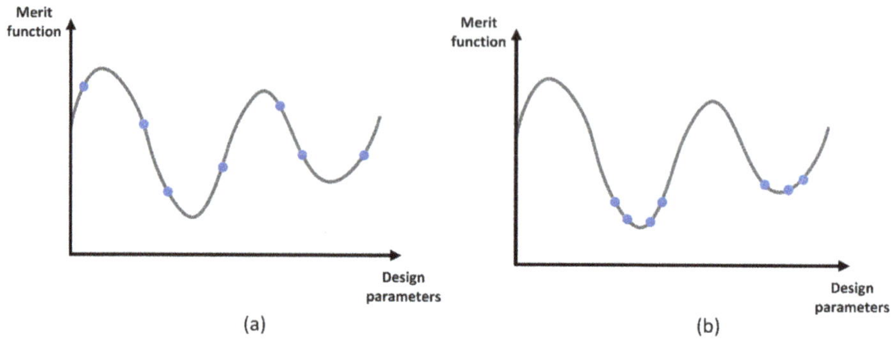

Figure 9.11. Hypothetical performance of the merit function versus the design parameters values for a monofocal lens design: (a) initial population distribution and (b) final population distribution achieved by GA method.

applied to the same monofocal lens design that had the merit function performance shown in figure 9.10. Figure 9.11(a) presents the set of solutions (blue dots) that corresponds to the initial population of candidate solutions of the GA. The initial population of possible solutions of the GA is randomly generated and aims to explore as much as possible the solution space as can be seen in figure 9.11(a). At each iteration of the GA, the population of solutions changes towards the region of the solution space that has the best alternatives (best fitness). At the end of the GA process, the solutions will be concentrated in the regions that resulted the best merit function values as presented in figure 9.11(b).

However, as the GA is a stochastic method, it does not guarantee that the global minimum of the problem can be found [22]. But this method, in general, is efficient to find the solutions that are in the neighbourhood of the global minima. Besides, depending on the search space region of the optimization problem, the GA might not find the convergence basin that contains the global minima, but it can find regions that yield promising solutions for the problem. Based on that, the second approach consists in applying the GA on the optical design optimization and using the best solution of each region found by the GA as a starting point for the DLS algorithm (figure 9.12). With this approach, the starting point selection for the DLS algorithm does not depend on the experience of the optical designer because it is determined by the GA.

There are several deterministic and evolutionary methods that can be applied and combined to optimize a lens design, other than the DSL and GA [22, 42]. However, no other determinist method was capable of outperforming the DLS algorithm in an overall basis, and the GA is one of the most investigated evolutionary algorithms in the field of lens design [16, 42]. An overview regarding some other promising hybridization techniques among evolutionary algorithms and local search strategies can be found in [46]. The next and last stage after the optimization process related to the computational model of a lens design is described in the next subsection.

9.2.3 Post-optimization phase

In the end of the optimization process a final design is found by the algorithm. This is the moment for the optical designer to analyse whether the parameters are within the

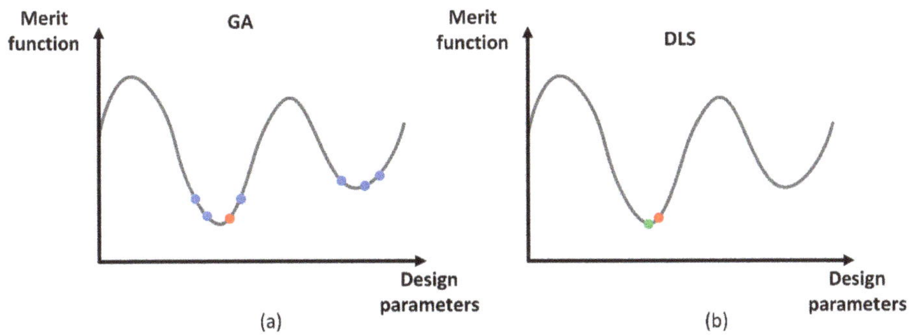

Figure 9.12. Hypothetical performance of the merit function versus the design parameters values for a monofocal lens design: (a) final population of solutions found by GA (all dots) and its best solution (red dot) and (b) starting point of DLS algorithm from GA best solution (red dot) and global minimum found by DLS (green dot).

specifications and the merit function corresponds to best possible solution or not. In the last case, the optical designer can make small adjustments on the design variables, design constraint limits and merit function weights. Once all adjustments are done, the optimization process is restarted to check if it is possible to achieve a better design [42]. At this stage it is important to consider that, although the final design will yield the best performance for the system, it is not possible to guarantee that the parameters of the manufactured design will have the same values as established in the computational model (ideal design). This is an intrinsic aspect of the manufacturing process and because of that the constructive parameters determined by the computational design will present a value within a range based on the manufacturing precision. An important step at this stage of the lens design process is the manufacturing tolerance analysis, which is discussed in detail in the following subsection.

An additional fact to be highlighted is that, even if the manufacturing process could guarantee a lens design with performance matching the nominal project requirements, there are several clinical factors after the implantation of the lens into the eye that would affect the final optical performance. Clinical aspects related to the patient biometry such as the cornea keratometry, pupil aperture, the anterior chamber depth and aspects related to the surgery such as incision size, centralization and rotation of the lens after implantation, capsulorhexis size, they are all important factors that could affect the nominal performance of a lens design.

9.2.4 Tolerance analysis

The tolerance analysis is an important tool for the optical designer to verify the impact of the manufacturing errors on the performance of the final manufactured product and what the sensitive parameters are in the optical system that most impact its performance [47, 48]. It is important, firstly, to define what is the acceptable variation in the best scenario of the optical performance of the system and to establish the tolerances of all optical and mechanical parts of the system. The tolerances of each parameter can be found in the technical and clinical literature or in the manual of the equipment (such as lathe manuals) [42, 47, 48].

There are cases where the tolerance of a certain parameter depends on the combined tolerance of other ones. For instance, the conic constant is an optical parameter that has its tolerance dependent on the combined tolerance of the clear optic diameter, thickness and radius of the curvature of the surface. In this case, to determine the conic constant tolerance it is necessary to apply a multidimensional tolerance analysis. Therefore, to determine the tolerance of the dependent variable (t_ζ), for example the conic constant, the following equation needs to be employed [49]:

$$t_\zeta = \sum_{i=1}^{n} | a_i t_{\xi_i} | \tag{9.27}$$

$$a_i = \frac{\partial \zeta}{\partial \xi_i} | \text{ nominal values} \tag{9.28}$$

where ζ is the mathematical description of the dependent parameter as a function of the independent parameters and ξ_i is the ith independent parameter.

Once the tolerance of all optical and mechanical parameters is defined, the next step consists in verifying how the variation of each parameter affects the optical performance of the system. The Monte Carlo tolerancing method is one of the approaches that allows the analysis of the combined impact of the tolerances of all the parameters on the optical performance of the system [48]. In the Monte Carlo method, the tolerance of each parameter is changed simultaneously and randomly in the optical system considering a variation distributed according to a statistical probability function that is in general a normal distribution [48]. For instance, considering the center thickness (c_t) as a parameter with tolerance defined within the range of ±0.01 mm and nominal value of 1.00 mm, its variation would be described by the normal distribution presented in figure 9.13 that is based in a level of confidence of 95%. The 95% confidence level corresponds to 95% of the total area under the probability distribution function and it is considered a reasonable margin of variation in a tolerance analysis [48].

In the end of the tolerancing process, it is possible to analyse the sensitivity of each design parameter in relation to the optical performance of the system. The most sensitive set of parameters are those that have the highest impact on the degradation of the performance of the system. This set of parameters requires a high precision manufacturing process to reduce the high impact of its tolerance on the performance [47]. Based on that, it is possible to notice that the tolerance analysis consists in balancing the trade-off between cost and performance. The tighter the performance,

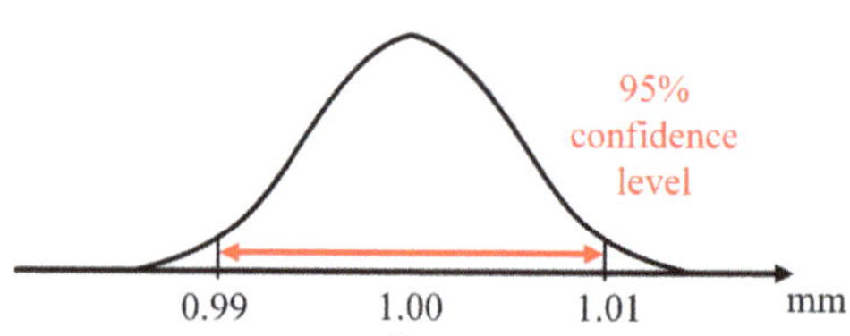

Figure 9.13. Normal probability function example describing the center thickness parameter variation in a Monte Carlo tolerance analysis.

the higher the cost due to the high required precision of manufacturing process. However, if the performance has some acceptable margin for variation, it is possible to manufacture the system at a more reasonable cost. At this stage, once the most sensitive set of parameters is identified and the manufacturing budget is defined, the tolerance process is repeated by tightening the sensitive parameters tolerance and loosening the non-sensitive parameters tolerance as a final check of the system performance achieved [47, 48].

9.3 Summary

This chapter presented the basics of optimization by means of the description of the mathematical formulation of an optimization problem and its elements such as design variables, design constraints and merit function. The classification of optimization problems was presented, and the optimization techniques recommended for each class of problem were addressed. Once the optimization concepts and techniques were detailed in section 9.1, section 9.2 showed how these elements are combined and applied to the optical lens design. An IOL was taken as baseline for the explanation and discussions about the decisions that are needed to be taken during the optimization process. The considerations and relevant aspects to be analysed before, during and after the optimization per se were addressed based on the perspective of an IOL. Finally, the basics of tolerance analysis were presented. In summary, the development of a lens design, more specifically of an IOL, is a complex process that requires multidisciplinary skills and a broad overview of the process to take into account all the aspects related to the physics of the design, optimization tools, computational design software, eye anatomy, manufacturing process and surgical implantation.

Chapter highlights

- The fundaments of optimization are discussed regarding the definition of an optimization problem and some related concepts.
- The principles of classical and modern optimization techniques are highlighted.
- Some of the main considerations applied to an IOL before, during and after the optimization process are addressed.
- Two optimization approaches for lens design using GA and DLS are presented.
- The basics of tolerance analysis applied to lens design are outlined.

References

[1] Rao S S 2009 *Engineering Optimization: Theory and Practice* 4th edn (New York: Wiley)
[2] Fletcher R 2000 *Practical Methods of Optimization* 2nd edn (Chichester: Wiley)
[3] Mokhtar C M S, Bazaraa S and Sherali H D 2006 *Nonlinear Programming: Theory and Algorithms* 3rd edn (New York: Wiley)
[4] Ravindran A, Ragsdell K M and Reklaitis G V 2006 *Engineering Optimization: Methods and Applications* 2nd edn (Hoboken, NJ: Wiley)

[5] Jorge S J W 2006 *Nocedal,* Numerical Optimization 2nd edn (Berlin: Springer)
[6] Bertsekas D P 1999 *Nonlinear Programming* 2nd edn (Nashua, NH: Athena Scientific)
[7] Vira Chankong Y Y H 1983 *Multiobjective Decision Making: Theory and Methodology* (Amsterdam: Elsevier Science Publishers)
[8] Miettinen K M 1999 *Nonlinear Multiobjective Optimization* (Dordrecht: Kluwer Academic)
[9] Collette Y and Siarry P 2003 *Multiobjective Optimization – Principles and Case Studies* 1st edn (Berlin: Springer)
[10] Meiron J 1965 Damped least-squares method for automatic lens design, *J. Opt. Soc. Am.* **55** 1105
[11] Sinclair D C 1995 Optical design software, *Handbook of Optics: Fundamentals, Techniques, and Design* 2nd edn (New York: McGraw-Hill) p 34
[12] Deb K 2008 Multi-objective optimization, *Multi-Objective Management in Freight Logistics* 1st edn (London: Springer) p 187
[13] van D A, Carlos Coello Coello V and Lamont G B 2007 *Evolutionary Algorithms for Solving Multi-Objective Problems* 2nd edn (New York: Springer)
[14] Eiben J E S A E 2015 *Introduction to Evolutionary Computing* 2nd edn (Berlin: Springer)
[15] Dréo E T J, Pétrowski A and Siarry P 2006 *Metaheuristics for Hard Optimization – Methods and Case Studies* 1st edn (Berlin: Springer)
[16] Höschel K and Lakshminarayanan V 2019 Genetic algorithms for lens design: a review, *J. Opt.* **48** 134–44
[17] Holland J H 1992 *Adaptation in Natural and Artificial Systems* (Cambridge, MA: MIT Press)
[18] Goldberg D E 1989 *Genetic Algorithms in Search, Optimization, and Machine Learning* (Reading, MA: Addison-Wesley)
[19] Malacara-Hernández Z M-H D 2013 *Handbook of Optical Design* 3rd edn (Boca Raton, FL: CRC Press)
[20] Laikin M 2006 *Lens Design* 4th edn (Boca Raton, FL: CRC Press)
[21] Smith W J 2007 *Modern Optical Engineering* 4th edn (Bellingham, WA: SPIE Press)
[22] Vasiljević D 2002 *Classical and Evolutionary Algorithms in the Optimization of Optical Systems* (New York: Springer)
[23] International Organization for Standardization – ISO, 'ISO – ISO 11979-5:2006 – Ophthalmic implants—intraocular lenses—part 5: biocompatibility' 24 2006
[24] Yu N, Fang F, Wu B, Zeng L and Cheng Y 2018 State of the art of intraocular lens manufacturing, *Int. J. Adv. Manuf. Technol.* **98** 1103–30
[25] Zhou T P R, Stephen Q, Sy J C and Berteig M A 1992 High refractive index silicone compositions *Patent* EP0593684A1
[26] Sergei E J A, Nanushyan R and Valunin I 2000 High refractive index silicone for use in intraocular lenses *Patent* WO 01/17570 A1
[27] Lehman C F C 2009 High refractive index ophthalmic device materials *US Patent* US 20090281209 A1
[28] Contamac, Contamac | World leader in contact lens and intraocular lens material technology www.contamac.com/ (accessed 14 Jul. 2020)
[29] Benz, Benz Research & Development. www.benzrd.com/ (accessed 14 Jul. 2020)
[30] Menapace R 2009 Mini-and micro-incision cataract surgery – a critical review of current technologies *Eur. Ophthalmic Rev.* **2009** 52–7
[31] International Organization for Standardization 2012 Ophthalmic implants—intraocular lenses— mechanical properties and test methods (ISO Standard No. 11979-3:2012) https://www.iso.org/standard/55681.html

[32] Simpson M J 1992 Diffractive multifocal intraocular lens image quality, *Appl. Opt.* **31** 3621

[33] Bellucci R 2013 An introduction to intraocular lenses: material, optics, haptics, design and aberration *Cataract* vol 3 (Basel: S. Karger AG) pp 38–55

[34] Jim Deacon M L P, Shepherd D J and Brady D G 2004 Foldable intraocular lens and method of making *US Patent* US 9737396 B2

[35] Barbero S, Marcos S, Montejo J and Dorronsoro C 2011 Design of isoplanatic aspheric monofocal intraocular lenses *Opt. Express* **19** 6215

[36] Yu-Chin Lai D P V and Ruscio D V 2004 Intraocular lens *US Patent* US 2004/0002757 A1

[37] Nanavaty M A, Spalton D J, Boyce J, Brain A and Marshall J 2008 Edge profile of commercially available square-edged intraocular lenses *J. Cataract Refract. Surg.* **34** 677–86

[38] Nanavaty M A, Zukaite I and Salvage J 2019 Edge profile of commercially available square-edged intraocular lenses: part 2 *J. Cataract Refract. Surg.* **45** 847–53

[39] Alió J L, Alió del Barrio J L and Vega-Estrada A 2017 Accommodative intraocular lenses: where are we and where we are going *Eye Vis.* **4** 16

[40] Qun Peng Y, Yang X and Zhang 2003 Accommodative intraocular lens *US Patent* US 2003/0204256 A1

[41] Daniel Brady R R and Silvestrini T 2016 Two-part accommodating intraocular lens device *US Patent* US 2016/0317286 A1

[42] Kingslake R and Johnson R B 2010 *Lens Design Fundamentals* 2nd edn (Amsterdam: Elsevier)

[43] Noll R J 1976 Zernike polynomials and atmospheric turbulence, *J. Opt. Soc. Am.* **66** 207–11

[44] Barbero S and Marcos S 2007 Analytical tools for customized design of monofocal intraocular lenses *Opt. Express* **15** 8576

[45] International Organization for Standardization 2014 Ophthalmic implants—intraocular lenses—optical properties and test methods (ISO Standard No. 11979-2:2014) https://www.iso.org/standard/55682.html

[46] Krasnogor N and Smith J 2005 A tutorial for competent memetic algorithms: model, taxonomy, and design issues *IEEE Trans. Evol. Comput.* **9** 474–88

[47] Shannon R R 1995 Tolerancing techniques, *Handbook of Optics: Fundamentals, Techniques, and Design* 2nd edn (New York: McGraw-Hill) p 35

[48] Robert P R Y, Fischer E and Tadic-Galeb B 2008 *Optical System Design* 2nd edn (New York: McGraw-Hill)

[49] Van Wyk D 1999 Multi-dimensional tolerance analysis (manual method) *Dimensioning and Tolerancing Handbook* 1st edn (New York: McGraw-Hill Education) p 704

Chapter 10

Design of diffractive multifocal intraocular lenses

Luiz Melk de Carvalho, Davies William de Lima Monteiro and Otávio Gomes de Oliveira

Diffraction phenomena are commonly used in intraocular lenses (IOLs) to provide two or more focal points and hence allow the eye to project on the retina the superimposed images of objects located at different distances from the observer, enabling gaze-independent multifocality. Such effect can be achieved through the use of diffractive optical patterns, which are microstructures engraved on the surface of the lens to split the light into the different focal points. The geometrical shape of the diffractive optical elements determines parameters of optical performance of the lens, such as the difference in optical power between the foci, whose diffraction orders are used for image formation and the amount of light energy associated with each of them. The design of diffractive ophthalmic lenses is a complex discipline that requires strong competences in mathematical and optical modelling, which are not always clearly compiled and structured in the available literature.

In this context, this chapter presents, from concept to computational model results, the method to design a multifocal diffractive intraocular lens. To illustrate the degree of complexity that can be imparted to the diffractive pattern, the lens herein presented features diffractive steps height modulated by the modulus of a shifted cosine function. Although the approach presented in this chapter is related to the cosine function, its application is not limited to it and provides the basis for deployment of other modulation functions.

The chapter is structured in three sections. The first section covers the mathematical model for light efficiency distribution and additional power in kinoform (sawtooth structure) diffractive surfaces. The second section presents the main characteristics of some diffractive multifocal intraocular lenses (MIOLs) available on the market and known worldwide. The third section describes the process to modulate a diffractive surface by a mathematical function; the optimization procedure of a diffractive MIOL, describing all specifications: algorithm selection,

doi:10.1088/978-0-7503-3263-7ch10

design and constraint parameters, merit function and solution selection; and, finally, a comparative analysis of the computational optical results with those of MIOLs present on the market. The presented method can be applied to any mathematical function and the parameters of each modulation function can be correlated to the metrics that describe the performance of the optical design. On this basis, future research can be conducted to verify, not only if there is any other mathematical function that can yield a better optical performance, but also how to mathematically correlate the metrics of the optical performance of the lens with the parameters of the modulation function.

10.1 Multifocal diffractive lens parameters and design

The working principle of diffractive ophthalmic lenses is based on an optical phenomenon called diffraction. The diffraction consists of a disturbance of the light wavefront when the light beam hits or passes through a geometrical artifact that has a dimension of the same order of magnitude of the light wavelength. The artifact can be an aperture, a sharp edge, a step, or any kind of topological feature with a usually sharp transition of shape or refractive index. The wavefront propagation due to diffraction was qualitatively described by a Dutch scientist called Christian Huygens [1]. According to Huygens, upon encountering an artifact, the wavefront propagation from that position onwards can be considered as that of punctual light sources, whose emitted waves propagate as secondary wavefronts (figure 10.1). The combination of the secondary wavefronts generates constructive and destructive interference, rendering a composite wavefront shape different from that of the original incoming wavefront [1, 2].

In diffractive lenses, small steps (see figure 10.2) are responsible for generating the diffraction effect. Each diffractive step on the lens surface splits the incoming light into numerous secondary beams called diffractive orders (figure 10.2) and at some points on the optical axis they cross over generating a constructive interference [3].

The diffractive step heights have their magnitudes on the order of μm or less. Thus, their size has been sketched out of scale in figure 10.2, in comparison to the

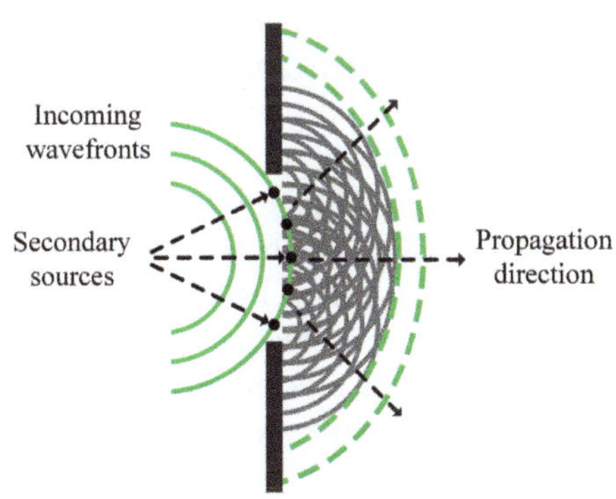

Figure 10.1. Wavefront propagation considering the diffraction phenomenon: incoming wavefront (green semi-circles), secondary light sources (black dots), propagation direction (dashed arrows), secondary wavefronts (gray semi-circles) and resulting wavefront (green dashed semi-circles) due to the interference of wavefronts.

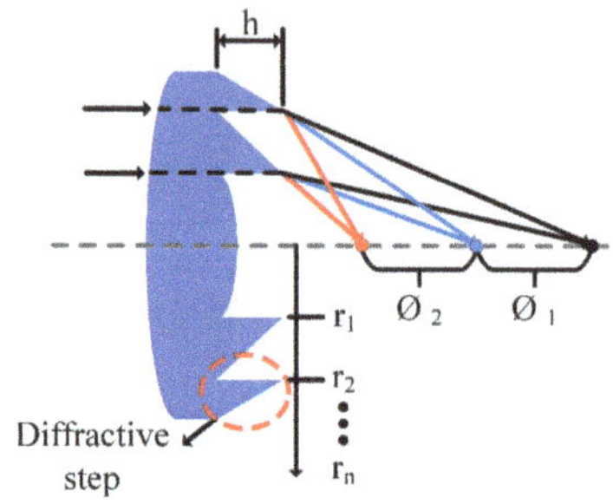

Figure 10.2. Diffractive lens parameters: step height (h) and step height radial position (r_1, r_2, …, r_n). Vergence of the diffractive orders: order 0 (black arrow), order +1 (blue arrow) and order +2 (red arrow). Vergence power difference between two consecutive diffractive orders (ϕ_1 and ϕ_2).

lens dimensions (diameter and center thickness), to more clearly present the constructive details.

Regarding the vergence of the diffractive orders, the refractive power of the lens (also known as base power) determines the dioptric power of the order 0 [1]. The power vergence of each diffractive order (see terms ϕ_1 and ϕ_2 in figure 10.2) relative to the previous order is known as the additional power or add power (ϕ_{add}), which is kept constant in a diffractive IOL and determined by the radial positions of the step heights (r_1, r_2, …, r_n—see figure 10.2). The mathematical relation between the add power and the step-height position (r_n) is given by equation (10.1) [1]:

$$r_n = \sqrt{\frac{2 \cdot n \cdot \lambda}{\phi_{\text{add}}}} \tag{10.1}$$

where λ is the light wavelength, n is an integer that corresponds to the nth diffractive step. The number of diffractive steps on the optical surface impacts the interference phenomenon at the foci created by a diffractive multifocal IOL and consequently the efficiency distribution [4]. Based on that, it is recommend that a diffractive surface of an IOL should have at least four diffractive steps to guarantee a well-defined intensity distribution at each focus [4]. On the other hand, the maximum number of diffractive steps on an optical surface, i.e. the upper limit for n in equation (10.1), depends on the optical area available for the construction of the steps and on the value of additional power (ϕ_{add}).

While r_n has impact on the additional power, the diffractive step height at each step position and its profile has an impact on the energy distribution throughout the diffractive orders. There are numerous possibilities to design a diffractive profile (figure 10.3) and each design yields a different energy distribution [5, 6]. In all of them, the distance between subsequent steps becomes smaller towards the periphery, as expected from equation (10.1) as n increases.

The light energy distribution (light efficiency) in a kinoform diffractive profile having a linear transition with constant step height (figure 10.3(b)) is given by equations (10.2)–(10.4) [7]:

$$\text{eff}_0 = \left[\frac{\sin(\sigma/2)}{\sigma/2}\right]^2 \tag{10.2}$$

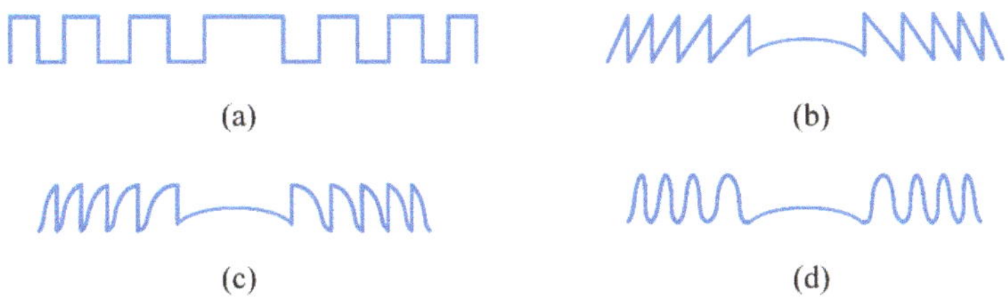

Figure 10.3. Different diffractive profiles: (a) binary profile, (b) kinoform profile with linear ramp transition, (c) kinoform profile with sinusoidal transition and (d) sinusoidal profile.

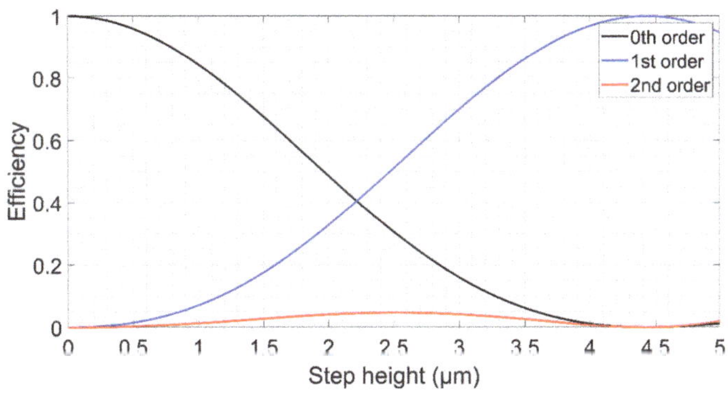

Figure 10.4. Efficiency variation of the 0th, 1st and 2nd order as a function of the diffractive step height.

$$\text{eff}_m = \left[\frac{\sin(\sigma/2)}{(2\pi \cdot m - \sigma)/2}\right]^2 \tag{10.3}$$

$$\sigma = 2\pi \cdot h \cdot (n_{\text{IOL}} - n_{\text{aq}})/\lambda \tag{10.4}$$

where eff_0 is the efficiency of the 0th diffractive order, eff_m is the efficiency of the mth order, h is the height of the diffractive step, n_{IOL} is the refractive index of the IOL material, n_{aq} is the refractive index of the surrounding medium and λ is the light wavelength.

Considering equations (10.2)–(10.4) it is possible to vary the efficiency of the diffractive orders by changing the diffractive step height (h). Figure 10.4 shows the efficiency variation of the 0th, 1st and 2nd diffractive orders as a function of the diffractive step height (h) and considering $\lambda = 550$ nm, $n_{\text{IOL}} = 1.460$ and $n_{\text{aq}} = 1.336$. It is possible to notice that at 2.2 μm the efficiency of the 0th and 1st order intersects at about 0.41 amplitude. Thus, a kinoform diffractive design with this step height creates two main foci (bifocal lens), because most part (82%) of the incoming light is directed to these two foci. On the other hand, if the step height of the design were 4.5 μm, the light efficiency would have been concentrated in the 1st order, which would characterize a monofocal lens.

Another important aspect to consider in a diffractive design is that the efficiency of each diffractive order also depends on the wavelength (λ) (see equations (10.2)–(10.4)). However, the visible spectrum is composed by a range of wavelengths (from 390 nm to 780 nm [8]). As an example of the impact of the wavelength on the diffractive performance, figure 10.5 shows the efficiency distribution of the 0th, 1st and 2nd diffractive orders as a function of the wavelength considering the previously mentioned diffractive design with h = 4.5 μm, n_{IOL} = 1.460 and n_{aq} = 1.336. The curves were determined by using equations (10.2)–(10.4). It is possible to notice that the diffractive profile directs 100% of the light efficiency on the 1st order for the wavelength (550 nm) considered in the design, but at wavelengths above and below 550 nm the efficiency presents a drop in the 1st order and an increase in the 0th and 2nd orders. Even so, the 1st diffractive order is still predominant, and the lens performs as a monofocal design, which means that the diffractive order of interest during the lens design continues to be prioritized.

10.2 State-of-the-art of diffractive multifocal IOLs

The previous section detailed the diffraction principles that describe the optical performance of diffractive lenses, where one of the most basic forms, the linear kinoform with constant step height, can be designed to generate a bifocal performance. This approach is employed, as an example, by the Tecnis ZM900 IOL (from the former Abbott Medical Optics and now Johnson & Johnson Vision), among many other diffractive bifocal lenses. Its diffractive profile, which is situated on the posterior surface, has 32 steps with the same height covering the entire optic surface [9, 10]. The bifocality is created through the 0th and 1st diffractive orders that correspond, respectively, to the far and near focus [9]. As the diffractive profile covers the entire optics and the step height is kept constant, the efficiency distribution is independent of the pupil aperture [10]. The additional dioptric power of the lens is +4 D and the lens is made of a silicone material [9]. Figure 10.6 illustrates the diffractive profile of the lens without the base curve and it is important to notice that the dimension scale does not depict the real size of the diffractive steps in comparison to the optic size. The modulation transfer function (MTF) of the lens at 3.0 mm aperture and 50 lp mm^{-1} of spatial frequency at far and near focus is 0.45 and 0.30, respectively [10]. The MTF at an aperture of 4.5 mm pupil diameter at far and near focus is 0.44 and 0.30, respectively [10]. The MTF values were measured in an optical bench that considers a model eye proposed by the ISO standard 11979-2 [10].

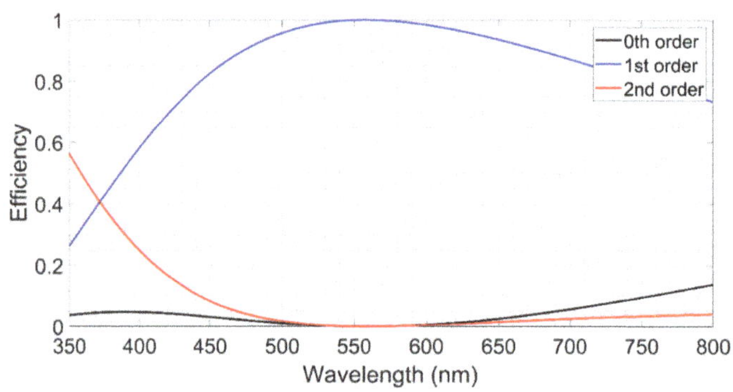

Figure 10.5. Efficiency variation of the 0th, 1st and 2nd order as a function of the wavelength.

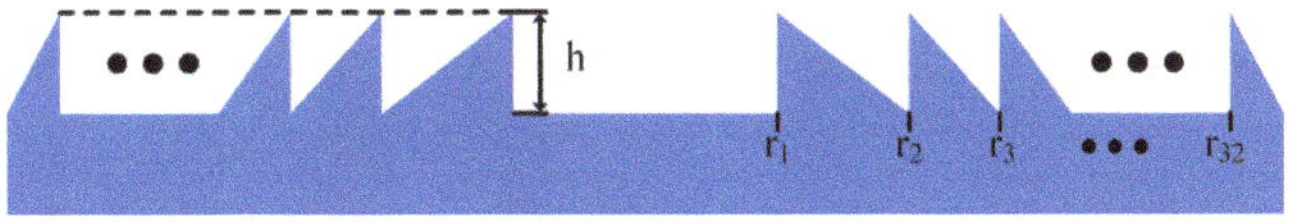

Figure 10.6. Diffractive profile of the Tecnis ZM900 IOL: constant step height (h) and 32 diffractive steps (from r_1 to r_{32}).

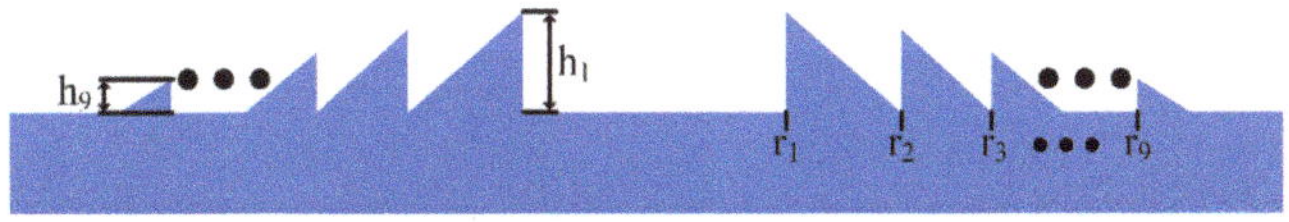

Figure 10.7. Diffractive profile of the Acrysof Restor SN6AD1 IOL: the apodized diffractive steps (from $h_1 = 1.3$ µm to $h_9 = 0.2$ µm) and nine diffractive steps (from r_1 to r_9) covering the 3.6 mm aperture diameter.

Another bifocal lens with a different profile than the Tecnis lens is the Acrysof Restor SN6AD1 IOL (from Alcon Laboratories, Inc.), which presents a diffractive profile having nine steps on a region between 0 mm to 3.6 mm of diameter of the optic surface (from 3.6 mm to the edge of the optic, it is a refractive lens) [10]. The bifocality is created by the 0th and 1st order, which correspond to the far and near focus, respectively [11]. The diffractive profile of the lens is apodized, which means that the step heights are gradually reduced from the center to the periphery. This lens features the first step with a height of 1.3 µm and the last one of 0.2 µm [12, 13]. Due to the apodization, the optical performance of the lens is pupil dependent. At about 1.8 mm pupil diameter the efficiency distribution at the two main foci is about 41%, but with the increase of the pupil aperture the far focus is prioritized and at 5 mm pupil diameter for example the distribution is about 10% at the near and 84% at the far focus [12]. The MTF performance of the lens at the far and near focus considering an aperture of 3.0 mm diameter and 50 lp mm^{-1} is 0.42 and 0.33, respectively [10]. The MTF at the same foci, but considering an aperture of 4.5 mm diameter, is 0.55 at far and 0.18 at near focus [10]. The MTF values were assessed in an optical bench that considers an eye model proposed by ISO standard 11979-2 [10]. The additional power of the lens is +3 D and the lens material is a hydrophobic flexible acrylic [13]. Figure 10.7 illustrates the apodized diffractive profile of the lens without the base curve. To represent the diffractive steps, the details of the diffractive steps in figure 10.7 have been magnified and do not correspond to the real scale of the lens.

One of the approaches described in the literature to achieve a trifocal optical performance using a diffractive profile consists in alternating the step heights between two values (see figure 10.8) [14]. With this design it is possible to create three main foci (far, intermediate and near) by using the 0th, 1st and 2nd diffractive orders and keeping an efficiency of 27% at each one of them [14].

The FineVision Micro F (from PhysIOL SA) is a diffractive trifocal lens with diffractive profile featuring 26 steps covering the entire optic surface [10]. The diffractive profile (figure 10.9(c)) consists on the combination of two apodized bifocal profiles (figure 10.9): one with +3.50 D additional power (figure 10.9(a)) between the 0th and 1st order and the other with +1.75 D additional power (figure 10.9(b)) between the 0th and 1st order [15]. In this design, the diffractive

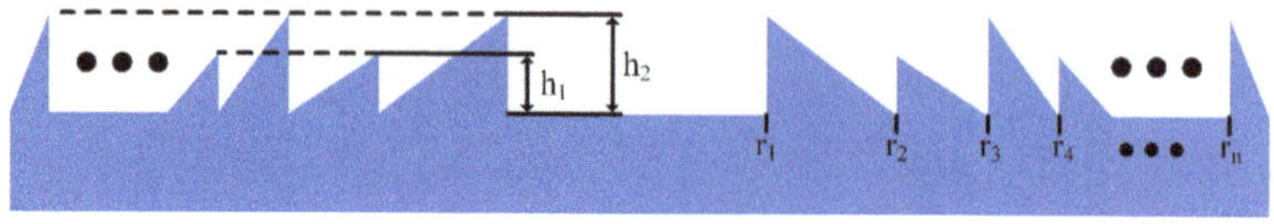

Figure 10.8. Diffractive profile of a trifocal lens considering the steps height with alternating values: h_1 (at the odd positions r_1, r_3, ..., r_n) and h_2 (at the even positions r_2, r_4, ..., r_{n+1}) [14].

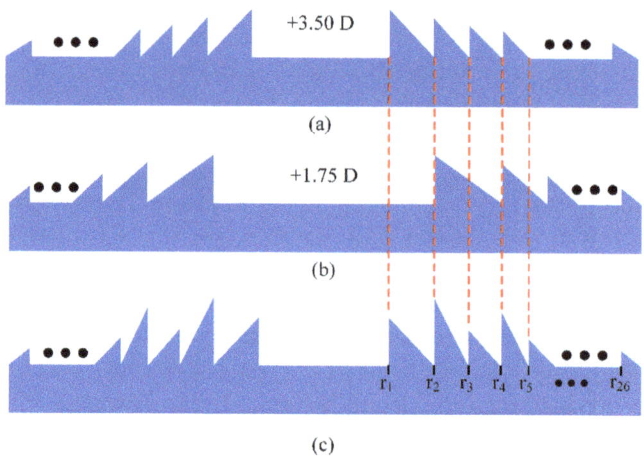

Figure 10.9. Diffractive profile of the FineVision IOL: (a) bifocal apodized diffractive profile with additional power of +3.50 D, (b) bifocal apodized diffractive profile with additional power of +1.75 D and (c) combination of both diffractive profiles previously mentioned to create the trifocal profile of the lens [15].

profile with +3.50 D addition directs most part of the incoming light energy onto the 0th and 1st diffractive orders (far and near focus, respectively), while part of the energy is lost on the other high diffractive orders (2nd, 3rd and so on) [15]. The diffractive profile with the +1.75 D addition directs most part of the incoming light energy onto the 0th, 1st and 2nd diffractive order (far, intermediate and near focus, respectively), while part of the energy is lost on the other diffractive orders [15]. As the second diffractive profile (with added +1.75 D) has an additional power that is half of the first diffractive profile (with added +3.50 D), it is possible to notice that the 2nd order of the second profile is directed onto the near focus which coincides with the 1st order of the first profile [15]. Due to its apodized profile the optical performance of the lens is pupil dependent and the lens material is a hydrophilic acrylic [15]. The MTF of the lens at far, intermediate and near focus considering a pupil aperture of 3.0 mm and a spatial frequency of 50 lp mm^{-1} is, respectively, 0.35, 0.16 and 0.30 [15]. The MTF at the same foci (far, intermediate and near) and considering an aperture of 4.5 mm pupil diameter is 0.42, 0.11 and 0.21 [15]. The MTF values were assessed in an optical bench that is in accordance to the model eye proposed by the ISO standard 11979-2 [15].

The AT LISA Tri 839 MP (from Carl Zeiss Meditec) is a diffractive trifocal lens with two diffractive profiles that cover the entire optic surface [16]. The first profile is responsible for generating the trifocal performance (far, intermediate and near focus) and it covers an optic area up to 4.34 mm diameter [16]. The second diffractive profile that covers the surface from 4.34 mm to 6.00 mm diameter consists of a bifocal design (far and near focus) [16, 17]. The trifocality is created by using the 0th, 1st and 2nd diffractive orders [18]. The lens material consists in a hydrophilic acrylic and the additional power between the far and intermediate focus is +1.66 D, while the additional power between the far and near focus is +3.33 D [16]. To the best of our knowledge no detailed information regarding the diffractive profile of the lens has been published by Carl Zeiss. However, there is a patent issued by Menicon [19] (a Japanese company) that had its rights transferred to Carls Zeiss. In this patent is described a range of possible modes to construct a diffractive profile to generate trifocality [19]. The MTF performance of the lens at 3.0 mm aperture and considering a spatial frequency of 50 lp mm^{-1} at far, intermediate and near focus is 0.38, 0.15 and 0.19, respectively [17]. At an aperture of 4.5 mm the MTF at far, intermediate and near is 0.27, 0.11 and 0.18, respectively [17]. The MTF values were acquired in an optical bench that considers a model eye proposed by the ISO standard 11979-2 [17].

The Acrysof IQ PanOptix (from Alcon Laboratories) is a diffractive trifocal lens with a non-apodized profile that covers an optical area up to 4.5 mm diameter (the remaining optical area from 4.5 mm to 6.0 mm is refractive) [14]. The trifocal optical performance is achieved by using the 0th, 2nd and 3rd diffractive orders to create the far, intermediate and near focus, respectively [11]. The diffractive profile of the lens is based on a quadrifocal profile (figure 10.10(a)) which consists in a diffractive structure with three step heights [16]. However, one of the steps of the quadrifocal profile is suppressed to redistribute the light energy onto the far focus (figure 10.10(b)) [16]. The additional power between the far and intermediate focus is +2.17 D and the additional power between the far and near focus is +3.25 D [16]. The lens is made of hydrophobic acrylate material [16]. The MTF of the lens considering an aperture of 3.0 mm diameter and a spatial frequency of 50 lp mm^{-1} at far, intermediate and near focus is 0.40, 0.15 and 0.18, respectively [18]. The MTF

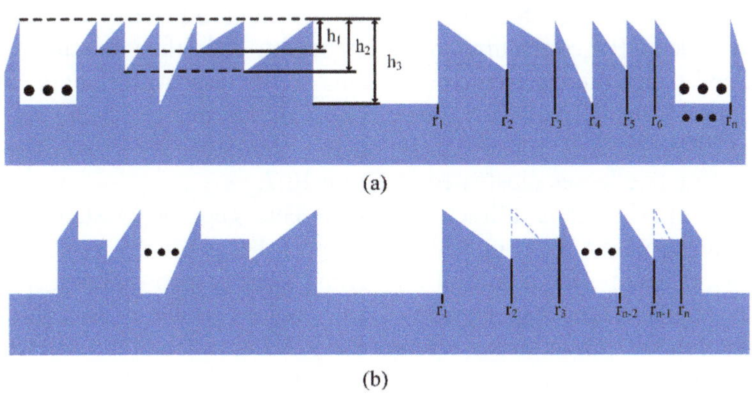

Figure 10.10. Diffractive profile of: (a) a quadrifocal design covering the entire optics and (b) Acrysof PanOptix IQ design covering the optic region up to 4.5 mm diameter [16].

Table 10.1. Some of the main optical features of the bifocal and trifocal diffractive IOLs mentioned in this section and present on the market.

IOL Parameters	Tecnis ZM900	Acrysof Restor SN6AD1	FineVision Micro F	AT LISA Tri 839 MP	Acrysof PanOptix IQ
Multifocality	Bifocal	Bifocal	Trifocal	Trifocal	Trifocal
Add power (near/inter.)	+4 D/NA	+3 D/NA	+3.50 D/ +1.75 D	+3.33 D/+1.66 D	+3.25 D/+2.17 D
Diffractive orders	0th, 1st	0th, 1st	0th, 1st, 2nd	0th, 1st, 2nd	0th, 2nd, 3rd
Diffractive optical size	Full optics	Up to 3.6 mm	Full optics	Up to 4.34 mm trifocal and from 4.34 to 6.00 bifocal	Up to 4.5 mm
MTF at 3 mm aperture and 50 lp mm^{-1} (far/inter./near)	0.45/NA/0.30	0.42/NA/0.33	0.35/0.16/0.30	0.38/0.15/0.19	0.40/0.15/0.18
Lens material	Hydrophobic	Hydrophobic	Hydrophilic	Hydrophilic	Hydrophobic

measurements were performed in an optical bench that follows the model eye proposed by the ISO standard 11979-2 [18].

Table 10.1 shows a compilation of the main features of the diffractive profiles of the previously mentioned off-the-shelf IOLs and some of their optical performance metrics, such as the MTF.

10.3 Diffractive multifocal lens modulated by a mathematical function

As explained in the previous section and noted from equations (10.2) to (10.3), the height of the diffractive steps determines the light efficiency of the different diffraction orders. By controlling the step height, it is then possible to control how light splits into the different focal points of the lens. Additionally, if the step height changes from the center to the periphery of the lens, the energy distribution among the different focal points can be made pupil dependent, as in the so-called apodized diffractive lenses illustrated in figure 10.7.

Regardless of whether a diffractive lens presents steps of constant height as in figure 10.6, an apodized behavior with step heights that gradually reduce towards the periphery as in figure 10.7, a mix of alternating steps of two different heights as in figure 10.8, or alternating step heights of an apodized profile as in figure 10.9, the distribution of the heights of the steps across the radial direction of the diffractive profile can be mathematically described. In other words, it is possible to define a mathematical function, periodic or not, that envelopes the diffractive profile of the lens and defines the height of each individual diffractive step at each respective radial position r_n.

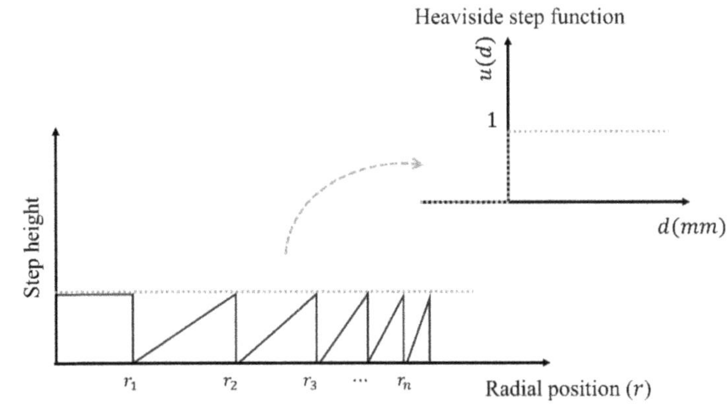

Figure 10.11. Diffractive profile with steps of constant height enveloped by a Heaviside step function.

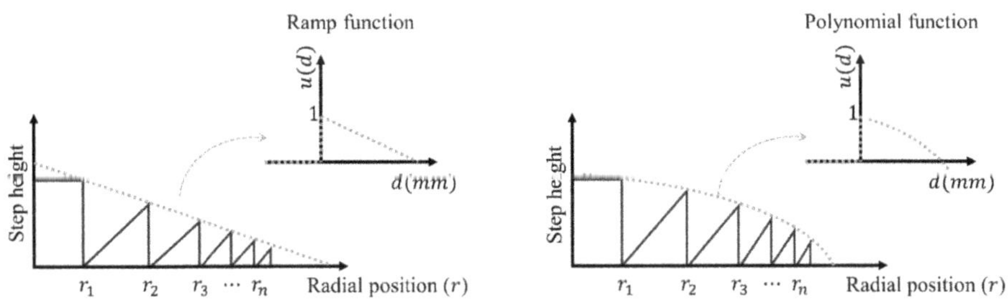

(a) (b)

Figure 10.12. Apodized diffractive profiles described by linear (a) and polynomial functions (b).

Consider a diffractive lens such as the bifocal Tecnis ZM900. For such a profile, each diffractive step, positioned at a given radial distance r_n from the center of the lens, presents the same height h. Therefore, this diffractive profile can be enveloped by a Heaviside step function $u(d)$, for instance, in which the dependent variable u represents the step height and the independent variable d is the radial distance from the lens center, such as illustrated in figure 10.11.

Similarly, for apodized lenses, such as Acrysof Restor SN6AD1, the gradual reduction of the height of the steps towards the periphery of the lens can be described by either linear (figure 10.12(a)), exponential or polynomial (figure 10.12(b)) functions that decay with increasing radial position. In this case, the choice of the function defines the optical behavior of the lens for different pupil sizes.

In a trifocal lens, the final diffractive profile is obtained from the combination of two independent profiles with different step heights. The resulting profile presents then a pattern of alternated lower and higher diffractive steps. Such pattern suggests that a periodic function can be used to envelope the diffractive profile and define the height of each individual step. The figure 10.13 illustrates the use of a square wave function with frequency that increases towards the periphery, in which the valleys define the height of the lower steps and the peaks define the height of the higher steps. The varying

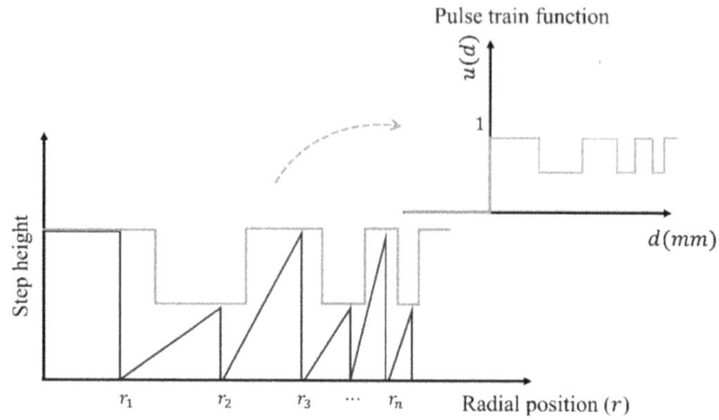

Figure 10.13. Diffractive profile of a trifocal lens described by a square wave function with increasing frequency.

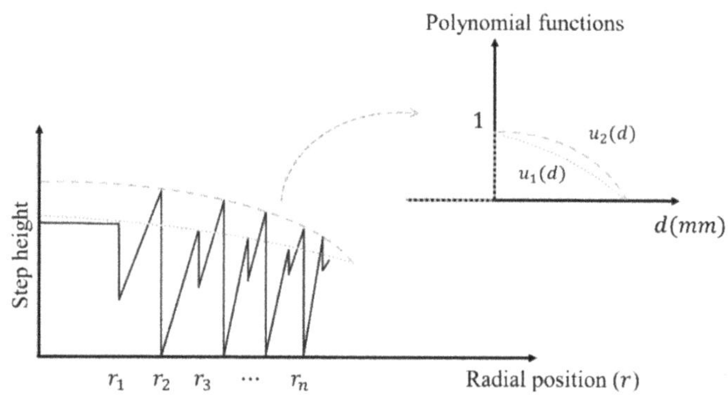

Figure 10.14. Apodized trifocal diffractive profile described by two polynomial functions.

frequency of this function is required to account for the gradual reduction in the distance between two subsequent steps, as established by equation (10.1).

If the trifocal lens includes the apodization feature, such as the FineVision Micro F, each of the individual diffractive component can be independently modulated by linear or polynomial functions. The final profile can then be described by two different functions that apply to either even or odd steps, as illustrated in figure 10.14.

In the context of the present text, the functions $u(d)$ will be called modulation functions in the sense that they can be used to define the shape of the diffractive profile by determining the height of each individual diffractive step.

10.3.1 Diffractive multifocal lens modulated by the modulus of a shifted cosine function

In the previous sections, it was demonstrated that the distribution of heights of the steps of a diffractive profile can alter significantly the optical behavior of the lens by controlling how light splits among the different focal points and across the pupil size. Moreover, it was also demonstrated that a mathematical function can be used to envelope the diffractive profile and hence define the height of each individual diffractive step, therefore referred to as a modulation function.

In this context, it makes sense to wonder if there is the possibility to correlate the analytical parameters of a given modulation function with the merit figures that characterize the optical behavior the lens, and specifically the light efficiency for the different diffraction orders. Then, for a given desired optical behavior, the modulation function could be promptly specified. Alternatively, the study of the variation of the function parameters can lead to optical performance that cannot be intuitively achieved with the current lens design methods. And finally, the said parameters could yet be defined through optimization algorithms so that the lens could meet the required optical performance. Therefore, this approach can represent a very useful tool for a lens designer, since it allows for automation of the process to define the diffractive profile and also opens up new possibilities to enhance the optical performance of a diffractive lens.

The alternated distribution of step heights in a diffractive profile of a trifocal lens suggests that a periodic function would be suitable as a modulation function.

As can be noted from equation (10.1), the distribution of the radial positions (r_n) of the diffractive steps is not linear. They get closer to each other towards the periphery of the lens. The combination of a periodic function to modulate the step heights with an aperiodic distribution of the radial positions of the diffractive steps (r_n), generates a diffractive pattern that appears to be randomized rather than alternated, as in conventional trifocal lens designs (figure 10.15).

A natural choice for a periodic function to modulate the step heights would be a cosine function. The function would have to be shifted upwards to yield only positive values, and a phase would need to be applied so that the function only starts in the radial position of the lens where the diffractive steps start. However, if a shifted cosine function is used, the steps that fall within the valley regions could be too small (even less than 0.1 μm), as is illustrated for steps r_3 and r_4 in the figure 10.15. With such small steps, the manufacturing of the diffractive profile could become unfeasible or too costly. So, a shifted cosine function would have limited applicability as a modulation function of the diffractive profile.

An alternative to reduce the likelihood of generating such small steps is to use the modulus of the cosine function. Such function reduces the sizes of the valley regions

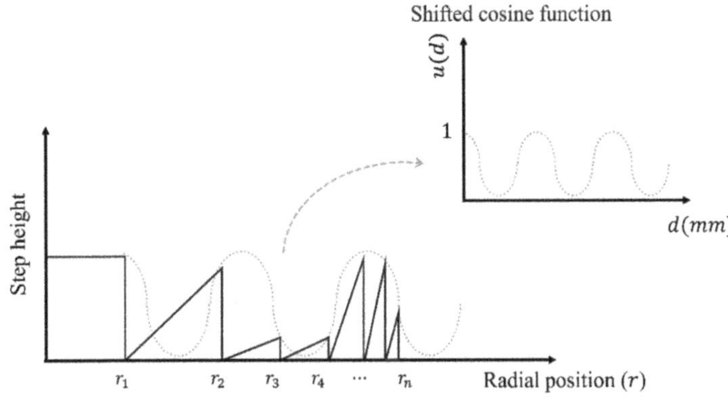

Figure 10.15. Diffractive pattern modulated by a cosine function.

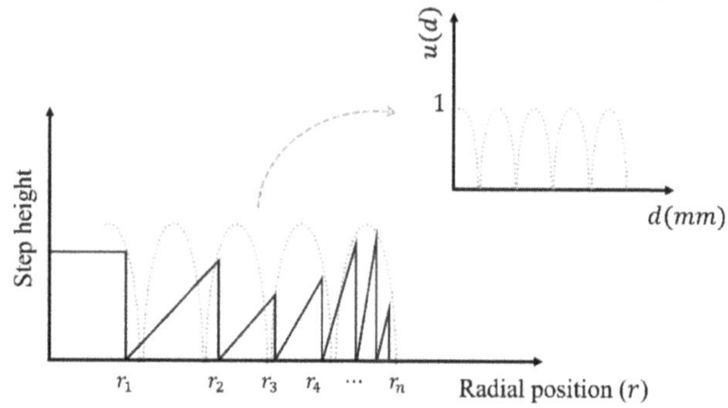

Figure 10.16. Diffractive pattern modulated by the modulus of a cosine function.

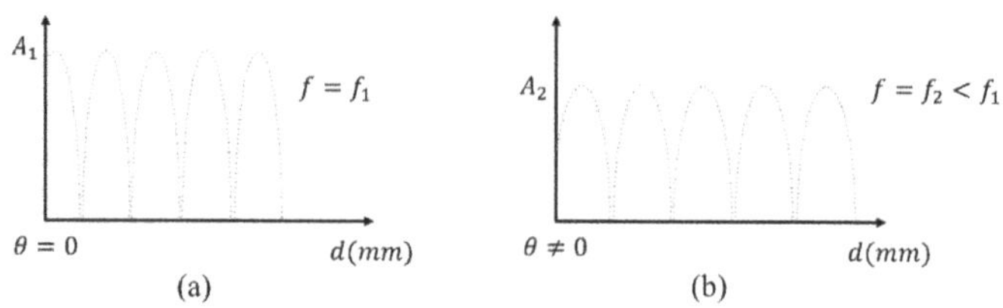

Figure 10.17. Impact of varying the parameters of the modulus of the cosine function.

and still provides a periodic behavior, as can be seen in figure 10.16. Therefore, the obtained step heights are high enough to be reliably manufactured.

The modulus of the cosine function can be mathematically represented by the equation (10.5),

$$m(d) = |A \cdot \cos(2\pi \cdot f \cdot d + \theta)| \tag{10.5}$$

where A is the amplitude of the cosine function in micrometers; f is the spatial frequency in mm^{-1}; d is the independent variable that describes the radial distance from the center of the lens and thus is associated with the position r_n of each diffractive step; and θ is a phase constant in rad.

Figure 10.17 illustrates the effect of varying the function parameters. The graph shown in figure 10.17(a) shows a phase constant equal to zero, a higher spatial frequency f_1 and a higher amplitude A_1, when compared with the graph in figure 10.17(b), in which the phase constant is non-zero and the spatial frequency and amplitude are represented by f_2 and A_2, respectively.

10.3.1.1 Mathematical model of the light efficiency

The calculation of the energy distribution among the different diffractive orders created by a diffractive structure is an essential step of the lens design process. Different articles in the literature approach the calculation of light efficiency of diffractive profiles [1, 7, 20–22], and consider profiles with steps of constant height to demonstrate that the energy efficiencies can be calculated following equations (10.2)–(10.4).

However, a recent study [14] proposes the calculation of the light efficiency of the diffractive orders based on the orthogonal expansion of the PSF produced by the lens. The adopted method requires that the steps have triangular shape, as is often the case, and that the annular surface area defined between two subsequent steps is constant, which is guaranteed by the radial positions as given by equation (10.1). Under such conditions, the light efficiency of a diffractive profile with any distribution of step heights can be calculated through the square modulus of the term b_{00}, which is defined by equation (10.6).

$$b_{00} = \sum_{n=0}^{N} (\rho_n^2 - \rho_{n-1}^2) \cdot \exp\left[-i2\pi\left(\beta_n + \frac{\alpha_n}{2} - \frac{W_{20}}{2}(\rho_n^2 + \rho_{n-1}^2)\right)\right] \cdot \text{sinc}\left[W_{20}(\rho_n^2 - \rho_{n-1}^2) - \alpha_n\right] \quad (10.6)$$

where ρ_n represents the position of the nth diffractive step (r_n) normalized by the semi-diameter of the lens optical zone ($r_{max} = 3.0$ mm); W_{20} is the defocus coefficient in waves; α_n and β_n describe the phase profile of the nth step, in waves, as shown in figure 10.18; and N corresponds to the total number of diffractive steps.

Equation (10.6) describes a generalist formulation for the calculation of the light efficiency of a diffractive profile. For a particular condition in which the step heights are modulated by a given mathematical function, equation (10.6) has to be manipulated so as to include the parameters of the modulation function. Different parameters of the argument of the function can be rewritten according to the following steps.

The defocus coefficient can be written as shown in equation (10.7) [14]:

$$W_{20} = \phi_{\text{base}} \cdot \frac{r_{\text{max}}^2}{2\lambda} \quad (10.7)$$

where ϕ_{base} is the nominal optical power of the lens; r_{max} is the maximum aperture size (3.0 mm) and λ is the wavelength used in the design (546 nm).

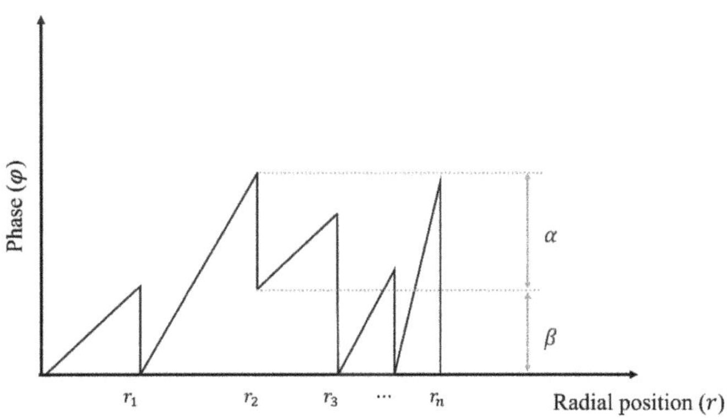

Figure 10.18. Graphical representation of the parameters α and β in equation (10.6).

Considering that the areas of the diffractive zones are always the same, the area of each zone can then be described as shown in equation (10.8):

$$A_{\text{zone}} = \frac{A_{\text{disk}}}{\text{number of zones}} = \frac{\pi\rho_{\max}^2}{N} \tag{10.8}$$

where A_{disk} is the normalized area of the optical surface limited by the normalized radius $\rho_{\max}$; and N is total number of diffractive steps on the optical surface of the lens.

An alternative way to calculate the normalized area of a given diffractive zone limited by two subsequent radial positions ρ_{n-1} and ρ_n is presented in equation (10.9):

$$A_{\text{zone}} = A_{\text{outer disk}} - A_{\text{inner disk}} = \pi(\rho_n^2 - \rho_{n-1}^2) \tag{10.9}$$

Combining equations (10.8) and (10.9) and considering that $\rho_{\max} = 1$, leads to equation (10.10):

$$\frac{\pi \cdot 1^2}{N} = \pi\left(\rho_n^2 - \rho_{n-1}^2\right) \longrightarrow \left(\rho_n^2 - \rho_{n-1}^2\right) = \frac{1}{N} \tag{10.10}$$

Using equation (10.1) to describe the radial position of each diffractive step, it is possible to rewrite the ρ_n^2 as shown in equation (10.11):

$$\rho_n^2 = \left(\frac{r_n}{r_{\max}}\right)^2 = \left(\frac{\sqrt{\frac{2n\lambda}{\phi_{\text{add}}}}}{r_{\max}}\right)^2 = \frac{2n\lambda}{r_{\max}^2 \cdot \phi_{\text{add}}} \tag{10.11}$$

Equations (10.7) and (10.11) can then be used to rewrite the argument of the sinc-function of equation (10.6), as follows in equation (10.12):

$$W_{20}\left(\rho_n^2 - \rho_{n-1}^2\right) = \phi_{\text{base}} \cdot \frac{r_{\max}^2}{2\lambda}\left(\frac{2n\lambda}{r_{\max}^2 \cdot \phi_{\text{add}}} - \frac{2(n-1)\lambda}{r_{\max}^2 \cdot \phi_{\text{add}}}\right) = \frac{\phi_{\text{base}}}{\phi_{\text{add}}} \tag{10.12}$$

Similarly, one of the parameters of the argument of the exponential function of equation (10.6) can be rewritten as shown in equation (10.13):

$$\frac{W_{20}}{2}\left(\rho_n^2 + \rho_{n-1}^2\right) = \phi_{\text{base}} \cdot \frac{r_{\max}^2}{4\lambda}\left(\frac{2n\lambda}{r_{\max}^2 \cdot \phi_{\text{add}}} + \frac{2(n-1)\lambda}{r_{\max}^2 \cdot \phi_{\text{add}}}\right) = \frac{\phi_{\text{base}}}{\phi_{\text{add}}}\left(n - \frac{1}{2}\right) \tag{10.13}$$

The phase parameter β is assumed as zero and α is defined by the modulation function described by equation (10.5), as follows in equation (10.14):

$$\alpha_n = |\, A \cdot \cos(2\pi \cdot f \cdot r_n + \theta)| \cdot \frac{(n_{\text{IOL}} - n_{\text{aq}})}{\lambda} \tag{10.14}$$

Equations (10.10) and (10.12) to (10.14) can then be substituted into equation (10.6) to yield the final version of b_{00}, as follows:

$$b_{00} = \sum_{n=0}^{N} \frac{1}{N} \cdot \exp\exp\left[-i2\pi\left(\frac{A \mid \cos\cos(2\pi \cdot f \cdot r_n + \theta) \mid \cdot \left(n_{\mathrm{IOL}} - n_{\mathrm{aq}}\right)}{2\lambda} - \frac{\phi_{\mathrm{base}}}{\phi_{\mathrm{add}}}\left(n - \frac{1}{2}\right)\right)\right] \cdot \mathrm{sinc}\left[\frac{\phi_{\mathrm{base}}}{\phi_{\mathrm{add}}} - A \mid \cos\cos(2\pi \cdot f \cdot r_n + \theta) \mid\right] \quad (10.15)$$

Finally, the light efficiency of the lens can be calculated from the square of the modulus of equation (10.15).

10.3.2 Optical design optimization process and solution selection

By means of the methodology and mathematical formulation presented in the previous sections, it is possible to establish an optimization process based on the parameters of the modulation function as design variables to develop a diffractive trifocal IOL [23]. The optimization process can be separated in two main stages. In the first one the base power of the lens is optimized by using a direct search method called damped least square (DLS), and in the second stage the heights of the diffractive steps of the lens are optimized considering the modulation function proposed in the previous section by using a stochastic method called non-dominated sorting genetic algorithm II (NSGA-II). By separating the optimization process in two stages it is possible to simplify the complexity of the entire problem because in the first stage a monofocal problem is solved by a well-known and efficient direct search method, and in the second stage the process is focused only in the design variables (step height) that directly affect the multiobjective merit functions (efficiency at each focus).

An example is outlined in this section to demonstrate how to optimize a diffractive multifocal IOL using the two-stage method previously mentioned. In the first stage of the optimization some pre-defined conditions must be established, such as the lens material, the model eye, the base power and the lens shape (aspherical or spherical). The eye model considered is the ISO eye model because it is commonly employed to compare IOLs with different optical designs and it is the model required by the ISO standard 11979-2 to evaluate the performance of IOLs [24]. The lens material selection is important to define the refractive index of the lens, which has an impact on the curvatures and thickness of the IOL for a certain base power. Among the possibilities for IOL raw material, it is chosen an hydrophilic option with refractive index of 1.461 [25]. The clear optical size of the IOL must be within the specifications of the applied standard [26] and considering off-the-shelf IOLs [27] a value of 6.00 mm diameter is chosen. The base power of the lens is defined as 20 D, which corresponds to a conventional IOL power. The lens shape is biconvex with an anterior aspheric surface to eliminate the spherical aberration of the lens. The minimization of the difference between the MTF curve at the focal plane of the system, the retina, and that of the diffraction limited MTF is taken as an objective function of the optimization process. The design variables, which are optimized to guarantee that the IOL meets the base power constraint (20 D) and the merit function requirement (MTF at the diffraction limit), are the following: the

Table 10.2. Pre-defined parameters of the lens.

Parameter	Value
IOL refractive index	1.461
IOL clear optics	6.00 mm
IOL base power	20 D

radius of curvature of the anterior surface (R_a), the conic constant of the anterior surface (k_a), the radius of curvature of the posterior surface (R_p) and the center thickness (t). Table 10.2 presents the pre-defined parameters of the lens.

The optimization problem and goal for this stage of the process is formulated in equations (10.16) to (10.21). The limits established for the design variables R_a, k_a, R_p and t were based on typical values found in IOLs on the market.

Minimize:

$$\Delta\text{MTF}(R_a, k_a, R_p, t) \tag{10.16}$$

Subjected to:

$$\text{IOL}_{\text{power}} = 20\text{D} \tag{10.17}$$

$$8\ \text{mm} \leqslant R_a \leqslant 100\ \text{mm} \tag{10.18}$$

$$-20 \leqslant k_a \leqslant 0 \tag{10.19}$$

$$-29\ \text{mm} \leqslant R_p \leqslant -6\ \text{mm} \tag{10.20}$$

$$0\ \text{mm} \leqslant t \leqslant 1.2\ \text{mm} \tag{10.21}$$

Once the curvatures of the lens and conic constant are optimized to guarantee the base power and a diffraction limited MTF performance at the focal plane, the second stage starts with the optimization of the diffractive steps height by employing the modulation function (see equation (10.5)) to achieve a trifocal optical performance.

Initially, a few decisions must be taken to define the optical properties of the diffractive structure and guide the optimization flow. As presented in section 10.2, there are diffractive lenses on the market that have diffractive steps on the entire clear optics, such as Tecnis ZM900 IOL model, and models that have diffractive steps only up to a portion of the clear optics (Acrysof Restor SN6AD1 model for example). As the diffractive steps get narrower towards the periphery of the lens, it has been established that the diffractive region covers the clear optics on a region up to 5.00 mm diameter, thus avoiding the additional hurdles and costs of attending to such fine diffractive steps in the manufacturing process. Although there are diffractive IOLs on the market with additional power between the far and near focus varying within a range of +3.00 D and

+4.00 D, studies based in clinical assessment have shown that diffractive MIOLs with +3.00 D additional power yield an near visual acuity with superior quality when compared to MIOLs with +4.00 D additional power [28, 29]. Based on that, the additional power between two consecutive diffractive orders for the lens in this study has been defined as 1.50 D, which corresponds to an additional power of 1.50 D between the far and intermediate focus (0th and 1st diffractive orders, respectively) and 3.00 D between the far and near focus (0th and 2nd diffractive orders, respectively). Considering a design wavelength (λ) of 546 nm, which is the wavelength required by the ISO standard 11979-2 to evaluate IOLs [24], it is possible to find from equation (10.1) all the diffractive-step positions (r_n), yielding 16 diffractive steps to cover an optical area up to 5.00 mm diameter.

The last step to finalize the structure of the optimization problem consists in defining the merit function. As the main objective of the optimization of the step heights is to achieve a trifocal performance, the efficiency at the three main foci (far, intermediate and near) is considered as merit functions. Having more than one objective function characterizes a multiobjective optimization problem and uses the stochastic method NSGA-II.

The ISO standard 11979-2 requires the evaluation of the MIOLs for two distinct pupil apertures: one between 2 mm and 3 mm diameter and the other between 4 mm and 5 mm diameter [24]. Based on that, the optimization of the diffractive surface is performed in two steps: in the first step the diffractive step heights covering the optics region up to 3 mm diameter are optimized by the modulation function, and in the second step the remaining diffractive steps (from 3 mm to 5 mm diameter) are optimized by a modulation function with the same mathematical form, but different parameters values. The segmentation of the optimization problem in two steps is convenient because the parameters of the modulation function that guarantee a good optical performance at 3 mm aperture do not necessarily yield a good optical performance at an aperture of 5 mm. Besides, this procedure enables the reduction of the number of parameters to be simultaneously optimized by the NSGA-II, and renders the flexibility to optimize the lens performance for different pupil apertures [23].

Considering the parameters of the modulation function (A, f and θ) as design variables, the optimization problem is formulated by equations (10.22) to (10.25) [23]:

Maximize:

$$\mathrm{eff}_{\mathrm{IOL}}(A, f, \theta) = \{\mathrm{eff}_{\mathrm{far}}(A, f, \theta), \mathrm{eff}_{\mathrm{intermediate}}(A, f, \theta), \mathrm{eff}_{\mathrm{near}}(A, f, \theta)\} \quad (10.22)$$

Subjected to:

$$0(\mu\mathrm{m}) \leqslant A \leqslant 5(\mu\mathrm{m}) \quad (10.23)$$

$$0(1/\mathrm{mm}) \leqslant f \leqslant 2 \times 10^6(1/\mathrm{mm}) \quad (10.24)$$

$$0(\mathrm{rad}) \leqslant \theta(\mathrm{rad}) \quad (10.25)$$

The limits established for the design variables (equations (10.23)–(10.25)) have been defined to guarantee a broad search space during the optimization process. For instance, the step height (A) upper limit guarantees a broad efficiency variation for the 0th, 1st and 2nd diffractive orders as shown in (10.4).

At the end of the optimization process by the NSGA-II algorithm, a set of estimated Pareto solutions are available as feasible and possible solutions for the aforementioned optimization problem (equations (10.23)–(10.25)). Each solution of the Pareto front corresponds to a different combination of efficiencies distributed on the far, intermediate and near focus. Based on that, a selection criterion must be defined to extract the solution that best attends to the required performance. From the optical performance of the trifocal IOLs presented in section 10.2, it is possible to notice that the far and near focus of the lenses are often prioritized in relation to the intermediate focus. Besides, there are models that present a pupil dependent performance (such as the FineVision Micro F model) and models that do not depend on the pupil aperture (model AT LISA Tri 839 MP). With the intent to develop a pupil independent trifocal IOL and prioritizing the far and near focus over the intermediate focus, the ratio between the efficiency at the far and near foci close to the unit is considered as selection criterion. It is possible to find a diversity of solutions that respect this criterion. Aiming at trifocality with optimal energy distribution among the foci, it is reasonable to introduce a second criterion that searches for the highest sum of efficiencies at the three lens foci. Therefore, in summary, these two criteria are applied in sequence to select the best solution. Firstly, the set of solutions with ratio between the far and near focus close to the unit is selected. Subsequently, within that set, the solution with the highest sum of the efficiencies at the main foci is chosen.

10.3.3 Design parameters and optical performance of selected solution

In the first stage of the optimization process, as mentioned in the previous section, the optical parameters R_a, k_a, R_p and t were optimized to guarantee the base power (20 D) and a diffraction limited MTF performance. The optimal parameters found by the DLS algorithm are presented in figure 10.19 [23].

In the second stage, the parameters of the modulation function (A, f and θ) were optimized to achieve a diffractive structure that creates a trifocal performance. At this stage, the optimization is performed in two regions of the diffractive surface: one from the center of the lens to 3 mm diameter and the other from 3 mm to 5 mm diameter. The set of estimated Pareto solutions generated by the NSGA-II algorithm for the first diffractive region (0 mm to 3 mm diameter) and the selected solution is presented in figure 10.20. The Pareto dominance plots (figures 10.20 and 10.21) show on the x axis the efficiency on the far focus, on the y axis the efficiency on the near focus, and the color bar indicates the efficiency on the intermediate focus. For the chosen solution, the efficiency at the far, intermediate and near focus is 0.34, 0.10 and 0.35, respectively [23].

The set of estimated Pareto solutions obtained for the second diffractive region (from 3 mm to 5 mm diameter) is presented in figure 10.21. It is important to mention that the chosen solution for the region from 3 mm to 5 mm diameter

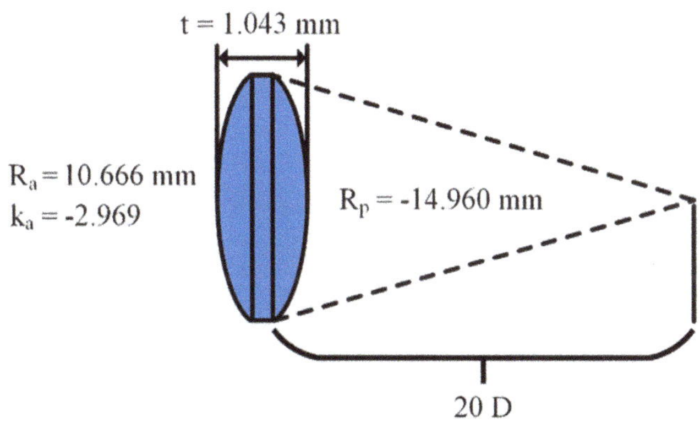

Figure 10.19. Optimized refractive optical parameters of the IOL found by the DLS algorithm: R_a—anterior radius of curvature, k_a—conic constant of the anterior surface, R_p—posterior radius of curvature and t—center thickness.

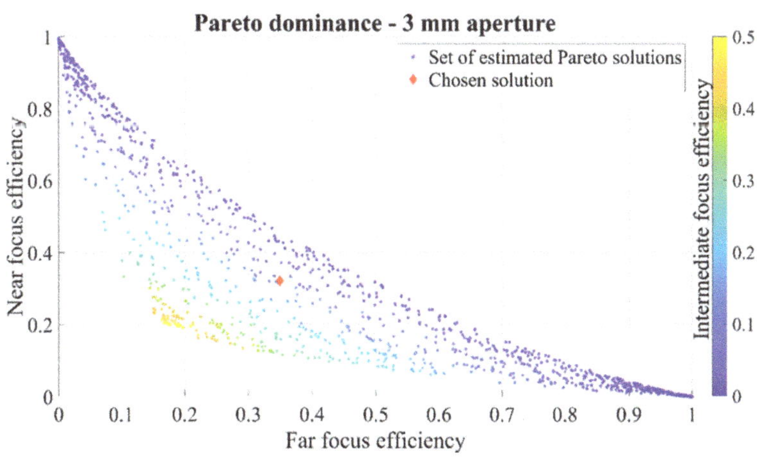

Figure 10.20. Set of estimated Pareto solutions and chosen solution obtained after the NSGA-II optimization process applied to the diffractive steps from 0 mm to 3 mm diameter. Reproduced from [23], copyright (2020) with permission from Elsevier.

considers a diffractive structure from 0 mm to 3 mm that had been chosen in the previous step. For the selected solution, the efficiency at the far, intermediate and near focus is 0.32, 0.10 and 0.32, respectively [23].

The parameters of the modulation function (A, f and θ) that correspond to the solution presented in figure 10.20 are: $A = 2.859$ μm, $f = 8.004\,857\,017\,75 \times 10^3$ mm^{-1} and $\theta = 6.148$ rad [23]. The optimum parameters that correspond to the solution highlighted in figure 10.21 are: $A = 2.862$ μm, $f = 1.362\,444\,7692 \times 10^6$ mm^{-1} and $\theta = 6.568$ rad [23]. The two set of optimized parameters are employed in the modulation function to determine the step height of each diffractive step at each radial position (r_n). At this stage, all the constructive parameters of the diffractive trifocal IOL have been defined by optimization. Considering those parameters, the optical performance of the lens was evaluated in a ray-tracing software (OpticStudio

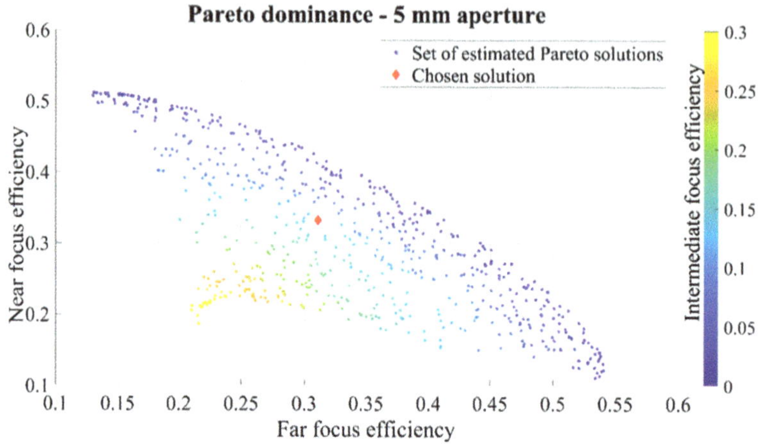

Figure 10.21. Set of estimated Pareto solutions and chosen solution obtained after the NSGA-II optimization process applied to the diffractive steps from 3 mm to 5 mm diameter. Reproduced from [23], copyright (2020) with permission from Elsevier.

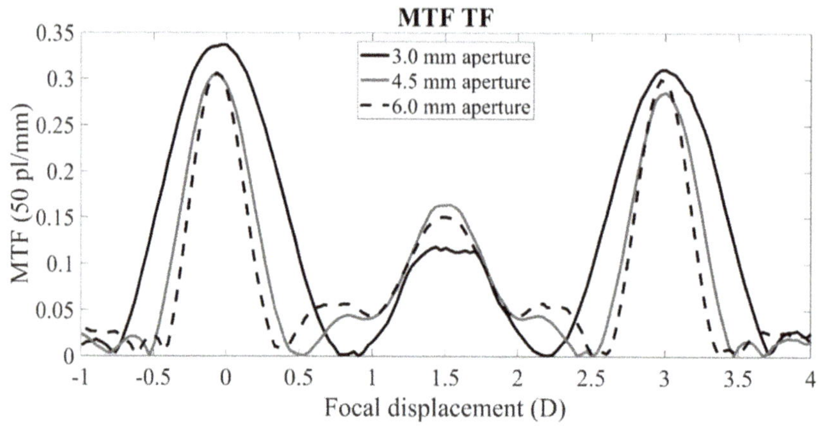

Figure 10.22. Simulated TF-MTF curves at 50 lp mm^{-1} for 3.0 mm, 4.5 mm and 6.0 mm aperture in the ISO eye model. Reproduced from [23], copyright (2020) with permission from Elsevier.

2013 version) in a model eye proposed by ISO standard 11979-2 [23]. The through-focus MTF (TF-MTF) curves at a spatial frequency of 50 lp mm^{-1} considering 3.0 mm, 4.5 mm and 6.0 mm aperture diameter were obtained and it is presented in figure 10.22 [23].

The plots presented in figure 10.22 indicate that the design guarantees an additional power between the far and intermediate focus and between the far and near focus of +1.5 D and +3.0 D, respectively, where the far focus corresponds to 0 D. The TF-MTF peaks at far and near focus for 3.0 mm, 4.5 mm and 6.0 mm aperture had a ratio between the far and near peak of 1.09, 1.07 and 1.02, respectively. This result shows that this ratio was close to the unit for all apertures, which represents a good pupil independent optical performance, with a slight absolute decrease in contrast and a decrease in depth of focus as the pupil increases. The MTF performance at far and near focus is significantly higher than the performance at the intermediate, as expected due to the criterion defined to select the best solution from the estimated set of Pareto solutions.

In order to further evaluate the optical performance of the optimized diffractive trifocal IOL, a comparison analysis is established considering two diffractive trifocal lenses present on the market: AT LISA Tri 839 MP and FineVision Micro F [23]. The simulated MTF results of the lens are compared with the *in vitro* measurements of the AT LISA Tri and FineVision lenses (both lenses measured considering the same model eye employed on the optimization of the lens presented in this section [10, 17]). Table 10.3 presents the MTF values at 50 lp mm^{-1} spatial frequency of the optimized IOL at far, intermediate and near distances considering 3.0 mm and 4.5 mm aperture. It is important to highlight that, as the trifocal IOLs considered in this comparison analysis present different additional powers, the MTF was evaluated at the focal plane corresponding to the additional power of each lens. Therefore, the comparison at each focus corresponds to focal planes in different positions.

Analysing table 10.3 for 3.0 mm aperture values, it is possible to notice that the optimized lens presents MTF values close (less than 4%) to the values of the FineVision Micro F model at far and near focus and a more significant difference (25%) at the intermediate focus, in which the optimized lens has a lower performance. In relation to the AT LISA Tri 839 MP, the MTF of the optimized lens is lower for the far and intermediate focus and higher for the near focus. Besides, the AT LISA Tri 839 MP presents the highest difference between the performance at the far and near focus, while the optimized lens and FineVision Micro F model present a performance at far and near focus with a ratio closer to the unit at a 3-mm aperture.

At 4.5 mm aperture, table 10.3 shows that the optimized lens presented a better optical performance at all the three foci when compared to the AT LISA Tri 839 MP model and it also presents a higher MTF value at intermediate and near focus when compared with the FineVision Micro F model. The optimized lens has a lower MTF value at the far focus when compared to the FineVision Micro F model, as expected, since the latter is apodized, which corresponds to a prioritization of the far focus

Table 10.3. Comparative table with the MTF values at 50 lp mm^{-1} and apertures of 3.0 mm and 4.5 mm diameter in ISO eye model at far, intermediate and near distances.

	IOL Model	Optimized IOL	AT LISA Tri 839 MP	FineVision Micro F
3.0 mm	MTF far	0.34	0.40	0.35
	MTF intermediate	0.12	0.15	0.16
	MTF near	0.31	0.18	0.30
4.5 mm	MTF far	0.31	0.27	0.42
	MTF intermediate	0.16	0.10	0.11
	MTF near	0.29	0.20	0.21

performance as the aperture increases, whereas the optimized lens was designed to have favor pupil independent optical performance.

In summary, the optimized lens has an optical performance comparable and competitive with the two trifocal lenses present on the market and used in the comparison study. Some differences are observed for different apertures and foci and this is due to the differences in the design process of each IOL such as additional power, apodization and diffractive step height combination. It is important to highlight, as an evident advantage of the method presented, that the use of analytical functions combined with optimization procedures broadens the range of possible diffractive solutions, sometimes not intuitive to the designer, that can be deployed in an IOL to deliver optical functions with the characteristics required by the project inputs.

10.4 Summary

This chapter presented the mathematical modeling, design and optimization process of a diffractive trifocal IOL with step height modulated by the modulus of a shifted cosine function. The main objective of the optimization process was to achieve a pupil independent trifocal optical performance with prioritization of the performance at the far and near foci.

From the computational optical results obtained through ray-tracing software, the optimized lens presented an optical performance in accordance with the criteria established: pupil independence; additional power of +1.5 and +3.0 D between far and intermediate and between the far and near focus, respectively; and prioritization of the far and near focus performance. Besides, the comparison analysis with two trifocal IOLs shows that the optical performance of the proposed lens is comparable with that for trifocal lenses currently available on the market.

By using the methodology presented in this chapter, it is possible to develop a broad range of IOLs with different efficiency distribution at each focus by selecting other solutions from those presented in the set of estimated Pareto solutions. Such solutions will provide different MTF profiles for the far, intermediate and near foci, i.e. each point present in the set of estimated Pareto solutions corresponds to an IOL with different optical performance characteristics.

Although the optimization process was developed considering a modulation function represented by the modulus of the cosine function, a variety of other mathematical functions can be employed in the same basis to achieve a different design.

Chapter highlights

- Optimization method applied to kinoform diffractive IOLs to achieve multifocality.
- Mathematical description of the diffraction phenomena present in kinoform diffractive structures.
- Mathematical description of energy efficiency for a diffractive profile modulated by cosine function.
- New method for designing diffractive structures of intraocular lenses.

References

[1] O'Shea D C, Suleski T J, Kathman A D and Prather D W 2003 *Diffractive Optics: Design, Fabrication, and Test* 1st edn (Bellingham, WA: SPIE)

[2] Daniel Malacara-Hernández Z M-H 2003 Diffraction in optical systems, *Handbook of Optical Design* 2nd edn (Boca Raton, FL: CRC Press) p 512

[3] Palmer C 2000 *Diffraction Grating Handbook* 4th edn (Newport, RI: Richardson Grating Laboratory)

[4] Lenkova G A 2010 Effect of the eye pupil diameter and the phase shift in the diffraction structure on bifocal properties of diffractive-refractive intraocular lenses *Optoelectron. Instrum. Data Process.* **46** 264–73

[5] Castignoles F, Flury M and Lepine T 2010 Comparison of the efficiency, MTF and chromatic properties of four diffractive bifocal intraocular lens designs *Opt. Express* **18** 5245

[6] Sokolowski M K-H M, Pniewski J and Brygola R 2015 Hybrid heptafocal intraocular lenses *Opt. Appl.* **45** 285–98

[7] Lenkova G A 2007 Methods for investigating optical characteristics of bifocal diffractive-refractive intraocular lenses *Optoelectron. Instrum. Data Process.* **43** 262–73

[8] David G S and Atchison A 2000 *Optics of the Human Eye* (Oxford: Butterworth-Heinemann)

[9] Akaishi P F T L, Vaz R, Vilella G and Garcez R C 2010 Visual performance of Tecnis ZM900 diffractive multifocal IOL after 2500 implants: a 3-year followup *J. Ophthalmol.* **2010** 8

[10] Gatinel D and Houbrechts Y 2013 Comparison of bifocal and trifocal diffractive and refractive intraocular lenses using an optical bench *J. Cataract Refract. Surg.* **39** 1093–9

[11] Lee S, Choi M, Xu Z, Zhao Z, Alexander E and Liu Y 2016 Optical bench performance of a novel trifocal intraocular lens compared with a multifocal intraocular lens *Clin. Ophthalmol.* **10** 1031–8

[12] Davison J A and Simpson M J 2006 History and development of the apodized diffractive intraocular lens *J. Cataract Refract. Surg.* **32** 849–58

[13] Toto L *et al* 2013 Comparative study of acrysof ReSTOR multifocal intraocular lenses +4.00 D and +3.00 D: visual performance and wavefront error *Clin. Exp. Optom.* **96** 295–302

[14] Schwiegerling J 2016 Diffraction efficiency and aberrations of diffractive elements obtained from orthogonal expansion of the point spread function *Proc. Optical Modeling and Performance Predictions VIII* 9953 995307

[15] Gatinel D, Pagnoulle C, Houbrechts Y and Gobin L 2011 Design and qualification of a diffractive trifocal optical profile for intraocular lenses *J. Cataract Refract. Surg.* **37** 2060–7

[16] Sudhir R R, Dey A, Bhattacharrya S and Bahulayan A 2019 Acrysof IQ panoptix intraocular lens versus extended depth of focus intraocular lens and trifocal intraocular lens: a clinical overview *Asia-Pacific J. Ophthalmol.* **8** 335–49

[17] Soo Son H *et al* 2017 *In vitro* optical quality measurements of three intraocular lens models having identical platform *BMC Ophthalmol.* **17** 9

[18] Carson D, Xu Z, Alexander E, Choi M, Zhao Z and Hong X 2016 Optical bench performance of 3 trifocal intraocular lenses *J. Cataract Refract. Surg.* **42** 1361–7

[19] Kobayashi I A A and Suzuki H 2011 Method for manufacturing aphakic intraocular lens *Patent* EP 2 377 493 A1

[20] Loewen E G and Popov E 1997 *Diffraction Gratings and Applications* (New York: Marcel Dekker)

[21] Levashov V E and Vinogradov A V 1994 Analytical theory of zone plate efficiency *Phys. Rev.* E **49** 5797–803

[22] Londoño C and Clark P P 1992 Modeling diffraction efficiency effects when designing hybrid diffractive lens systems *Appl. Opt.* **31** 2248–52

[23] de Carvalho L M, de Lima Monteiro D W and de Oliveira O G 2020 Computational modeling and optimization of a diffractive trifocal intraocular lens with step heights modulated by a cosine function *Opt. Commun.* **476** 126325

[24] International Organization for Standardization – 2014 Ophthalmic implants—intraocular lenses—optical properties and test methods (ISO Standard No. 11979-2:2014) https://www.iso.org/standard/55682.html

[25] O&O 2020 O&O mdc | IOL Polymers Hydrophilic. http://oo-mdc.com/oomdc-IOL-Polymers-Hydrophilic.html (accessed 29 Aug. 2020)

[26] International Organization for Standardization – 2012 Ophthalmic implants—intraocular lenses—mechanical properties and test methods (ISO Standard No. 11979-3:2012) https://www.iso.org/standard/55681.html

[27] Bellucci R 2013 An introduction to intraocular lenses: material, optics, haptics, design and aberration *Cataract* vol 3 (Basel: S. Karger AG) pp 38–55

[28] Kohnen T, Nuijts R, Levy P, Haefliger E and Alfonso J F 2009 Visual function after bilateral implantation of apodized diffractive aspheric multifocal intraocular lenses with a +3.0 D addition *J. Cataract Refract. Surg.* **35** 2062–9

[29] Maxwell W A, Cionni R J, Lehmann R P and Modi S S 2009 Functional outcomes after bilateral implantation of apodized diffractive aspheric acrylic intraocular lenses with a +3.0 or +4.0 diopter addition power. Randomized multicenter clinical study *J. Cataract Refract. Surg.* **35** 2054–61

Chapter 11

Intraocular lens with sinusoidal patterns: design assisted by a classification algorithm

Diogo Ferraz Costa and Davies William de Lima Monteiro

The industrial sector for intraocular lenses (IOLs) has been constantly developing new models with improved performance and attending to different focal classes, such as multifocal, or enhanced monofocal or extended-depth-of-focus lenses. A tailored performance and the IOL classification within a certain class usually rely on a suitable choice and combination of the lens topographical features. Recently, there has been a special interest in lenses with an extended depth of focus. This chapter presents the design of a biconvex IOL with a bidimensional refractive sinusoidal profile distributed over its posterior surface in an orthogonal grid, where the configuration of different amplitudes and frequencies of the sinusoidal functions lead to distinct optical performance figures. The lens is simulated in a modified Liou–Brennan eye model and, conveniently, the choice of the sinusoidal parameters could enable the IOL to behave as either monofocal (MONO), or multifocal (MIOL), or extended depth of focus (EDoF). A custom methodology then proposed and employed to verify into which class a lens falls when designed with a certain set of parameters. The performance of some of the designed IOLs is presented and discussed, while verified with Snellen visual acuity charts and estimated preclinical defocus curves. This methodology can be used as a reliable method to assist lens manufacturers to investigate optimal design parameters targeting a predefined performance metric, especially when tailored to consider the much richer defocus curves to obtain the score, instead of the through-focus modulation transfer function (MTF) curves.

11.1 Lens design

The proposed lens in this case study is a refractive biconvex lens with a smooth orthogonal sinusoidal pattern distributed over its posterior surface [1, 2]. The sinusoidal pattern allows the introduction of additional design parameters, namely,

doi:10.1088/978-0-7503-3263-7ch11

the amplitude (A) and frequencies (α, β) along the x- and y-axis on the plane of the lens. It can be viewed as an alternative to the designs presented in the literature and in the market. This undulatory pattern 'mimics' a lenslet array, where the small pseudo aperture of each lenslet intrinsically offers a larger depth of field. This also introduces the ability to slightly redirect light rays refracted by the base surface of the lens. This lens design is henceforth referred to as periodic.

The eye model used to simulate the lens performance was a modified Liou–Brennan (L–B) model. The lens emulating the cornea consists of an anterior and a posterior surface separated by 0.5 mm in the center. It features most of the refractive power of the eye, i.e., around 40 D. After the cornea, the anterior chamber is filled by an aqueous material with refractive index of 1.336 and has a depth of 3.16 mm along the main optical axis. Then, the pupil is emulated by an optical clear aperture of a given diameter. The aperture was set to 3.0 mm of diameter for all the simulations presented herein. The original L–B crystalline lens has been replaced by a custom designed IOL with a refractive power of around 20.0 D. The L–B eye features an axial length of 23.95 mm, extending from the anterior cornea to the retina. Considering the IOL central thickness of 1.59 mm, the distance between the center of the posterior IOL surface and the retina, in the vitreous cavity, is 18.70 mm. The vitreous humor is modelled as having a 1.336 refractive index for a 550 nm wavelength. The modified L–B eye model has the parameters presented in table 11.1, where the IOL is to be inserted.

A simple base design could have been that of a plano-convex IOL, where the posterior flat surface would have been modified to include the orthogonal bidimensional sinusoidal characteristics. However, it would have required a significantly high curvature of the anterior surface to account for the necessary ray bending to achieve the base refractive power. This would be particularly detrimental for lenses with high powers. Therefore, the base design was instead chosen to have a biconvex symmetrical structure, maintaining the sinusoidal pattern on the posterior surface (figure 11.1). All the lenses presented in this chapter have an outer optic diameter of 6 mm. The simulations were performed using ZEMAX OpticStudio.

The base design comprises two convex aspheric surfaces whose parameters are presented in table 11.2. They were optimized for emmetropia with a pupil of 6.0 mm, a central lens thickness of 1.59 mm, and an effective lens position of 0 mm from the

Table 11.1. Modified Liou–Brennan eye model specification (adapted from [3]).

	Anterior radius (mm)	Posterior radius (mm)	Anterior conic	Posterior conic	Central thickness (mm)	Refractive index (550 nm)
Cornea	7.770	6.400	−0.180	−0.600	0.50	1.376
Anterior chamber	—	—	—	—	3.16	1.336
IOL	—	—	—	—	1.59	—
Posterior chamber	—	—	—	—	18.70	1.336

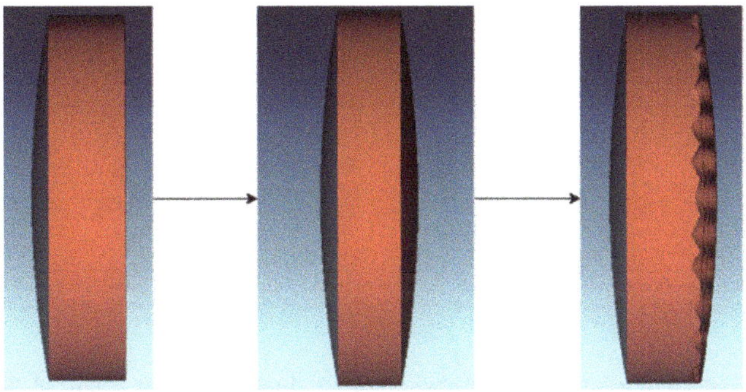

Figure 11.1. IOL design choices.

Table 11.2. Parameters of the designed 20 D lenses (adapted from [3]).

	Aspheric reference IOL
Anterior radius (mm)	15.607
Anterior curvature (mm^{-1})	64.07
Posterior radius (mm)	−15.607
Posterior curvature (mm^{-1})	−64.07
Anterior conic constant	−10.255
Posterior conic constant	−18.010
Central thickness (mm)	1.59
Refractive index (@550 nm)	1.492

pupil location (inside the 3.0 mm capsular bag). It resulted in a lens with refractive power of 19.88 D. The sinusoidal pattern is then added to the posterior convex surface of this lens.

The description of the compound sinusoidal posterior surface of the periodic IOL is given by equations (11.4) and (11.5), where the latter indicates the profile of the base surface

$$z_{\text{sin}} = z_{\text{asf}} - A\left\{\frac{1}{4}[1 + \cos(2\pi\alpha x)][1 + \cos(2\pi\beta y)] - 1\right\} \tag{11.4}$$

$$z_{\text{asf}} = \frac{cr^2}{1 + \sqrt{1 - c^2 r^2}} \tag{11.5}$$

where z_{sin} is the elevation in the direction along the IOL optical axis; A is the amplitude of the sinusoidal pattern; α is the frequency of peaks along the x-axis; β is the frequency along the y-axis, both in cycles/mm; c is the curvature of the posterior surface, corresponding to the inverse of the radius of curvature R_c; and r is the radial coordinate orthogonal to the lens optical axis. The lens profile is shown in different views in figure 11.2, with a disproportionally magnified wave amplitude for the sake of a clearer visualization.

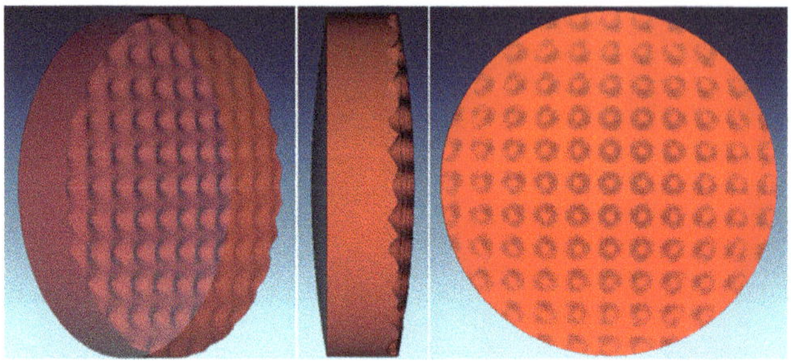

Figure 11.2. Diagonal view (left), lateral view (center) and posterior view (right) of the periodic IOL.

Table 11.3. Periodic IOL range restrictions and simulation step.

	Minimum	Maximum	Step
Amplitude A (μm)	0.25	1.25	0.25
Frequency α (cycles/mm)	0.25	2.00	0.25

Although one can attribute different frequencies, α and β, to the periodic function along the x- and y-axes, the results presented henceforth consider the sinusoidal frequency in both directions equal to α. The simulations are performed for monochromatic light with a wavelength of 550 nm (green). A population of 40 periodic IOLs has been designed with a systematic variation to the amplitude (A) and the frequency (α) in the range presented in table 11.3. The limits of these parameters were chosen so as to maintain an MTF of at least 0.43 at the spatial frequency of 100 lp mm^{-1}. This is the specified minimum contrast of a MONO IOL in accordance with the ISO standard [4].

Since the proposed IOL design is entirely refractive, it is expected that it will exhibit reduced positive dysphotopic phenomena, such as halo and glare. Both effects are more significantly associated with diffractive IOLs, since light diffraction can generate interference patterns and wavefront distortions that lead to such optical conditions. Halo can, nevertheless, also be experienced with refractive multifocal lenses due to the superimposed images from different foci.

11.2 Classification algorithm

An algorithm has been developed with Python programming language to enable the batch simulation of series of lens designs with different values for the periodic parameters in order to provide a general overview map of the simulated performance trend. This approach greatly increases the efficiency when serially evaluating many topologies. It automates the availability of design parameters to be used in the optical simulations; it systematically reads, manipulates and stores simulated data; and it evaluates the figures of merit of interest. The goal of this algorithm is to classify the simulated lenses according to their focal performance in three categories, namely, MONO, MIOL and EDoF. The algorithm relies on the interpretation of the through-focus MTF (TF-MTF) curve of the simulated lens for its classification.

The first step in the algorithm, whose flow is presented in figure 11.3, is to specify the base lens parameters for the periodic IOL (step 1). Then, lower and upper MTF threshold values are set to determine the presence of meaningful peaks and valleys (step 2). They are input parameters used by the algorithm to aid in the classification process and will be further discussed. In this procedure, the TF-MTF curves are compared with preset thresholds that, although arbitrarily chosen, are based on the

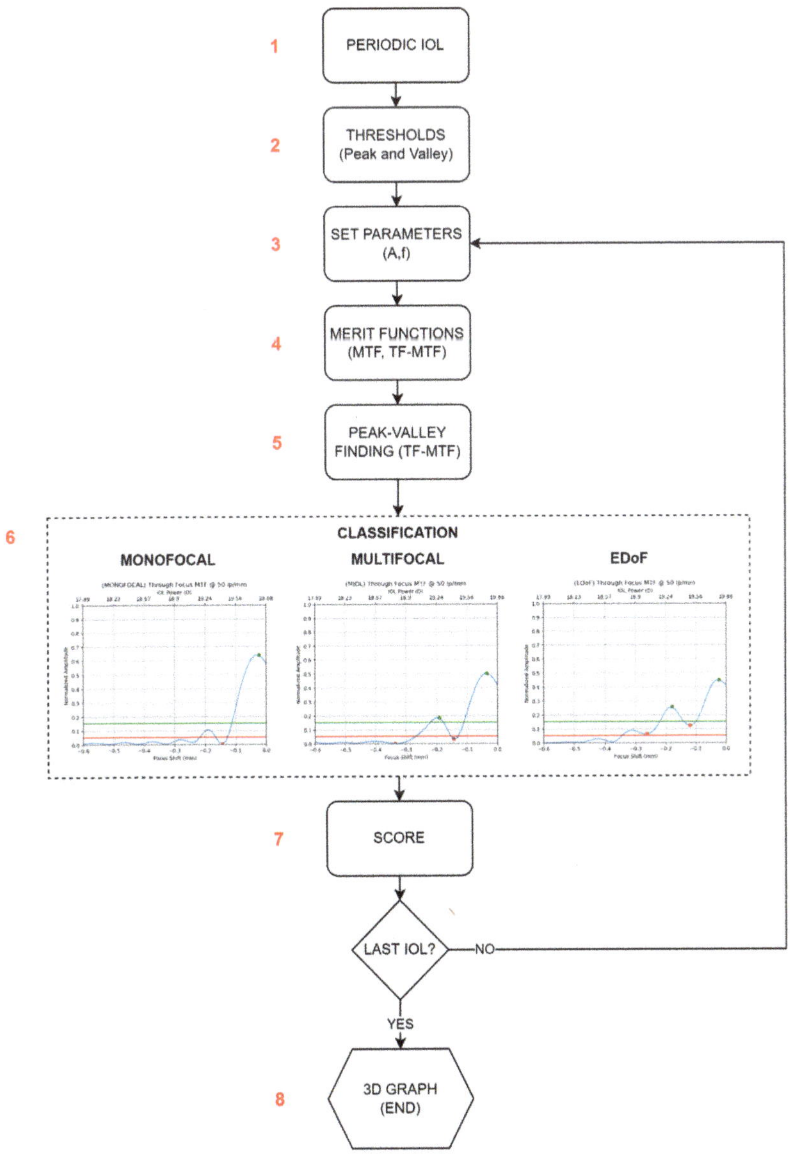

Figure 11.3. Classification algorithm.

values shown in an extensive clinical trial done by Alcon about the Vivity IOL [5, 6]. The next step is to modify the frequency (α) and amplitude (A) values of the IOL periodic surface (step 3) within a predefined parameter range, as was shown in table 11.3. In step 4, the merit functions (MTF and TF-MTF) are extracted from each individual configuration and saved to the database. In each TF-MTF curve a peak detection function (PDF) is executed, where all the peaks are found (step 5). The valleys are found by using an artifact, which consists in first vertically flipping the TF-MTF curve, turning valleys into peaks, and then applying the same PDF function as already used for peak detection. Peaks and valleys are only counted as such if they extrapolate the threshold MTF values previously set. After that, the algorithm compares each individual TF-MTF curve with some preset conditions, classifying the lens (step 6) as MONO, MIOL or EDoF. Besides the classification, a score value is attributed to each periodic IOL configuration (step 7), which is based on physical characteristics that are desired for the IOL performance (further discussed in this section). Subsequently, the classification algorithm ends by plotting a 3D bar chart with the population of simulated lenses with their scores and indication of class as a function of the design parameters (step 8) (section 11.3.1).

The TF-MTF curve presents the behavior of the normalized MTF amplitude at planes shifted from the focal plane (retina) for a specific spatial frequency. Figure 11.4 presents an example of a TF-MTF curve. In this particular case, and throughout the analyses, the TF-MTF curve is limited to the contrast evaluation at 50 lp mm^{-1} and a focus range from −0.6 mm to 0.6 mm, centered at the retina. The higher the MTF value at a certain position off the focal plane (focus shift), the better the contrast of an observed pattern located at a corresponding position on the object side. The image with a spatial frequency of 50 lp mm^{-1} on the retina matches an

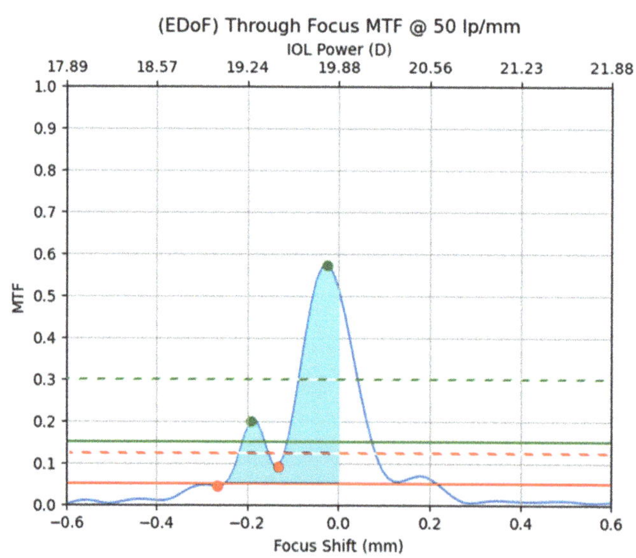

Figure 11.4. Example of the threshold choices and implications on the classification methodology based on the TF-MTF curve.

object represented by a Snellen letter inscribed in a cone with an opening angle of 10 arcmin, i.e. a visual acuity of 20/40. Very low MTF values, on the other hand, suggest a poor ability to distinguish the object features.

The curves that are evaluated in the algorithm are obtained for an object located at infinity, therefore, with its image plane corresponding to the retina at a focus shift of 0 mm. Nevertheless, far vision is sometimes assessed when viewing objects at distances down to 6 m, whose image plane is beyond the retina but close enough to it to yield a contrast similar to that of an object at infinity. Lenses designed with a target distance of 6 m, on the other hand, project objects at that distance on the retina, and objects at infinity at a distance slightly before the retina.

Figure 11.4 shows an arbitrary TF-MTF curve. Let us first consider the solid horizontal lines representing the peak threshold (green) and the valley threshold (red). Since two peaks (green dots) in the blue curve were detected as coinciding or lying above the peak threshold, the methodology rules out the MONO class. The algorithm must decide if the lens is either MIOL or EDoF. When considering the valley between the two detected peaks (central red dot), note that it lies above the valley threshold. This means that the secondary peak is not completely separated from the main one, therefore, we instructed the algorithm to classify this lens configuration as EDoF.

If the valley threshold had been chosen at a higher value (the red dashed line), the algorithm would have considered that the secondary peak is disconnected from the main peak, classifying this lens configuration as MIOL. In another example, if the peak threshold had been changed to a higher value (green dashed line), the algorithm would have taken only the central peak as having a relevant contrast value and therefore would classify the lens configuration as MONO.

There is one important factor worthy of attention when evaluating the TF-MTF curve. Imagining the scene on the left and the eye on the right, as objects get closer to the eye than infinity, they are imaged at best focus on a plane increasingly farther beyond the retina to the right. In the curve shown in figure 11.4, 0 mm corresponds to the retinal plane, the object is at infinity and the focus shift corresponds to the MTF registered at planes at different distances from the retina to either side. In this particular curve the maximum point, related to the best focus, lies on a plane slightly to the left of the retina. This indicates the lens that generated this curve has been designed with a target far vision at a point closer to the eye than infinity, as will be explained in sequence. To understand it, it is important to know that keeping the object fixed and moving the image plane left and right is equivalent to keeping the image plane fixed and moving the object in the opposite direction, respectively, and registering what happens to the MTF at the fixed image plane. If we consider that the retina is always fixed at 0 mm, as the object approaches the eye, the entire TF-MTF curve shifts to the right and if the object moves away from the eye, the curve shifts to the left. For a lens designed for best focus at, say, 6 m, when the object is at that distance, the highest peak of the TF-MTF curve will shift to the right and coincide with the retina. Therefore, for a TF-MTF curve obtained in the usual way, i.e. for an object fixed at infinity, it only makes sense to use its left-hand side (negative focus shift), as those are the MTF values that, in practice, will represent objects closer to the eye than infinity. The object will only have closer distances to the eye from there, whose respective image planes are located beyond the retina, unless additional aiding spectacles are used. In this latter case, not only will the curve shift, but it will also change its shape due to the contribution of the MTF of the spectacles or contact lenses. In the case of MIOLs, we should expect multiple peaks of the TF-MTF curves to the left; the farther left the peak is, the closer the additional focus represented by it.

The algorithm first employs its PDF to search all TF-MTF crests and troughs, among which actual pertinent peaks and valleys are selected based on upper and lower MTF threshold values, respectively.

The upper threshold specifies the minimum MTF value that still yields an acceptable contrast. If only the central peak, corresponding to the focal plane, exceeds the peak threshold, the IOL is classified as MONO. If there are two or more peaks, it can be classified as MIOL or EDoF, and further distinction is made by evaluating the curve with respect to the lower threshold. If multiple peaks have an amplitude above the peak threshold, the troughs that occur between those peaks are considered for the classification between these two classes. If any trough between two adjacent peaks falls below the lower threshold, i.e., if it is considered a valley, the IOL configuration is classified as MIOL. Otherwise, the lens is labeled as EDoF.

The preset upper and lower MTF threshold values chosen are 0.15 and 0.05, respectively. These are somewhat arbitrary values that were based on prior observation of the focal behavior of some simulated IOLs and have been chosen so that the peak threshold attempts to select peaks considered significant, while the valley threshold limits the width of a given peak, even setting boundaries to whether two adjacent crests are to be considered separate peaks or simply oscillations of a single peak. A more thorough choice should be based on the standards to which the designed IOLs must comply.

A monofocal aspheric IOL usually has a narrow and high central peak, while an EDoF IOL often has a wider but lower central peak. In contrast, a multifocal IOL can have multiple peaks with varying heights and widths, performing as bifocal, trifocal, or even quadrifocal. Some lenses within a certain class can perform better than others and an additional score is useful to rank them. If the area under the TF-MTF curve is used as the score parameter it will not be able to properly distinguish between, for instance, a curve with a single elevated central peak or a curve with lower but multiple peaks. Hence, a score parameter is chosen that partly mitigates this deficiency and that attributes a higher weight to curves with a higher central peak, meaning a high contrast associated with the far focus. The score value proposed uses the area under the TF-MTF multiplied by the MTF value at the retina, i.e. at 0 mm focus shift. Only the portion of the TF-MTF curve to the left of the retina is used, for the reasons explained above. The choice of score-value computation can be easily adapted to account for different priorities or merit functions.

One limitation is that if the algorithm selects only one peak in the TF-MTF, but that peak is wide, which typically characterizes an EDoF IOL, it will classify the IOL as being MONO but with a large score value, which could be an indication to yet another category, that of enhanced monofocal.

The classification algorithm maps out a panorama of possible lens classes and performance trends within certain boundaries. A useful future addition would be the subsequent deployment of an optimization algorithm to find amplitudes and frequencies around those of singled out candidates of a certain class to fine tune the desired optimal performance, using additional figures of merit for that purpose, such as the preclinical visual acuity and image quality. These two merit functions are discussed in the next section and evaluated for some selected lens models, but they have not been used in the algorithm.

11.3 Merit functions

In this section the resulting bar chart simultaneously exhibits the classification and the score given to each simulated periodic IOL. A few of the simulated IOLs are selected, one being monofocal, one multifocal and all three EDoF IOLs, and their TF-MTF curves are presented and evaluated. The image simulation of a 20/40 Snellen letter located at five different distances from the eye is presented for these five lenses and for the best reference aspheric lens, for comparison. Lastly, the preclinical visual acuity charts are produced and discussed for the most suitable EDoF with respect to the best aspheric IOL, for different pupil apertures. The image simulations and the preclinical visual acuity results were not used as criteria in the classification algorithm, which currently only considers the TF-MTF behavior at 50 lp mm^{-1}.

The results have been obtained for a wavelength of 550 nm (green color); an effective lens position of 0 mm from the pupil location, inside the 3 mm capsular bag; and a 3 mm pupil aperture for the TF-MTF curves and image simulations. The effective focal length of all the periodic IOLs is 17.0118 mm in the model eye and that of the best aspheric lens is 16.8494 mm.

11.3.1 3D bar chart

The generated three-dimensional bar chart shown in figure 11.5 exhibits the classification and score results for the forty nominal 20 D periodic IOLs, with sinusoidal amplitudes and frequencies within the pre-established range. The color of each bar in the 3D graph represents the classification result *per se*, whereas each respective height represents the calculated score. This graph is intended to aid the lens designer in the identification of the most suitable parameters to yield a given class of lenses. One of the main goals was to observe if any combination of the parameters would indeed result in EDoF lenses. With the upper and lower MTF thresholds set at 0.15 and 0.05, and a depth-of-focus interval of 0.6 mm, 33 lenses have been classified as MONO, 3 as MIOL, and 4 as EDoF. For comparison, the score of the reference best aspheric IOL is 0.03447.

The graph suggests that intermediate to high amplitudes and intermediate frequencies, within the respective selected ranges, yield either EDoF or MIOL lenses. All others result in MONOs. High frequencies and high amplitudes present the worst scores. Regarding exclusively the scores, within this population, the best MONOs have a high frequency and low amplitude, the best MIOLs and EDoFs feature amplitudes towards the lower end and intermediate frequencies. It is

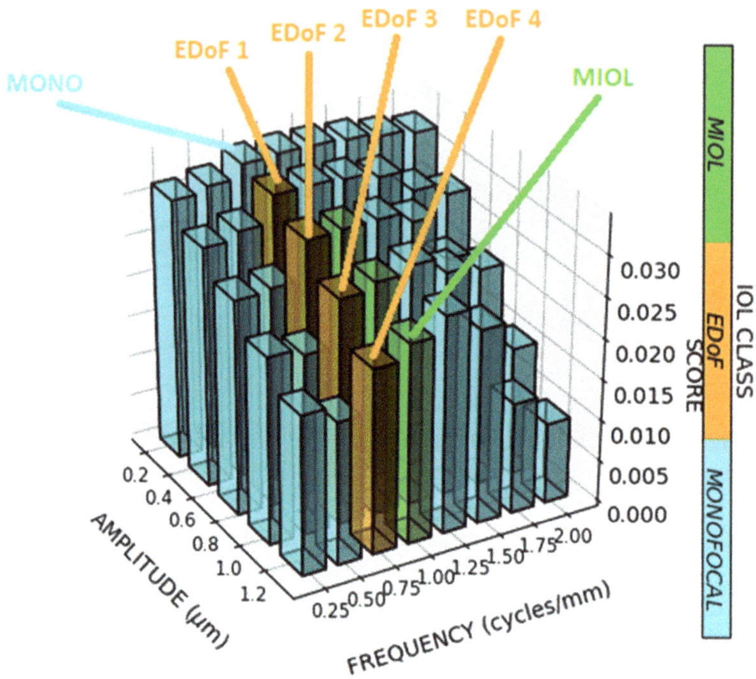

Figure 11.5. IOL score and classification as a function of the sinusoidal amplitudes and frequencies imparted to the posterior surface of a 20 D IOL.

important to reiterate that changes in the base lens or in the MTF thresholds can yield a distinct classification of these lenses. The method is also limited to the analysis in the spatial frequency of 50 lp mm^{-1}.

This algorithm highly relies on a suitable and justifiable selection of the upper and lower MTF thresholds. For instance, if the upper threshold is too high, many IOLs are classified as MONO, because most of the secondary crests will not reach the threshold. If this threshold is low, the procedure tends to classify most IOLs between EDoFs or MIOLs. If the lower threshold is high, most troughs between peaks will occur below it, therefore the algorithm will classify some potential EDoFs as MIOLs. If the lower threshold is too low, many IOLs will be classified as EDoFs, since then neighboring peaks are considered as being extensions of the main central peak, even with oscillations between them.

Five IOLs were chosen for the purpose of comparison. The four EDoF samples have been selected, as well as one MONO and one MIOL, which is bifocal. These latter have been arbitrarily chosen among those with the highest scores and with parameters close to each other in the map, featuring reasonably high and equal sinusoidal amplitudes (A) between them, and also a close value for the sinusoidal-pattern frequency (α). The sinusoidal parameters for the selected IOLs are shown in table 11.4. The depth of focus is measured at the position the TF-MTF curve crosses the lower threshold. The nearest field is calculated as the corresponding nearest distance the object can be to the eye to still yield the minimally acceptable MTF

Table 11.4. Parameters of the selected IOLs and the best aspheric (ASPH).

IOL	ASPH	MONO	MIOL	EDoF 1	EDoF 2	EDoF 3	EDoF 4
Score	0.03447	0.03398	0.02435	0.03332	0.03114	0.02739	0.02257
A (mm)	0	0.25	1.25	0.50	0.75	1.00	1.25
α (cycle/mm)	0	0.75	1.00	0.75	0.75	0.75	0.75
MTF 1st peak	0.65	0.64	0.49	0.62	0.57	0.52	0.45
MTF 2nd peak	—	—	0.18	0.16	0.20	0.24	0.26
Position 2nd peak (mm)	—	—	−0.18	−0.18	−0.19	−0.17	−0.17
Depth of focus (mm)/ Nearest field (m)	0.14/1.8	0.14/1.9	1.05 m*	0.24/1.6	0.25/1.5	0.33/1.2	0.35/1.1

*Object distance associated with the second peak.

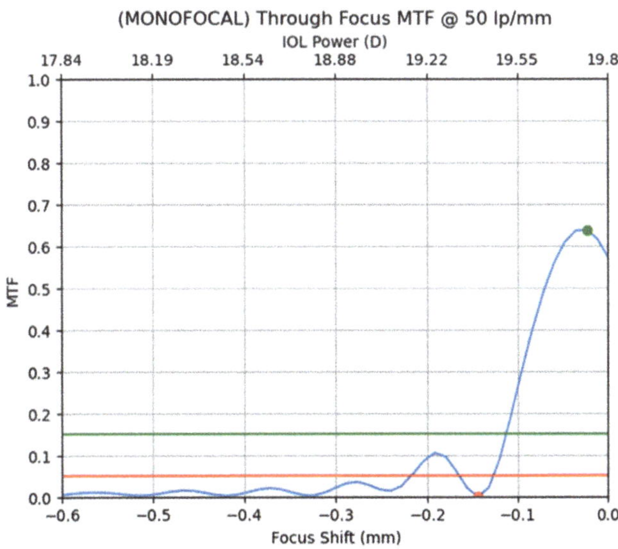

Figure 11.7. TF-MTF curve for monofocal IOL.

corresponding to the lower TF-MTF threshold. The MONO lens, with sinusoidal features, presents no advantage compared to the reference aspheric lens (ASPH), from the parameters in the table. The only possible minor advantage is the appearance of a low secondary peak as will be shown in figure 11.7. The value indicated for the nearest field of the MIOL deems its nearest focus distance, corresponding to the position of the secondary peak.

The TF-MTF curve for the best aspheric IOL is shown in figure 11.6.

First, it is important to note that the TF-MTF curve is shown with an object placed at an infinite distance from the eye, so that the main peak is visible. This happens because the base IOL is optimized for an object placed at 6 m from the eye. In order to find the values of depth of focus, nearest field, peak positions and amplitudes (shown in table 11.4), the object distance from the eye was manually

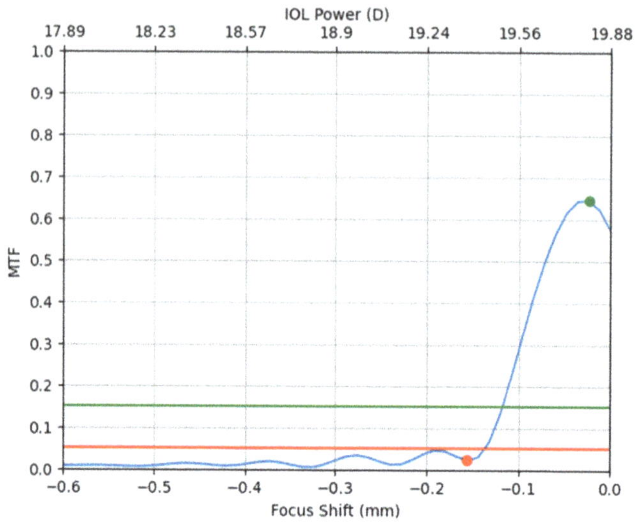

Figure 11.6. TF-MTF curve for the best reference aspheric IOL.

changed and measured directly from the ZEMAX interface. It is possible to notice that the TF-MTF curve for the aspheric IOL has only one main peak, that is narrow and exhibits a relatively high contrast value (>0.6).

The TF-MTF curve for the monofocal IOL (MONO) is shown in figure 11.7. The TF-MTF curve for this selected IOL configuration exhibits only one central peak above the peak threshold. In this case, the algorithm ignores both the valley threshold and the adjacent valleys found, classifying the lens as MONO.

The TF-MTF curve for the multifocal IOL is shown in figure 11.8. The MIOL exhibits two peaks above the peak threshold, and the detected valley between those peaks is below the valley threshold. Therefore, the algorithm considers the secondary peak as being separated from the main peak, classifying this IOL configuration as multifocal, namely a poor bifocal with an add IOL power of approximately 0.5 D.

The TF-MTF curves for the EDoFs are presented in figures 11.9–11.12 and they have similar behaviors. In all of them, the second peak appears around −0.19 mm of focus shift and their MTF lie between 0.15 and 0.25. The trough between that peak and the central one does not fall under the lower threshold, therefore is not counted as a valley. The position of the second peak corresponds to an object located at roughly 1.4 m from the eye. The point where it crosses the lower threshold lies around −0.26 mm and corresponds to a nearest acceptable field of about 1.0 m. EDoF 1 features the best compromise of score, peak MTFs, position of the second peak and nearest field.

It is possible to notice that for each of the EDoF TF-MTF curves, the valleys (red dots) between the two peaks do not fall below the valley threshold value (red horizontal line). Therefore, the algorithm considers that the secondary peak provides an extended range of focus and not a secondary isolated focus, classifying the IOLs as EDoFs instead of MIOLs.

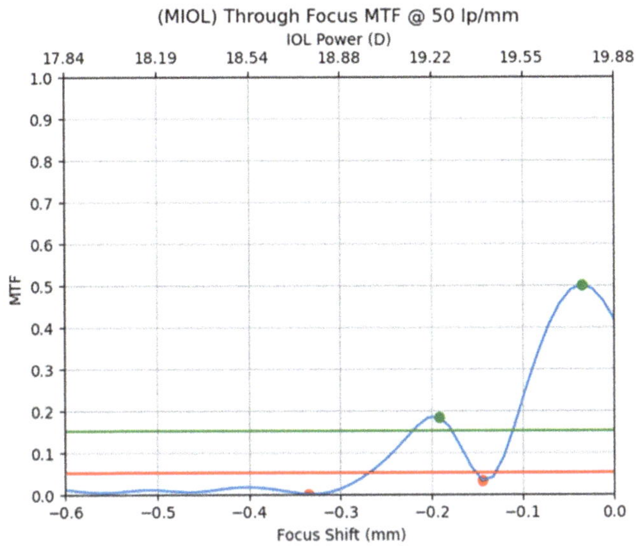

Figure 11.8. TF-MTF curve for multifocal IOL.

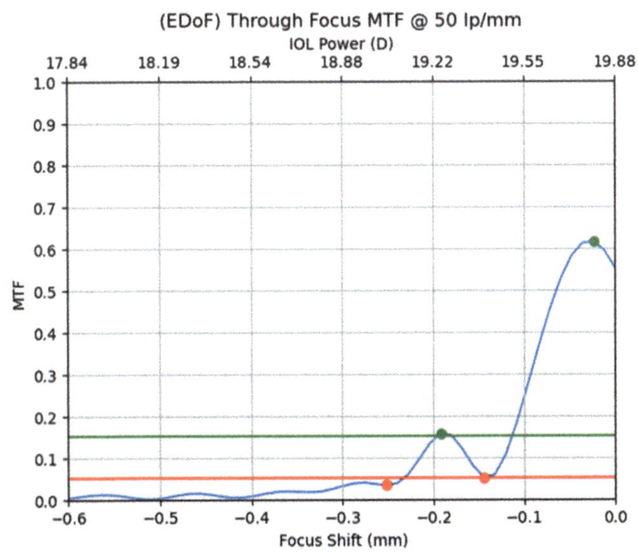

Figure 11.9. TF-MTF curve for EDoF 1.

According to [7], the values of contrast exhibited in table 11.4, and the chosen lower MTF threshold, are within the range of the TF-MTFs of commercial EDoF IOLs, such as the TECNIS Symfony ZXR00 (Johnson & Johnson) and the AT LARA 829 MP (Carl Zeiss Meditec). The TF-MTF response at 50 lp mm^{-1} is shown in figure 11.13.

It is important to indicate that the ISO eye model was used for the IOL comparison. Also, the *x*-axis scale is measured in diopters, possible at the cornea

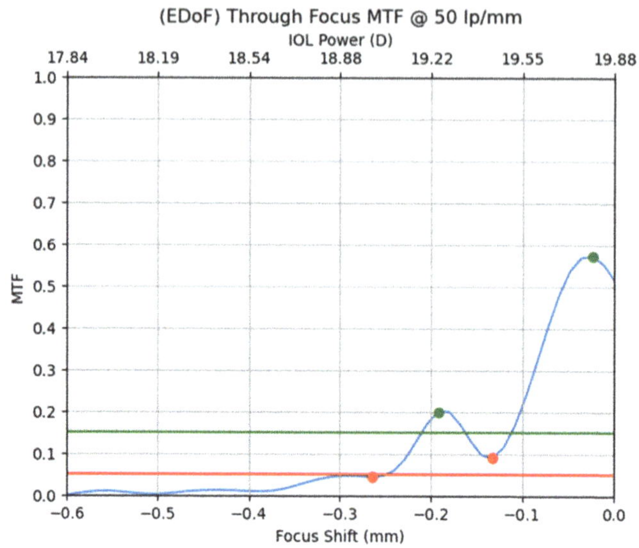

Figure 11.10. TF-MTF curve for EDoF 2.

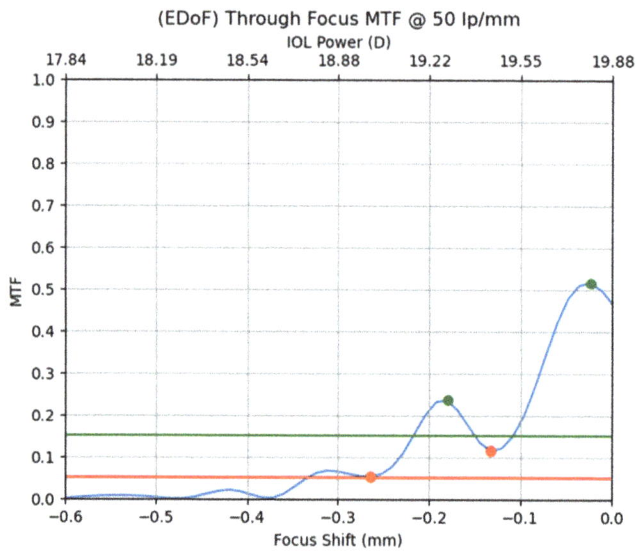

Figure 11.11. TF-MTF curve for EDoF 3.

vertex plane, and not in focus shift, in millimeters. The mathematical parameters of both IOLs are not known, therefore it is impossible to estimate the effective focal length (***EFFL***) and, consequently, the accurate conversion from diopters to focus shift. In our TF-MTF curves, however, we indicate on the upper axis the corresponding refractive power at the IOL plane. What is known is that both

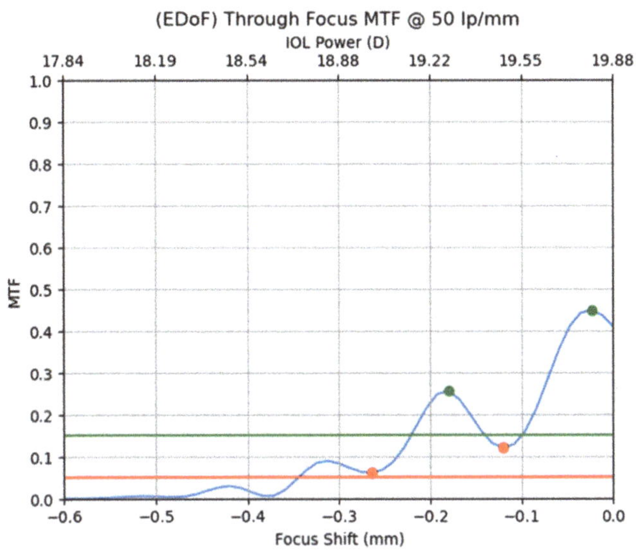

Figure 11.12. TF-MTF curve for EDoF 4.

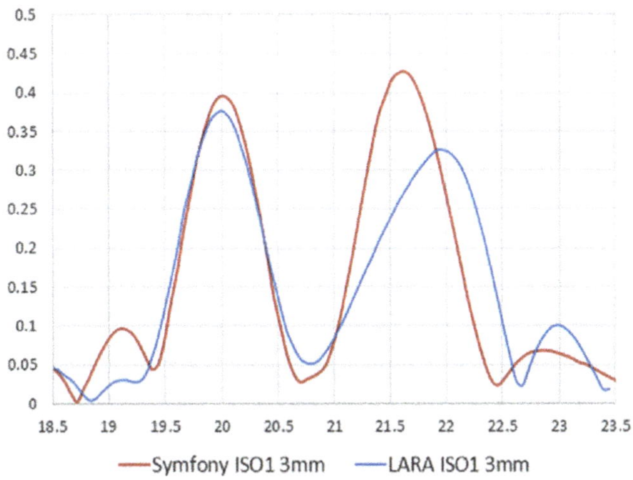

Figure 11.13. Comparison between the TF-MTF curves of the Symfony and AT LARA IOLs [7].

IOLs were designed with 20 D and a secondary focus around +1.5 or +2 D. If only the refractive powers are compared, it is possible to notice that both the main central peak amplitude (@ 20 D) and the secondary peak amplitude (@ ~22 D) are between 0.33 and 0.43. The periodic EDoFs we found feature higher main peaks, lower second peaks, and limited extended focus compared to these commercial models. The simulation and classification methodology, however, allows one to further exploit different parameters, different ranges and to base the analysis on a different performance metric.

11.3.2 Image simulations

The results shown in tables 11.5–11.10 are the visual representations of the Snellen-chart letter E, corresponding to a visual acuity of 20/40, which represents a minimum quality threshold (logMAR 0.3), according to some standards, in the evaluation of IOLs in clinical defocus curves. Each table presents the image results at a certain object distance for the reference best aspheric (ASPH), and the best extended-depth-of-focus lens found (EDoF 1).

The image simulations were done for three different pupil diameters 3.0 mm, 4.5 mm and 6.0 mm, relating to photopic, mesopic and scotopic illumination conditions. For each pupil aperture and object distance the result comprises the image projected on the retina for the object with fading contrast levels with the following values: 100%, 77%, 55%, 33% and 11%. Different user activities and corresponding scenes entice a distinct combination of illumination condition, object distance and presented contrast. In the simulations, the object size was calculated for each distance to represent the Snellen visual acuity of 20/40, which leads to an image height on the retina of 49.62 μm for an ***EFFL*** of 17.0118 mm. The object sizes are,

Table 11.5. Visual acuity of a 20/40 letter at 1 km from the eye. Object size 2.33 m.

	Pupil 3.0 mm	Pupil 4.5 mm	Pupil 6.0 mm
ASPH			
EDoF 1			

Table 11.6. Visual Acuity of a 20/40 letter at 6 m from the eye. Object size 13.96 mm.

	Pupil 3.0 mm	Pupil 4.5 mm	Pupil 6.0 mm
ASPH			
EDoF 1			

Table 11.7. Visual acuity of a 20/40 letter at 2 m from the eye. Object size 4.65 mm.

	Pupil 3.0 mm	Pupil 4.5 mm	Pupil 6.0 mm
ASPH			
EDoF 1			

thus, proportionally altered to maintain the same reference visual acuity for 1 km, 6 m, 2 m, 0.66 m and 0.35 m, which are, respectively, 2.33 m, 13.96 mm, 4.65 mm, 1.54 mm and 0.81 mm. The letter E contains 2.5 line pairs vertically, which divided by its height on the retina yields a spatial frequency of 50 lp mm^{-1}. The presented images provide an objective indication of the image quality to be optically projected on the user's retina, where no subjective assessment is considered, as it would involve neural processing, including adaptation and resignation.

As the base geometry of the lenses was designed for far vision, they exhibit good results for an object located at 1 km. It would correspond to a letter E about the size of an ordinary door observed from that distance. In that situation the reference aspheric IOL, without any sinusoidal pattern, is overall better. The periodic EDoF 1 lens presents a blur around them for larger pupils.

Table 11.6 presents the letter E closer to the eye, namely at 6 m, corresponding to the length of an average meeting room. The object now has approximately the height of a pen. For practical purposes, it still represents far vision. This object distance of 6 m, roughly 20 ft, was used because it represents the distance at which the test using the Snellen eye chart is usually performed in clinical practice.

Little difference is noted for the 3 mm pupil, compared to the observation at 1 km. Also, the advantage of the periodic IOL starts to show, presenting approximately the same behavior as before for larger pupils, whereas the aspheric lens presents a degradation in the sharpness of the letter.

Table 11.7 presents the letter E at 2 m from the eye and its height is equivalent to that of a small matchbox in a supermarket shelf across a wide aisle.

At this distance, the performance of the aspheric lens decreases further, even for a small pupil. The periodic lenses also exhibit dissatisfying behavior for the 3 mm pupil. However, for larger apertures, they show a less sharp object than before, but still a recognizable letter profile with no considerable increase in the blur around the letter. One would still be able to read the letter E on the matchbox.

If the object is placed at 1.5 m from the eye, it is possible to notice the advantage of the selected IOL when compared to the aspheric, as presented in table 11.8. As the pupil size increases, it is possible to distinguish the simulated letter which is not possible using the aspheric IOL.

Table 11.9 shows the image results for a 1.5 cm object at 0.66 m from the eye, which is basically the separation of the lips in a smile when looking at someone across the dinner table.

Table 11.8. Visual acuity of a 20/40 letter at 1.5 m from the eye, corresponding to the second focus of the multifocal IOL. Object size 4.36 mm.

	Pupil 3.0 mm	Pupil 4.5 mm	Pupil 6.0 mm
ASPH			
EDoF 1			

Table 11.9. Visual acuity of a 20/40 letter at 0.66 m from the eye. Object size 1.54 mm.

	Pupil 3.0 mm	Pupil 4.5 mm	Pupil 6.0 mm
ASPH			
EDoF 1			

Table 11.10. Visual acuity of a 20/40 letter at 0.35 m from the eye. Object size 0.81 mm.

	Pupil 3.0 mm	Pupil 4.5 mm	Pupil 6.0 mm
ASPH			
EDoF 1			

The aspheric IOL would not allow the user to distinguish between adjacent teeth in the smile either in a well lit room (small pupil), or in a darker room (larger pupils). The EDoF 1 also perform poorly for the small and medium pupils, but offers remarkably acceptable results for a large aperture. This phenomenon can be explained by the gradual insertion of the sinusoidal peaks as the optical aperture increases, forcing part of the refracted rays to converge around the focal point. None of them present acceptable results for objects with the lowest contrast (11%) for any of the apertures.

When the object is placed at a distance closer to the eye (35 cm), it is not possible to distinguish a 20/40 letter with any IOL configuration, as seen in table 11.10. This agrees with the results of the TF-MTF curves, since a focus shift of around 1.0 mm would be necessary to be able to see the simulated letter.

11.3.3 Preclinical visual acuity-defocus curve

The analyses presented so far are limited to very specific spatial frequencies. The through-focus curve only contains the MTF signature at 50 lp mm^{-1} and the evaluation carried out with the images of the letter E are limited to 50 lp mm^{-1}. A real scene, nevertheless, contains features with multiple spatial frequencies that are perceived with different weights by patients. Therefore, a more suitable method to evaluate the lens performance should take that into account and the data from the simulations was used to generate preclinical defocus curves [8].

The preclinical defocus curve is a more reliable merit function for performance evaluation since it can be obtained by correlating the area under the MTF curve with the psychophysical subjective perception of target scenes by a population of users [8]. This estimate considers not only a single frequency of 50 lp mm^{-1} but all the spatial frequencies from 0 to 50 lp mm^{-1}. This is done for various object distances from the eye. The preclinical study exhibits a correlation between a purely theoretical evaluation and the psychophysical perception of a population of patients.

The curve indicates the visual acuity the subject is expected to experience, in logMAR units, as a function of additional refractive power on the IOL plane. 0 D corresponds to object at infinity and positive additional powers correspond to distances closer to the eye. The usual threshold for the minimum acceptable visual acuity lies on logMAR 0.2, which relates to 20/32 in the Snellen scale.

Figure 11.14 presents the defocus curve for the best aspheric IOL and the EDoF 1 lens for a 3 mm pupil, for three different wavelengths (red, green and blue).

Both IOLs yield a visual acuity at far vision (0 D) of approximately 20/16 (logMAR –0.1). For planes closer to the eye, represented by an increasing additional power, the acuity provided by the aspheric lens drops faster than that of the EDoF 1. The best aspheric only maintains the visual acuity better than 20/32 (logMAR 0.2) up to 1.5 D, whereas the EDoF 1 extends this range up to roughly 1.75 D.

The requirements for a lens to be considered EDoF by the ANSI standard Z80.35–2018 [9] are that for a 3-mm pupil (i) it must have a range extension of at least 0.5 D at 0.2 logMAR, when compared to the reference monofocal aspheric IOL (photopic condition, monocular, distance corrected visual acuity); (ii) it must feature no less than 0.2 logMAR for an object placed at 66 cm from the eye; (iii) the acuity at 0 D must not be lower than that of a reference monofocal lens by 0.1 logMAR; (iv) the defocus curve must be monotonically descreasing, i.e. there should not be an increasing trend larger or equal to 0.04 logMAR; (v) for a 4.5-mm pupil, it ought to demonstrate a depth of focus of at least that of the reference monofocal aspheric lens at 0.2 logMAR.

Considering the graph in figure 11.14, the EDoF 1 IOL meets the last two criteria. The EDoF 1 curve crosses 0.2 logMAR at 0.25 D further than the curve for the aspheric;

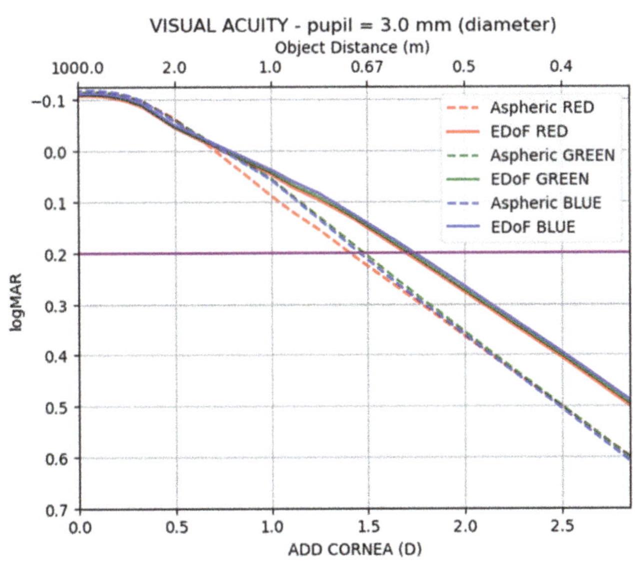

Figure 11.14. Defocus curve, in logMAR units, for pupil diameter of 3.0 mm.

but the EDoF 1 curve keeps above the threshold only until 1.75 D, which corresponds to an object distance of 0.57 m, which is closer to the eye than 0.66 m (~2.0 D), but does not yield the 0.5 D extension compared to the reference aspheric. This makes it an enhanced monofocal lens rather than an EDoF one. At 0 D the logMAR score difference to the aspheric IOL is lower than 0.0125 logMAR. The fourth criterion is to be verified for a 4.5 mm aperture and is evaluated from figure 11.15.

For this larger pupil the acuity at 0 D becomes lower for the EDoF 1 but is maintained high for the aspheric. The EDoF 1 defocus curve also crosses the threshold at a larger defocus value, namely, 2.0 D, whereas the aspheric IOL features a reduced range to 0.9 D, therefore with a difference of 1.1 D, showing the superiority of the EDoF 1 far beyond the requirements of the aforementioned standard. The aspheric lens was expected to have a lower crossing point at logMAR 0.2 mm for a 4.5 mm aperture, as the lower depth of focus is then dominated by the intrinsic effect of the increased pupil size. The EDoF 1, on the other hand, congregates more peaks of the sinusoidal pattern when the pupil increases, causing a larger number of rays to converge to a region closer to the optical axis, and dominating the effect of the depth of focus.

Figure 11.16 shows the results for a 6 mm pupil opening, where there is a further reduction of the focal range of the aspheric lens and a maintenance of the range for the EDoF 1 IOL, where the additional portions of the sinusoidal pattern now only counterbalance the reducing effects of a larger pupil. For this pupil, the performance of the EDoF 1 at 0 D is considerably lower, dropping down to an acuity of 0.1 logMAR (20/25 Snellen), but still almost reaching 20/20 for planes about 2 m from the eye (0.5 D).

The logMAR score at 0 D for the aspheric is always higher, independent of the pupil aperture, since this lens has the best contrast for the distant focus and was optimized for a 6 mm pupil, covering the whole range presented herein. The EDoF 1

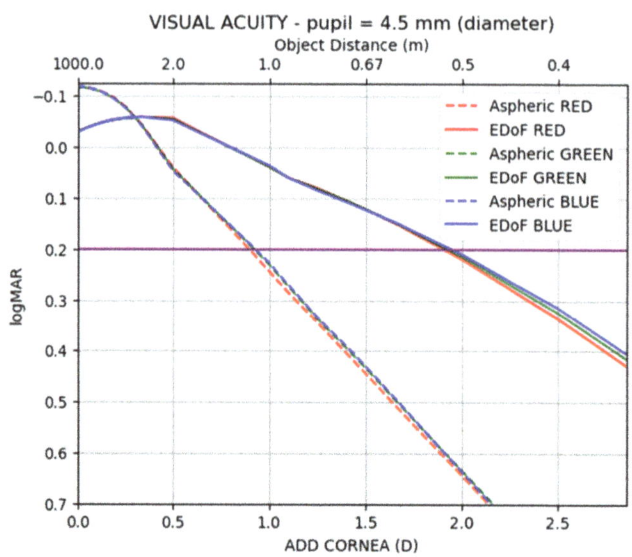

Figure 11.15. logMAR plot for pupil diameter of 4.5 mm.

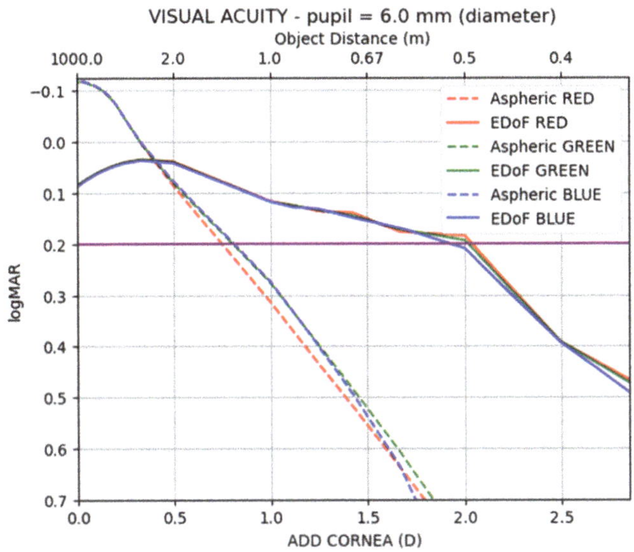

Figure 11.16. logMAR plot for pupil diameter of 6.0 mm.

IOL follows the tendency of the aspheric lens that constitutes its base profile. Although the general methodology is useful, the results from the defocus curve suggest that assessing whether a lens is EDoF based solely on the TF-MTF results can be misleading. An IOL that was considered EDoF by the classifier was indeed an enhanced monofocal lens. A more accurate classification algorithm can keep the base methodology herein presented but must be based on scores derived from the defocus curve and attending to the specifications of the ANSI standards.

11.4 Summary

Through this study it was possible to investigate the different types of IOLs present on the market, as well as the procedure on how to design a lens using the geometric and material characteristics for any given refractive power. It also presented an insight on features of different eye models. The raytracing process and useful merit functions were investigated to allow the comparison of the optical performance of distinct IOLs.

An algorithm was proposed to help the classification of different periodic IOL configurations among different types of optical performances. This classification occurs by comparing the peaks and valleys of an important optical merit function, the TF-MTF, which is the most common curve used to assess multifocality and extension of focus in current IOL design.

After classifying the various simulated topologies, they were compared with some well-established off-the-shelf IOLs. Some of the interesting IOL topologies from different classes were chosen and a visual acuity analysis was made, which was represented by the preclinical defocus curve and image simulations. The preclinical defocus curve gives an excellent method to evaluate the overall IOL performance, since it is based on the MTF curve for many spatial frequencies and not one specifically, in contrast to the TF-MTF curve.

Although the algorithm could sort the lens performance as MONO, MIOL or EDoF, resorting to the results of preclinical defocus curves, it has been concluded that the through-focus curve does not provide an accurate means to classify the lens. IOLs identified as EDoF by the through-focus analysis, as a matter of fact behaved as enhanced monofocal lenses, according to its defocus curve.

The sinusoidal parameter range from the lens design could be further discretized to increase accuracy and/or explore different regions of the 3D map. The lens design could be refined so that the sinusoidal pattern would have different amplitudes and/or frequencies along the x- and y-axis. Also, the amplitude could be modulated by another function, as well as suffer apodization from the center to the edge of the base lens. The proposed algorithm could be adapted to incorporate the preclinical visual acuity charts. It could also contain a polychromatic analysis to further refine the decision-making process.

This study was important to promote expertise in lens design for the ophthalmic optics field and to investigate the specifics of IOL performance analysis. It was possible to understand the parametric influences on different merit functions, enabling a systematic analysis of optical performance, as practiced in the ophthalmic scenario for IOLs.

Chapter highlights

- Design of a refractive intraocular lens with a complex geometry that allows the configuration of additional parameters.
- Creation of a methodology for the classification of such a lens according to its optical performance.
- Comparison of the proposed methodology with the preclinical visual acuity-defocus curve.
- The through-focus MTF curve is a limited resource for the classification of an intraocular lens and can be misleading. The use of the preclinical defocus curve is advised.

References

[1] Costa D and Monteiro D Wd L 2020 *Methodology for the Classification of an Intraocular Lens with an Orthogonal Bidimensional Refractive Sinusoidal Profile*. Congresso Brasileiro de Engenharia Biomédica (CBEB)

[2] Costa D and Monteiro D Wd L 2020 Classification and performance estimation of an intraocular lens with an orthogonal bidimensional refractive sinusoidal profile *SPIE Photonics East. Optical Design and Testing X* **11548** 446–59

[3] Almeida M S de and Carvalho L A 2007 Different schematic eyes and their accuracy to the *in vivo* eye: a quantitative comparison study *Braz. J. Phys.* **37** 378–87

[4] No, I.S. 11979-2 2014 Ophthalmic implants-intraocular lenses—part 2: optical properties and test methods *International Organization for Standardization, Geneva, Switzerland*

[5] Alcon Inc 2009 Extended depth of focus (EDOF) lens to increase pseudo-accommodation by utilizing pupil dynamics *US Patent* Reissue No. USRE45969E1
[6] https://www.accessdata.fda.gov/cdrh_docs/pdf/P930014S126C.pdf (accessed 30 Nov. 2020)
[7] Chae S H, Son H S, Khoramnia R, Lee K H and Choi C Y 2020 Laboratory evaluation of the optical properties of two extended-depth-of-focus intraocular lenses *BMC Ophthalmol.* **20** 1–7
[8] Alarcon A *et al* 2016 Preclinical metrics to predict through-focus visual acuity for pseudophakic patients. *Biomed. Opt. Express* **7.5** 1877–88
[9] ANSI Z80.35–2018 2019 American Vision Standards Institute, Inc., The Vision Council, The Accredited Committee Z80 for Ophthalmic Standards

Chapter 12

Eye-Fi and electronically equipped lenses

Andrés Felipe Vasquez Quintero and Davies William de Lima Monteiro

This chapter tackles developments on electronically equipped lenses turning them from conventional passive components into actively functional devices intended to either assist vision, embed health diagnostic features, or even foster augmented reality. As wearable devices, besides caring for user comfort and endorsement, their development faces several challenges, from miniaturization of features, to energy transfer and storage, to wireless data communication. As medical devices they are also subjected to strict regulations in terms of biocompatibility, electrical safety, usability, and clinical validation. We will briefly describe heads-up displays on spectacles, review current developments on smart contact and intraocular lenses (IOLs), and provide a glimpse on proposals for actuating retinal implants. A somewhat deeper focus will then be spent on stretchable technology (enabling technology to realize smart contact lenses), wireless power transfer by means of radio frequency and iris control (for vision correction and light intensity modulation). Later, we discuss the visionary prospects of Eye-Fi, where functions and data communications to and from the eye could be incorporated into an IOL and its surroundings, as a permanent implant, benefiting from the rich environment of the ocular globe to monitor physiological parameters.

12.1 Introduction

Spectacles, contact, and intraocular lenses are primarily used to guarantee that images of the world are properly formed on the retina. However, technology evolution in optics and electronics enables more functions to be embedded in them. Smart lenses have once been imagined within the realm of science fiction, allowing documents to be scanned or scenes to be inconspicuously recorded when viewed. To the despair of misoneists and the thrill of tech enthusiasts, lenses endowed with novel functions have recently gained a tint of reality, targeting at presenting the user with data or with the experience of augmented reality (AR). Data can be text, picture, or video, as they would have been seen on a smartphone display,

doi:10.1088/978-0-7503-3263-7ch12

for instance. AR, on the other hand, virtually overlays pertinent interactive information onto the scene the user sees. Among a plethora of possibilities, deploying real-time object and face recognition algorithms, these extended functions could mean informing price tags on products, revealing a person's name and ID, applying custom subtitles to a movie, or presenting traffic directions while driving. Also, some smart lenses are equipped with sensors and propose to collect personal health data or to track the environment where the user wanders. These developments are poised to become ubiquitous and need to be carefully implemented and used under the auspices of a strict ethical conscience.

Numerous smart glasses have been developed. Morrow, for instance, is a Belgian company that developed glasses that embed a thin liquid-crystal layer between two optical lenses. With a small dose of electrical power activated at the touch of a button, the user can switch between near and far vision. Most smart spectacle developments, however, incorporate near-eye displays (NEDs) directly on one or both lenses of the spectacles, or as an add-on heads-up screen. They have been designed to incorporate data and AR for either personal or professional purposes. Advances in microLEDs and lasers make possible compact displays with a high density of pixels or high-rate beam steering directly to the retina. The technology of waveguides and bird's bath optics[1] allow information to be viewed as if directly on the lens, with a short eye relief[2]. The display itself is located somewhere off the lens, usually on the lateral support and at a distance compatible with the focusing capability of the eye. The miniaturization of electronic components, the increased computational capacity of processors and the low-power commercial integrated-circuit modules enable a high degree of information processing directly incorporated on the glass frame with acceptable additional weight and reduced delay, avoiding eventual vestibular discomfort. Yet, several manufacturers opted to have the processing power delegated to an accompanying pod, or even to a smartphone or notebook. Energy is usually stored in batteries also packed with the glass frame. Data transmission and remote controls can be executed by means of standard electronic communication modules, such as low-energy Bluetooth, ZigBee and WiFi, or by more recent wireless solutions, such as LoRa or LiFi.

Some models of these spectacles are commercial off-the-shelf gadgets while others are claimed to be developed for research purposes only. While Google Glass was one of the first fashionable and complete models to be announced, in 2013, and came equipped with camera, microphone, speakers, processor and wireless communication, it is claimed to have caused uncomfortable eye strain due to the chosen upper diagonal location of the heads-up display. Besides, it raised privacy issues as the wearers could be inadvertently recording and broadcasting scenes of any public and private spaces they visited. Aware of this criticism, at the time of writing this chapter, Apple has been preparing to launch its

[1] It is a concave semi-transparent optical surface that, used together with a beam splitter, combines the image of distinct object fields (e.g. microdisplay and ambient view) in a single superimposed field as viewed by the observer. Its shape, drawn along with central and marginal rays, resembles that of a typical bird's bath.

[2] The relief is the distance between the effective projection plane and the apex of the anterior cornea. The image then appears to the viewer as if on the projection plane closer to the eye than where it actually is.

own yet unnamed glass with no camera but, surprisingly, with a high-resolution LiDAR (Light Detection and Ranging System), capable of tracking the topography of the surroundings in real time, including body and facial three-dimensional profiles. The data processing capability is going to happen in the highly advanced chipset of the accompanying iPhone. It has been rumored that the frame and projection will be such that different corrective lenses can be fitted and that the microdisplay employed is a Sony OLED model with a resolution of 3300 dpi. In the meantime, Facebook has been investing in a project they named Aria, with the motto to put people first and never to surprise anyone. It is supposed to distribute a limited number of units for the sake of research, where the glass manufactured by Oculus comprises two inertial measurement units (magnetometer, barometer, and GPS), and three cameras: two monochrome 640×480-pixel 90 fps, 150° field-of-view cameras and one RGB 8-Mpix camera, 30 fps, 150° field-of-view camera. Microsoft comes with the Hololens 2 glasses, Intel with its Recon JetPro, Epson with Moverio and Toshiba with the DynaEdge glass. Several other companies have been promoting their products as well, some for use in specific environments and applications, such as Magic Leap, advertised as a wearable spatial computer; LUMUS, with a family of smart lenses featuring reflective waveguides; Optinvent Clear Vu, for functional performance in various professional sectors; Solos, providing visual information for athletes on the go; North Focals, with a less conspicuous design; GWD HiiDii that directly interacts with the computer hands-free and is operated by blinking; Zungle, with bone conduction speakers and a grip on entertainment; and Vuzix Blade for entertainment and businesses.

Conferring the functions implemented on smart glasses to contact lenses faces a limited power budget and presents multiple other difficulties. The whole available area and volume are much smaller; the lens rests directly on the cornea (or 50 μm to 100 μm above it for scleral lenses), leaving basically no eye relief (the distance between the apex of the cornea and the display optics); the contact lens can move around the optical axis; oxygen transmission needs to be ensured; contact lenses have a more limited expiration date (from daily to yearly use in some special cases). Despite these challenges, several solutions have been proposed, less for AR and more oriented towards medical purposes. As of displays, in January 2021, Mojo Vision was awarded the Last Gadget Standing Prize at the Consumer Electronics Show (CES), for the development of its scleral lens that vaults over the cornea and features a GaN LED-array display on a silicon complementary metal-oxide-semiconductor (CMOS) backplane, occupying less than 0.5 mm across the center of the lens, where each pixel is only 1.3 μm wide. In 2022 Mojo also introduced an accelerometer, a gyroscope and a magnetomter that continuously track eye movements, one step further in what has been called 'invisible computing'. Even cameras have been proposed on contact lenses by patents from Google, Sony, and Samsung; none of which has announced a physical development to date. Regarding medical application, contact lenses are foreseen to continuously and non-invasively access biomedical data using physical (intraocular pressure, temperature) and chemical sensors (lactate/glucose monitoring and even some cancer biomarkers). Biochemical sensors in contact with the tear fluid are actively researched and developed thanks to the presence of all the biomarkers found in blood (although in different concentrations). A famous development was studied by Google and Novartis with a glucose sensor embedded on a soft contact lens. Although the project was officially stopped (around January 2018), the technology development and concept proposition are still valid today and gave rise to a few exciting realizations for the medical community. Next to sensing, active vision correction is perceived as one of the pinnacles of smart contact lenses thanks to their natural extension of conventional lenses available today. These new devices should enable high optical quality and comfort compared with current commercial devices to be accepted by the public. Such devices mainly enable multifocality, variable depth of field and active light management. These developments will be useful for a large part of the population (already more than 1 billion people suffer from presbyopia and need multifocality).

IOLs introduce yet another layer of challenges. They are to be implanted into the eye, preferably through a small incision, which inevitably means they must have a rather small body diameter (<8 mm) and they need to be manufactured with a foldable material. This will require contactless charge of its embedded electronic components and all additional components to the lens optics need to be biocompatible or hermetically encapsulated with no possibility of chemical leakage or diffusion whatsoever. Power is limited due to storage capacity, but also due to caution with intraocular temperature increase. IOLs with ingenious mechanical mechanisms have been reported to account for vision accommodation and some of them will be discussed in chapter 13. No IOL has yet been reported, to our knowledge, with embedded electronic functions. Charging of an energy storage element, such as a battery or a micro-supercapacitor, can be accomplished by coil induction with an external primary coil, or by locally harvesting energy available by means of vibration, temperature gradients or light. Advantages are that an IOL does not move about the optical axis and is expected to be permanent. As the fabrication technologies for micro-supercapacitors and energy harvesting devices evolve and observing that highly integrated and tiny low-power microelectronic circuits are already available, smart IOLs are poised to become a reality in a tangible future. Later in this chapter, Eye-Fi, a visionary concept of such a lens will be introduced, devised to monitor health parameters in elderly patients and exchange data by means of light signals.

12.2 Smart contact lenses

A smart contact lens is a device with integrated electronics in direct contact with the eye, which provides sensing, actuation and/or wireless communication [1, 2] and offers both remarkable opportunities and challenges in a wide range of ocular applications: (i) active vision correction [3–8], (ii) biomedical sensing [9–11] and (iii) augmented reality [12, 13]. These applications can be attained thanks to important breakthroughs in miniaturized stretchable systems and hybrid integration of a variety of components onto flexible platforms [14–20]. The interaction between such platforms and the eye is crucial to develop safe and comfortable devices. This is particularly important when considering the oxygen flow required by the cornea directly from the environment. The cornea is the only organ to get oxygen in this way in contrast to by means of blood oxygenation (any blood vessels in the cornea would hamper vision). The oxygen permeability of smart contact lenses needs to be as high as possible to avoid corneal edema due to lack of oxygen. The thickness and material properties of the lens as well as its design play an important role in this optimization to prevent serious complications. The concept of a platform embedded inside a contact lens allowing oxygen flow has been recently explored [21], proving interesting advantages and increasing the possibilities of diversification. Hereafter, we present an overview of the major breakthroughs and concepts on smart contact lenses, including future perspectives.

12.2.1 Vision correction

Vision correction in a dynamic manner is considered the direct evolution of conventional contact lenses to improve vision taking into account the environmental conditions and commands from the users. It can be categorized in multifocal correction, depth-of-field correction, and light management.

12.2.1.1 Multifocal correction

Multifocal correction is mainly intended for patients suffering from presbyopia, a highly prevalent condition starting after 45 years of age (affecting more than 1 billion people worldwide) [22, 23]. It is related to the increasing stiffness of the crystalline lens in the eye, which rather consistently provokes the need of an additional power of approximately +1.00 D at age 45, +2.00 D at age 55 and reaching values of +3.00 D after 60 years old [24, 25]. There is currently no cure and people normally use different sets of reading glasses or bifocal/multifocal ophthalmic devices (glasses, contact lenses and/or IOLs), typically adapted in one eye (monovision correction). Monovision and static multifocal correction introduce different optical powers to stereoscopic vision system at the same time, which requires a certain degree of neural adaptation or resignation. Although the latter condition is debatable, implying eventual acceptance of the undesirable effects, it is known that a fraction of the implanted patients do not get used to it at all. Smart lenses with dynamic change of focal power have been proposed to alleviate the previously presented problems [26]. One of the most studied methods is the use of liquid crystals (LCs) due to their low voltage requirement and high refractive index changes (+2 D or higher) when actuated. LCs, found both in the natural world and artificial technological applications, are intermediate phases of matter which flow like a liquid and have optical properties of solids such as ordered arrangements of their molecules, whose most common types are cholesteric (aka chiral with molecules oriented in a helical structure), smectic (molecules oriented in parallel and in well-defined planes) and nematic (molecules oriented in parallel but not arranged in well-defined planes). The LC molecules can be aligned by an electric field in order to change their orientation and optical properties. Regarding the optical power design, LC-based devices can be designed following four main approaches, being concave/convex solid LC lenses, diffractive LC lenses, Fresnel shape LC lenses and electrode patterned LC lenses. These types are explained below:

(a) Concave/convex solid LC lenses: the LC layer is shaped into a curve inside the contact lens between two substrates with different internal radii of curvature [27–29]. The difficulty is to have relatively thin layers and to shape the LC layer to generate the wanted optical power correction [30]. The best performance was found with a homeotropic alignment of the LC because of reduced scattering and fast reaction time; [31].

(b) Diffractive LC lenses: a diffraction grating between curved substrates can reduce the overall thickness of the LC layer and provide a high optical power change [32, 33]. However, this type of LC cell configuration can provoke high chromatic aberrations since the variable cavity thickness is designed for a single wavelength. Manufacturing of such structures has been demonstrated [34] by means of photolithography resulting in cured/uncured zone rings enabling the diffractive behavior;

(c) Fresnel shape LC lenses: the use of Fresnel lenses can reduce the cell gap considerably from tens of microns to just a few, which enables their integration in contact lenses with fewer chromatic aberrations [35]. In this case, the LC layer is shaped as a Fresnel lens with either positive or negative focal power. The Fresnel structure is index-matched to the ordinary refractive index of the LC, which can be switched by applying an electric field to dynamically change the optical power;

(d) Electrode patterned LC lenses: the idea here is to also use the diffractive concept but this time with individual electrodes for each diffractive ring or sub-domain (groups of several rings). This translates into a variable electric field over the lens to enable the optical correction properties. This type has the most complex design and fabrication process [36].

12.2.1.2 Depth-of-field correction

Patients suffering from iris disorders such as aniridia (congenital and traumatic) and coloboma, and transilluminance disorders such as ocular albinism have short depth-of-field and light sensitivity due to the relatively large pupil diameter [37]. Guest–host liquid crystals (GH-LCD) have been demonstrated as a potential solution to mimic the functionality of the pupil/iris arrangement in an automatic manner, creating an artificial iris [38]. GH-LCD uses a combination of LC and dichroic dyes in order to change the transmittance of the cell by changing the orientation of the molecules when an electric field is applied. The dyes have a distinct transmittance depending on their orientation and, due to their geometry, they follow the LC orientation. GH-LCD are intended to be combined with electronics and on-lens light sensors to actuate independent concentric rings on the LC cell and change the effective size of the pupil, thus varying the depth-of-field according to the ambient light. A contrast of 1:2, effectively blocking 50% of the incoming light, was proven to enable the pin-hole effect for patients suffering from aniridia, where the pupil size could be bigger than 8 mm [37].

12.2.1.3 Light management

Photophobia, very high light sensitivity, affects a large percentage of the population suffering from different disorders such as chronic migraine, dry-eye syndrome, albinism and aniridia, among others [39]. These patients could have their pain alleviated with a device that could effectively reduce the incoming light to the retina at the precise moment and with appropriate reaction speed. Two technologies have been recently presented, namely: GH-LCD [37] and electrochromic displays [40]. GH-LCD devices function under the same principle as the artificial iris presented in the previous section, with the difference that the LC cell is not necessarily partitioned in rings, but it features full area actuation. The electrochromic display was based on electrodeposited Prussian blue and actuated between −0.5 V and 0.5 V. Relative contrast, which is the ratio between OFF and ON transmittance, still has to be increased in order to fully protect the patients from extra light. Currently, the contrast ratio is 1:2 for GH-LCD devices, and 1:1.5 (at 500 nm) and 1:2.6 (at 700 nm) for electrochromic devices. At this transmittance levels, such devices have been presented as light attenuation displays in order to present critical information to the user, for instance about a danger in the way, or to communicate by means of Morse code [40]. This GH-LCD technique is being developed and prepared for clinical studies by the company Azalea Vision in Belgium, including an autonomous electro-optical system, on-chip photodiodes to measure the environmental light conditions and powered by a rechargeable micro battery. Azalea Vision is a startup company spin-off from Ghent University and IMEC, which focuses on vision correction by means of LC cells embedded inside a contact lens. Their innovative system classifies as a medical device for a range of visual disorders and patients with high light sensitivity. The device works fully autonomously along the day and can be recharged during the night. Active light management, ALMA for short, is intended to mimic the functionality of the iris while providing an uncompromising level of light adaptation and comfort.

12.2.2 Biomedical diagnosis

Biomedical diagnosis and healthcare monitoring is of great importance to identify and control different types of disorders. This is even amplified if the testing can be done continuously and in a non-invasive manner. This is the main purpose of a smart contact lens with biomedical diagnostic capabilities. In general, these devices can be divided into physical and chemical sensors in order to determine the respective biomarkers to monitor [41].

Physical sensors: monitoring of intraocular pressure (IOP) can potentially help to diagnose and control glaucoma, the number one cause of irreversible blindness worldwide [42]. The standard of care procedure consists in using a tonometer only once or a couple of times per appointment at the ophthalmologist's office. This methodology fails to continuously measure the IOP, especially at night when it is known to be at its highest [43]. It is reported that a change of 1 mmHg in IOP corresponds roughly to a 3 μm change in the central corneal radius of curvature, after considering all the forces acting on the eye [44], which makes it possible to track it by following the forces at the cornea. Next to IOP, eye movement tracking is very useful to identify problems of the nervous system. For instance, diagnosing concussions and even more severe traumatic brain injuries [45]. Finally, electroretinography, which investigates the proper function of the retina, can help to diagnose problems with the light receptor cells, namely rods (for night blindness) and cones (for color blindness) [46]. A three-electrode system on the eye, forehead and side of the eye, can provide such a measurement of the electric signal amplitude as a function of time. Different types of physical sensors are presented next.

Capacitive sensors: they rely on the change of pressure that provokes changes in parallel-plate capacitors embedded in soft contact lenses [47]. However, most prototypes presented interfere with the line of sight since the capacitive transducer is located on top of the center part of the cornea for maximum sensitivity [48]. Developments with transparent conductive layers were proven to have sensitivities of 240 ppm mm^{-1} Hg between 7.8 and 42.8 mmHg [49] on eyes of pigs, which have similar dimensions and pressure ranges to those of humans.

Strain sensors: they transduce changes in the curvature of the cornea into electrical signals by using a strain gauge [50]. The Wheatstone bridge is used to increase the sensitivity of the sensor while temperature compensation resistors are used to keep the device independent from changes in temperature [51]. Ren's group made a strain sensor with Ti/Pt on PET substrate with high sensitivity of 20 μV mm^{-1} Hg in the range of 10–30 mmHg [52]. The company Sensimed introduced the product Triggerfish to measure the IOP from a soft contact lens with Ti/Pt strain sensors [51, 53]. Although a correlation has been shown between the pressure sensor and the IOP, the lens is not fitted for each patient's eyes, which diminishes its ability to measure very small changes in the eye's curvature and thus IOP [54].

Microfluidic sensors: they rely on the volume change of liquid inside small channels in the contact lens. The channels are based on PDMS and transparent glues using techniques of embossing, lamination, and soft lithography. Some groups have developed prototypes with sensitivities of 600 mmHg for channels 20 μm wide [55]. The changes in pressure can be immediately observed without any electrical power, however, it is difficult to quantify the effect with a proper readout system inside the lens.

ERG sensors: they are important to detect genetic anomalies and the functionality of photoreceptors by means of measuring reaction voltages at the cornea after light impulses [56]. Gold rings in contact with the cornea are commercially available, and new approaches using graphene electrodes have been demonstrated [56]. Graphene is very interesting because it is highly transparent, thus it is possible to place it in the whole area of the cornea to obtain more precise readings.

Chemical sensors: many of the constituents and markers present in blood are also present in tears in different concentrations [57, 58] with numbers of proteins ranging from 54 to 1543 [59]. The tear–blood barrier filters the plasma and enables direct access to its constituents in a non-invasive manner with high correlation. Concentrations on tears can be lower (glucose) [2] or higher (lactate) [60] and varies depending on the type of tears, which can be classified as basic functional tears to keep the cornea lubricated and free of dust, known as basal [59], reflex and psycho emotional [61]. Some chemical sensors will be discussed next.

Electrochemical sensors: they use an electrochemical transducer to identify biological analytes through a biological recognition element. The latter can be an enzyme, protein or antibody in order to produce the respective reaction with the target analytes [62]. The electrical signal is measured by a working electrode where the reaction occurs, while a reference electrode maintains the electrical potential constant. A counter electrode ensures the current flow through the working electrode. Up until now, the most common analytes for smart contact lenses have been glucose and lactate, as presented hereafter. An electrochemical sensor based on the field-effect transistor (FET) was demonstrated on a contact lens to detect glucose. The FET device presented electrodes based on graphene and silver nanowires, while the channel was made of graphene. The sensor showed high responsiveness for glucose concentrations between 0.1 mM and 0.6 mM, even after 24 h of usage [47]. Parviz's group presented a lactate sensor on a PET substrate with Pt electrodes [63]. The substrate was thermoformed to fit the curvature of the eye, while the electrodes were exposed to the analytes for characterization. The sensing structures were functionalized by cross-linkage of lactate oxidase with glutaraldehyde and bovine serum albumin, and presented an average sensitivity of 53 $\mu A\ mM^{-1}\ cm^{-2}$. This results in sufficient resolution within the physiological range of lactate concentrations in tear fluid.

Fluorescence sensors: these are based on the emission of photons with longer wavelengths by excitable synthetic molecules (fluorophores) when absorbing a specific electromagnetic wave [64]. Prototypes on silicone-hydrogel lenses have been presented to detect pH on the tear fluid. The lenses were mixed with hydrophobic ion-sensitive fluorophores (H-ISFs) and characterized by means of UV illumination emission spectra. The fluorescence intensity was decreased as the pH increased [65]. Other studies have presented similar concepts for glucose measurement. In that case, the fluorescence was observed to decrease as the glucose increased [66].

12.2.3 Augmented reality

AR on a contact lens aims to provide visual information primarily to low vision patients, by means of simple LED signals, and, in general, to provide complex information by means of nanoLED technology projected directly to the retina [67]. The latter is the concept of the company Mojo Vision (US) that has developed a nanoLED projector with a resolution of 14k PPI as the smallest pixel size and highest pixel density display, with around 300 times higher density than current smartphone displays [67]. Currently, the projector is monochrome on the green wavelength. This early technology demonstration is intended to be placed inside a rigid contact lens to project images directly to the retina. The display resting on the cornea is intended to project images directly to the fovea while the eye moves in order to maintain an overlay image seamlessly combined with the scene image. Currently, the power consumption and data communication of information have not been reported. Additionally, aspects such as oxygen permeability and contact lens fitting need to be addressed to envision a feasible application.

12.3 Current developments in smart contact lenses

Smart contact lenses present different challenges regarding their full autonomy (electrical power and decision making) and integration. The embedded devices need to be stretchable/flexible to conform with the shape of the eye, oxygen permeable and optically transparent [2]. Their integration techniques need to result in high manufacturing yields and be compatible with high volume production complying with medical regulations and quality assurance [2].

12.3.1 Thin-film technology and hybrid integration

One of the approaches used to fabricate smart contact lens platforms is by combining thin-film circuits and hybrid integration of active components. On one hand, the thin-film technology is based on micro-electro-mechanical systems (MEMS) processes including photolithography, wet/dry etching, electro-plating, etc. On the other hand, hybrid integration makes use of custom-made flip-chip techniques, lamination steps, ani-/isotropic conductive adhesives, and thermoplastic materials as substrates. The latter allows the use of thermoforming techniques to shape the smart circuit to match the shape of the eye or the contact lens body where it is embedded. In literature one finds devices based on the thermoforming of PET [13, 68] and parylene-C [1, 69] substrates. Although PET-based substrates can be formed at relatively low temperatures (around 85 °C due to its glass transition temperature), parylene-C ones require elevated temperatures and long processes (up to 48 h at 200 °C) [69]. The work presented by A Vásquez Quintero *et al* [14] makes use of thermoplastic polyurethane (TPU) layers as mechanical carriers hosting the thin-film electronics with active components. This configuration allows for stretchability, low temperature integration processes and relatively elastic platforms (TPU hardness Shore A87—ASTM D2240) [14], comparable with the mechanical properties of conventional soft contact lenses. The planar structure embedded inside the TPU carrier is then thermoformed into a spherical cap in order to fit the shape of the eye and it is compatible with vision correction modules on the line of sight (i.e., liquid crystal-based cells). This process gives rise to a soft, stretchable, wrinkle-free and oxygen permeable platform with electronic components, adequate to be inserted into contact lenses, rendering them suitable for daily use.

Figure 12.1(a) presents the stretchable platform including the thin-film based circuitry, silicon chip, mockup gold RF antenna, gold interconnections and TPU-based mechanical carrier. The latter enables the spherical thermoforming, with a radius of curvature between 8 and 9 mm to fit the cornea [70], including the supporting meanders from the RF antenna to the inner circuit, as shown in figure 12.1(b). The thin-film based circuitry (figure 12.1(c)) was fabricated by: spin-coating of polyimide (PI) layers (5.5 μm thick), sputtering of gold layers (150 nm-thick), photolithography and wet etching to pattern the gold layers and reactive-ion etching (RIE) to pattern the PI layers, making use of a hard evaporated aluminum mask. The gold circuitry is fully embedded inside the PI layers, which render it flexible and stretchable with high mechanical reliability. The thin-film circuit is then transferred from the rigid processing carrier to the bottom TPU substrate for the subsequent integration of the silicon chips. Figure 12.1(d) shows the integrated platform with a daisy chain chip (DCC) inside two laminated layers of TPU (100 μm-thick). The DCC incorporates several on-chip resistive paths connected in sequence which allow the measurement of contact resistance and integration yield. The latter is laser ablated with predefined meander shapes to mechanically support the components. The DCC was integrated by a custom-made flip-chip technique and laser-induced forward transfer (LIFT) of indium (In) bumps. At this point, the self-standing platform is ready to be thermoformed by means of

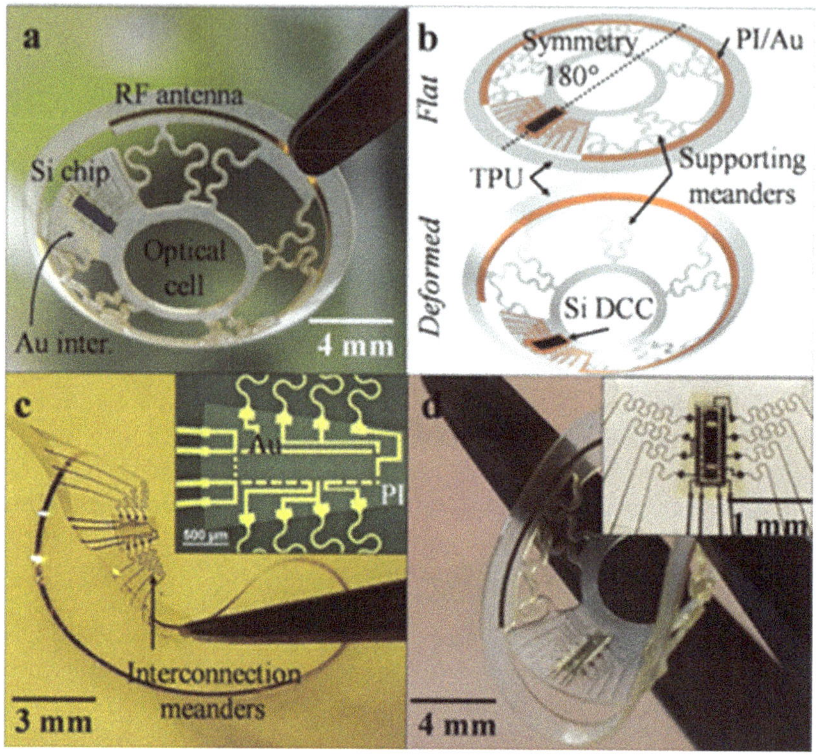

Figure 12.1. (a) Picture of the thermoformed stretchable platform. (b) Schematic of the planar and thermoformed platform. (c) Self-standing thin-film circuitry. (d) Picture of the platform bent by tweezers. Reproduced with permission from [14] John Wiley & Sons, copyright 2017 WILEY-VCH Verlag GmbH & Co. KGaA, Weinheim.

applied pressure with a spherical aluminum mold with convex and concave radii of curvature of 8 mm and 8.2 mm, respectively. The gap of 200 μm left between the surfaces protects the platform from being crushed in the process. The procedure lasts about 10 min at 120 °C.

The thermoforming of the self-standing platform into a spherical cap consists of the transformation of a developable surface (e.g. circular plane) into a non-developable or zero-Gaussian surface [71–73]. The modeling and optimization of the process was supported by finite element models (FEMs) implementing the generalized Maxwell viscoelastic model [74] combined with the William–Landel–Ferry thermal shift function [75]. Ultimately, the objective is to avoid high compressive strain at the outer edge, which can provoke failure by buckling. Compressive strains are present due to the reduction in the disk perimeter of the self-standing platform during the thermoforming. At the same time, a reduction in compressive strain would protect the mechanical integrity of the active components and interconnections. The principal hoop (or circumferential) strain, shown in figure 12.2, corresponds to $\varepsilon_{\mathrm{h}} = \frac{\sin(\Theta)}{\Theta - 1}$, where $\Theta = \frac{R}{r}$ is the ratio between the disk radius R and its radius of curvature r. The patterning of the TPU-based mechanical

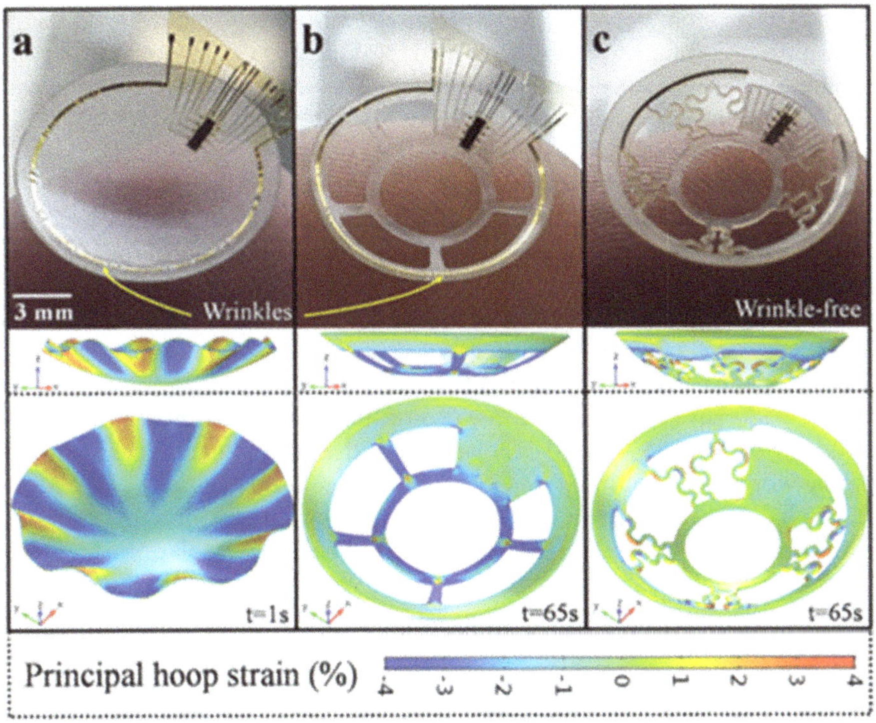

Figure 12.2. Pictures and images of simulation models of the thermoformed platforms indicating the hoop strain during the thermoforming process. (a) No patterning. (b) Straight supports. (c) Meandered supports without hoop strain along the outer edge of the platform. Reproduced with permission from [14] John Wiley & Sons, copyright 2017 WILEY-VCH Verlag GmbH & Co. KGaA, Weinheim.

carrier in the shape of meanders as stress release patterns, considerably reduced the hoop strain avoiding wrinkles along the RF antenna. Additionally, the perforated areas allow oxygen to traverse freely through the platform. Functional implementations of this concept are presented in the next two sections.

12.3.2 Power transfer through radio frequency (RF)

On-lens electrical power is indispensable for the proper functionality of any active application of a smart contact lens, either to sense, to actuate, to communicate and/or provide power autonomy. Several ideas have been proposed to provide energy to the integrated electronics, either directly or through a primary or secondary rechargeable battery. These works include photovoltaics [8, 76], tribological energy from blink events [77], transduction of chemical energy from the tear fluid [78] and RF wireless energy [68]. The latter is the only one that provides both high transfer efficiency and data communication through the same architecture, including the RF transceiver and receiver. All the systems mentioned above are potentially compatible with secondary rechargeable batteries in order to provide extended power autonomy during operation. Different RF resonating frequencies have been explored for both wireless power transfer

(WPT) and data communication, mainly focused on high frequency (HF—13.56 MHz [38]) and ultra-high frequency (UHF—900 MHz [1] and 2.45 GHz [79]). Main challenges lie on maintaining adequate form factors and thicknesses, as well as compatibility with saline solutions and contact lens daily use. An RF antenna at 2.4 GHz was fabricated directly on a thermoplastic PET carrier for WPT functions [68]. The antenna made out of copper powered up a red-light emitting diode (LED) but failed to comply with long-term exposure to saline solution, in direct contact with the eye. Another antenna was designed to resonate at 2.45 GHz combined with off-the-shelf components [79]. However, it failed to comply with contact lens standards regarding thickness and conformability. Systems compatible with near-field communication (NFC) protocols at 13.56 MHz combined with spectacle-based power transceivers provided a high-power transfer efficiency [80], but the RF antenna was a solid 100 μm thick gold wire protruding at the back side of the CL. Other works have introduced the use of ultra-thin antenna configurations by means of depositing silver nanowires [81]. However, the design presented blocked the line of sight. The work presented by A Vásquez Quintero *et al* [21] considers the previously described challenges and goes beyond the state-of-the-art to provide a thin and foldable platform compatible with soft hydrogel-based contact lenses. The platform incorporates an RF system (antenna and Si chip) designed for NFC protocols (data and power). As proof-of-concept, the design powered-up a blue micro-LED even when the lens was placed inside its conventional cleaning saline solution. The fabrication of the platform, comprising thin-film circuitry, lamination, and thermoforming, is the same as the one described in section 12.3.1. (figures 12.3(a)–(c)). The RF antenna was realized by means of pattern plating of a thin seed layer of gold (10 μm-thick and 20 μm-wide), as shown in figure 12.3(d). This structure was shown to have an impedance of 5.2 μH and a quality factor of at least 8.6. The quality factor quantifies the bandwidth of the antenna, and it is the ratio between the radiated and dissipated power at the resonance frequency. Therefore, the quality factor is desired to be as high as possible to increase the efficiency of the energy transfer and reduce the distance between the lens and the reader antennas. For the geometry presented in figure 12.3, a quality factor higher than 8 enables high enough magnetic coupling between the lens and reader antennas to generate 8 V_{pp} at the NFC chip. This voltage turns on the internal energy harvesting module of the chip, allocating DC current to the micro-LED. The gold antenna is fully embedded inside PI layers in order to increase their mechanical robustness [82]. The embedding of the platform on the hydrogel lens (2-Hydroxyethyl methacrylate—HEMA) was done by a custom-made double casting process.

Figures 12.4(a) and (b) show the flexibility of the smart contact lens by comparing it to a conventional pHEMA contact lens. Flexibility and mechanical robustness are required during insertion and removal from the eye. The smart platform, including the NFC antenna, chip and LED, conforms to the shape of the molded soft contact lens thanks to the similar mechanical properties of the mechanical carrier (TPU) with respect to the cured pHEMA [83]. By integrating the smart platform on the periphery of the contact lens, the visual path is not obstructed, and conventional vision correction is enabled for myopia, hyperopia, astigmatism, presbyopia, etc. Additionally, oxygen to the cornea is ensured through the center part and bevel edge (landing on the eye). Special attention was given to the bevel area in order to comply

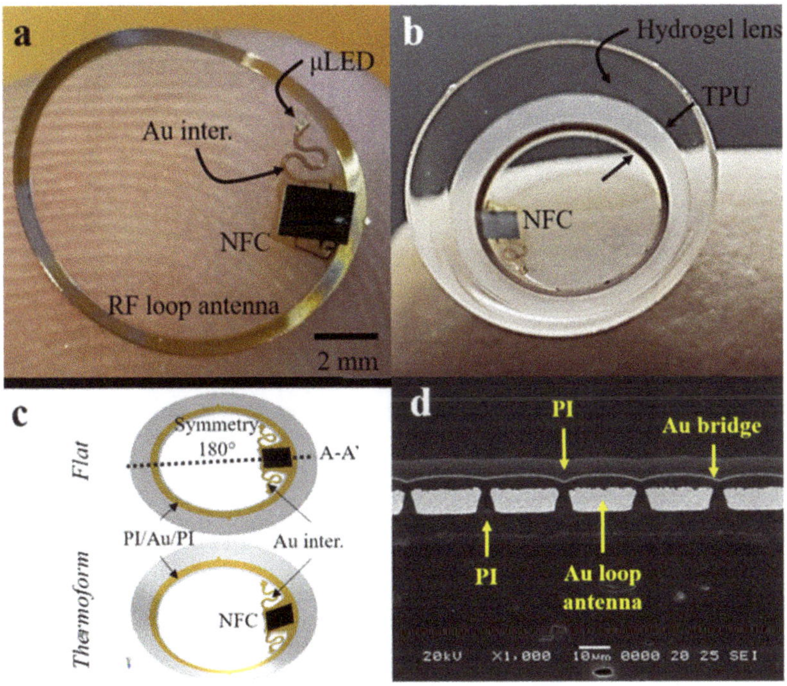

Figure 12.3. (a) Picture of the thin-film circuitry enabling NFC-based WPT. (b) Thermoformed platform fully embedded inside a hydrogel lens. (c) Schematics of the planar and thermoformed FEM models of the platform. (d) Cross-section of the ultra-compact antenna and second gold layer inside the PI layers. Reproduced with permission from [21] John Wiley & Sons, copyright 2020 Wiley-VCH GmbH.

with commercial hydrogel-based contact lenses. To test the WPT under typical conditions used for lens cleaning, the smart lens was placed in a glass container full of saline solution while the NFC chip was accessed wirelessly by the reader antenna. This process was repeated for at least 25 weeks of full immersion (figures 12.4(c) and (d)), at which point the distance between the lens and reader antennas dropped by half from 1 cm to 0.5 cm. This reduction was attributed to a worsening contact resistance between the chip and the gold antenna. Further encapsulation of the platform, for instance by means of parylene-C, is required to protect the electrical functionality of the circuit. This implementation paves the way towards autonomous smart contact lenses with excellent eye fitting and comfort.

12.3.3 Artificial iris with liquid crystals

As described in section 12.2.1, smart contact lenses with guest–host liquid crystals (GH-LCDs) are suitable for active depth-of-field (DoF) and light management [5, 8, 37, 84]. These developments are particularly interesting for patients suffering from poor visual acuity, glare, halos, optical aberrations, and photophobia. The latter, in particular, is originated from a wide range of indications and it is known to trigger pain-sensing neurons connected directly to the retinal photoreceptors [85, 86]. The GH-LCD device can be operated in two different modes depending on the protocol

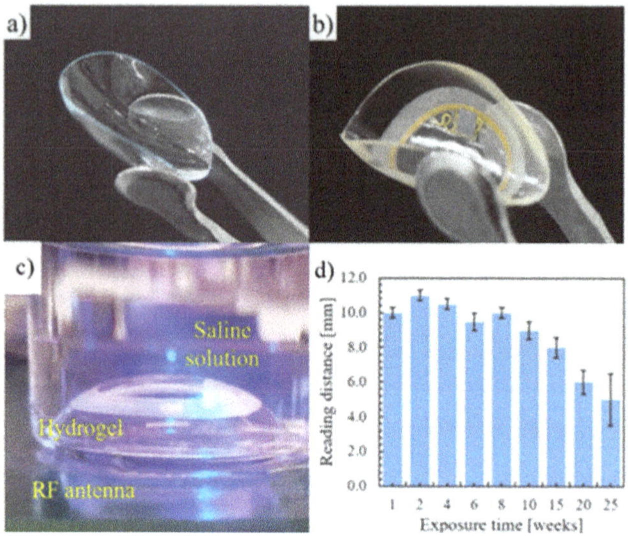

Figure 12.4. (a) Picture of a conventional pHEMA contact lens. (b) Picture of the smart contact lens between tweezers. (c) Picture of the smart contact lens inside a saline solution demonstrating the wireless power transfer and the blue micro-LED emitting light. (d) Measured reading distance versus immersion exposure time. Reproduced with permission from [21] John Wiley & Sons, copyright 2020 Wiley-VCH GmbH.

used to actuate the concentric rings. On one hand, the consecutive actuation of the rings, starting with the outer one at low light intensities and ending with the inner one at high intensities, has been demonstrated to mimic the functionality of a healthy iris. This mode, known as artificial iris, is intended to provide a variable DoF and a controllable illumination into the eye by changing the effective pupil size. This solution is suitable for patients suffering from congenital/traumatic iris disorders (such as aniridia and coloboma) [87] and transillumination (ocular albinism) [88], where the function of the pupil/iris has been impaired. On the other hand, the simultaneous actuation of all the rings in hundreds of milliseconds [38],at a particular light intensity threshold, reduces the overall light intensity reaching the retina. This solution is intended for patients suffering from photophobia to extreme photophobia due to various distinct causes, either ocular (dry-eye syndrome) [85], neurologic (chronic migraine) [89], traumatic (brain injury [85]), achromatopsia [90] or others (medication [85]). The work presented by A Vásquez Quintero demonstrated the optical benefits of an artificial iris embedded in a scleral contact lens. The light was filtered by a GH-LCD using the artificial iris mode with a contrast of 1:2 (ration between the OFF/ON transmittance states). The DoF increment, and light reduction was simulated by means of an optical model (supported by ZEMAX, Focus Software) user real patient data and the model of the smart contact lens.

Figures 12.5(a)–(c) show the transmittance attenuation of the GH-LCD with respect to a reference image with the lower-case letter 'e'. The visible-light transmittance (VLT) values were calculated as an average of the RGB (red, green, and blue) color code from the raw images of a CMOS digital camera. The contrast ratio between the OFF/ON state, and reference/ON were shown to be 1:2 and 1:3.3,

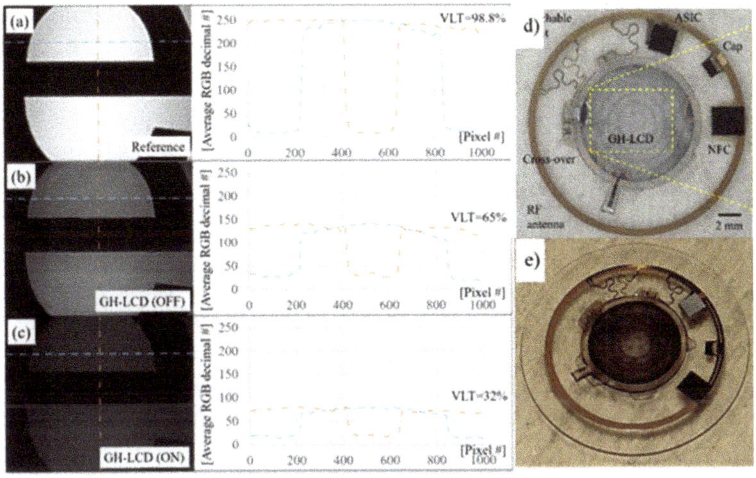

Figure 12.5. Pictures from CMOS camera and visible light transmittance (VLT). (a) Reference image without GH-LCD. (b) GH-LCD OFF state. (c) GH-LCD ON state. The values for the vertical and horizontal VLT values are represented on the plots to the right. Reproduced from [37]. (d) Picture of the smart contact lens platform with GH-LCD OFF state. (e) ON state. Figure 12.5(d) is reproduced with permission [38], copyright 202o, IEEE.

respectively, both measured around 530 nm, where the human eye has its highest sensitivity. Diffraction effects were observed around the border of the high contrast letter caused by intrinsic haze promoters, due to the multilayer stack of the substrate, the transparent conductor, and the alignment layer; and by diffraction of small spacers that had been used to keep the top and bottom substrates equally distanced. The GH-LCD was composed of PET substrates (50 μm thick), PEDOT:PSS (from Sigma Aldrich) as organic transparent conductor, SU-8 (form MicroChem) as spacers (10 μm thick and 40 μm wide) UVS 91 (from Norland Products) as outer gasket, and SiO_2 as evaporated alignment layer (6 nm thick oblique evaporation) [4, 5]. The GH-LCD was composed of the liquid crystal MLC-6608 (from Merck), the chiral dopant S811 (from Merck) and neutral black dichroic dye (at 3 wt%). At the surface, the LC was homeotropically aligned (vertically) in order to provide a transparent OFF state, representing a safe OFF option for wearable devices. The cell was electrically driven with a square signal at 1 kHz at 20 V_{pp}. A practical implementation was demonstrated with the integration of the GH-LCD (with four concentric rings) with an Application-Specific Integrated Circuit (ASIC) capable of actuating the LC rings automatically depending on the light conditions [38], as shown in figures 12.5(d) and (e). The ASIC included a range of photodiodes to detect the light from very dark (1.2 lux) to very bright (190 klux), an LCD driving voltage of 13 V_{pp}, an LCD smart flicker removal (during a blink event) and a blink detector (65% sensitivity). All these building blocks were integrated in a chip with an area of 1.45 mm^2 fabricated using a hybrid 0.18 μm BCD (Bipolar-CMOS-DMOS) technology with a supply range of 3–3.6 V and power consumption of 0.93, 1.9, 2.2, 4.2 μW when the rings 1, 2, 3 and 4 are ON, respectively.

The through-focus plot of Visual Strehl (figure 12.6) visualizes the variation of visual quality of an eye with aniridia (without iris) depending on the artificial iris aperture, optical aberrations compensation and light levels. The geometrical model of the aniridia eye was measured by optical coherence tomography (OCT) images of an 18-year-old female patient, with a 16 mm diameter in scanning range. The refractive index of the parts of the eye were as follows: cornea (1.3769), aqueous humor (1.3369), crystalline lens (1.4199) and vitreous humor (1.3359). The artificial iris aperture was simulated from 6 mm to 2 mm (smart CL), represented by solid lines in the figure. A diameter of 2 mm resulted in higher vergence range for the same Visual Strehl and light conditions. This range, representing the visual quality, can be found by finding the two intersection points of a horizontal line with the simulated Visual Strehl. The visual quality is higher for all the artificial iris apertures compared to the other cases. The higher visual quality is attributed to the pin-hole effect provided by the artificial iris aperture. The other cases include: the eye model with astigmatism and high order aberrations (HOAs) without any corrective lens (blue dotted-line), the eye model with HOAs without any corrective lens (red dotted-line), the eye model with a spherical contact lens correction (green dotted-line) and the eye model with a toric best design (purple dotted-line).

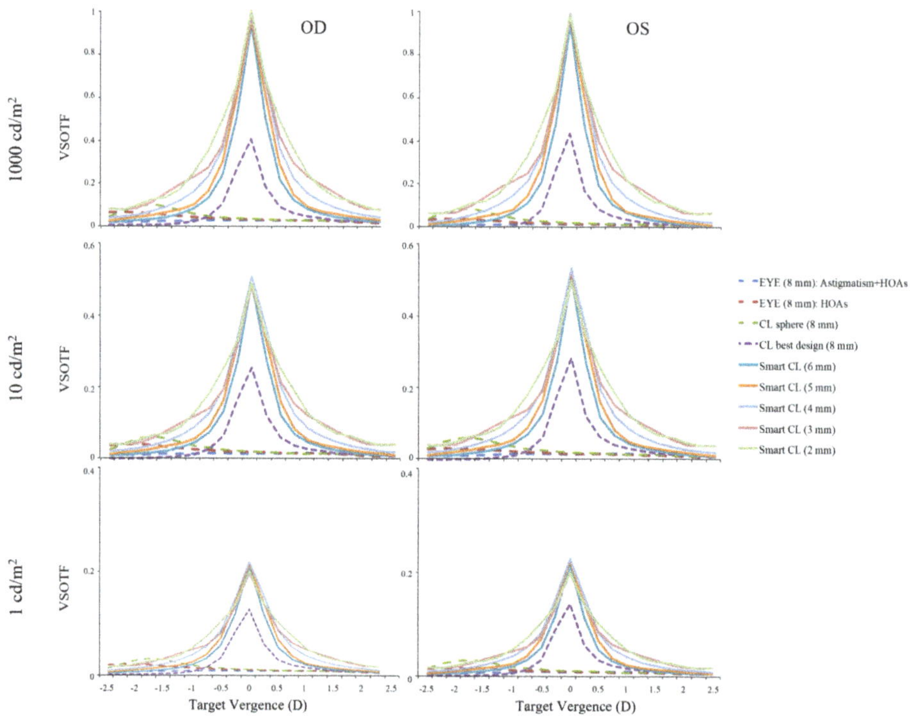

Figure 12.6. Simulated through-focus plots of Visual Strehl for different GH-LCD apertures and light level conditions (high photopic—1000 cd m^{-2}, low photopic—10 cd m^{-2} and mesopic—1 cd m^{-2}), for the left and right real eyes of a patient with aniridia. Reproduced from [37], copyright 2020, The Authors, with permission of Springer.

The light conditions were simulated for high photopic (1000 cd m^{-2}), low photopic (10 cd m^{-2}) and mesopic (1 cd m^{-2}). It was found that the highest DoF was provided by the artificial iris model with a pupil diameter of 2 mm with values of 3 D, 2 D and 0.75 D for each light condition, respectively. The current designs and implementations still allow too much light entering the eye when compared to a healthy eye, which blocks almost 100% of the light passing through the annular ring it covers. The protection provided by the GH-LCD starts to compare to the healthy iris for contrast values above 1:4, especially for the high/low photopic condition. Future studies with patients (under clinical trials) will help to validate the functionality of such devices especially covering the wide range of luminance variations across the day (as much as 100 000:1).

12.4 Eye-Fi: smart intraocular lens

The eye presents itself as a promising environment for implanted healthcare solutions. Sensors, electronic circuits, and non-volatile memory arrays could be embedded in an IOL to monitor health parameters such as IOP, intraocular temperature, glucose levels, oxygenation and cardiac cycle. Energy harvesting and storage modules could collect, store, and provide electrical power for the circuits to operate. Wireless transceiver modules would ensure data communication with a gateway external to the eye. This possibility of data exchange, especially with light, to and from the eye, involving an ensemble of integrated technologies implanted into the eye has been coined as Eye-Fi [91]. In this section, we explore further some of these possibilities.

12.4.1 The potential in healthcare

The increasing longevity of the population and the maintenance of an active lifestyle of elderly individuals present a backdrop that calls for the development of smart intraocular lenses (SIOLs). Several diseases, besides cataracts, tend to appear or become more critical at an older age, such as diabetes, glaucoma, cardiovascular and breathing conditions. These often need to have diagnostic data frequently monitored. The possibility of advancing forgetfulness, not to mention dementia and other afflictions of the mind, on top of the disadvantage of constant voluntary self-tests, data logging and sharing, must be considered as well. Hence, a patient undergoing cataract surgery could benefit from an IOL that combines vision restoration to an implanted test facility, comprising an *in situ* point-of-care solution. Also, for their own comfort, younger individuals have been prematurely choosing to have multifocal or extended-depth-of-focus lenses implanted at the onset of presbyopia, which suggests that a multi-functional lens could have a high adoption rate in that group too.

Each parameter to be monitored requires a distinctive signal or marker to be detected. For example, type-2 diabetes accounts for the overwhelming majority of the cases and it is mostly observed in patients older than 40 years. The hyperosmolar hyperglycemic state (HHS) commonly occurs in patients with this type of the disease and it can evolve over several weeks presenting a mortality rate of around 15%. Glucose concentration is an important marker to reveal this state and is ordinarily monitored in blood plasma. Contact lenses developed with a glucose sensor track the concentration of this marker in the tear film, which varies from 1% to 2% of the concentration in the plasma [92]. In the case of deployment in an SIOL, the vitreous humor is a good carrier of such a marker, where the glucose concentration is as high as 85% of that found in the plasma. It has been found in post-mortem samples that glucose concentration in the vitreous humor higher than 6.9 mmol l^{-1} (125 mg dl^{-1}) is abnormal and might suggest the diagnostic of a potentially incapacitating illness. The aqueous humor also has a diffusion mechanism of elements of blood plasma [93]. Additionally, hormones such as catecholamines and cortisol are known to promote the increase of glucose levels that, in turn, can reveal the stress level of the patient [94].

A significant advantage of the SIOL is that it brings additional features to a lens that is already being implanted, whose main purpose is to restore or enhance vision. Besides, it could be implanted in a single eye, whereas its corresponding pair receives a regular IOL. Also, the SIOL is expected to accompany the patient for decades. Considering the parameters of electronic circuits and sensors are often dependent on the temperature, the regulated temperature inside the eye offers an excellent stable environment with high reliability to the expected performance of these elements. Also, as the system will only operate intermittently, self-heating is not expected to be an issue.

12.4.2 The Eye-Fi concept

Eye-Fi technology will be integrated to a smart lens involving a set of monolithic microdevices interconnected. Figure 12.7 shows the concept for a smart pseudophakic intraocular lens with some example components that could compose it.

Most elements integrated to the lens are located along its peripheral region and are shielded by the iris from direct incidence of light (microLED2, micro-supercapacitor, RF

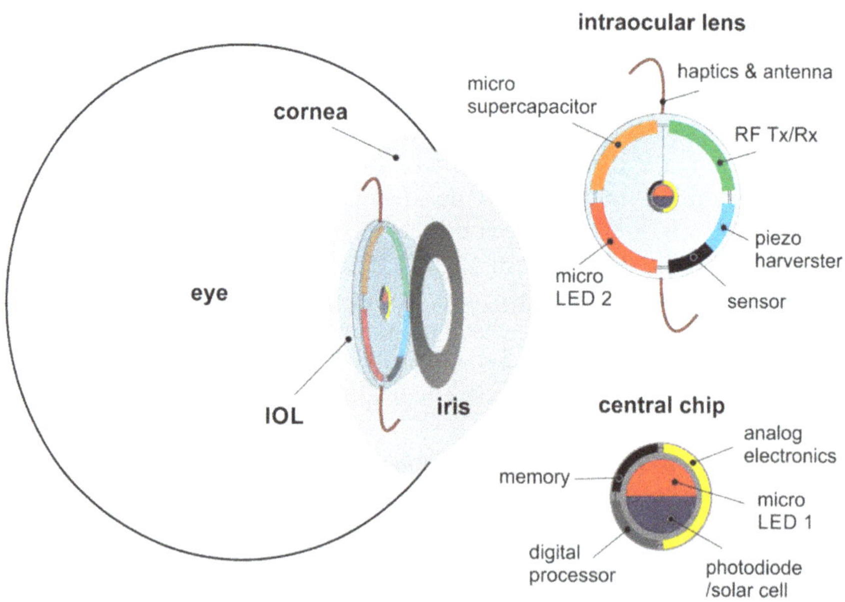

Figure 12.7. Eye-Fi artistic concept based on a smart pseudophakic intraocular lens.

transceiver and antenna, vibration micro-energy harvester and sensor). Some or all integrated circuits (ICs) can be peripheral as well. There is also a microchip depicted in the very center of the lens that connects to the peripheral elements by means of a bus of thin flat electrical interconnecting wires. The chip contains analog and digital integrated circuits, non-volatile memory, microLED1 and a photosensitive element doubling as a photodiode and as a photovoltaic solar cell. The reason for this chip to be positioned in the center is the need for its exposure to light and ability to send light signals off the eye via microLED1, regardless of the pupil-aperture size. Contrary to what might be intuitively thought, the chip neither blocks part of the scene observed by the user, nor casts an opaque shadow spot on the image on the retina. According to the principles of wave optics, provided the chip is put in the optic center of the eye and does not occupy the whole clear aperture, the primary consequence it imparts to image formation is the reduction of the overall intensity level of the formed scene image, proportional to the ratio of the chip surface area to the area of the pupil aperture. It is, therefore, desirable to have it occupy as small an area as possible. The chip edges will also introduce diffraction effects, but no more than those already present in IOLs with diffractive elements or even those due to the pupil itself. If the perimeter of the chip is made circular, rather than rectangular, the perception of these effects is expected to be lower.

Next, the several parts that could potentially compose the smart Eye-Fi system are discussed, along with a brief discussion of the underlying technologies and some state-of-the-art solutions.

12.4.2.1 Integrated circuits

The most ubiquitous and dominating IC technology available for microchip design and fabrication is silicon CMOS, accounting for the overwhelming majority of the chips in electronic devices and equipment. Several to many thousands of chips can be simultaneously fabricated on a single silicon wafer, whose diameter ranges from 100 to 450 mm, depending on the industrial facility. Each chip, also called die, is then separated from the wafer by diamond scribing or sawing, or by laser cutting.

The technology node is a label that indicates the minimum feature size that can be fabricated in a certain factory. The industrial production nodes have evolved from tens of micrometers in the 1970s to as low as 3 nm in 2021. It enables the integration of millions of nanometer-scale transistors per mm^2 on a thin silicon chip, a density soon to be increased to up to a billion with the projected advent of the 1.5 nm node in 2026, reaching material atomic-scale boundaries [95]. The actual chip size depends on the technology node and on the number of integrated components, internal interconnection wires and contact pads needed to execute its proposed functions. Depending on its purpose and complexity a chip can typically measure from half a millimeter to several centimeters across its diagonal. The clustering of so many electronic components in such a small silicon patch area is referred to as very large-scale integration (VLSI) or ultra large-scale integration (ULSI).

A CMOS process consists of a large series of interdependent steps involving the doping of the semiconductor substrate according to a specific topographical pattern, the deposition of extremely thin films of isolating and conductive materials, and the etching of those layers to create topological patterns. Most of these steps are combined with photolithography, a method using light, usually ultraviolet (UV), and a photosensitive polymer to create the desired patterns on the chip surface. Although the pipeline of a complete CMOS run takes at least several weeks, it is a batch production process with a large throughput and high yield. The quick evolution of any new CMOS node has been pulled by an immense market drive, attracting resources and governmental subsidies to foster steady research and innovation from materials to equipment.

The FET is a semiconductor device that basically allows an electric current to be controlled through it by means of voltages applied to its three active terminals, most distinctively to its isolated gate terminal. The amount of current can be regulated, interrupted, or reversed, which enables the execution of many electronic functions, from simple signal on and off switching to amplification. There are fundamentally two types of these transistors (NMOS and PMOS) with slightly different but complementary characteristics. The combination of different sizes and types of transistors yield all kinds of electronic circuits in the digital and analog domain.

Digital circuits often rely on the smallest possible transistors and operate based on binary logic, zeros and ones, the driving force of data computation. They are the core of computer microprocessors, whose building blocks are smaller cells called logic gates. These are usually circuits with a high transistor density, which depends on the computational efforts required.

Analog circuits generally employ transistors larger than the minimum size feasible and function, for example, as signal amplifiers, signal filters, voltage converters and comparators. They are necessary to directly handle the raw signals from sensors and data receivers, which are sometimes weak, noisy, or incompatible with the signal requirements demanded by the processor. They are also necessary to power a sensor, or to handle the necessary signal to light up an LED, or to drive an antenna that transmits data. Additionally, analog circuits are used to couple a micro-energy harvester to the energy storage unit or to the circuit to be supplied with power. Analog integrated circuits, along with transistors, often need resistors, capacitors and sometimes inductors. These latter components have a much larger footprint than minimum-size transistors. Digital and analog circuits are combined in analog-to-digital (ADC) and digital-to-analog (DAC) signal converters. These are important circuit blocks in the communication and compatibilization of data between digital and analog parts.

IC designers count on accurate and reliable device models to propose and simulate circuits to custom needs, using advanced software platforms to that end. Non-volatile memory arrays are also possible in some versions of CMOS technology. They are useful for storing data locally without the need of a power supply during the retention period.

A very advantageous trend of CMOS circuit design is that there has been an increasing effort to develop reliable ultra-low-power circuits aiming at the rapid growing rate of adoption of portable devices, with 21.7 billion connected wireless devices in 2020 and a projected number of 41.2 billion in 2025, with more than half being devices for the Internet-of-Things [96]. Another important benefit of CMOS technology is the availability of non-proprietary foundries, meaning that anyone who designs an IC can have the chip fabricated in one of those facilities, only paying for the processing steps and retaining the rights to the intellectual property for its design, deployment, and commercialization. Since in one particular CMOS process the materials and the sequence of the fabrication steps remain the same, some of these foundries offer the possibility to have one's design fabricated alongside other designs on the same wafer, in the same run, in what is termed a multi-project wafer (MPW). It delivers a limited amount of chips per design but cuts the cost significantly, which is shared among participants.

To bring a sense of proportion of chip size, number of transistors and clear-optic size, if a 1 mm^2 chip is fabricated, using an advanced but now conservative 32 nm digital CMOS technology node, 20 million minimum-size-cell transistors could be implemented, about the number of transistors found in a processor (CPU—central processing unit) of a modest smartphone model. Considering it is a chip with digital and analog blocks (amplifiers, voltage converter, ADC), a memory array integrated and that several contact pads are necessary for the interconnection with peripheral devices, it could still have a transistor count for digital purposes around 1 million, which can enable a reasonable number of digital-processing custom features. A 2 mm pupil offers a clear aperture area of roughly 3 mm^2, whereas a 6 mm pupil encompasses about 28 mm^2. This means that millions of transistors can be placed in the center of an IOL, executing plenty of electronic functions, and still guaranteeing a considerable amount of clear area, 67% to 96%, depending on the pupil aperture. For analog circuitry, in contrast to purely digital circuitry, it is usually preferred to choose a higher and more traditional CMOS node, usually larger than 90 nm. This is justified by the larger supply voltage-to- threshold voltage ratio, enabling a wider signal range and signal-to-noise ratio; and by the higher current handling capability, necessary when coupled to some types of sensors. This comes at the cost of a lower transistor density for the implementation of digital functions.

The printed conductive ribbons connecting the chip to other elements in the periphery of the IOL could be either metallic (Cu, Ag, Au), or made of a liquid metal alloy (EGaIn, eutectic gallium–indium; Galinstan, gallium–indium–tin), or polymeric (PEDOT: PSS, poly(3,4-ethylenedioxythiophene) polystyrene sulfonate), transparent conductive oxide (ITO, indium tin oxide), graphene or silver nanowires (AgNWs). To align with the requirements of a foldable, or partly foldable IOL, a good compromise of electrical conductivity and tolerance to strain must be sought. Modern flexible IOLs are folded and loaded into an injector that dispenses it into the eye through an incision in the cornea. The IOL then unfolds and its haptics are brought to their intended fixed position. To resist folding of the IOL, in the injector, and its unfolding in the eye, upon implantation, a more stretchable geometry is recommended, as a serpentine, instead of a straight wire.

Lastly, it is also possible to have a lower level of electronic integration directly on substrates other than silicon, and even flexible and curved ones, by employing thin-film transistor (TFT) technology, as has been common in displays for decades, using amorphous or polycrystalline thin silicon layers. There is also the alternative of organic electronic devices, which are often compatible with polymers and flexible curved substrates. Their demanded fabrication infrastructure is orders of magnitude cheaper than that of a CMOS facility. These devices are based on organic materials, their footprint is much larger than those of silicon devices, their reproducibility is poorer, they age faster, thus their functional lifetime tends to be shorter.

12.4.2.2 Sensors and actuators

Microsensors have been developed for various domains: optical, mechanical, magnetic, chemical, thermal, etc. Usually, for the measurement of a certain quantity, there are different proposed methods and devices. For instance, short-wave infrared radiation (SWIR) can be detected using either a microbolometer or a quantum-well photodetector. The first operates on the heating of a thin film of vanadium oxide, whereas the latter works by having electrons being promoted to a current through a sandwich of heterogeneous layers of a semiconductor. These microdevices are often produced using process steps similar to those employed in a CMOS process. Photodiodes that detect visible or near-infrared light, for example, can be produced alongside CMOS electronic circuits, with basically no intervention to the process steps. Camera image-sensor chips of the CMOS type rely on this to create the megapixel arrays.

Some sensors and most microactuators need moving parts, as cantilevers, bridges, membranes, plates, ratchets, wheels, or sliding beams. Some of these parts can be made suspended on top of the substrate, some can be suspended on a hole in the substrate, and some can simply slide, with some friction, against another layer. These can be fabricated using any of a group of processes that are coined as MEMS technology. MEMS is not as standard as CMOS in terms of having the same type of device as the building block or following the same kind of steps and materials. Most MEMS processes are based on silicon substrates but vary across foundries and are often tailored to a specific kind of microdevice structure. These processes often allow scalable fabrication of devices. Some MEMS processes can be combined to a CMOS chip in a post-processing step. An ISFET (ion-sensitive field-effect transistor), for instance, requires few post-processing steps to work as a chemical sensor. Earphone actuating membranes, inkjet-nozzles for printing cartridges, projector micromirrors and piezo-rod retinal implants are examples of MEMS microactuators. Accelerometers, pressure sensors and flow sensors are some types of MEMS microsensors.

Given the plethora of choices and ample possibility for customization, it is promising that microdevices could be designed to attend to the needs of Eye-Fi as, for instance, the IOP sensor, the glucose sensor, the intraocular temperature sensor, the fall sensor, the heartbeat sensor and the optical retina scanner. Looking yet one level further, sensors to collect surface biopotentials, based on skin-tattoo electrodes, could be deployed to monitor vital body activities by means of *in situ* EEG (electroencephalography), ECG (electrocardiography) and EMG (electromyography) [97]. In all these cases, the sensors need to have a small footprint, they should not be power hungry, and applied voltages should be as low as possible. These microdevices can be positioned at the periphery of the IOL and, depending on its function, can be completely encapsulated by the lens. However, if a chemical glucose sensor is used, it needs to have an access window for its active surface to be in contact with the vitreous. This could be assisted by microfluidics technology, where a network of micro-channels, valves and chambers can be fabricated on polymer or glass. Nevertheless, photonic sensor solutions usually avoid contact with the analyte.

12.4.2.3 Energy harvesters

It is possible to harvest energy with microdevices from energy sources in various domains. A photovoltaic solar cell converts light into electrical charges that can be stored or immediately used to supply energy to circuits and sensors. Conveniently, any silicon diode, discrete or integrated, works as a solar cell when exposed to visible or near-infrared light. Therefore, a simple photodiode in a CMOS chip can double as a solar cell and could be placed in the optic area of an IOL [98]. Hence, while the user has the eye open, not only is the image of the scene formed on the retina, but also the solar cell generates energy. The amount of energy produced depends on the illumination level of the scene reaching the eye, on the exposed area of the solar cell and on its efficiency. An additional generation boost can be accomplished by placing within the visual field an infrared charger with a wavelength ranging from approximately 740 nm to 1100 nm, which is invisible to the human eye. Although this range is not hazardous to the eye in low intensities, care must be taken with the designed luminance of the source and its proximity to the eye. The cell can be integrated on-chip with other functions. Auxiliary circuits, as a charge controller, are necessary to manage the generated energy with the storage element and the circuits to be fed. DC–DC converters can also be deployed to adjust the generated voltage level to that of each specific integrated element. Also, because the illumination level varies with the scene and time of the day, it is convenient to employ a maximum-power-point-tracker (MPPT) algorithm, running on hardware on-chip, to identify and enforce the most energy profitable operating point of the solar cell for a given illumination level. This strategy will be further discussed in the next subsection.

Vibration energy from heartbeat, breathing, micro-saccadic movements, ciliary muscle contraction or body motion could also be converted to electrical energy by means of moving cantilevers with piezoelectric film patches applied to their anchor points. Such a piezo material generates electrical charges across its terminals when subjected to mechanical stress and the anchor point of a suspended cantilever is the region where most of the stress occurs. The cantilever can be designed with its length, width and tip mass tailored to resonate at a specific vibration frequency, which accounts for the highest vibration amplitude and maximum stress, therefore maximum generated energy. These micro-energy harvesters can be fabricated using MEMS bulk micromachining of silicon with the addition of an aluminum nitride (AlN) thin film where applicable. If encapsulated inside the IOL, the cantilever needs to have some gap around it to enable the vibrations and it should preferably be encapsulated in vacuum to avoid oscillation damping due to air or other media. It is important that the structure is robust enough to remain operational for a lifecycle of several decades. An accompanying rectifier circuit is necessary to convert the alternate voltage into a continuous one before the energy is stored and used. When a voltage difference is applied across a piezoelectric layer, its thickness changes. Micropumps or micro-pistons can be devised using this electrostrictive phenomenon.

The generated power depends on the intensity of the energy source and on the area of the harvester, among other specific features of each type of device. The produced power is typically rather low and lies in the pico-watt to the nano-watt range. The generated energy depends on the power and the amount of time the device has been exposed to the source. In one hour a micro-harvester could generate energy from nano-joules to tens of micro-joules. Since most of the healthcare functions will be intermittent, sensors and electronics will only enter operation for a short period of time, fractions of a second to a few seconds, and will stay in stand-by for all other periods, during which energy is generated and stored. Electronic circuitry and sensors, nevertheless, should consume as low power as possible.

Harvesting solutions can co-exist in the eye, guaranteeing redundancy and contributing to the accumulated energy budget. A few other microgenerators could be listed: thermoelectric harvesters based on a Peltier device, using bismuth telluride (Bi_2Te_3) and the Seebeck effect, could be considered if there were reasonable temperature gradients in the eye; enzymatic biochemical harvesters, as the glucose bio-battery, are still at their dawn, and only accommodate few charging and discharging cycles; and coils around the lens periphery could have power transferred wirelessly by radio frequency induction from a primary coil external to the eye.

12.4.2.4 Maximum power-point tracking of solar cells

When a solar cell is used as an integrated energy harvester, the amount of energy generated depends on the illumination level available in the scene. Sometimes there will be bright daylight outdoors and other times dim light indoors. To ensure the maximum power is collected, the solar cell needs to operate close to its maximum power point and an MPPT needs to be employed. The MPPT must guarantee a maximum power transfer between the photodiode, operating as a solar cell, and the storage element or the load, i.e. circuit or sensor to be powered. Several methods have been developed for this purpose [99, 100]. A usual method for MPPT for solar cells is the perturb-and-observe (P&O) algorithm. This method monitors the output voltage and current at frequent intervals and calculates the corresponding power. A circuit named charge controller varies the cell operating point to track the point of maximum power. This method guarantees high efficiency but requires a large quiescent power, as the MPPT is continuously monitoring and adjusting the PV operation point. It is, therefore, not viable for the SIOL. Other methods, such as the Hill Climb and Incremental Conductance, are based on similar principles and are incompatible with ultra-low power applications. A promising compliant alternative that requires a lower quiescent power is the fractional open circuit voltage (FOCV). It works based on the assumption that the open circuit voltage (V_{OC}) is approximately linearly related to the maximum power voltage (V_{MP}) at varying illumination level and temperature. The FOCV method measures the V_{OC} of the cell and calculates the V_{MP}. Even though it does not verify the actual maximum-power point *per se*, its low cost of implementation, fast tracking and low power consumption makes it the most suited to be implemented in an SIOL [101]. The challenge with the FOCV method is to measure the V_{OC} of the cell while the system is in operation. Usually, this is done by briefly disconnecting the cell terminals from the load, which causes some power loss. This can be mitigated by using an adjoining pilot cell, additional to the main cell, to determine the V_{OC} [102]. An important aspect of the power management circuitry is that it should employ topologies that favor the use of capacitors instead of inductors. Most of the classical DC–DC converter topologies, used for voltage compatibility, rely on the use of inductors, which are difficult to be properly fabricated in a standard CMOS process. Topologies based on switched capacitors are more compliant choices [103, 104].

12.4.2.5 Energy storage

It is important to have a rechargeable element to store the energy generated by the harvesters and to retain this energy for a period at least as long as the time between two wake-up calls of the electronics. The storage capacity should also be compliant to the amount of energy that is expected to be generated within that period. Although an estimation of the necessary power and energy depends on the number and type of electronic devices, given the need for brief power consuming activities, it is probably safe to assume that a high-power density is more important than a high energy density. With that in mind, microsupercapacitors are promising candidates. They offer power densities higher than those of batteries and higher energy densities than those of traditional capacitors. They are also rechargeable and can store charge according to different principles, depending on the microsupercapacitor type [105]. In electrochemical double-layer supercapacitors, the electrodes feature a structure whose effective surface area is much larger than the apparent surface area, due to creases, pores, or the presence of a mesh of nanometric elements. The large effective area is important to store charge electrostatically. Conventional electrodes and electrolyte materials are conducting polymers, metal oxides and activated carbon, where the key performance is intimately dependent on the structure and composition of the electrode material [106]. Currently, there is research on electrodes of carbon nanotubes, nanoporous and microporous carbon films, graphene, transition metal oxides, conducting polymers and combined composites. The microsupercapacitor should be resistant to some degree of bending and twisting if it is to be incorporated into a foldable IOL. The components can be suitably placed in the IOL opposite to the folding axis, where strain is lower. There has been investigation of self-healing supercapacitors aided by a PVP–H_2SO_4 gel electrolyte [107].

Batteries are less attractive because of the hazardous materials they usually employ, the risk of substance leakage and volume change when aged. Solid-state polymer-lithium batteries on silicon substrates seem to be promising candidates in exhibiting low cytotoxicity. The energy density of batteries is high, but the power density is often lower than that of the supercapacitor counterpart. Nevertheless, for wearable applications, zinc-ion batteries look promising, claimed to be flexible, safe, non-flammable, durable and non-toxic [108]. They have been studied in different varieties as Zn–carbon, Zn–air and Zn–MnO_2 [109, 110].

12.4.2.6 RF transceiver

Radio frequency (RF) comprises a band of the electromagnetic spectrum ranging from tens of hertz to hundreds of gigahertz, encompassing the microwave and the millimeter-wave spectra. The higher the signal frequency, the lower its reachable distance. The RF spectrum serves a vast field of communications, as long-range AM radio, FM radio, television open-air broadcasts, 4G mobile-phone communications, LoRa, 5G multi-band technology; and short-range communications, such as wireless local networks (WiFi), Bluetooth, Low-Energy Bluetooth, and ZigBee.

An RF transmitter and receiver, named transceiver, can be implemented on-chip. It relies basically on an antenna, an impedance matching module, a rectifier circuit, and an electronic signal filter. A transceiver is used to both receive and transmit signal data wirelessly and it can also be designed to double as an energy harvesting module, where it is usually termed as a rectenna [111]. The electrical matching module is used to maximize either the signal transfer or the power transfer, depending on the application. When charge carriers oscillate in RFs in the conductive rods of an antenna, energy can radiate off the antenna to the medium around it in the form of radio waves. Likewise, when radio waves reach an antenna, electric charges in the material are brought to oscillation, inducing a measurable alternate electrical signal, whose energy could also be alternatively stored when the rectenna serves as an energy harvester. An IOL prototype based on a modal LC spatial phase modulator has been devised to restore accommodation. The control signal is delivered to the lens wirelessly by means of a 6 MHz RF carrier wave, whose amplitude is modulated by a low-frequency signal up to 50 kHz [112], see chapter 13.

The antenna length depends on the intended wavelength of the signal, which is calculated by dividing the speed of light in the medium by the signal frequency. In the half-wave dipole architecture, there are two conducting parallel wires or ribbons with one quarter of the wavelength each, considering the wavelength in the medium where the antenna is implemented. At an RF normally used for WiFi, 2.45 GHz, the length of each wire would have to be 30.6 mm in air, and about 20.4 mm in a polymer with a refractive index like that of PMMA, and 9 mm if implemented in a silicon chip. If the communication frequency increases to 10 GHz, the antenna length decreases to 7.5 mm in air and 5 mm in the polymer. Efficient antennas are usually implemented off the chip and can be placed along the peripheral region of the IOL. A lens with a 6.5 mm optic zone and 0.5 mm rim to accommodate the antenna, results in an outer perimeter of 20.4 mm, therefore, matching the requirements for antenna lengths for frequencies down to 2.45 GHz. Perhaps, even the haptics could be used as antennas.

When on-chip integration is desired, the following modules are more likely to be integrated: the rectifier, the filter and the signal-processing circuitry, or the load, in the case of the rectenna working as a harvester; whereas the antenna and matching structures tend to be external to the chip. There are several RF transceiver topologies in the literature that feature low power consumption and fulfill the low area requirements for an SIOL [113–117].

An SIOL with the antenna serving both for energy harvesting and for communications has been proposed by Boysset [118]. This electronic lens is supposed to restore accommodation by changing the curvature of a central optic zone with the controlled injection of a liquid by means of a micropump. The lens is meant to be inserted in the anterior chamber and to complement the natural crystalline or another implanted IOL within the capsular bag. It is supposed to accommodate vision based on the direction of gaze of the user, where the best contrast of the target object in the scene is assessed by an integrated autofocus system (see chapter 13).

Some disadvantages of radio frequencies when applied to IOLs are the already densely populated and highly regulated communication bands, the wide spatial power distribution of the emitted radiation, and its high absorption by aqueous media and conductive media. Besides having a high-water concentration, the electrical conductivity of the tear film, the cornea and the aqueous humor increase with the increase of the RF [119]. In addition, RF is transmitted through walls and information could be easily tapped by unwanted parties.

12.4.2.7 LiFi transceiver

A different type of data communication can be achieved with light to and from the eye. When visible or infrared light is used for communications without any guided media, fiber optics or waveguides, it is referred to as optical wireless communications (OWC) [120]. If the exchange of data happens in the visible spectrum it is called visible light communications (VLC) [121]. More recently, especially targeting high data rates, the term LiFi has been used, meaning light fidelity, as opposed to WiFi (wireless fidelity) [122]. The advantages of the optical spectrum, especially in the visible range, are the absence of strict band allocations; the much wider band than the RF spectrum; beam directionality, therefore increasing data security; compatibility with lighting, displays, and cameras; compatibility with silicon technology for the photodetector; compatibility with LEDs. LiFi has been gaining momentum in recent years due to the massive adoption of LED lighting, which can be adapted to become data access points in indoor as well as in outdoor environments, from medical to industrial, from residential to commercial applications.

On top of the higher energy efficiency of this type of light source, an important feature for data communications is that they can be electronically modulated by a driver circuit. It means that a custom LED lamp can both illuminate the room and simultaneously transmit data. Data can be coded as small amplitude variations of the intensity level, imperceptible to the individuals in the environment. This is called amplitude shift keying (ASK), of which on–off keying (OOK) is the most usual variation, working in a way similar to Morse code. More advanced schemes have been developed, allowing increased data rates, more robustness to interference and reflection, and using multiple colors simultaneously.

Besides ambient lighting, LEDs have been included in most electronic home appliances, featuring one or more pilot lights or indicator lights. More vehicles sport head and tail LEDs. Traffic lights and billboards too. LED displays, either in pixel or as backlighting, are the dominant screen technology from TVs, to computers, to smartphones. LED stands for light emitting diode. It is a diode made of a special combination of semiconductor materials and dopings. When the diode is forward biased and an electrical current above a certain threshold flows through it, it emits photons, and the number of emitted photons increases as the current magnitude rises. There are inorganic and organic LEDs, and both can be discrete devices or assembled in arrays. Their dimensions can range from several microns to dozens of centimeters [123]. MicroLEDs in AMOLED displays are about 15 μm small. LEDs are intrinsically monochromatic; however, they can be clustered to provide polychromatic light, including warm and cold white. The white light results from the perceived weighted combination of primary red, green, and blue colors by the human eye. White LED modules are either phosphor coated, where a phosphor layer receives photons of a given wavelength directly from a certain diode and reconverts them to a different color; or they are RGB, containing three diodes of different materials each emitting at red, green and blue, respectively. Higher data rates are obtained with RGB LEDs, as direct light emission is faster than photon absorption and conversion in phosphor LEDs.

A microLED could be integrated to the SIOL to transmit the collected data to an external receiver unit (gateway). If the receiver is put close enough to the eye, the LED power can be kept to a minimum. Data codification could be implemented with a simple OOK scheme, where light is emitted in the near-infrared (NIR), which is invisible to the human eye, but detectable by a silicon photodetector. Driver circuits, to control the currents through microLEDs, can be implemented on a CMOS chip. An additional implanted microLED could operate in the visible range and face the retina. It could shine a warning signal programmed according to the user needs or to a predefined threshold to a critical health parameter.

The data receiver is a photodetector. Silicon photodetectors are ubiquitous as discrete devices or as arrays of pixels in cameras. They can detect light in the visible spectrum and in the near-infrared till 1100 nm. Although there are different possible topologies for a photodetector, a simple and reliable one is an ordinary photodiode. It is simply a monolithic surface junction of two opposite parts of silicon doped differently and is compatible with CMOS chips. When light shines on the surface of the photodiode, photons penetrate the material and generate charge carriers, which result in an electrical current at the terminals of the photodiode. The current is proportional to the rate of photons reaching the device. To turn on and off the photodiode and to improve the quality of the signal, an electronic circuit can be directly connected to it. The complexity of the electronics can range from a simple electrical switch to that of a circuit that amplifies the signal and converts it to a digital number. The combination of the detector and the immediate proxy circuit is called a pixel. A receiver pixel can be integrated into the IOL in its central clear optic to receive control data signals from an external gateway. If equipped with proper switches, the same photodiode can operate as a solar cell when it is not receiving data.

The external gateway unit can be designed to work in a full-duplex operation, both receiving data from the eye and instructing the SIOL what data to collect and when. It can also be used to routinely check and calibrate the smart systems after implantation. The gateway can be a custom hardware unit, or an available device with at least one photodetector and one LED and control software. A smartphone, a smartwatch, a tablet, or a notebook could be suitable candidates. Additionally, the system can be designed in such a way that the digital processor implanted in the lens can execute simple functions when the user deliberately performs a certain blink pattern.

Photodetectors, solar cells and the microLEDs do not necessarily need to be in the exposed clear optic zone of the pseudophakic lens. They can be placed in a rim along the perimeter of the lens provided the lens features a waveguide combiner that works as a light-beam splitter, redirecting part of the incoming light to the peripheral region where the solar cell is located, and letting the rest through towards the retina. Moreover, this same element can guide part of the light from a microLED, at the periphery of the lens, off the eye towards an access point. Several different types of waveguide combiners have been developed for head-mounted and near-eye displays for augmented reality [124]. The relative dimension of the waveguide combiner and the ratio of light intensity that is transmitted or that is deflected can be tailored to each specific need.

12.4.2.8 Overall system requirements

The integration of such a system in the eye is visionary and challenging, presenting many issues that need to be carefully assessed. The system must be reliable and durable with biocompatible materials and, otherwise, hermetically encapsulated. No leakage of substance or charge is acceptable; it must not be flammable, and the operational temperature must be as close as possible to the body temperature while dissipated heat must consider the thermal conductivity and specific heat of the media in the eye. Voltages and currents must have carefully limited amplitudes and must be strictly applied to the elements of the system, without any chance of electrical arcs or charge flow to other elements of the eye. The magnitude and frequency of the electric fields must not be hazardous to any of the tissues in the eye. The illumination levels and wavelengths of the implanted microLEDs must not be damaging to the retina nor detrimental to vision.

It is also important to guarantee that the implanted microprocessor, that will manage all the data processing and traffic on-chip, can be reprogrammed, for instance, with the update delivered by a light data stream. Also, the user must have a reliable and uncomplicated mechanism to mechanically shut down the system, for instance, by means of a wireless switch externally actuated by a magnetic coupled transceiver. Cybersecurity must be ensured too. A promising solution called LiFiX proposes both signal-frequency modulation and spatial color mosaicking to create a coded key of sequential data transmission and reception [125].

12.4.3 Preliminary Eye-Fi developments

A silicon microchip has been designed and fabricated containing a module with digital and analog integrated circuits, and an optical receiver, to serve as a preliminary demonstrator, aligned with the proposal of an energy autonomous system. The operation at low power also keeps the generated heat as low as possible, preventing IOL deformation and a rise in the intraocular temperature. The embedded digital circuit works in two stages: sensing and transmitting, for example. The circuit changes states when the optical receiver detects a pattern of three eye blinks, represented by abrupt changes in illumination. The optical-receiver module consists of a photodiode, a signal amplifier and a comparator.

Figure 12.8 shows a block diagram for the integrated module. The system operates by sensing the light intensity with the photodiode, where its current is converted into a

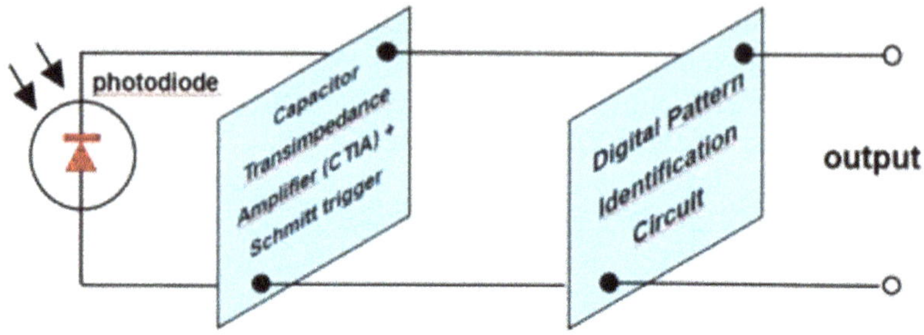

Figure 12.8. Block diagram for the integrated module that detects the blinking pattern.

voltage by a capacitive transimpedance amplifier (CTIA). This amplifier uses the photo-generated current to charge a capacitor for a given time interval and produces a voltage signal at its output that corresponds to the charge level of the capacitor. This voltage signal is proportional to the amount of light received by the photodiode during the specified time interval. The voltage is then fed to a signal comparator (Schmitt trigger circuit). This comparator sets a threshold value that relates to the expected presence of light and outputs a significant signal whenever there is a transition from a low light level to a high light level, representing a blink.

The final block is a digital pattern recognition circuit that identifies the pre-configured blinking pattern. The circuit uses a combination of logic gates to implement a finite-state machine (FSM) with two output states. An FSM is simply a code that has an initial state and a finite number of output states, where the transition between any two output states is triggered by an external input signal. An output state corresponds to a certain function to be executed, for instance, to instruct an appended temperature sensor to collect data for 30 s and store them to the memory. In this FSM the initial state of the circuit is sleeping. Meanwhile, the optical receiver module is constantly detecting light intensity. Abrupt changes thereof are constantly monitored by the input digital circuit of the FSM. Once the expected serial pattern, represented by three consecutive blinks is identified, this serves as a trigger to the FSM that changes the output state. This alternate output state could be the instruction for an appended transmitter to transmit wirelessly the data collected in the previous output state until the memory is empty. The blinking time window can be adjusted by choosing an appropriate clock frequency to the circuit. A 1 kHz clock generates a 1 s window, meaning that the circuit will only change its output if it detects three blinks within one second. By blinking in this predetermined pattern, the user can define when data is collected and when data is transmitted.

Figure 12.9 presents the flowchart of the detection scheme. The initial sleep state consumes minimum power, and the first detected blink is used to wake up the circuit. In terms of general power budget, it was not a suitable choice, considering that involuntary blinks happen quite frequently and would, therefore, often wake up the digital pattern recognition circuit unnecessarily. After the first blink, the circuit starts a timer implemented by a 10-bit counter, which counts from 0 to 1023 in one second. Every subsequent blink detected is registered. If three blinks are detected before the timer goes off, the output changes and the circuit resumes the sleep state.

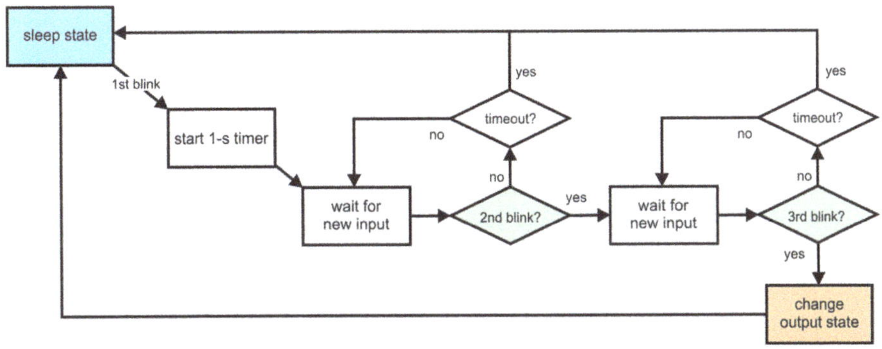

Figure 12.9. Flowchart of the blink detection scheme used in the FSM implemented with integrated logic gates.

If, however, the three blinks do not happen before the time is up, the output does not change, and the circuit is brought back to sleep.

The proposed CMOS chip was fabricated using Austria Microsystem (AMS) 0.35 μm technology, featuring two poly-silicon and three metal layers. The layout of the physical implementation of the smart module is presented in figure 12.10. The circuit area is approximately 26 500 μm^2, having a length of 165 μm and width of 160 μm. When external pads are considered, the area almost doubles to 52 000 μm^2. Yet, this represents only 1.66% of a 2 mm pupil area.

The nominal supply voltage for circuits in CMOS AMS 0.35 μm is 3.3 V, however, to save power, the operating voltage was deliberately reduced to 1 V. The main disadvantage of this voltage reduction is the higher propagation delay in the logic gates, limiting the maximum clock possible in the circuit. Both simulation and experimental tests show that it was possible to drive the circuit with the desired clock of 1 kHz with no adverse effects. The use of a relatively slow clock was also a design decision to save power, as the dynamic power of the logic cells is proportional to the clock frequency.

The proposed circuit was tested using a simple setup where the chip was placed in a dark box with a white LED that emitted light pulses with several different patterns to validate the pattern detection capabilities of the circuit. The LED driving current was adjusted so that the illuminance would be about 50 lux on the photodiode plane, compliant with medium to low illumination levels indoors. The blinks were modeled by turning off the LED for 100 ms. The integrated system worked successfully.

The simulated photo-generated current for a 1 V input voltage was 952 nA, resulting in a 952 nW power consumption, for the analog and digital parts combined. Although the current circuit has been designed to detect only relatively slow blinking patterns, a similar concept can be implemented to detect rapid changes in light intensity at frequencies compatible with a custom external controller. One such device could be the display of a smartphone running an app to generate modulating patterns at frequencies up to 60 Hz, which is the maximum refresh rate of most smartphone displays. This would enable the transmission of much more complex commands to the SIOL, enabling its use in a wider range of applications.

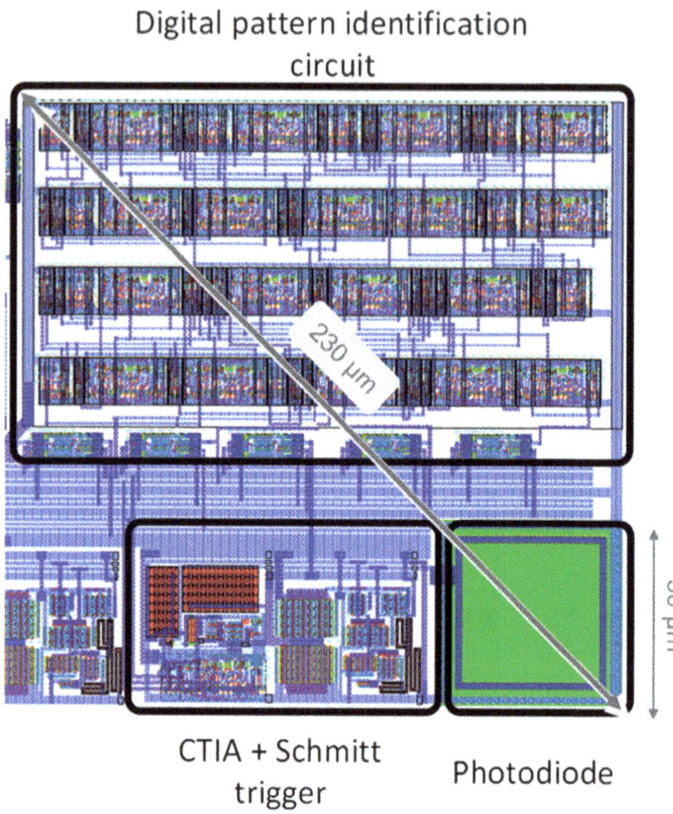

Figure 12.10. Layout of a CMOS integrated smart module comprising digital pattern-identification circuit, a capacitive transimpedance amplifier (CTIA), a Schmitt trigger comparator and a photodiode.

12.5 Summary

Already for some time, electronic-based solutions for ophthalmic devices have been developed and presented for diverse applications. Typically, components, such as microchips, RF antennas, sensors and actuators can be integrated in so-called smart platforms, compatible with contact lenses or IOLs. The incorporation of these platforms in ophthalmic devices is possible thanks to the continued progress on miniaturization, hybrid integration, ultra-lower power electronics and innovative wireless power transfer. Such smart devices are foreseen to be used in a range of medical applications encompassing biochemical sensing for healthcare monitoring; augmented/enhanced reality for real-time supplementary information overlaid on the scene; and vision correction, for the restoration or the improvement of the functionality of the eye (e.g. iris and pseudophakic lens).

The electronic design of the platform needs to consider very high energy efficiency and relatively small form factors, in the order of a few square millimeters. Such miniaturization is only possible thanks to application-specific integrated circuits

(ASICs), which allow for a very high density of small electronic components, such as transistors, diodes, photodiodes, microLED, capacitors and resistors, among others. A custom integrated circuit is designed and verified by accurate simulation models, after which a layout is generated for the chip fabrication by an independent CMOS foundry. The fact that a complete electronic system can be designed and manufactured in a single silicon die, gives rise to ultra-compact and ultra-low power architectures compatible with smart ophthalmic devices.

The form factor and location of the electronics and active components are key in the realization of devices with high visual quality and high oxygen transmission. First of all, the components are desired to be sufficiently small where possible oxygen paths are devised. For smart contact lenses, the concept is to place the main active components in the periphery of the lens, away from the visual path, unless vision enhancement modules (e.g., LCD, LED) are used. Here, oxygen flow is devised to be around the active components through several perforations of the smart platform. For the SIOL it has been shown that a silicon chip of about 1 mm^2 placed in the pupil center only covers 3% to 4% of the optical region, depending on the pupil aperture. This level of dimensions ensures a clear vision with only a small reduction in illumination reaching the retina.

The mechanical properties of the electronic-based devices need to be as close as possible to the properties of the eye, especially for soft contact lenses and foldable IOLs. These include the ability to withstand stress (strength), to bend without plastic deformation (rigidity/flexibility), and to resist pressing forces without breaking (hardness). This gives rise to mechanically invisible platforms compatible with soft contact lenses and foldable IOLs. In the case of soft contact lenses, the mechanically invisible platform would allow for seamless wearing with much lower rejection by a foreign object and bending cycles during insertion and removal of the lens. For IOLs, it would allow a foldable design compatible with current implantation procedures. And for rigid contact lenses (corneal and scleral), it would permit the seamless integration with their predetermined curvatures.

The mechanical design of the platform needs to accommodate for mechanical forces to protect the robustness of the electronic components. Generally, this is achieved by means of meandered designs instead of straight lines for the electrical interconnections. The meanders allow for stretchability and in/out of plane deformations. Such interconnections are then attached between several rigid islands where the main active components are located. This technology allows the realization of non-flat and curvilinear devices compatible with spherically shaped optical devices.

The add-on functions of the smart contact lens or SIOL, beyond passive vision correction, need to be powered up by electrical energy which is naturally not available on the lens. Thanks to current advances in energy storage and WPT, it is possible to accumulate energy either continuously by specific embedded harvesting techniques, or on demand, by electromagnetic induction, at relatively high efficiencies and short times. Micro batteries and supercapacitors have been demonstrated as suitable candidates to be used in medical devices for visual optics with high biocompatibility scores. Their capacity to store energy could be used for some applications with low power consumption, for instance, some types of LC technologies and low frequency biomarker monitoring, among others. However, more power demanding applications such as continuous biosensing or active vision correction would require higher energy capacities and more efficient harvesters to provide sufficient energy for at least one day before a recharging cycle. Images are formed on the retina after light is properly focused. Besides this fundamental functionality, light can also be used as a main energy supplier and, even more, as a communication channel by means of the Eye-Fi technology. This fosters interesting user cases where on-lens solar cells (photodiodes) can be used to harvest the electromagnetic spectrum in need of providing continuous energy to the smart application. Additionally, a predetermined combination of microLEDs and on-lens photodetectors can be used to realize a fast and safe communication protocol to reach the inner eye ensuring its wellbeing.

To sum up, the merger of passive ophthalmic devices and smart platforms with electronic functionality guarantees innovative solutions in the medical sector, which will surpass current solutions. After proper validation through preclinical trials, biocompatibility and, later, clinical trials, this technology promises to increase the quality of life of millions and the diagnosis and treatment of many visual and neurologic pathologies.

Chapter highlights

- Smart ophthalmic devices are foreseen to be used in medical applications ranging from biochemical sensing to augmented/enhanced reality and to vision correction.
- Ultra-compact and ultra-low-power electronic architectures are possible thanks to the adequate design of advanced ASICs in a single silicon die.
- Micro batteries and supercapacitors combined with energy harvesting techniques are suitable candidates to provide energy to continuous monitoring and actuation applications.
- Besides image formation, light can be used as a main power supplier and as a communication channel by means of the Eye-Fi technology.
- The hybrid assembly of microelectronics, optoelectronics, and visual optics can lead to innovative ophthalmic devices (contact lenses and IOLs) that promise improved quality of vision and timely healthcare monitoring for millions of people.

References

[1] Chiou J-C, Hsu S-H and Huang Y-C *et al* 2017 A wirelessly powered smart contact lens with reconfigurable wide range and tunable sensitivity sensor readout circuitry *Sensors* **17** 108

[2] Farandos N M, Yetisen A K and Monteiro M J *et al* 2015 Contact lens sensors in ocular diagnostics *Adv. Healthcare Mater.* **4** 792–810

[3] Senior M 2014 Novartis signs up for Google smart lens *Nat. Biotech.* **32** 856

[4] de Smet H, de Smet J and de Backer P *et al* 2014 *Curved Guest-Host LCD Used as a Dynamic Artificial Iris (Society for Information Display, Mid-Europe Chapter, Spring Meeting).* p 22

[5] Vanhaverbeke C, Verplancke R and de Smet J *et al* 2017 Microfabrication of a spherically curved liquid crystal display enabling the integration in a smart contact lens *Displays* **49** 16–25

[6] Carpi F, Frediani G and Turco S *et al* 2011 Bioinspired tunable lens with muscle-like electroactive elastomers *Adv. Funct. Mater.* **21** 4152–8

[7] Woltman S J, Jay G D and Crawford G P 2007 Liquid-crystal materials find a new order in biomedical applications *Nat. Mater.* **6** 929–38

[8] de Roose F, Steudel S and Myny K *et al* 2016 An active artificial iris controlled by a 25-mu W flexible thin-film driver *IEDM 2016 (San Francisco, CA, Dec. 3–7)* pp 798–801

[9] Alvarez-Lorenzo C, Hiratani H and Concheiro A 2006 Contact lenses for drug delivery *Am. J. Drug Delivery* **4** 131–51

[10] Mansouri K and Shaarawy T 2011 Continuous intraocular pressure monitoring with a wireless ocular telemetry sensor: initial clinical experience in patients with open angle glaucoma *Br. J. Ophthalmol.* **95** 627–9

[11] Leonardi M, Pitchon E M and Bertsch A *et al* 2009 Wireless contact lens sensor for intraocular pressure monitoring: assessment on enucleated pig eyes *Acta Ophthalmol.* **87** 433–7

[12] de Smet J and de Backer P E I *et al* 2013 A spherically shaped display for use as an artificial iris *EuroDisplay* **2013** 61–4

[13] Ho H, Saeedi E and Kim S S *et al* 2008 *Proc. IEEE Int. Conf. on Micro Electro Mechanical Systems (MEMS).* pp 403–6

[14] Vásquez Quintero A, Verplancke R and de Smet H *et al* 2017 Stretchable electronic platform for soft and smart contact lens applications *Adv. Mater. Technol.* **2** 1700073

[15] de Rossi D 2007 A logical step *Nat. Mater.* **6** 328–9

[16] Someya T 2010 Tiny lamps to illuminate the body *Nat. Mater.* **9** 879–80

[17] Stoppa M and Chiolerio A 2014 Wearable electronics and smart textiles: a critical review *Sensors* **14** 11957–92

[18] Ahn J-H, Je J H and Carta R *et al* 2009 Design and implementation of advanced systems in a flexible-stretchable technology for biomedical applications *IEEE Trans. Compon. Hybrids Manuf. Technol.* **104** 1350–3

[19] Kim J, Banks A and Xie Z *et al* 2015 Miniaturized flexible electronic systems with wireless power and near-field communication capabilities *Adv. Funct. Mater.* **25** 4761–7

[20] Fan J a, Yeo W-H and Su Y *et al* 2014 Fractal design concepts for stretchable electronics *Nat. Commun.* **5** 3266

[21] Vásquez Quintero A, Arai R and Yamazaki Y *et al* 2020 Near-field communication powered hydrogel-based smart contact lens *Adv. Mater. Technol.* **2000702** 1–13

[22] Morgan P B and Efron N 2009 Contact lens correction of presbyopia *Contact Lens Anter. Eye* **32** 191–2

[23] Madrid-Costa D, García-Lázaro S and Albarrán-Diego C *et al* 2013 Visual performance of two simultaneous vision multifocal contact lenses *Ophthalm. Physiol. Opt.* **33** 51–6

[24] Antona B, Barra F and Barrio A *et al* 2008 Comparing methods of determining addition in presbyopes *Clin. Experiment. Optom.* **91** 313–8

[25] Blystone P A 1999 Relationship between age and presbyopic addition using a sample of 3,645 examinations from a single private practice *J. Am. Optom. Assoc.* **70** 505–8

[26] Guoqiang L 2010 Adaptive lens *Progress in Optics* ed E Wolf (Amsterdam: Elsevier) pp 199–283

[27] Milton H E, Morgan P B and Clamp J H *et al* 2014 Electronic liquid crystal contact lenses for the correction of presbyopia *Opt. Express* **22** 8035–40

[28] Syed I M, Kaur S and Milton H E *et al* 2015 Novel switching mode in a vertically aligned liquid crystal contact lens *Opt. Express* **23** 9911–6

[29] Kaur S, Kim Y-J and Milton H *et al* 2016 Graphene electrodes for adaptive liquid crystal contact lenses *Opt. Express* **24** 8782–7

[30] McEldowney S C, Shemo D M and Chipman R A 2008 Vortex retarders produced from photo-aligned liquid crystal polymers *Opt. Express* **16** 7295–308

[31] Bailey J, Kaur S and Morgan P B *et al* 2017 Design considerations for liquid crystal contact lenses *J. Phys. D: Appl. Phys.* **50** 485401

[32] Valley P, Mathine D L and Dodge M R *et al* 2010 Tunable-focus flat liquid-crystal diffractive lens *Opt. Lett.* **35** 336–8

[33] Fan Y-H, Ren H and Wu S-T 2003 Switchable Fresnel lens using polymer-stabilized liquid crystals *Opt. Express* **11** 3080–6

[34] Li L, Bryant D and Bos P J 2014 Liquid crystal lens with concentric electrodes and inter-electrode resistors *Liquid Cryst. Rev.* **2** 130–54

[35] Sato S, Sugiyama A and Sato R 1985 Variable-focus liquid-crystal Fresnel lens *Jpn. J. Appl. Phys.* **24** L626–8

[36] Rossi M, Kunz R E and Herzig H P 1995 Refractive and diffractive properties of planar micro-optical elements *Appl. Opt.* **34** 5996–6007

[37] Vásquez Quintero A, Pérez-Merino P and Smet H d 2020 Artificial iris performance for smart contact lens vision correction applications *Sci. Rep.* **10** 14641

[38] Raducanu B C, Zaliasl S and Stanzione S *et al* 2020 An artificial iris ASIC with high voltage liquid crystal driver, 10-nA light range detector and 40-nA blink detector for LCD flicker removal *IEEE Solid-State Circuits Lett.* **3** 506–9

[39] Katz B J and Digre K B 2016 Diagnosis, pathophysiology, and treatment of photophobia *Survey Ophthalmol.* **61** 466–77

[40] Kim M, Jung I D and Kim Y *et al* 2020 An electrochromic alarm system for smart contact lenses *Sens. Actuators B: Chem.* **322** 128601

[41] Kim J, Cha E and Park J U 2020 *Adv. Mater. Technol.* **5** 1900728

[42] Chen G-Z, Chan I-S and Lam D C C 2013 Capacitive contact lens sensor for continuous non-invasive intraocular pressure monitoring *Sens. Actuators* A **203** 112–8

[43] Mosaed S, Liu J H K and Weinreb R N 2005 Correlation between office and peak nocturnal intraocular pressures in healthy subjects and glaucoma patients *Am. J. Ophthalmol.* **139** 320–4

[44] Hjortdal J Ø and Jensen P K 1995 In vitro measurement of corneal strain, thickness, and curvature using digital image processing *Acta Ophthalmol. Scand.* **73** 5–11

[45] Samadani U, Farooq S and Ritlop R *et al* 2015 Detection of third and sixth cranial nerve palsies with a novel method for eye tracking while watching a short film clip *J. Neurosurg. JNS* **122** 707–20

[46] Niggemann B 2015 Ophthalmological investigations in toxicity studies *The Nonhuman Primate in Nonclinical Drug Development and Safety Assessment* (San Diego, CA: Academic) pp 543–50

[47] Kim J, Kim M and Lee M-S *et al* 2017 Wearable smart sensor systems integrated on soft contact lenses for wireless ocular diagnostics *Nat. Commun.* **8** 14997

[48] Chiou J-C, Huang Y-C and Yeh G-T 2015 A capacitor-based sensor and a contact lens sensing system for intraocular pressure monitoring *J. Micromech. Microeng.* **26** 15001

[49] Pierscionek B K, Asejczyk-Widlicka M and Schachar R A 2007 The effect of changing intraocular pressure on the corneal and scleral curvatures in the fresh porcine eye *Br. J. Ophthalmol.* **91** 801–3
[50] Piso D, Veiga-crespo P and Vecino E 2012 Modern monitoring intraocular pressure sensing devices based on application specific integrated circuits *J. Biomater. Nanobiotechnol.* **3** 301–9
[51] Leonardi M, Leuenberger P and Bertrand D *et al* 2004 First steps toward noninvasive intraocular pressure monitoring with a sensing contact lens *Invest. Ophthalmol. Vis. Sci.* **45** 3113–7
[52] Pang Y, Li Y and Wang X *et al* 2019 A contact lens promising for non-invasive continuous intraocular pressure monitoring *RSC Adv.* **9** 5076–82
[53] Leonardi M, Pitchon E M and Bertsch A *et al* 2009 Wireless contact lens sensor for intraocular pressure monitoring: assessment on enucleated pig eyes *Acta Ophthalmol.* **87** 433–7
[54] Dunbar G E, Shen B Y and Aref A A 2017 The Sensimed Triggerfish contact lens sensor: efficacy, safety, and patient perspectives *Clin. Ophthalmol.* **11** 875–82
[55] Yan J 2011 *IEEE Eng. Med. Biol. Soc.* 2522
[56] Yin R, Xu Z and Mei M *et al* 2018 Soft transparent graphene contact lens electrodes for conformal full-cornea recording of electroretinogram *Nat. Commun.* **9** 2334
[57] van Delft J L, Meijer F and van Best J A *et al* 1997 Spatiotemporal electrochemical sensing in a smart contact lens *Mediat. Inflamm.* **6** 438257
[58] Donora M, Vásquez Quintero A and de Smet H *et al* 2020 Spatiotemporal electrochemical sensing in a smart contact lens *Sens. Actuators, B: Chem.* **303** 127303
[59] Zhou L and Beuerman R W 2012 Tear analysis in ocular surface diseases *Prog Retin Eye Res* **31** 527–50
[60] Mitsubayashi K and Arakawa T 2016 Cavitas sensors: contact lens type sensors & mouthguard sensors *Electroanalysis* **28** 1170–87
[61] Harvey D, Hayes N W and Tighe B 2012 Fibre optics sensors in tear electrolyte analysis: towards a novel point of care potassium sensor *Contact Lens Anter. Eye* **35** 137–44
[62] Thévenot D R, Toth K and Durst R A *et al* 2001 Electrochemical biosensors: recommended definitions and classification *Anal. Lett.* **34** 635–59
[63] Yao H, Shum A J and Cowan M *et al* 2011 A contact lens with embedded sensor for monitoring tear glucose level *Biosens. Bioelectron.* **26** 3290–6
[64] Pickup J C, Hussain F and Evans N D *et al* 2005 Fluorescence-based glucose sensors *Biosens. Bioelectron.* **20** 2555–65
[65] Badugu R, Jeng B H and Reece E A *et al* 2018 Contact lens to measure individual ion concentrations in tears and applications to dry eye disease *Anal. Biochem.* **542** 84–94
[66] Badugu R, Lakowicz J R and Geddes C D 2004 Noninvasive continuous monitoring of physiological glucose using a monosaccharide-sensing contact lens *Anal. Chem.* **76** 610–8
[67] Martin P S 2020 Mojo vision nanoLEDs for invisible computing *Proc.SPIE* **11302** 1130204
[68] Pandey J, Liao Y-T and Lingley A *et al* 2010 A fully integrated RF-powered contact lens with a single element display *IEEE Trans. Biomed. Circuits Syst.* **4** 454–61
[69] Kim B J, Chen B and Gupta M *et al* 2014 Formation of three-dimensional Parylene C structures via thermoforming *J. Micromech. Microeng.* **24** 065003
[70] Olsen T 1986 On the calculation of power from curvature of the cornea *Br. J. Ophthalmol.* **70** 152–4
[71] Rogers J A, Someya T and Huang Y 2010 Materials and mechanics for stretchable electronics *Science* **327** 1603–7

[72] Ko H C, Stoykovich M P and Song J *et al* 2008 A hemispherical electronic eye camera based on compressible silicon optoelectronics *Nature* **454** 748–53

[73] Cerda E and Mahadevan L 2005 *Proc. R. Soc. A* **461** 671

[74] Lan S, Lee H J and Lee S H *et al* 2009 Experimental and numerical study on the viscoelastic property of polycarbonate near glass transition temperature for micro thermal imprint process *Mater. Des.* **30** 3879–84

[75] Williams M L, Landel R F and Ferry J D 1955 The temperature dependence of relaxation mechanisms in amorphous polymers and other glass-forming liquids *J. Am. Chem. Soc.* **77** 3701–7

[76] Lingley A R, Otis B P and Shen T T *et al* 2012 A contact lens with integrated micro solar cells *Microsyst. Technol.* **18** 453–8

[77] Li J, Wang Y and Liu L *et al* 2019 A biomimetic soft lens controlled by electrooculographic signal *Adv. Funct. Mater.* **29** 1903762

[78] Falk M, Andoralov V and Silow M *et al* 2013 Miniature biofuel cell as a potential power source for glucose-sensing contact lenses *Anal. Chem.* **85** 6342–8

[79] Yuan M, Das R and Ghannam R *et al* 2020 Electronic contact lens: a platform for wireless health monitoring applications *Adv. Intell. Syst.* **2** 1900190

[80] Takamatsu T, Chen Y and Yoshimasu T *et al* 2019 *Adv. Mater. Technol.* **4** 1800671

[81] Ku M, Kim J and Won J E *et al* 2020 *Sci. Adv.* **6** 1–10

[82] Verplancke R, Bossuyt F and Cuypers D *et al* 2012 Thin-film stretchable electronics technology based on meandering interconnections: fabrication and mechanical performance *J. Micromech. Microeng.* **22** 15002

[83] Vásquez Quintero A, Verplancke R and de Smet H *et al* 2017 Stretchable electronic platform for soft and smart contact lens applications *Adv. Mater. Technol.* **2** 1700073

[84] de Smet J, Avci A and Joshi P *et al* 2013 Progress toward a liquid crystal contact lens display *J. Soc. Inf. Disp.* **21** 399–406

[85] Digre K B and Brennan K C 2012 Shedding light on photophobia *J. Neuroophthalmol.* **32** 68–81

[86] Cortez M, Digre K and Uddin D *et al* 2019 Validation of a photophobia symptom impact scale *Cephalalgia* **39** 1445–54

[87] Tripathy K and Salini B 2020 Updated 2020 Aug 10

[88] Hong E, Zeeb H and Repacholi M 2006 Albinism in Africa as a public health issue *BMC Public Health* **6** 212

[89] Wu Y and Hallett M 2017 Photophobia in neurologic disorders *Transl. Neurodegen.* **6** 26

[90] Aguilar M C, Gonzalez A and Rowaan C *et al* 2018 Automated instrument designed to determine visual photosensitivity thresholds *Biomed. Opt. Express* **9** 5583

[91] Vecchia V, Silva M F and de Lima Monteiro D W 2018 Eye-Fi: integrating optical data communication into intraocular lenses *Proc. of the 1st Global LiFi Congress* vol 1 (Piscataway, NJ: IEEE Press) pp 189–95

[92] Peng B, Lu J, Balijepalli A S, Major T C, Cohan B E and Meyerhoff M E 2013 Evaluation of enzyme-based tear glucose electrochemical sensors over a wide range of blood glucose concentrations *Biosens. Bioelectron.* **49** 204–9

[93] Sears M L 1975 The aqueous *Physiology of the Eye – Clinical Application* ed R A Moses and R A Adler (St. Louis, MO: Mosby) ch 8 pp 232–51

[94] ⊘ ; Hockenhull J, Dhillo W, Andrews R and Paterson S 2012 Investigation of markers to indicate and distinguish death due to alcoholic ketoacidosis, diabetic ketoacidosis and hyperosmolar hyperglycemic state using post-mortem samples *Forens. Sci. Int.* **214** 142–7

[95] Jone S SemiWiki *Intel Node Names* https://semiwiki.com/semiconductor-services/295767-intel-nodes/ [accessed 26 June 2022]

[96] IOT Analytics *State of the IoT 2020: 12 billion IoT connections, surpassing non-IoT for the first time* https://iot-analytics.com/state-of-the-iot-2020-12-billion-iot-connections-surpassing-non-iot-for-the-first-time/ [accessed 26 June 2022]

[97] Alberto J, Leal C, Fernandes C, Lopes P A, Paisana H, de Almeida A T and Tavakoli M 2020 Fully untethered battery-free biomonitoring electronic tattoo with wireless energy harvesting *Nat. Res.—Sci. Rep.* **10** 5539

[98] Cevik I, Huang X, Yu H, Yan M and Ay S U 2015 An ultra-low power CMOS image sensor with on-chip energy harvesting and power management capability *Sensors* **15** 5531–54

[99] Esram T and Chapman P L 2007 Comparison of photovoltaic array maximum power point tracking techniques *IEEE Trans. Energy Convers.* **22** 439–49

[100] Husain M A, Tariq A, Hameed S, Arifa M S B and Jain A 2017 Comparative assessment of maximum power point tracking procedures for photovoltaic systems *Green Energy Environ.* **2** 517

[101] Hart G W, Branz H M and Cox C H 1984 Experimental tests of open loop maximum-power-point tracking techniques *Sol. Cells* **13** 1850195

[102] Kobayashi K, Matsuo H and Sekine Y 2004 A novel optimum operating point tracker of the solar cell power supply system *Proc. 35th Annual Conf. of the IEEE Power Electronics Society* (Piscataway, NJ: IEEE Press) pp 2147–51

[103] Chang M H, Wu J Y, Hsieh W C, Lin S Y, Liang Y W and Wei H 2010 High efficiency power management system for solar energy harvesting applications *Proc. of the 2010 IEEE Asia Pacific Conf. on Circuits and Systems* pp 879–82

[104] Chen G *et al* 2011 A cubic-millimeter energy autonomous wireless intraocular pressure monitor *Proc. of the Int. Solid-State Circuits Conf.* (Piscataway, NJ: IEEE Press) pp 310–2

[105] Zhang J, Zhang G, Zhou T and Sun S 2020 Recent developments of planar micro-supercapacitors: fabrication, properties, and applications *Adv. Funct. Mater.* **30** 1910000

[106] Bocchetta P, Frattini D, Ghosh S, Mohan A M V, Kumar Y and Kwon Y 2020 Soft materials for wearable/flexible electrochemical energy conversion, storage, and biosensor devices *MDPI Mater.* **13** 2733

[107] Niu Z Q, Dong H B, Zhu B W, Li J Z, Hng H H, Zhou W Y, Chen X D and Xie S S 2013 Highly stretchable, integrated supercapacitors based on single-walled carbon nanotube films with continuous reticulate architecture *Adv. Mater.* **25** 1058–64

[108] Pan J, Xu Y Y, Yang H, Dong Z, Liu H and Xia B Y 2018 Advanced architectures and relatives of air electrodes in Zn–air batteries *Adv. Sci.* **5** 170

[109] Fu J, Hassan F M, Li J, Lee D U, Ghannoum A R, Lui G, Hoque M A and Chen Z 2016 Flexible rechargeable zinc-air batteries through morphological emulation of human hair array *Adv. Mater.* **28** 6421–8

[110] Zeng Y, Zhang X, Meng Y, Yu M, Yi J, Wu Y, Lu X and Tong Y 2017 Achieving ultrahigh energy density and long durability in a flexible rechargeable quasi-solid-state Zn–MnO_2 battery *Adv. Mater.* **29** 170

[111] Gasulla M, Ripoll-Vercellone E and Reverter F 2019 A compact thévenin model for a rectenna and its application to an RF harvester with MPPT *MDPI Sensors* **19** 1641

[112] Simonov A, Vdovin G and Loktev M 2007 Liquid-crystal intraocular adaptive lens with wireless control *Opt. Express* **15** 7468–78

[113] Wang P P, Zhang C, Yang H, Bharadia D and Mercier P P 2020 20.1 A 28 μW IoT tag that can communicate with commodity WiFi transceivers via a single-side-band QPSK

backscatter communication technique *Proc. of the IEEE Int. Solid- State Circuits Conf. (ISSCC)* (Piscataway, NJ: IEEE Press) pp 312–4
[114] Roberts N E *et al* 2016 A 236 nW −56.5 dBm-sensitivity bluetooth low-energy wakeup receiver with energy harvesting in 65 nm CMOS *Proc. IEEE Int. Solid- State Circuits Conf. (ISSCC)* pp 450–1
[115] Wang P-H P *et al* 2018 A 6.1 nW wake-up receiver achieving −80.5 dBm sensitivity via a passive pseudo-balun envelope detector *IEEE Solid-State Circuits Lett.* **1** 134–7
[116] Roy A *et al* 2015 A 6.45 μW self-powered SoC with integrated energy-harvesting power management and ULP asymmetric radios for portable biomedical systems *IEEE Trans. Biomed. Circuits Syst.* **9** 862–74
[117] Zhang Y *et al* 2013 A batteryless 19 μW MICS/ISM-band energy harvesting body sensor node SoC for ExG applications *IEEE J. Solid-State Circuits* **48** 199–213
[118] Boysset M, Pascarella A, Jacot J and SAV-IOL S A 2020 Ophthalmic assembly for implantation in an anterior chamber of an eye of a patient and method for accommodating the vision of the patient *US Patent Specification* 2020/0397563 A1
[119] Ramirez J A, Carvalho D F Jr and Silva E J 2015 Determination of the temperature increase in the human eye due to electromagnetic fields *Int. J. Comput. Math. Elec. Electron. Eng.* **34** 1489–500
[120] Koonen T 2018 Indoor optical wireless systems: technology, trends, and applications *J. Lightwave Technol.* **36** 1459–67
[121] Lian J, Vatansever Z, Noshad M and Brandt-Pearce M 2019 Indoor visible light communications, networking, and applications *J. Phys.: Photonics* **1** 012001
[122] Haas H, Yin L, Wang Y and Chen C 2016 What is LiFi? *J. Lightwave Technol.* **34** 1533–44
[123] Wu T, Sher C-W, Lin Y, Lee C-F, Liang S, Lu Y, Chen S-W H, Guo W, Kuo H-C and Chen Z 2018 Mini-LED and Micro-LED: promising candidates for the next generation display technology *Appl. Sci.* **8** 1557
[124] Kress B C and Chatterjee I 2021 Waveguide combiners for mixed reality headsets: a nanophotonics design perspective *Nanophotonics* **10** 41–74
[125] Souza R A, Salles L P, de Lima Monteiro D W, Rocha D, Bataillou B and Mérelle T 2019 CMOS-based single-channel circuit prototype for bidirectional transmission via visible light for LiFiX *Proc. 2nd Global LiFi Congress* pp 1–6

IOP Publishing

Advances in Ophthalmic Optics Technology

Davies William de Lima Monteiro and Bruno Lovaglio Cançado Trindade

Chapter 13

Mechanically adjustable lenses

Thiago Daniel de Oliveira Moura, Davies William de Lima Monteiro, Jhonattan Córdoba Ramirez, Andrea Chiuchiarelli and Diego Miranda Bruno Lovaglio Cançado Trindade

The human eye has a formidable tunable lens that allows one to deliberately focus on far and near objects with a refocusing time that can be faster than one-third of a second and a perceived accommodation amplitude better than 10 D in the very first years of life [1, 2]. Presbyopia, however, is a condition that limits the ability of the ageing crystalline lens to accommodate, restricting one's aptitude to focus on objects at closer distances in an ametropic eye. It affects a significant part of the world population, and those who have the crystalline replaced by an implant almost entirely lose their dynamic refocus ability. To overcome this, accommodative intraocular lenses (AIOLs) have been constantly improving and are certainly one of the advances that is most awaited in the ophthalmic industry.

This chapter presents an overview of the types and actuation principles of mechanically adjustable and liquid-crystal lenses that allow some degree of visual adjustment in spectacles or IOLs, including solutions that extrapolate ophthalmic applications, such as to achieve auto-focus or angle of gaze in equipment. Although some of the contents of this chapter might seem a little far-fetched, and perhaps still a long way from being applicable, they are intended to highlight important technological advances in several engineering fields that might, already or at some point, be deployed not only to ophthalmic solutions but also to bionics and robotics.

Most of the actuation mechanisms for intraocular use require some degree of miniaturization of the devices. Hence, this chapter also includes a brief overview on micromachining technologies and active materials that could enable dynamic operation and low power consumption of components with reduced physical dimensions. Micromachined structures integrated with microelectronic circuits, and the deployment of soft materials, could meet the need for folding of the IOL and its implantation through smaller incisions in the cornea. To illustrate the interdisciplinary nature of such a development, these latter topics will be presented

doi:10.1088/978-0-7503-3263-7ch13

alongside the implementation of a double-optic electromechanical lens system designed to be dimensionally compatible with implantation in the human eye.

13.1 Adjustable lenses for vision

Today, the mimetism of vision accommodation reaches the most diverse pieces of equipment, ranging from traditional photographic cameras, to miniaturized cell phone optical systems, to advanced machine-vision modules and to microscopy.

The first generation of mobile devices, for instance, had only fixed lenses, but the increasing demand for different photography modes at various distances led to the development of miniaturized autofocus systems that encompass lens movement and stabilization [3].

The autofocus systems adjust the lenses to have an object at a certain distance in focus, traditionally by looking for the best contrast response measured by the camera sensor. In some of the most recent systems, phase detection is used, which is a faster mechanism for autofocus [4]. Whichever mechanism is employed to determine the focus, an actuator, usually a motor, moves the lens, or lenses, to the position where the sharpest image of the object is achieved. Most strategies used to move the lenses are aimed at low energy consumption and high speed. Traditional voice coil actuator (VCM) was one of the first approaches deployed, using electromagnetic force. It consists of fixed coils and magnets in the lens that respond to the magnetic field generated by the coil drives [3].

Micro-electro-mechanical systems (MEMS) refer to devices with integrated mechanical and electrical functions in micrometric scale. MEMS modules are two to four times faster than VCM and can be produced to confer mechanical actuation to microstructures. Besides, low-energy alternatives have been proposed to replace VCM modules, like ionic polymer–metal composite (IPMC), which is one of the promising electroactive polymers being investigated today.

Traditional optics are based on glass or plastic lenses, which are moved back and forth to focus or zoom. The so-called tunable optics, on the other hand, can change the shape and behavior of lenses. For example, varifocal microlenses have been developed, whose focal length varies along with the deformation of a transparent elastomer membrane under hydraulic pressure tailored by electroactive polymer actuators [5].

Xiaomi® was one of the first companies to make liquid lens technology available in mobile products. It can change its shape according to the need, called 'bionic photography'. The lens consists of a drop of liquid that can move and change mechanically when electrically actuated. The advantage of this type of technology is that it is fast, and it is possible for a single lens to function as both macro and telephoto.

Corning® Varioptic® lenses rely on the electrowetting principle where the interface of two liquids is modified when a voltage is applied. This eliminates moving parts, bringing a unique competitive advantage to the market, and enabling fast focus. Other companies like Samsung®, Apple® and Philips® are also betting on this technology in their most recent smartphone releases.

Through-focus (TF) or volumetric type of optical imaging has gained momentum in several areas such as biological imaging, microscopy, adaptive optics, material processing, optical data storage, and optical inspection. This scanning method provides three-dimensional information that can also be used in virtual reality (VR). For this application, flexible membranes are varied to make the membranes inflate or deflate. This process alters the radius of curvature of the membranes and thus changes the effective focal length of the lens, enabling the collection of TF images. Other common examples of adaptive optics elements are deformable mirror devices or liquid-crystal spatial light modulators (SLMs). These elements can also correct high-order residual aberrations as well as perform the TF scanning [6].

Adjustable eyeglasses are also a promising area, with solutions across a wide cost range. Companies like EYEJUSTERS® already make glasses available with the so-called eyeDialS®. This approach customizes the two lenses for each eye to achieve a good performance with a more discreet design, following the Alvarez principle, where a lateral shift between the two complementary components enables a change of focus. More advanced solutions using liquid lenses appear as new perspectives, as in the developments by Joshua Silver on self-refraction using adaptive spectacles. There, the spectacles can be easily adjusted by the wearer to yield acceptable defocus refractive correction at low cost. This fluid-filled lens technology takes advantage of the incompressibility of liquids to change the physical shape of the lens. The lens is constructed with two flexible membranes on the optical surfaces. The power of the lens is changed by pumping fluid into or out of the central lens reservoir. Another interesting development is that by the company Morrow, where a liquid-crystal layer lies within two optical lenses and can be electrically activated to change refraction to guarantee near or distant vision at the touch of a button.

In addition to the challenges related to self-focusing, consideration of the angle of sight is sometimes needed, related in the human eye to the angle of gaze. In many applications the gaze direction of the eyes must be known. Works have developed a statistical model that used Gaussian process regression to estimate horizontal gaze direction based on pupil position relative to the corners of the eyes and use mapping of the gaze angle to automatically update the focal length of a focus-tunable lens [7, 8]. This development is particularly interesting in devices that use displays close to the eyes, such as VR glasses. In this situation, users are forced to accommodate to a single plane, decoupling it from their other depth cues such as vergence, resulting in a vergence–accommodation conflict [9].

New optical device control perspectives focus on human–machine interaction (HMI) based on electrophysiological signals, including electroencephalography, electromyography, and electrooculography (EOG). EOG signals reflect the change of electrical potential difference between the cornea and the ocular fundus of the eyes, which is closely associated with the eye movements. When the eyes move from a primary central position to any direction (right, left, up, or down) and then move back, a peak with a subsequent valley successively (or a valley with a subsequent peak) is observed in the EOG signal. There are signs indicating if the eyes stay at the gaze direction or if they move back to the primary position. The HMI interacts with the soft tunable lens through drives and a microcontroller [10]. The same work shows a design of the soft tunable lens composed of electroactive polymer films. The change of the focal length and the motion of the soft lens closely resembled those of human eyes, which were achieved by the electrical-potential-induced actuation of a dielectric elastomer (DE), composed of a soft dielectric layer sandwiched between two compliant electrodes. When an electrical potential is applied to the two electrodes, the soft layer can instantly expand. The motion and deformation of the soft lens was synchronized with the movements of the eyes [10].

13.2 A glimpse of accommodative IOL technologies

This section presents a summary of accommodative intraocular lenses (AIOL), sometimes referred to as accommodating lenses, and highlights the advantage of dual-optic systems compared to single-optic lenses in yielding a wider accommodation range. Regardless of the many attempts to investigate and commercially promote AIOLs, in practice, the issue with the visual accommodation loss has not yet been satisfactorily addressed, leaving plenty of room for more advanced and reliable solutions [11–14].

Natural accommodation as that of a 10-year-old individual (>10 D), or even that of a 40-year-old subject (>4 D), has not yet been convincingly achieved by any reported AIOL, despite some IOL models having even reached the commercial market in the previous decades [15]. The principle of accommodation varies between these lenses, and it is important to compare the amount of accommodation provided by each lens at the same specific optical plane. Corneal apex is the usual reference, but there is to date no consensus in commercial leaflets or scientific publications at which plane this additional refractive power is reported.

Post-operative vision at different distances can be achieved with multifocal IOLs, extended-depth-of-focus IOLs (EDoF IOLs), pinhole IOLs or, to some degree, with pseudoaccommodation. The latter can be imparted by some degree of spherical or coma aberration on the cornea and IOL; or by a smaller pupil, with an efferent reflex to focus on a closer object (accommodative miosis) [16]. The American Academy of Ophthalmology (AAO) has issued a special report to assess accommodation produced by IOLs, where they recommend the use of methods to measure the objective accommodation in clinical studies. Although the performance of multifocal and EDoF IOLs can be evaluated by subjective tests, such as that of clinical defocus curves, the AIOL requires objective measurements to prove that the near-vision quality can be achieved by the actual change in the optical power of the eye, and not by pseudoaccommodation mechanisms. To verify this, additional testing was recommended by the AAO. These include not only the more commonly used autorefraction and aberrometry, but also measuring changes in biometric parameters such as axial distances, thicknesses and surface curvatures using the optical biometers, OCT and UBM [17]. Thus, the accommodation achieved by an AIOL can be either directly measured as a refractive change, or the biometric parameters can be recorded, and the corresponding refractive power changes can be inferred from them, using ray tracing in schematic eyes or standard power formulas. The AAO Task Force report recommends that the subjective and objective results are recorded for the same accommodative stimulus amplitude, and then plotted and fitted with a linear regression to demonstrate correlation between them.

A true solution that mimics the physiological accommodation would relieve the patient from adaptation or resignation to int*er*ocular or int*ra*ocular rivalry. The first is due to focal disparity between eyes in monovision or mini-monovision, when lenses with different focal ranges are implanted in opposite eyes. Usually, one eye is deliberately made more myopic to offer binocular extension of focus to nearer vision. Intraocular rivalry, on the other hand, results from superimposed images from different foci in the same eye in individuals implanted with multifocal lenses. The accommodation of refractive lenses can be achieved by a limited number of options, such as changing the effective lens position in the eye, or the relative position between two lenses in line, or the lens curvatures, or the refractive index of the lens. Relatively fast, adjustable, and reversible changes to the refractive index can be accomplished by electrically biased nematic liquid crystals [18–21]. The other methods require mechanical actuation to carry out the desired degree of change. In the simplest case, it is merely a switch between two distinct states, for instance, one that favors far vision and another that prioritizes near vision. In terms of dynamic accommodative response, two figures of merit are important: latency, which is the delay between a target change and the onset of a response, and time constant, which is the time for the response to achieve 63% of the new steady state. The reaction time is the sum of both. It is desirable that the latency and time constant with an AIOL are close to those of an average healthy eye, ~350 ms and ~200 ms, respectively, amounting to a reaction time of ~550 ms [2].

A natural choice, literally, for the mechanical actuation is the ciliary muscle, which is natively employed to change the shape of the crystalline lens and the corresponding tension and elasticity of the capsular bag. Helmholtz theory, dating from the mid-19th century, states that by the contraction of the ciliary muscle, most of its mass shifts anteriorly and crucially inwards, reducing the diameter of the muscle collar and relaxing the tension on the anterior and posterior zonular fibers, thus allowing the crystalline lens to adopt a more convex shape with a higher refractive power [22]. Schachar questions that accommodation mechanism and proposes that the much finer equatorial zonules play a more decisive role, also better explaining the change in the spherical aberration in the negative direction upon accommodation [23]. Despite the recurrent rebuttals and rejoinders in this topic, whichever is the most relevant mechanism in natural accommodation, there is consensus about the underpinning function of the ciliary muscle in the process.

Accommodation loss may result from several combined factors, including the progressive hardening of the natural lens structure and the reduced ability of the zonular fibers to release tension. According to Schachar, however, the decline of the amplitude of accommodation with age is mostly due to the equatorial crystalline growth [23]. Despite the ongoing controversy, the most crucial element for the development of several AIOL models is the ciliary muscle, and not the crystalline or the zonules. As will be later discussed, those AIOLs change their focus because of the contraction of the ciliary muscle. Magnetic resonance imaging (MRI) has demonstrated that this muscle maintains its ability to contract throughout life, even in pseudophakic subjects [24]. This confirms that, after the lens implantation, it can keep being employed as the voluntary activation mechanism, reacting to visual cues sent to the visual cortex.

Relying on the ciliary muscle to control accommodation can be tricky. Clinical studies of muscle actuated IOLs indicate a great variability of results [25, 26]. The post-operative behavior of the ciliary muscle does not work in a very predictable way for all patients, depending on the outcome of the surgery, proper lens positioning and the specific anatomy of each eye. Hyperopic patients, on average, have stronger ciliary muscles than myopic patients, for instance [22]. Besides, the lens can be designed to be implanted in the capsular bag or in the ciliary sulcus, where the use of the forces created by the muscle can be optimized. Hermans *et al* estimated that for a normal eye, the strength of the ciliary muscle responsible for accommodation reaches 0.06 N [27], which is equivalent to the weight of two US dime coins. If the muscle movement does not result in a correct distribution of forces on the lens structure, it could introduce unwanted effects, such as a skew displacement of the lens in the capsular bag, resulting in deviating optical effects. Over the years there might also be a reduction of the contraction force of the muscle and the development of capsular fibrosis, reducing its elasticity, altogether affecting the amplitude of accommodation of some AIOL models.

It has been claimed that the first focusable pseudophakic intraocular lens was proposed in a patent by Chauncey F Levy in 1981 [28], where a rigid monofocal lens with radial flexible struts, terminating at blunt abutments, was to be implanted in the capsular bag. The lens is first implanted touching the posterior capsule, corresponding to far vision. The struts respond to the contraction of the ciliary body, moving the lens closer to the cornea, thus favoring near vision. The same patent offers the addition of a second fixed IOL at the posterior capsule, suggesting an enhancement of the accommodation range. As an example, a single-optic AIOL with a refractive power of 19 D, can contribute to up to about 1.2 D when longitudinally shifted forward by 1 mm in the capsular bag, whereas a double-optic AIOL, with approximately the same power, designed with 32 D and −12 D anterior and posterior lenses, respectively, can achieve an increased accommodation amplitude of 2.2 D with a 1 mm displacement between lenses [16].

Next, several categories of AIOLs are presented with their respective lens models: the single-optic IOLs that rely on the axial shift of one lens in the posterior chamber; the double-optic IOLs, in which two lenses are axially connected by spring-like articulations; and deformable lenses, whose shape or structure is changed by mechanical or electrical stimulation. Estimated accommodation amplitude and visual acuity depend on many factors and test conditions. It is, therefore, essential to emphasize that for detailed information on the conditions and statistical variations of the mean clinical figures herein presented, one should refer to the respective literature where data have been originally published.

13.2.1 Single-optic AIOLs

BioComFold—Morcher, Germany—This is claimed to have been the first commercialized AIOL, in 1996, with a newer model introduced in 1998. It is a forward shifting lens with the action of the ciliary muscle on its distinctively robust perforated haptics. It has been reported that it featured a large forward shift of 710 μm, based on subjective tests. However, other reports indicate 116 μm or 222 μm based on objective measurements [29].

1CU—Human Optics, Germany—This lens is sometimes referred to as ICU. It has a 5.5 mm optic zone and measures 9.8 mm across including the haptics, its overall diameter. Its achieved accommodative effect ranges from 0.5 D to 1 D, with a mean forward shift of 314 μm [30]. According to Uthoff *et al* the one-year results in 553 eyes suggested a minor statistical advantage of 0.25 D compared to monofocal IOLs measured with subjective methods using an accommodometer, defocusing curve, and Nieden charts. The average near visual acuity with distance BSCVA (best spectacle-corrected visual acuity) was 0.41 [31].

Crystalens AT-45—Bausch & Lomb, USA—This lens was earlier designed by C&C Vision and later commercialized by Eyeonics before it became Bausch & Lomb. There were the sequel models AT-50, HD and AO. It entered the market in 2002 and was approved by FDA. It is a biconvex lens that has a 4.5 mm high-refractive index silicone elastomer optic and flexible haptics that allow it to move forward under increased vitreous cavity pressure due to the redistribution of the ciliary muscle mass when constricted. In some cases, it was found that the optic moved backwards, instead [30]. The mean add power found for this lens was 1.24 D [24]. The earlier models have been discontinued but the two AO models are still in the company catalogue, claiming to provide 1 D of monocular accommodation [32]. The Trulign Toric is a similar IOL by Bausch & Lomb, also with hinges connecting the optic to the haptics, but with the additional capacity to compensate for corneal astigmatism [33].

Tetraflex KH-3500—Lenstec, USA—It works based on the forward shift of the lens in addition to some flexure of the optic. It is a 5.75 mm lens, with an accomplished add power between 1 D and 3 D. The distant corrected near vision acuity was at least 20/40 for 85.3% of patients. A mean forward shift of 337 μm strongly correlates with an accommodation amplitude of 0.94 D [34].

OPAL-A—Human Optics, Germany—It features a 5.5 mm optic and four flexible haptics devised to move the lens forward under the ciliary muscle contraction. It features a mean forward shift of 306 μm, corresponding to 0.36 D, which has been observed to decline with implantation time [35].

Magnet-driven IOL—This lens employs a pair of repulsing magnets sutured under the superior and inferior rectus muscles, while two other magnets are inserted in the capsular bag periphery. A ring structure with mini-magnets supports an IOL in its center and is inserted in the capsular bag. It prevents capsular shrinkage and when the zonules relax, because of the ciliary muscle constriction, the ring structure is pushed forward due to the repulsive magnetic forces, thus accommodating the system to near vision [36]. A later adaptation of the surgical procedure made the ring unnecessary and used magnets integrated to the optic periphery [37].

Although not dynamically accommodative in nature, besides the refractive-index shaping described in chapter 6, another light adjustable lens (LAL) has been introduced by RxSight, in which the shape and power can be modified in the doctor's office by UV light exposure after implantation.

13.2.2 Dual-optic AIOLs

This alternative can offer a wider accommodation range than the single-optic models. The most common configuration is that of an anterior positive lens connected by articulations to *a posterior* negative lens. The ensemble is placed in the capsular bag and the interoptic space is filled with the aqueous humor. Therefore, it is of uttermost importance to remove the viscoelastic material, used during the implantation, from between the lenses. When the ciliary muscle is rested, the capsular bag has tension in its walls and the two optics are compressed together, yielding unaccommodation, hence far vision. When the muscle is contracted, the capsule releases tension and the anterior lens moves forward, favoring nearer vision.

Synchrony—Visiogen, Abbott Medical Optics, Johnson&Johnson, USA. It was designed by Visiogen, and next became Abbott Medical Optics, which was later acquired by Johnson&Johnson. It has been a commonly implanted silicone lens, CE approved in 2006 in Europe, that requires a 3.8 mm incision [29]. The visual acuity of 0.15 and 0.07 logMAR has been reported after 1 year and 2 years, respectively, offering improved reading ability [38, 39]. In a comprehensive study by Ossma *et al* among other results, best corrected distance visual acuity was observed for all eyes, and with distance correction, at near the vision acuity of 20/40 or better for 96% of the eyes [40]. The mean estimated accommodation is 3.22 D [41]. In the spirit of constructive scientific assessment, the accuracy of some of these results have been questioned by Sergienko [42] and clarified by McLeod *et al* [43].

Sarfarazi (EAIOL)—Bausch & Lomb, USA. This lens is named after its creator F Mona Sarfarazi at Shenasa Medical. The rights to the lens were bought by Bausch & Lomb but it does not currently feature on their catalog. The EA stands for Elliptical Accommodative. There are two optics whose distance in between is changed by the flexure of three wide curved bands connecting them. The bend of the bands, when looked axially, is elliptical [12]. This shape better fits the elliptical longitudinal profile of the capsular bag. Preliminary tests in monkeys in 2003 suggested an accommodation amplitude up to 6 D, which converted to the human-eye model indicated up to 4 D [44].

Lumina—AkkoLens International, The Netherlands. This is an ingenious lens where the anterior and posterior lenses slide transversely in opposite directions according to the principles of an Alvarez compound lens [45]. In an early prototype, the anterior surface of the anterior lens has an aspheric function for the base power and a cubic function. The posterior lens has a posterior surface with a cubic function. The inner surfaces of the compound lens are flat. The focal length is modified by changing how the two cubic functions transversely combine. A HEMA/MMA (hydroxyethyl methacrylate-methyl methacrylate) copolymer prototype with 5.7 mm of diameter of the clear optic was tested in air and has shown a 40 D accommodation amplitude for a 0.75 mm relative lateral shift, corresponding to approximately 4 D in the eye [46]. The solution that hit the market uses a hydrophilic acrylic polymer (Contaflex CI26) and requires a corneal incision from 2.8 mm to 3.0 mm. It is designed to be positioned in the sulcus plane and has part of the base power on the aspheric anterior surface of the anterior lens, and part on the aspheric posterior surface of the posterior lens. The inner surfaces of the anterior and posterior lenses, respectively, exhibit complementary cubic functions. A clinical study by Alió *et al* found a mean subjective accommodation of 3.87 D with a visual acuity of 0.2 logMAR [47].

Juvene – LensGen, USA. Another accommodating 2-piece lens with soft material and based on capsule filling to change the curvature, where the pieces are inserted into the eye separately through a small incision and assembled inside. Good preliminary results have been recently shown, with stable ELP (effective lens position), minimal rotation and virtually no disturbance of photic phenomena.

13.2.3 Deformable and electronic lenses

Fluid Vision—PowerVision, USA (acquired by Alcon in 2019). As the name implies, it is a lens with a liquid filling (silicone oil), index matching the hydrophobic acrylic material of the haptics and optic. The haptics consist of chambers that serve as a liquid reservoir that dispenses the liquid through channels into the central optic when pressed by the contraction of the ciliary muscle. The additional liquid volume in the lens changes the curvature of its deformable surface, favoring near vision [48, 49]. It can be implanted through a 4 mm incision and can accomplish a mean accommodation amplitude up to 5.6 D [29].

NuLens Dynacurve [50]—Nulens, Israel. The lens has a fixed volume of a silicone gel between two flat, rigid and parallel PMMA (polymethyl methacrylate) plates. The anterior one has a circular aperture covered with a flexible polydimethylsiloxane (PDMS) membrane that retains the gel within the plates [51, 52]. The IOL is implanted in the sulcus, minimizing capsular fibrosis and tilt. When the ciliary muscle is relaxed, the bag-zonular complex acts as a stretched diaphragm acting on the posterior plate [53]. It, thus, pushes the gel against the anterior plate and creates a bulge in the membrane, favoring near vision. When the muscle is contracted, disaccommodation occurs, setting far vision [54]. This works in reverse to natural accommodation. The lens requires a 9 mm incision and has been claimed to yield a mean accommodation amplitude of up to 10 D [12].

Wichterle Continuous Focus (WIOL-CF)—Gelmed International, today Medicem International, Czech Republic. This is a flexible lens with a biomimetic design devised by Professor Otto Wichterle and his team. It is made of a synthetic hydrogel from cross-linked metacrylic copolymer (WIGEL®) with high water content (44% H_2O), and is coined as a polyfocal bioanalogic hydrogel. This IOL features biconvex and hyperbolic anterior and posterior surfaces, and no haptics. It is intended to touch most of the posterior capsule without altering its shape [55]. The ciliary muscle and vitreous pressure alter the shape of the capsular bag and concomitantly that of the lens, therefore providing mean accommodation up to 2.5 D. The lens has a diameter of 8.9 mm and requires a 2.8 mm incision.

Medennium Smartlens [24, 56]—Medennium, USA. The lens to be implanted is a thermoplastic thin solid piece at room temperature, whose shape evolves to a specific size and power when under the body temperature, filling the capsular bag [57]. The resulting material is an elastic soft hydrophobic gel that deforms under the action of the ciliary muscle and the capsular-bag consequent deformation [58, 59].

Lens refilling—Several proposals have been presented for lenses that use the posterior capsule and the defining lens geometry. The strategies are based on having the bag filled with a silicone gel, with sealing membranes and plugs, and sometimes with an endocapsular balloon, combined or not with an IOL at its anterior face [29, 60]. The principle relies on the fact that the ciliary muscle change alters the pressure in the capsule, modifying the shape of the sealed bag. The best accommodation has been found when the lens is not completely filled (60% to 70%) [61, 62]. An undesirable event is leakage of the intracapsular fluid or fibrotic capsule opacification [63].

Sapphire Electronic AutoFocal (AF-IOL)—Elenza, USA. This is an ingenious acrylic refractive lens that can provide up to 3 D accommodative amplitude and does not rely on the ciliary muscle for the accommodation mechanism. Instead, it optically detects the involuntary pupil constriction due to visual cues revealing the intention to focus on a close object, and it actuates a liquid-crystal cylindrical cell, encapsulated in an aspheric monofocal optic, to change the refractive power accordingly. The light transmission through the liquid-crystal cell is greater than 90%. Photosensors are placed behind the iris plane and they detect and interpret the dynamics of the pupillary movement and ambient illumination, with the aid of an integrated processor. The pupil diameter responds to light intensity, whose variation could be a confounder to the accommodative miosis, however, it is claimed that minute pupil constriction movement patterns and constant levels of illumination are a pathognomonic sign of the intention to accommodate. Several other factors are involved with the pattern and amplitude of pupillary change, as age and even the level of difficulty while solving a problem, and they might all lead to a wrong interpretation of the sensor [64]. To mitigate this, the lens runs a self-learning algorithm in the processor that can be customized to the pupillary dynamics of each individual and has been validated with 300 pseudophakic patients [65]. A solid-state ultra-thin gold-encapsulated lithium-ion battery is wirelessly recharged by electromagnetic induction over up to 20 cm [66]. The micro-coils can double as an antenna for two-way data communication between the lens and an external device. The implant has an approximate rectangular shape (9 mm × 3 mm × 0.6 mm) and has been designed to be inserted through a 3.9 mm incision. Despite having had several patents on the various aspects of the lens and having been on the verge of settling this technology on the market around 2016, due to crucial issues with the investment commitment of large stakeholders, they seem to have discontinued activities since then [20, 21, 67, 68].

R-TASC Electronic IOL—SAV—Swiss Advanced Vision, Switzerland. This is a lens intended to be positioned in front of the iris, in the anterior chamber. It can be implanted with the crystalline preserved, to mitigate presbyopia, or in combination to a fitted monofocal IOL. It can detect the distance of objects in the visual field by an autofocus system with embedded hardware and software. The device is meant to comprise photosensors, signal processing, a photovoltaic solar cell, induction coils, an energy storage element, power management, control unit, a varifocal cell and a lens actuator mechanism [69]. When the signals indicate a close element in the scene, the control unit triggers micro-pumps to alter the curvature of an optical membrane with liquid displacement from a storage reservoir to the optic central region. The process occurs in 200 ms, equivalent to the reaction time of a healthy-eye accommodation process. The reference power of this add-on lens, to be added to that of the crystalline or pseudophakic IOL, can be adjusted by the amount of liquid left in the central region when there is no actuation. There has been reported the intention to develop an app to calibrate and pre-set accommodation modes or to eventually adjust the reference emmetropic condition after implantation [70].

Liquid-Crystal Adaptive-Optics IOL—TU-Delft, The Netherlands. This is a 9 mm prototype of a biconvex intraocular lens with a 5 mm clear optic. It has been tested *in vitro* in an optical bench and features a 40 μm nematic liquid-crystal layer between two plano-convex glass pieces coated with transparent electrode layers [71]. The anterior piece has a low resistive electrode layer (ITO—indium tin oxide) and the posterior piece features at its inner flat surface a high-ohmic film, an n-type doped silicon carbide layer deposited by plasma enhanced chemical vapor deposition (PECVD) [72]. The combined curvatures of the glass parts determine the base refractive power of the lens. A control voltage can be applied between the electrodes and the modulation of its amplitude and frequency enables adaptive modal correction of ocular aberrations. The control signal is wirelessly received by a coil antenna on the rim of the IOL. Because the material of one of the electrodes presents high resistivity, a voltage applied to its outer end suffers a distributed reduction towards its center, creating a different voltage difference to the opposite low-resistive electrode, hence causing different degrees of reorientation of the liquid-crystal molecules from the edge to the center, and an equivalent continuous phase profile across the lens. The phase profile created by the liquid crystal adds to the converging wavefront created by the refractive optics of the glass pieces. The lens was found to achieve a mean accommodation amplitude of ~2.5 D, which would be sufficient to reading at near distance, and a correction of root-mean-square spherical aberration between −0.43 μm and 0.36 μm. The settling time was, however, still rather long (~4 s) compared to average human accommodative response. Another important issue is that linearly-polarized light is necessary for the adequate operation of the liquid crystals used in the lens. As in a scene the light does not feature that oriented polarization, the addition of a linear polarizer would result in the loss of half of the light impinging on the lens.

Except for the Crystalens and the Syncrony IOLs, experience with all the other ones has been limited to simulations, testbenches or small trials. The quest for a clinically available accommodative lens with full refractive aspheric optics, clinically meaningful accommodative power, biologically inert material, and with a reasonably small incision required, remains largely unmet. Recent improvements and new technologies for EDoF, enhanced monofocal and multifocal IOLs have been representing an investment shift away from the development of a true accommodative lens.

13.3 Dual-optic design and accommodation amplitude

The general structure of a dual-optic accommodating IOL usually consists of a moving positive lens attached to a lower power static negative lens by a spring articulation, thus providing accommodation between far and intermediate/near vision. The design of dual-optic AIOLs is critical, as it must ensure an adequate set of parameters that need to be preserved for as long as possible after the implantation surgery. It is therefore important to provide a thorough analysis of the underlying optical model to clarify the most important aspects that need to be addressed during the design stage. However, if all the high number of variables and degree of freedom were considered, the resulting analysis would have a high level of complexity, which we believe is beyond the scope of this chapter. In this section we first present a simplified analytical model, based on the linear model proposed by Lin *et al* [73], to allow the reader to have a general understanding of a specific AIOL optical system. The results obtained with the analytical model are then combined with a more extensive analysis, based on the traditional ray tracing method, to define the parameters that ensure the manufacturability of a given lens design.

Figure 13.1 shows the basic geometric disposition of the optical elements of the system in two different configurations. We limit our analysis to the case of static posterior lens in the dual-optic AIOL, as this is the most common configuration. Both systems are composed of thin lenses, which means that the lens thickness can be ignored.

The most important parameter of such a dual-optic IOL is its accommodation rate M, which quantifies the change in the refractive power of the eye with respect to the AIOL movement between positions related to far and near vision. Based on the reference model shown in figure 13.1, it is possible to highlight the main variables affecting M, such as the overall refractive power of the AIOL and the geometric arrangement, as shown in figure 13.2. These parameters will be explored in the analytical model reported below.

According to Lin's model [73], the accommodation rate, in diopter/mm, for the forward movement of the AIOL is defined as $M = dC/dS$, where C is the power of the cornea and S is the depth of the anterior chamber. The axial movement dS is defined as $dS > 0$ when moving towards the cornea.

In an eye with a single-optics AIOL, considering an emmetropic state, i.e. the state in which far-distance objects are focused, the vergence relations can be described by equation (13.1) [73]:

$$C = \frac{1}{\frac{1}{\frac{n}{X} - P} + \frac{S}{n}} \tag{13.1}$$

where P is the optical power of the single intraocular lens, n is the refractive index of the medium, and X is the depth of the posterior chamber, related to the axial length L and the length of the anterior chamber S.

To obtain the accommodation rate, we must take the derivative of C with respect to S. From equation (13.1) we obtain:

$$M = \frac{P}{1336}\left(1 - \frac{S.\,C}{1336}\right)\left(2C + P - \frac{S.\,C.\,P}{1336}\right) \tag{13.2}$$

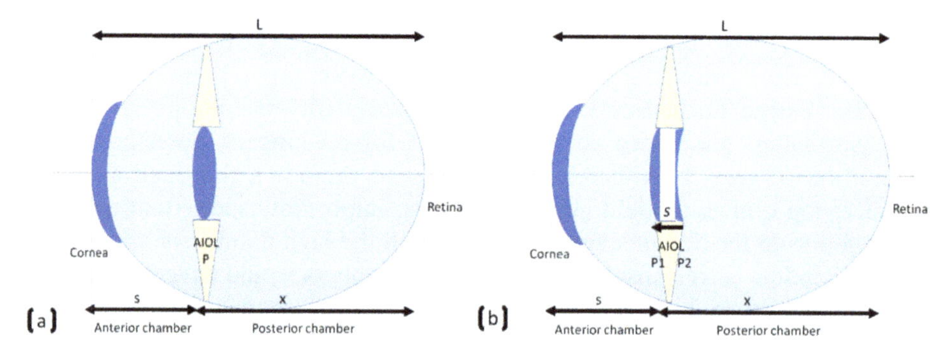

Figure 13.1. Simplified geometrical description of the optical system with (a) single-optic AIOL and (b) dual-optic AIOL.

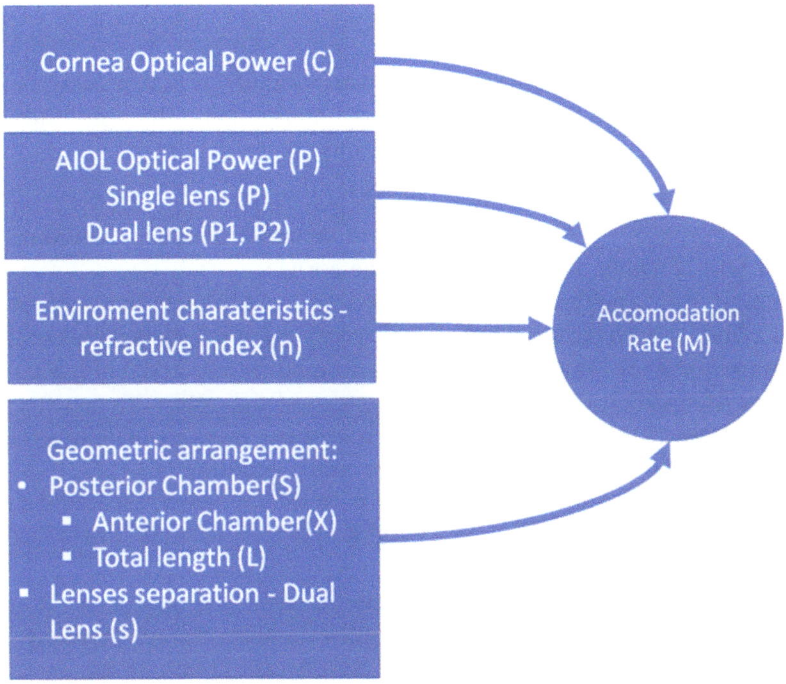

Figure 13.2. Mapping of the main variable affecting the accommodation rate (M) of the simplified eye model.

The refractive index is considered to be $n = 1.336$ for both the aqueous and the vitreous media. Since X and S are conventionally given in mm in such formulas, and the refractive powers C and P are given in diopter (or 1/m), both lengths need to be divided by 1000 in equation (13.1) for the coherence of units. This renders 1336 (i.e. 1000×1.336) in equation (13.2).

Extending the previous results to the case of a dual-optic AIOL is not straightforward, as a dual-optic lens presents a higher level of complexity with respect to a single-optic one. To simplify the analysis, it is useful to define the effective dual-optic AIOL dioptric power as a combined function of the optical powers, P_1 and P_2, of its anterior and posterior lenses, respectively (equation (13.3)):

$$P = P_1 + P_2 - \left(\frac{s}{1336}\right)P_1 .\ P_2 \tag{13.3}$$

where lower-case s is the separation between the two optics of the intraocular lens, given in mm.

Assuming the case of a static posterior lens, the accommodation rate due to the movement of the anterior lens (M_1) is given by equation (13.4):

$$M_1 = gM - \frac{1}{1336}\left(1 - \frac{SC}{1336}\right)^2 \cdot P_1 \cdot P_2 \tag{13.4}$$

where

$$g = \frac{P_1}{P}\left[1 + \frac{2s \cdot P_1 \cdot P_2}{1336.\ P}\right] \tag{13.5}$$

The geometric factor g quantifies the influence of the separation between lenses and the ratio between P_1 and P on the accommodation rate. The model reduces to the single-optic lens case (equation (13.2)) by imposing $P_2 = 0$, $P = P_1$ and $s = 0$. For the case of fixed posterior lens and moving anterior lens, the amplitude of total accommodation (A) is given by equation (13.6):

$$A = M_1(dS_1) \tag{13.6}$$

where dS_1 is the amount of axial movement of the anterior optics, assuming that the values are positive when moving towards the cornea.

It is important to highlight that the expression for M_1 given in equation (13.4) is based on a linear approximation, which holds only for small movements of the optics ($dS < 1$ mm). For $dS > 1$ mm, nonlinear variations have to be taken into account. This can be done by dividing the interval of movement of the optics in segments of 1 mm length, and applying the linear model for each segment, after recalculating the corneal power changes from one segment to another, using equation (13.1) [73].

The results of several different simulations are shown below, assuming a constant direct sum of $P_1 + P_2$ of 20 D, $L = 23.8$ mm and $S = 3.5$ mm. The effective total optical power of the dual optics AIOL is given by equation (13.4).

Figure 13.3 shows the accommodation rate M_1 as a function of the anterior lens optical power P_1 for different values of optics separation s, considering different values of the cornea optical power in a range between 38 and 48 D.

From figure 13.3 it is possible to verify that, for $P_1 = 30$ D and $P_2 = -10$ D, the accommodation rate is the highest for all configurations. Figures 13.4–13.6 highlight the effects of the distance s between the lenses, its optical power and the cornea power on the accommodation rate M_1. Figure 13.4(a) shows the accommodation rate as a function of the anterior lens optical power P_1 for different values of the cornea power, for $s = 2$ mm, while figure 13.4(b) shows the geometric factor g as a function of P_1 for different values of distance s. It is possible to notice that g exhibits a higher variability for higher absolute values (above 10) of the optical power P_1, and such variability increases with s.

Figure 13.5 shows the accommodation rate M_1 (a) and the geometric parameter g (b) as a function of the distance s between the AIOL lenses, for different values of the cornea power and for $P_1 = 30$ D. It can be noticed that M_1 is highly sensitive to the corneal power C, whereas its dependence on the distance s is reduced for $s > 0.5$ mm. The same behavior with respect to s can be observed for the geometric parameter g. With the help of figures 13.4(a) and 13.5(a) we can see that the accommodation rate M_1 is more affected by the anterior lens power (P_1) than the distance. It is also possible to observe from figures 13.5(a) and (b) that the first 1 mm displacement has a higher effect than the subsequent variations of s on both M_1 and g.

Figure 13.6 shows the final accommodation amplitude A as a function of the separation s between the two optics, for different values of the anterior lens power P_1. It is possible to notice that the sensitivity of A to the lens separation s increases with increasing values of P_1.

Based on the above results, we can now move to the second part of our analysis, which consists of a discussion on the geometrical characteristics of a dual-optic AIOL capable

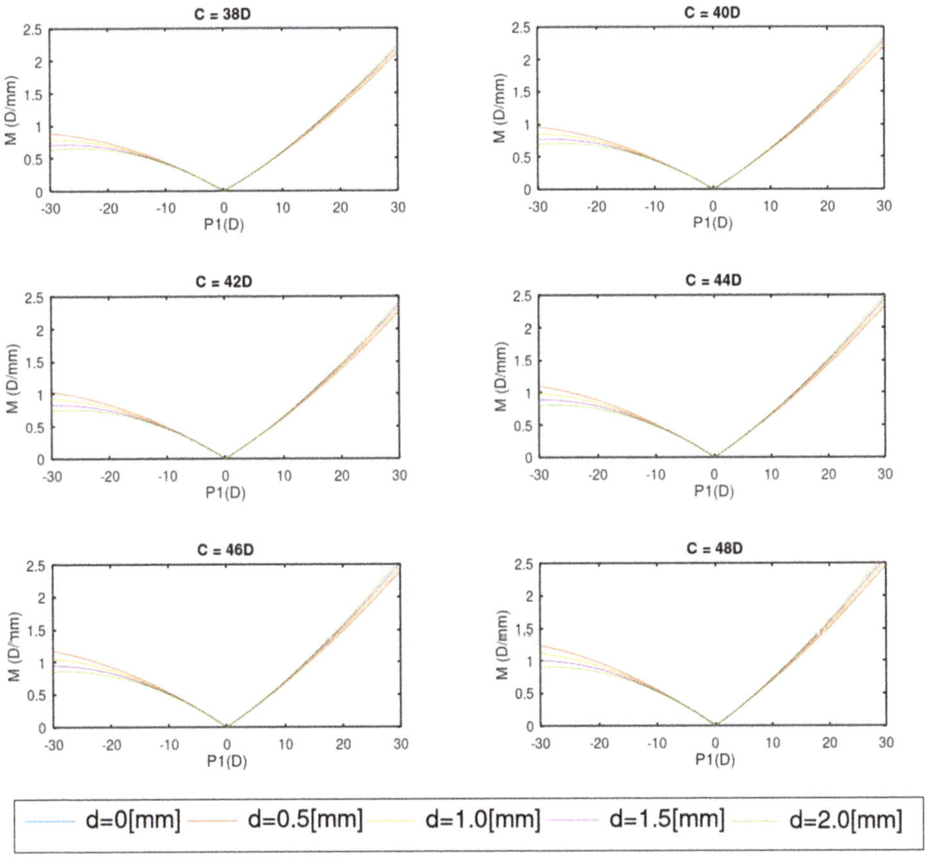

Figure 13.3. Accommodation power (M) as function of P_1, C and s.

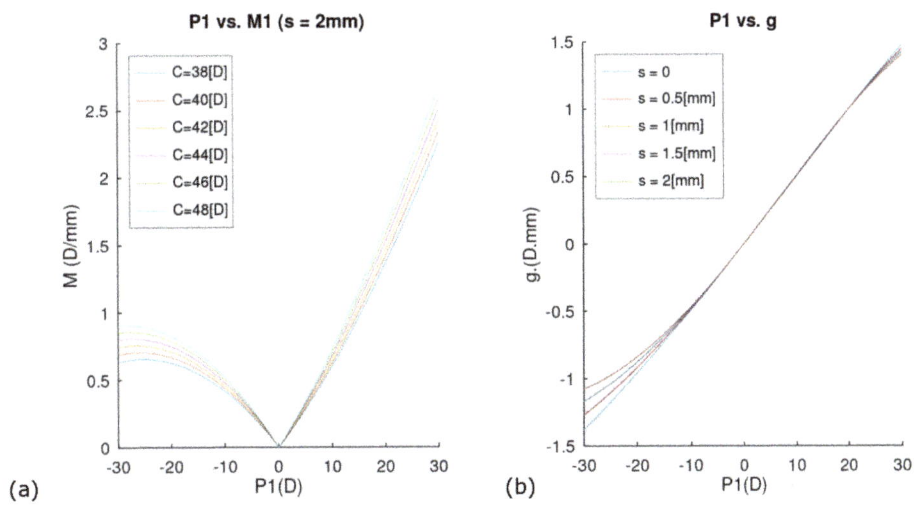

Figure 13.4. Accommodation rate (a) and geometric factor g as a function of P_1 (b).

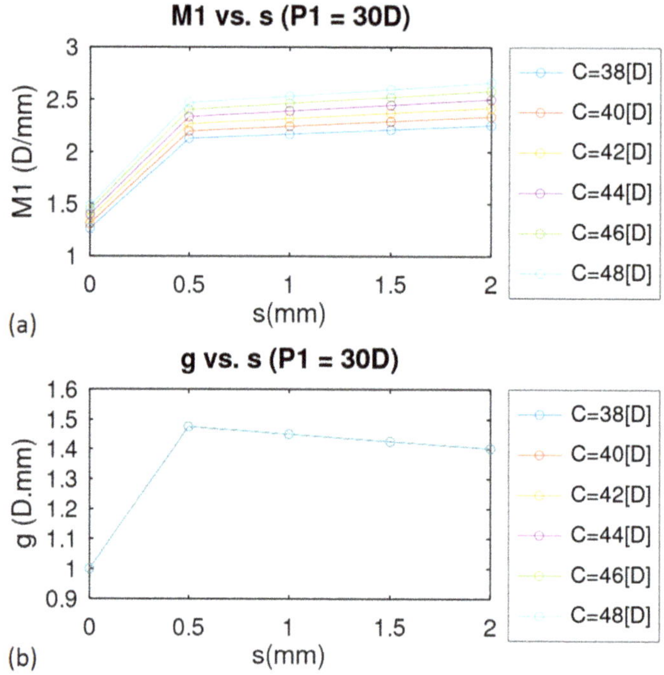

Figure 13.5. (a) Accommodation rate for $P_1 = 30$ D, as a function of the distance d for each of the cornea optical power C; (b) geometric factor for each of the distance d when $P_1 = 30$ D.

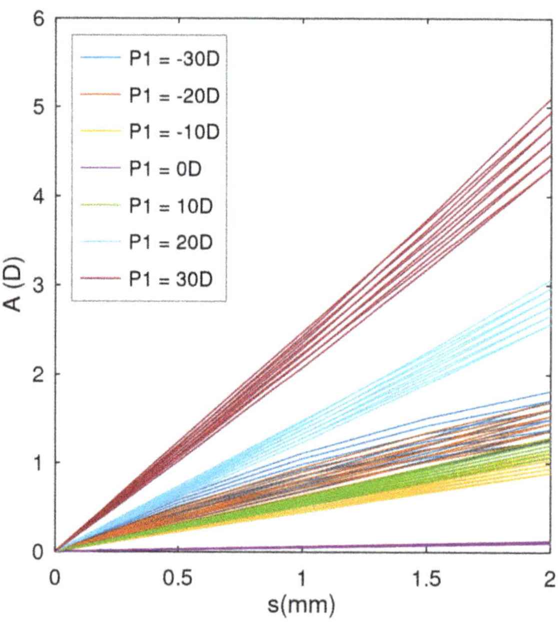

Figure 13.6. Accommodation amplitude A versus distance, for different values of optical power P_1.

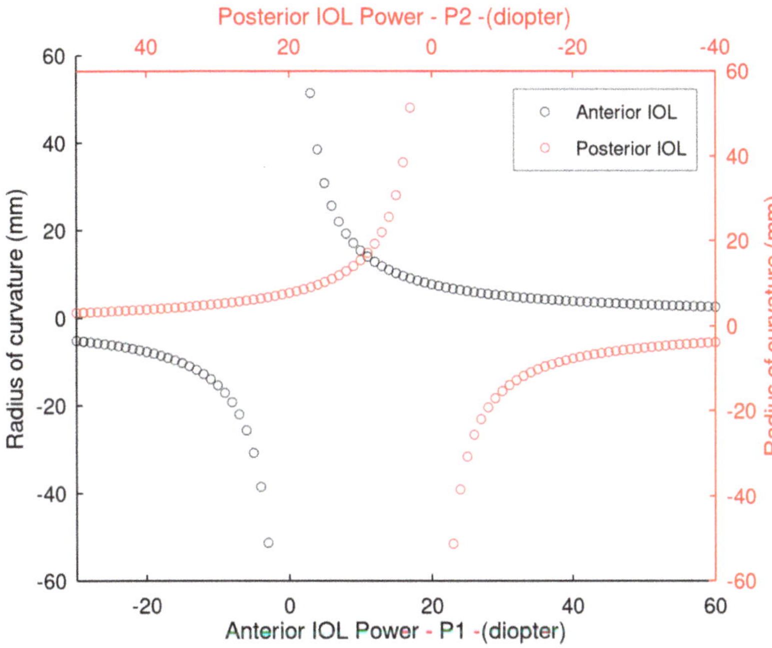

Figure 13.7. Radius of curvature of the L_1 and L_2 lenses satisfying the constraint $P_1 + P_2 = 20$ D.

of ensuring a given accommodation. Figure 13.7 shows the radius of the anterior and posterior lens as a function of its dioptric power, under the constraint that $P_1 + P_2 = 20$ D. It is possible to observe that for negative values of P_1 and P_2, the respective radius of curvature is also negative. The absolute value of the radius of curvature is inversely proportional to the refractive power, with the former increasing rapidly as the latter tends to 0 D. These values are used to perform ray tracing analysis, which is carried out using the ZEMAX Optics Studio software. The results obtained with the analytical model are used as a starting point for the analysis based on the ray tracing method. However, not all the solutions of the analytical model can be physically implemented, as it is necessary to consider the physical constraints imposed by the dimensions of the human eye. This represents the biggest advantage of associating the two methods of analysis, that is, the flexibility to explore the effect of multiple parameters offered by the analytical method and the higher accuracy of the ray tracing method.

Figure 13.8 shows two examples of infeasible configurations. The red lines in the picture indicate a physical inconsistency, since one of the lenses should overlap to the other to achieve the desired accommodation, making its manufacturing unfeasible. Figure 13.9 shows a feasible lens configuration, for comparison.

All the configurations considered in the analytical model were validated by ray tracing, where a short-eye model, whose axial length is equivalent to 22.2 mm, was assumed in the simulations, as shown in figure 13.9. The posterior lens was kept fixed while the anterior lens moved axially to promote the desired accommodation. To determine the required accommodation range, the reference distances consider a focus

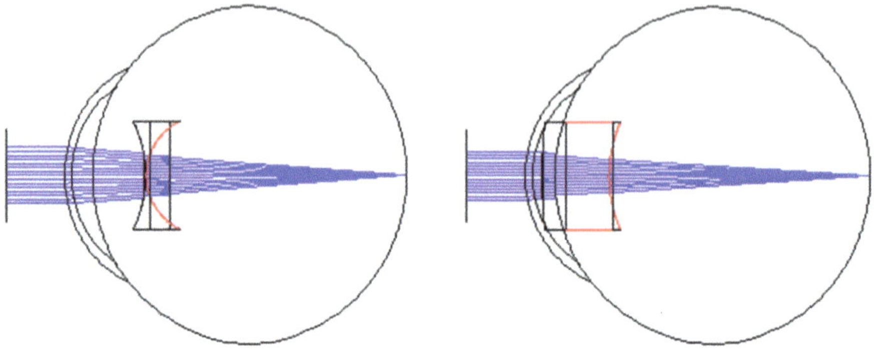

Figure 13.8. Examples of unfeasible lens compositions (a) $r_1 = -7.70$ mm, $r_2 = 3.85$ mm, (b) $r_1 = 7.70$ mm, $r_2 = 8.56$ mm.

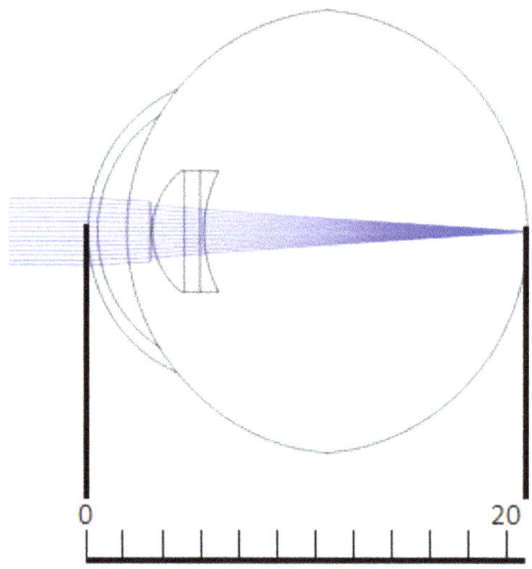

Figure 13.9. Example of a feasible geometrical layout for a 22.2 mm axial length.

distance range shifting from the situation when an object is located at an average reading distance (400 mm) to the situation of far distance vision, namely infinity (figure 13.10).

At this stage, the lens thickness was also considered since the manufacturability of the device and its compactness would play a relevant role on the incision size in the case of a real implant. The dioptric power of the anterior lens (P_1) is fixed at 32 D while the dioptric power of the posterior lens P_2 varies from −18 D to −13 D. Simulation results are shown in figure 13.11, where the red dots represent the lens separation when the object at the infinity plane is in focus while the black dots refer to focus at reading distance. The total thickness is given by the sum of the individual lens central thicknesses, which is fixed at 1.8 mm, plus the separation s between the lenses. The difference between red and black dots represents the separation range.

The modulation transfer function (MTF) was evaluated for the same group, and it is shown in figure 13.12. It gives an idea of the required dimensions for the physical

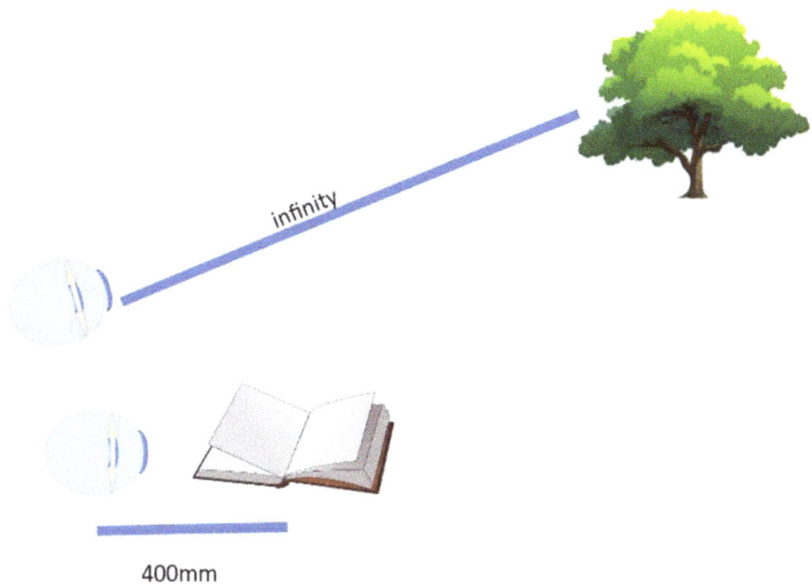

Figure 13.10. Object plane definition.

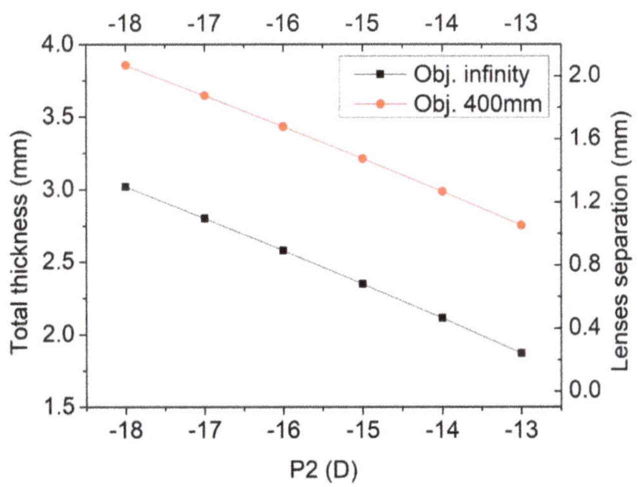

Figure 13.11. Total thickness and lens separation with anterior lens fixed with 32 D and posterior lens varying between −18 D and −13 D.

implementation of the device and the expected visual quality, with the aim of achieving the desired accommodation with the smallest possible displacement of the anterior lens L_1 with respect to the posterior lens L_2.

The MTF is one of the most used methods to assess the performance of optical systems. It determines how much contrast in the original object is maintained by the detector. In other words, it characterizes how faithfully the spatial frequency content of the object gets transferred to the image [74].

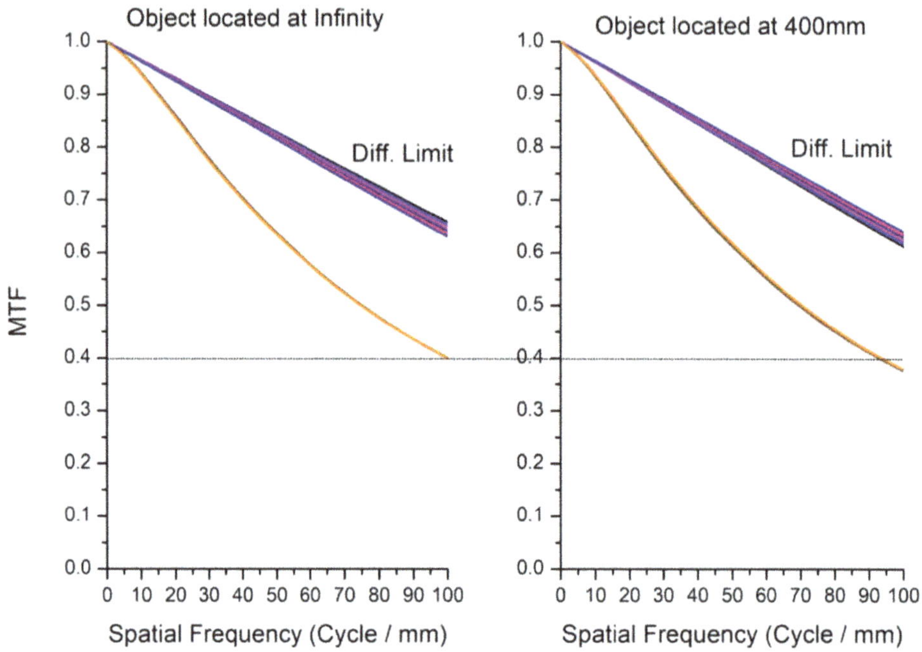

Figure 13.12. MTF for an optical system with anterior lens with 32 D and posterior lens varying between −18 D and −13 D.

The MTF graph plots the percentage of transferred contrast versus spatial frequency (cycles/mm) of the lines. The spatial frequency is expressed in terms of cycles or lp (line-pairs) per millimeter, which physically represents the density of the lines.

The MTF in figure 13.12 covers the spatial frequencies up to 100 cycles/mm. The TS Diff Limited (Tangential Sagital) set of curves represents the situation when the optical system is limited only by diffraction, which is the best-case scenario, while the other curves derive from the configuration and topology of the lenses. The minimum value of the MTF that ensures a sufficiently sharp image detection, at both infinite and reading distance, is 0.4, as indicated by the ISO standard [75]. For the considered case, the MTF satisfies this condition in the whole spatial frequency range between 0 and 100 cycles/mm.

13.4 Dual-optic MEMS IOL

In the past decades, medical robotics has been advancing in applications such as high-precision and remotely operated surgery, targeted therapy, and prosthetics. Particular focus has also been directed to implantable devices to assist or restore physiological processes [76]. There are ongoing challenges in this area, the most critical being biocompatibility of the materials used for the implants, their packaging, power efficiency, and energy harvesting capability. In this trend, the development of biocompatible electroactive polymers (EAPs) together with shape memory alloys (SMAs) have enhanced the potential for the design of MEMS as a viable solution for biomimetic applications [77].

A biomimetic signal-controlled soft lens was proposed in [10], showing the feasibility of MEMS for ophthalmic applications. The following sections present a brief introduction to EAPs, MEMS, and shape memory alloys followed by a proposed MEMS dual-optic lens structure that is evaluated for the case of intraocular deployment.

13.4.1 Electroactive polymers (EAPs)

EAPs are materials that can change shape when stimulated by an electrical signal. EAPs with an ionic character are polymers with ionizable groups in their main chains, and naturally show such electroactive response. These polymers are often known as artificial muscles due to their similarities to natural muscles. As a glimpse of their potential to impart displacement amplitudes compatible to AIOL's applications, figure 13.13 shows the displacement of a responsive hydrogel as a function of time stimulated by low voltage magnitudes.

This material responds to the voltage applied between two electrodes in a liquid medium and the performance registered in figure 13.14 is illustrated in the image taken during the experiment and presented in figure 13.14, which shows a hydrogel EAP, darker structure in the center, in its initial and final states, under an applied voltage.

The general performance of an EAP compound is highly dependent on several aspects such as the composition and the morphology. The composition and the

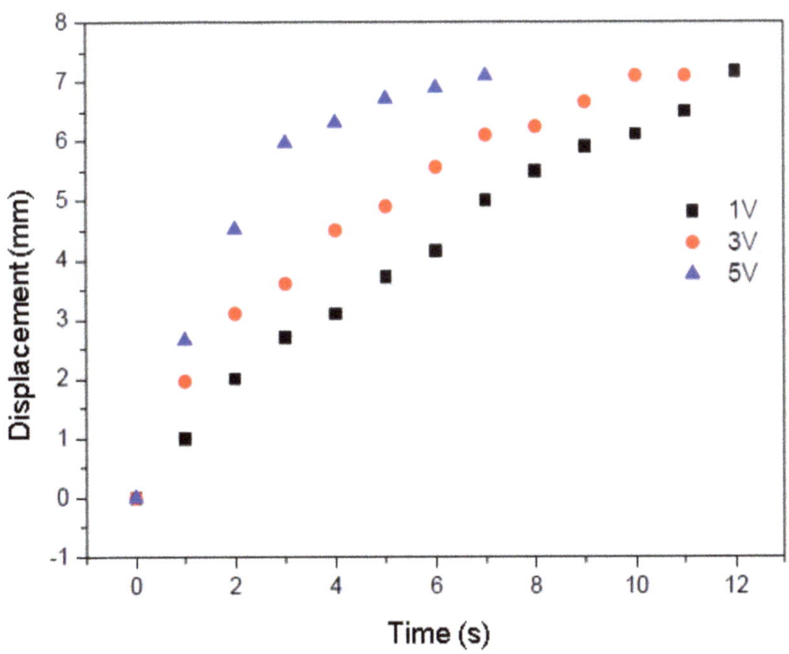

Figure 13.13. Displacement versus time for different voltages [78].

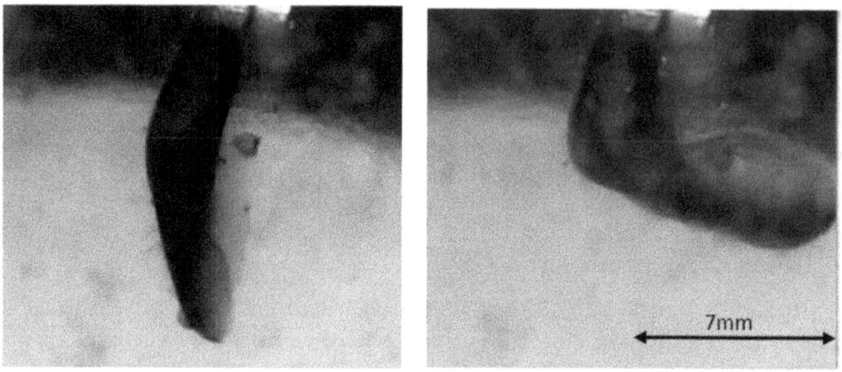

Figure 13.14. (a) Ionic EAP initial state without a voltage and (b) final state after applying a voltage [78].

Figure 13.15. EAP hydrogel nanofibers [78].

functional group in the main chain of the active monomers, may induce the EAP compound to respond either to voltage, temperature, magnetic field, or light, which gives such material a wide range of potential applications [79]. For the morphology, it has been seen that an organization in nanofibers shows a significant improvement in the performance of these materials. EAP nanofibers (figure 13.15) increase the surface area of active monomers allowing it to respond more quickly to stimuli that change its shape [78].

Some successful compositions are those blends of EAP with conductive polymers. A well-known conductive polymer is the polyaniline (PANI), and it has been studied for several applications such as supercapacitors, flexible electrodes, solar cells, sensors and actuators [80–82]. PANI is a conducting polymer of simple synthesis, low cost and high conductivity. It can be synthetized with different nanostructures which is the focus of intense research seeking improvements in processability and performance of devices using such materials. In general, the introduction of conductive polymers to EAP hydrogels may allow them to increase their actuation range by reducing the required voltage and increasing the response speed.

Figure 13.16 shows an image made by atomic force microscopy (I-AFM) that shows a surface map of an EAP/PANI sample. The measurement is done by a current flowing from the polarized sample in contact with a platinum tip (cantilever) used as a probe. From the image it is possible to see a structure of self-assembled nanorods of PANI in the hydrogel structure; it is a strategy for improving the final properties of EAP actuators [83].

EAPs are smart materials, many of which are suitable for deployment in actuators due to their high strain rate, reliability, and quick response [84]. These polymers can be produced in a lens shape and can feature different refractive indices and high transparency, thus being qualified to integrate the autofocus mechanism to the lens itself [85]. Some groups demonstrated different types of tunable liquid microlens devices, using stimuli-responsive polymers as actuators, inspired by the ciliary muscle in the human eye, to tune the shape of curved liquid–liquid interfaces and hence change the focal length of the resultant liquid micro-lenses [86].

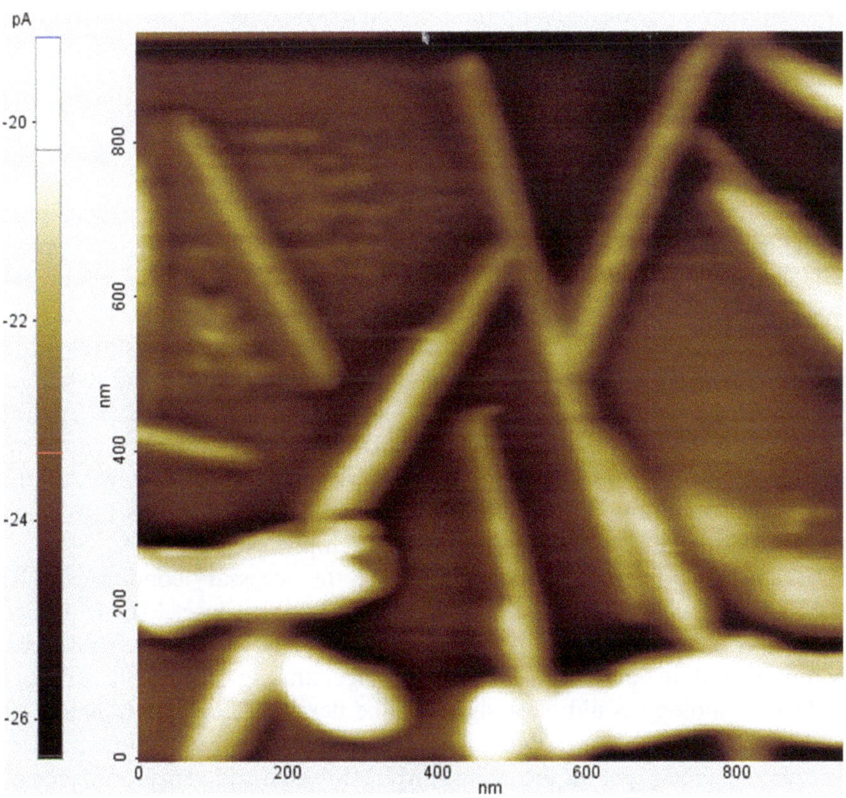

Figure 13.16. I-AFM Images of nanorods of PANI [78].

13.4.2 Micro-electro-mechanical systems (MEMS)

MEMS are micro devices, some also with nanometric dimensions, that often feature moving parts and are primarily based on silicon technology, although ubiquitously expanding to other base materials. Even though the term may sound new for some of the target audience of this book, most people have already been in contact, even if inconspicuously, with this technology. For example, an airbag system has a key component called accelerometer which is responsible for sensing the abrupt deceleration of the vehicle in an accident. Also, the gyroscope, which is present in most of the smartphones, allows one to rotate the screen to horizontal or vertical position and the image follows the movement. Numerous other examples are available in daily life such as certain kinds of pressure sensors, inkjet printer heads, projection displays, microphones, among others. In general, the actuation or sensing principle of MEMS devices relies on similar physical principles of macro devices, such as capacitive and resistive (ohmic) electrical changes. The MEMS process technology compatibility with electronic microchips allowed a rapid growth and deployment of the technology. The complexity of applications has increased with the maturity of the technology and more interdisciplinary applications have been found in biomedical and microfluidics. Even though MEMS technology is based on well-established principles and processes, its fabrication is not rigidly restricted to specific steps, equipment, or materials, and can be tailored to attend to diverse needs. Bringing up this topic in this chapter can inspire and enlighten curious and creative minds to search for disruptive applications in their fields. MEMS devices can be designed to be used as sensors and actuators, and it enables the design and fabrication of components capable of measuring or imparting, with great precision, angular movement, pressure, acceleration, electromagnetic fields, among others.

Today, remote-controlled surgeries are a reality, implementing fiber-optic camera systems, 3D rendering and augmented reality, all with nanometric precision, in part due to the deployment of MEMS devices in equipment used in invasive procedures [27, 87]. This represents a great leap compared to the need for large incisions in the operating room to have better access and greater visibility of the area being treated, which then was only possible using antiseptic, as introduced by Lister in 1867 [87].

Glucose meters, blood pressure monitors, electrocardiogram and more recently, DNA analyzers using a few drops of the patient's blood, are other applications suitable for MEMS in healthcare, providing direct interaction between the patient and the device. This concept constitutes one of the great challenges of modern bioengineering, obtaining fully functional platforms for detecting diseases efficiently, and in real time, which is known as a lab-on-a-chip (LOC) [88, 89].

AIOLs would optimally provide the eye with the necessary conditions to be able to focus at any distance with high visual acuity. In this sense, MEMS integration with AIOLs could provide additional benefits such as potential independence of the ciliary muscle and fast actuation for focus change, among others [90]. AIOLs based on MEMS technology could be designed to be flexible, biocompatible and electrically efficient, aiming to solve ophthalmological issues, such as the one proposed by Zhao *et al* [91], which uses an intraocular MEMS-based coil array, acting as a high-efficiency retina prosthesis, reaching a Q factor of 12.5 MHz at 30 MHz.

In this sense, recent research proposes a prototype of a MEMS AIOL, whose general concept integrates a photo-driven microchip with the function of sensing the user's command for triggering a focus change and managing the actuation mechanism of a silicon structure integrated with flexible parylene springs and SMA actuators. The microchip uses a photosensitive area, responsible for the sensing and generating photocurrent through a photovoltaic process based on blinking; an energy management block; and a logic circuit. The sensing block aims to translate a pre-established blinking sequence into a command to trigger the actuators and change the lens state from far to close focus, or vice versa [92–94].

13.4.3 Shape memory alloys (SMAs)

Shape memory alloys (SMAs) are a class of metallic materials that can exhibit the shape memory effect (SME) and the superelasticity effect (SE). The shape memory effect is a property where the material is initially deformed under low temperature and recovers its original shape when heated. As a pictorial example, imagine a paper sheet initially flat that you crumple completely until it turns to a paper ball. If you want to reverse the process and stretch and flatten the paper again, the fibers from the paper will be already damaged and it may never come back to the initial state anymore. If the same happens to an SMA material, as soon as you introduce heating to it after the deformation, the atoms tend to restore its original shape. Of course, the deformation cannot be so extreme as to induce irreversible hysteresis. These phenomena have been well understood as the result of thermoelastic phase transformations between the high temperature phase, known as austenite phase and a low temperature phase, known as martensitic phase [95].

Figure 13.17 shows a simplified illustration of the deformation process of the material structure. (a) its initial shape, (b) the deformed state and (c) the restored initial shape.

In summary, the SME is a macroscopic effect of thermally induced crystallographic phase change. The electrical resistivity of the material will also change during the deformation. The electrons will have less mobility to flow when the phase state of the material is such as in figure 13.17(a) than when it is on state (b). Since higher resistivity induces a material heating (ohmic losses) when traversed by an electrical current, it turns out one way to induce the phase restoring is by inducing an electrical current throughout the material by means of an applied voltage.

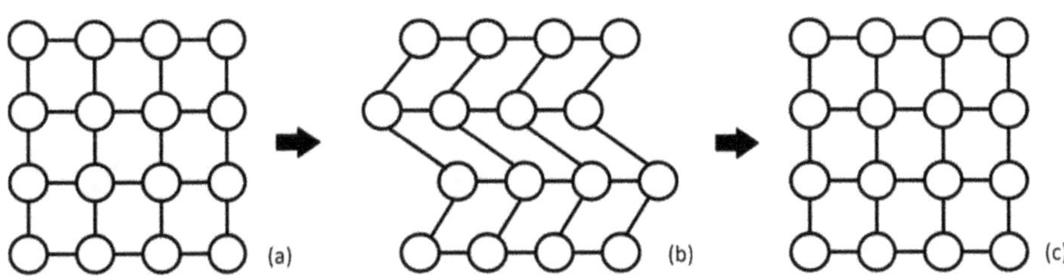

Figure 13.17. SME mechanism. (a) Original state, (b) deformed state and (c) restoring initial shape.

An important challenge in the development of microsystems is the implementation of the actuation principles in a miniaturized scale, and SMAs have big potential to provide this solution. Bellouard, presents many studies using such actuators for the development of micro clamps, micro endoscopes, micro valves, micro-pumps, among many other examples [96]. The authors note that SMA actuators do not operate at frequencies above 100 Hz, which is not a major problem, considering the eye is not sensitive to image transitions above 30 Hz.

13.4.4 Design of the dual-optic MEMS structure

The design of the actual dual-optic MEMS structure was guided by the results of the optical design and simulations, and it considered the compatibility of the final dimensions with those of the internal structures of the eye, although the type and size of the implant incision has not been considered at this stage. The thickness of the final structure was designed to be around 2.8 mm (figure 13.18(a)), including the required thickness of the lenses, with a total optical power of approximately 20 D and it considers the maximum relative displacement to reach sufficient accommodation range to cover objects located between 400 mm (acceptable reading distance) and far distance (infinity). From the optical simulations (section 13.3), the required displacement between lenses to achieve such accommodation lies between 0.6 mm and 1.0 mm.

Figure 13.18 illustrates a cross-section of the overall device. It is composed by an anterior lens L1 with 32 D, a posterior Lens L2 with −12 D separated by SMA springs. In the non-actuated state, the lenses remain on the axially most distant position, and the actuation induces a contraction. More details on the overall mechanism will be given below. The overall structure is encapsulated by a PMMA cylindric structure which has mainly three functions: it gives a mechanical stability for the lens system; protects the optical system from potential interferences from the ciliary muscles; and reduces the degree of freedom of the entire system sitting the

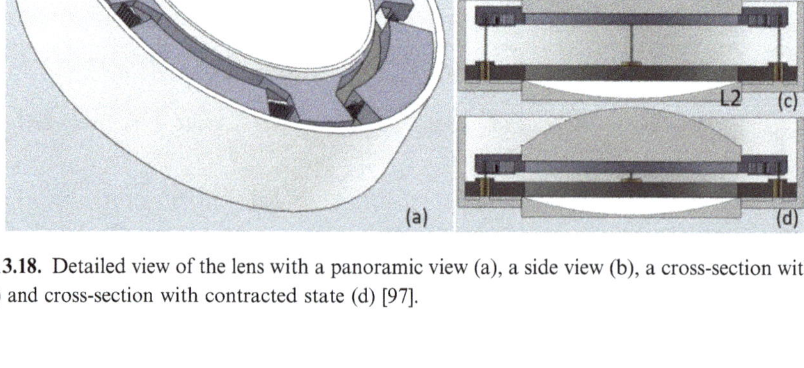

Figure 13.18. Detailed view of the lens with a panoramic view (a), a side view (b), a cross-section with relaxed state (c) and cross-section with contracted state (d) [97].

posterior lens on the back of the capsular bag in such a way that the accommodation is mainly driven by the anterior lens.

The device was designed to have an axial displacement. Until this development stage only two states were allowed to simplify the complexity and reduce the energy consumption. Both states are, respectively, the relaxed state (standby), and the actuated state. The first, with the lenses further apart, allows maintaining focus on objects located at long distances, on the other hand, the second one allows maintaining focus on objects located at short distance (reading distance 400 mm). The system is designed to require actuation energy only during the transition due to the Si shoe/latching mechanism.

Figure 13.19 introduces a detailed view of the proposed lens system.

The proposed device is composed of a PMMA cylinder serving as a sleeve that involves a fixed posterior Si base and an anterior movable Si structure responsible for accommodation. The anterior microfabricated structure, key element for the latching mechanism, has four Si shoes supported by parylene springs and actuated by the circular SMA crossing through all of them. The relative movement between the lenses is performed by the axial spring assembly composed by a metal/SMA spring. The latching mechanism holds the position of the anterior lens on both axial positions (figure 13.20(a) for axial spring stretched and (d) for axial spring contracted). For the axial spring state transition, the microfabricated Si shoes will retract and allow the axial movement of the anterior lens, either backwards or forward (figures 13.20(b) and (c)).

The shoe/latching mechanism requires actuation energy only during the state transition and avoids a large consumption of energy, and it requires a balance of forces on the system. When the axial actuator is in the contracted state, a simplified force balance evaluation considers for the vertical direction a spring force acting from the posterior to the anterior part to reestablish its relaxed state ($\vec{F}_{\text{spring}}$), in opposite

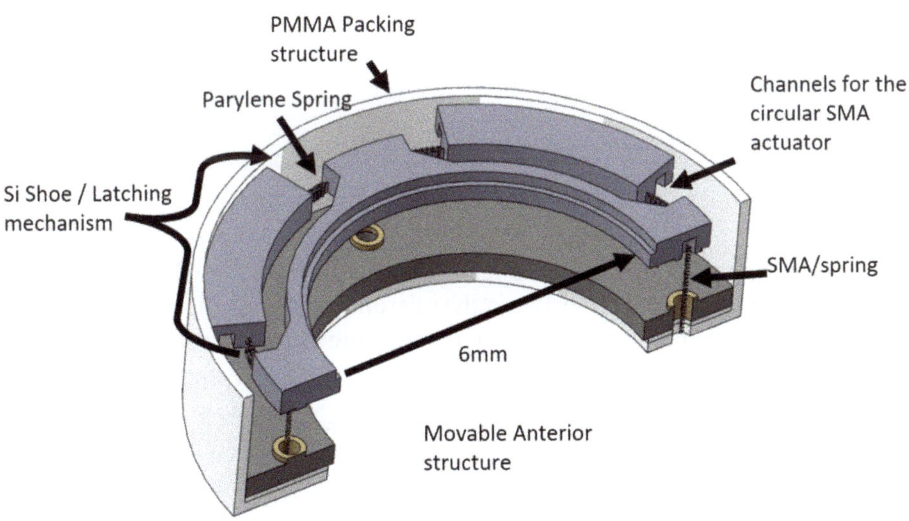

Figure 13.19. Schematic view in perspective of the proposed dual optics system [97].

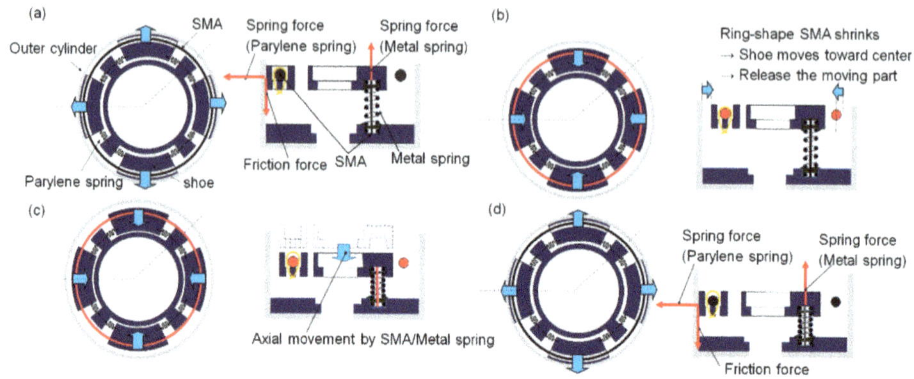

Figure 13.20. latching mechanism. (a) The friction and the axial spring forces are balanced (b) for state change, the radial SMA is activated and releases the outer cylinder shoes, followed by the (c) axial SMA spring shrinkage. When the radial SMA is disabled, (d) the silicon shoes are pushed against the cylinder and the position is maintained without power consumption. Reprinted from [94], copyright (2016), with permission from Elsevier.

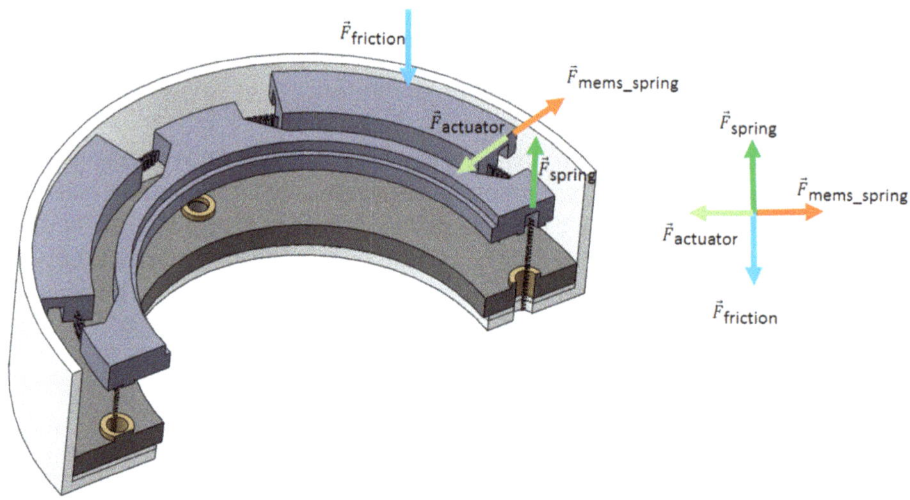

Figure 13.21. Force balance [97].

direction from the axial movement, intended by the spring force, there is the friction force created from the interface between the microfabricated Si-shoes and the PMMA wall ($\vec{F}_{friction}$). For the horizontal direction the pressure of the microfabricated Si-shoes against the PMMA wall ($\vec{F}_{mems\_spring}$) in a static condition and during the state transitions, the force exerted by the actuator ($\vec{F}_{actuator}$) which is responsible for releasing the Si-shoes, breaking the balance of forces, then allowing the axial movement due to the spring force ($\vec{F}_{spring}$). Figure 13.21 illustrates those forces.

In a more advanced development stage, the position could have more than two stages, by enabling an analogical reading of the internal ocular pressure to control the actuation, since it is known that even after surgery, the ciliary muscles can still be working [98], but it is still necessary to find a correlation between the pressure change with the accommodation level, which represents a challenging process.

13.4.5 Fabrication of the dual-optic MEMS structure

In previous section, we stated that a dual-optic accommodative intraocular lens as herein proposed has three separate major structures. We have posterior and anterior sections, the latter is responsible for the displacement of the lens, and a cylindrical structure that packages the two elements previously described.

However, to achieve this kind of structure, perfectly executed steps in the fabrication processes must be carried out. These processes can be classified into three large groups: deposition, etching and packaging.

In this section, we will describe some of the manufacturing processes that can be used to manufacture AIOL, the list of these techniques can be very extensive, for that reason we will only mention the most used.

The photolithography procedure in all stages of the manufacturing process in CMOS technology (complementary metal-oxide-semiconductor), integrated photonics, and MEMS, is required because it allows the patterning of the required structures with great precision. The standard photolithography process can be seen in figure 13.22.

The photoresist is a polymer that is sensitized by exposition to an appropriate light through a mask, usually with its wavelength in the ultraviolet regions, therefore, replicating the mask directly or inversely, depending on the type of photoresist used. The polymer is deposited as drops on the substrate and spread out through the spin coating technique. The thickness of the photoresist film directly depends on the speed of rotation. In some situations, the photoresist itself can be used as a protective layer to the subsequent substrate processing. However,

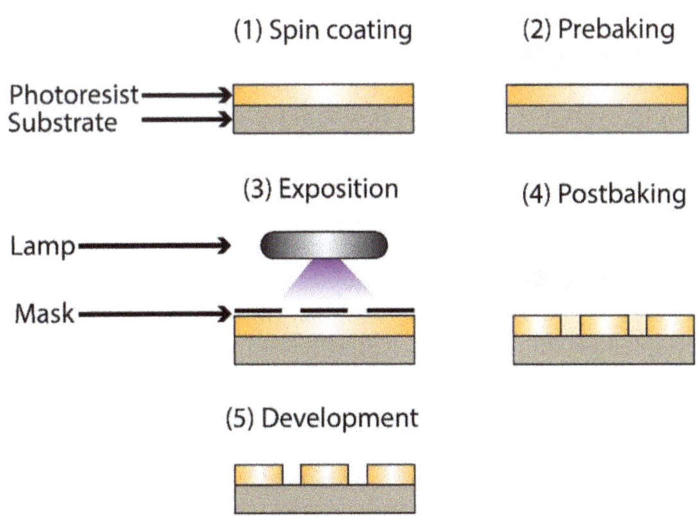

Figure 13.22. Schematic description of the photopolymerization process.

depending on the type of process to be executed, a layer of another material (e.g. SiO_2) might need to be grown or deposited prior to the photoresist coating, guaranteeing a more reliable barrier to etching or doping. Subsequently, the sample is pre-baked at a temperature that depends on the type of polymer (positive or negative) and the expected thickness.

The light exposition is carried out through a mask at wavelengths in the UV (ultraviolet) or in the violet region of the visible spectrum. Some procedures require a post-bake, but it is not essential. In figure 13.22, for simplicity, the contact method is depicted, where the mask is in direct contact with the photoresist. More reliable methods are proximity and projection printing. In the first the mask is held at proximity to the wafer, with a small gap in between. In the second, a lens system is positioned between the wafer and the mask, and the mask pattern is imaged on the photoresist layer covering the wafer.

There are different methods to deposit materials on a substrate, the most used in the semiconductor industry is chemical vapor deposition (CVD), which provides ultra-thin films. Among the materials that can be deposited with the CVD technique, there are monocrystalline, polycrystalline, and amorphous [99, 100]. These materials include silicon and derivatives, carbon and derivatives, and metals, such as tungsten.

However, this is not the most common method for depositing noble metals such as gold, silver, platinum, copper, among others. In this sense, thermal evaporation, sputtering and electron beam physical vapor deposition (EBPVD), are commonly used to deposit the aforementioned materials, allowing the obtention of high purity thin metallic films.

As previously presented, CVD is carried out at high temperatures, unlike CVD techniques such as ICP-PECVD and molecular capor deposition (MVD), in which films can be deposited as low as 100 °C and 80 °C respectively, for this reason, the corrosive gases used in the process can leave residues and some impurities on the surface. On the other hand, physical vapor deposition (PVD) is carried out at low temperatures and does not use corrosive gases in the process, affecting the deposition rate, reaching speeds between 0.1 and 100 $\mu m\ min^{-1}$.

There are essentially two methods to etch a material, dry etching or wet etching. Perhaps the most common are the wet etching processes, where through acetone, hydrofluoric acid (HF), and other acids, it is possible to etch materials such as polymers, silicon, silicon nitride, silicon dioxide, among others.

Other etching processes use gases, i.e., dry etching, where equipment such as reactive ion etching (RIE) uses a combination of gases to remove with chemical reactive plasma, materials previously deposited on the wafers, obtaining high-quality structures, with high aspect ratios, regardless of the size [101]. The inductively coupled plasma (ICP) is a variation of the RIE that has become popular in microfabrication because of the high-quality devices that can be obtained, because unlike the conventionally used RIE, the ICP generates plasma through RF powered magnetic fields.

In accordance with what was discussed above, we present an efficiently developed fabrication process for the AIOLs previously described. It is composed of an anterior part, a posterior part and a cylindrical structure that will bring together all the parts. The proposed device is based on Si wafers with 300 μm and 500 μm thickness, in addition, the surface with crystal orientation (1 0 0) was used.

The proposed structure is composed of an anterior structure which is responsible for the movements that induce the accommodation, on the other hand, the posterior structure basically holds the posterior lens. Similarly, the implementation of a PMMA structure not only protects the entire component, but it also accomplishes an important role in avoiding external interference of the ciliary muscle.

The step-by-step fabrication process, illustrating the steps [97] in a cross-section of each part is presented in figures 13.23 and 13.24, followed by the assembly of the separate parts (figure 13.25).

Assembly: After manufacturing each of the previously described structures separately, both were joined as demonstrated in this section, consolidating a single structure.

The development of the processes previously described, present in a sequential and organized way, the necessary steps to obtain a prototype of an AIOL, as will be further presented in the section to follow.

13.4.6 Results and future challenges of the dual-optic MEMS structure

Figure 13.26 presents the morphology of the parylene spring for the several initially proposed designs. The parylene deposition process guarantees a good repetition rate

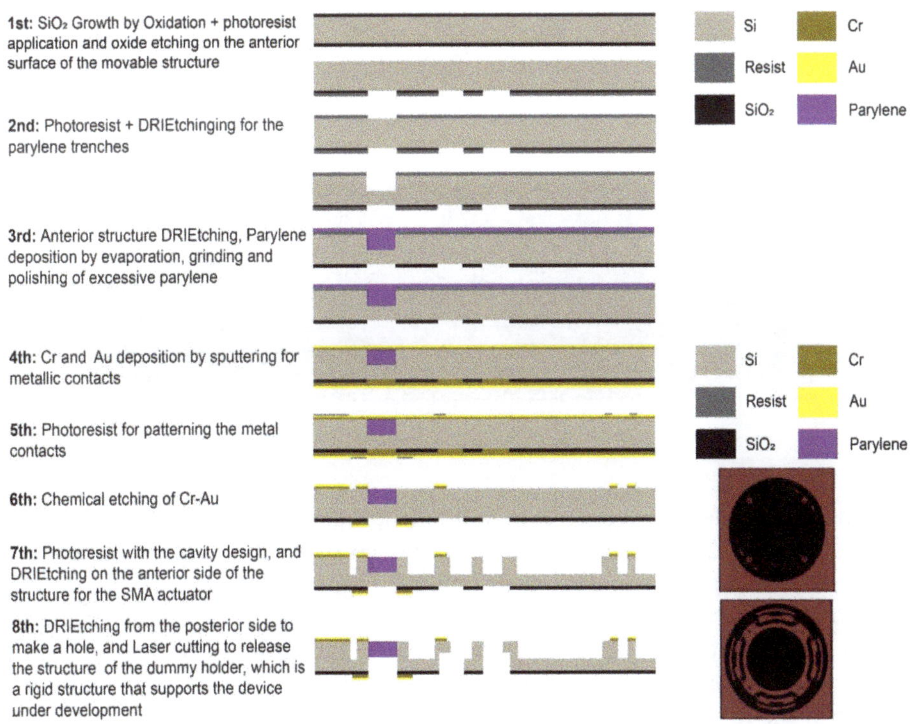

Figure 13.23. Fabrication steps for the anterior structure.

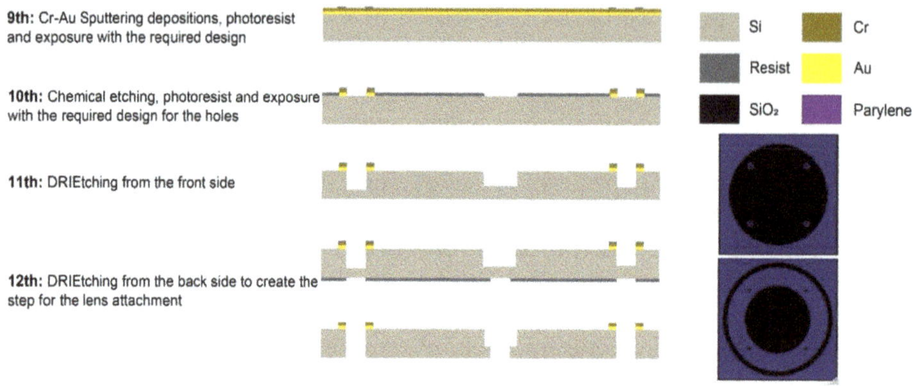

Figure 13.24. Fabrication steps for the posterior structure.

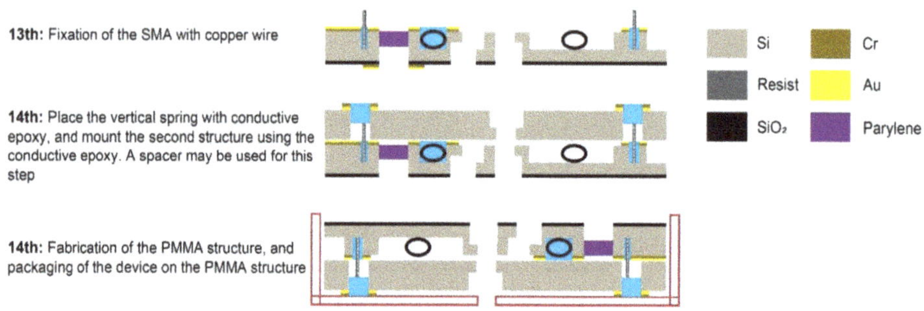

Figure 13.25. Assembly steps.

Figure 13.26. Fluorescence parylene spring [97].

and tends to be quite homogeneous even for critical aspect ratios between depth and thickness. However, large aspect ratios can induce structural failure and make the springs more fragile, affecting its performance. Figure 13.26 shows a fluorescence image for the aforementioned design. The central dark line represents a lower material density and the overall shape complies to the design.

To explore the mechanical stability of the device, the SMA described was made available in the form of a spring-shaped actuator along the outer circular cavity of the device. Figure 13.27 shows the experimental apparatus where it is possible to

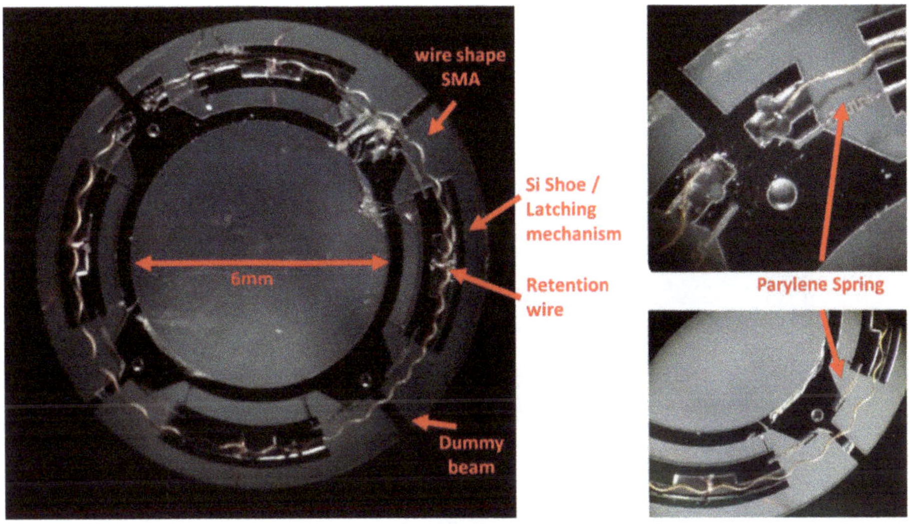

Figure 13.27. Complete assembly of the structure [97].

appreciate the composition of the device obtained in each of the stages in the manufacturing process, described in the previous section. Similarly, figure 13.27 shows a picture of the anterior structure assembly (in black against a gray background) still without the retention/locking piece on top. The Si shoes are still held by the dummy beam structure to the base to avoid damage of the parylene springs while assembling the SMA wire and the retention wires. Those beams are removed by laser cutting during the final assembly.

It is worth noting that these commercial SMA wire and springs were not designed for this application. The overall displacement was mostly not homogeneous and wire-shaped actuators induced greater mechanical stability. It is highly recommended the introduction of a dedicated electrochemical process for the SMA structure to avoid manual steps, such as the introduction of the SMA wires on the Si shoes, and to improve mechanical stability and allow further reduction of the dimensions,

The metal spring connecting the anterior and posterior supporting parts was the weakest point on the development, but the overall results show a good potential to reach high actuation displacement amplitudes between them. The metal spring was also not specifically designed for this project, which introduced a higher stiffness than desired. We propose the replacement of the previously mentioned metal structure by an EAP structure that connects both elements, serves as an actuation element and as mechanical support. Additionally, it can be made of a biocompatible polymer. However, there are many more challenges ahead.

Finally, the AIOL system was optically evaluated using the setup shown in figure 13.28 using an eye model based on [75].

Figure 13.29 presents the image focus change when the anterior and posterior lenses are shifted between 0 and 0.6 mm.

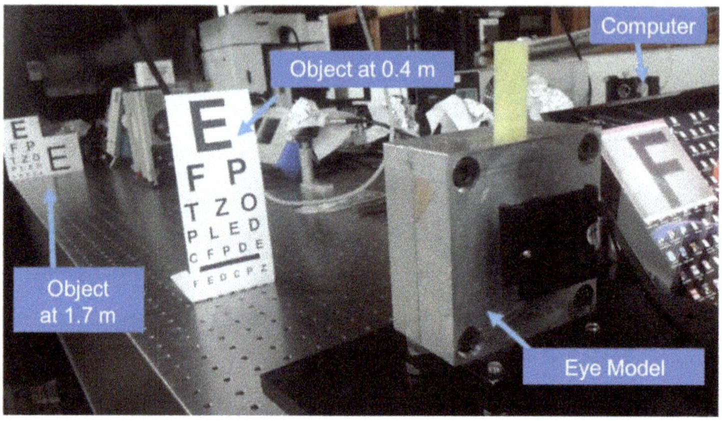

Figure 13.28. Experimental setup to assess the lens accommodation [97].

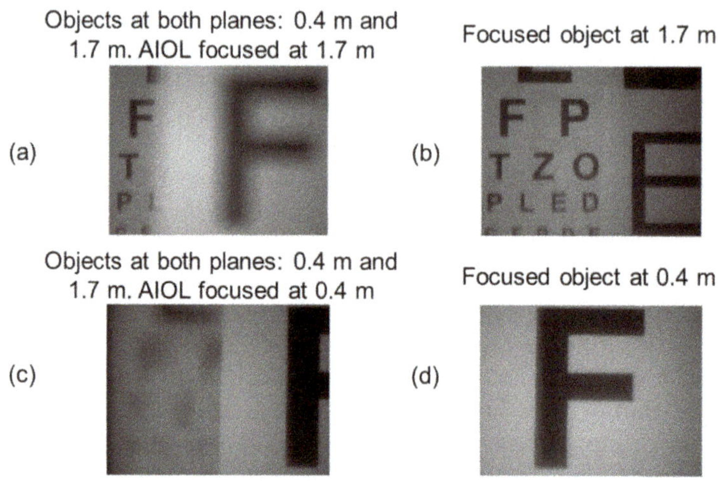

Figure 13.29. Images from planes at 1.7 m and 0.4 m, respectively [97].

13.5 Summary

This chapter presented an overview of a variety of mechanically adjustable lenses, especially in the context of AIOLs, covering single- and double-optic lenses featuring multiple designs and technologies. The ultimate target of these lenses is the full restoration of visual accommodation, after the explantation of the crystalline lens, free of undesired dysphotopic effects intrinsically caused by multifocal IOLs. A comparison between single- and dual-optic designs indicates that the extent of pseudophakic lens accommodation due to the action of the ciliary muscle may vary considerably among different patients. Lenses with other accommodation triggering and actuation mechanisms, with some degree of embedded electronics have also been addressed. From the optics perspective alone, the dual-optic approach provides higher accommodation amplitude within the feasible displacement range inside the eye.

To date, although reliable and high-performing AIOLs would be the holy grail of intraocular lenses, experience with most of the models developed so far has been limited to simulations, testbenches or small trials. The quest for a clinically available accommodative lens with full refractive aspheric optics, clinically meaningful accommodative power, biologically inert material, and compliant with a small incision, remains largely unmet. Recent improvements and new designs for EDoF, enhanced monofocal and multifocal IOLs have been shifting the investment away from the development of a true accommodative lens.

This chapter also presented a MEMS structure for an AIOL that can still benefit from many potential improvements that would not only enable a more compact version, but also higher mechanical stability and independence of the ciliary muscle. Further development of accommodative lenses will necessarily need a combination of multiples fields of science and engineering, as microelectronics, material engineering, optical engineering, and ophthalmic science, which have independently achieved impressive evolution in the past decades. Multidisciplinary development will, therefore, be the key to successfully address the ambition for accommodative lenses.

Chapter highlights

- Lenses with adjustable focus have been part of both traditional and modern devices and equipment featuring different mechanical and electrical actuation mechanisms.
- A predictable, reliable and high-performance IOL solution that mimics the natural physiological accommodation is yet to be made widely accepted and largely available.
- A variety of AIOLs have appeared in labs or on the market with either single or double-optic designs, whose focus adjustment relies on the movement of the ciliary muscle or on other vision cues.
- State-of-the-art microtechnologies and materials, such as microelectronics, micro-optics, MEMS, EAPs and SMAs, have been combined into promising prototype solutions.
- Simpler monolithic and static solutions as multifocal and EDoF IOLs have currently been the mainstream choices, partly countering the loss of the eye accommodative function.

References

[1] Kasthurirangan S and Glasser A 2006 Age related changes in accommodative dynamics in humans *Vision Res.* **46** 1507–19

[2] MORDI J A and Ciuffreda K J 2004 Dynamic aspects of accommodation: age and presbyopia *Vision Res.* **44** 591–601

[3] Hsieh C L, Chang Y H and Chen Y T *et al* 2018 Design of VCM actuator with L-shape coil for smartphone cameras *Microsyst. Technol.* **24** 1033–40

[4] Kobayashi M, Ohmura M, Takahashi H, Shirai T, Sakurai K, Ichikawa T, Yuzurihara H and Inoue S 2018 High-definition and high-sensitivity CMOS image sensor with all-pixel image plane phase-difference detection autofocus *Jpn. J. Appl. Phys.* **57** 1002B5

[5] CHOI S T, Lee J Y, Kwon J O, Lee S and Kim W 2011 Varifocal liquid-filled microlens operated by an electroactive polymer actuator *Opt. Lett.* **36** 1920–2

[6] Attota R K, Kang H, Scott K, Allen R, Vladar A E and Bunday B 2018 Nondestructive shape process monitoring of three-dimensional high aspect ratio targets using through-focus scanning optical microscopy *Meas. Sci. Technol.* **29** 125007

[7] Zhang Y, Bulling A and Gellersen H 2014 Pupil-canthi-ratio: a calibration-free method for tracking horizontal gaze direction *Proc. 2014 Int. Working Conf. on Advanced Visual Interfaces (AVI'14)* (New York: Association for Computing Machinery) pp 129–32

[8] Wojke N, Hedrich J and Droege D *et al* 2016 Gaze-estimation for consumer-grade cameras using a Gaussian process latent variable model *Pattern Recognit. Image Anal.* **26** 248–55

[9] Padmanaban N 2020 *Enabling Gaze-Contingent Accommodation in Presbyopia Correction and Near-Eye Displays* (Ann Arbor, MI: Stanford University. ProQuest Dissertations Publishing) 28104064

[10] LI J, Wang Y, Liu L, Xu S, Liu Y, Leng J and Cai S 2019 A biomimetic soft lens controlled by electrooculographic signal *Adv. Funct. Mater.* **29** 1903762

[11] Tonekaboni K and Whitsett A J 2005 The IOL horizon: accommodative intraocular lenses *Optometry* **76** 185–90

[12] SHEPPARD A L, Bashir A, Wolffsohn J S and Davies L N 2010 Accommodating intraocular lenses: a review of design concepts, usage and assessment methods *Clin. Exp. Optom.* **93** 441–52

[13] Qiao W, Johnson D, Tsai F S, Cho S H and Lo Y H 2009 Bio-inspired accommodating fluidic intraocular lens *Opt. Lett.* **34** 3214–6

[14] Bilbao-Calabuig R, González-López F, Amparo F, Alvarez G, Patel S R and Llovet-Osuna F 2016 Comparison between mix-and-match implantation of bifocal intraocular lenses and bilateral implantation of trifocal intraocular lenses *J. Refract. Surg.* **32** 659–63

[15] Alió J L, Alió del Barrio J L and Vega-Estrada A 2017 Accommodative intraocular lenses: where are we and where we are going *Eye and Vis.* **4** 16

[16] McLeod S D, Portney V and Ting A 2003 A dual optic accommodating foldable intraocular lens *Br. J. Ophthalmol.* **87** 1083–5

[17] Glasser A *et al* 2017 Special report: American Academy of Ophthalmology Task Force recommendations for test methods to assess accommodation produced by intraocular lenses *Ophthalmology* **124** 134–9

[18] Vdovin G, Loktev M and Naumov A 2003 On the possibility of intraocular adaptive optics *Opt. Express* **11** 810–7

[19] Lin H C, Chen M S and Lin Y H 2011 A review of electrically tunable focusing liquid crystal lenses *Trans. Elec. Electron. Mater.* **12** 234–40

[20] Holladay J T 2009 Multi-focal intraocular lens system and methods *Patent* US20090032679A1

[21] Gupta A, Mazzocchi R, Trip R, Schnell U and Fehr J N 2016 Electro-optical monofocal intraocular lens *Patent* US20160324628A1

[22] Helmholtz H L F 2000 *The Helmholtz's Treatise on Physiological Optics* 3rd edn vol 1 (New York: Thoemmes Continuum) pp 226–376

[23] Schachar R A 2006 The mechanism of accommodation and presbyopia *Int. Ophthalmol. Clin.* **46** 39–61
[24] Cumming J S, Colvard D M, Dell S J, Doane J, Fine I H, Hoffman R S, Packer M and Slade S G 2006 Clinical evaluation of the crystalens AT-45 accommodating intraocular lens: results of the U.S. food and drug administration clinical trial *J. Cataract Refract. Surg.* **32** 812–25
[25] Küchle M, Seitz B, Langenbucher A, Gusek-Schneider G C, Martus P and Nguyen N X 2004 Comparison of 6-month results of implantation of the 1CU accommodative intraocular lens with conventional intraocular lenses *Ophthalmology* **111** 318–24
[26] Mastropasqua L, Toto L, Falconio G, Nubile M, Carpineto P, Ciancaglini M, Di Nicola M and Ballone E 2007 Longterm results of 1 CU® accommodative intraocular lens implantation: 2-year follow-up study *Acta Ophthalmol. Scand.* **85** 409–14
[27] HermanS E A 2008 Geometry, optics and mechanical modelling of the humans lens during the accommodation (PrintPartners Ipskamp) Enschede ISBN: 978-90-8659228-9
[28] Chauncey F L 1983 Focussable intraocular lens *Patent* US4409691A
[29] Artal P, Findl O and Hirnschall N 2017 Accommodating intraocular lenses *Handbook of Visual Optics, Volume Two: Instrumentation and Vision Correction* (Boca Raton, FL: CRC Press—Taylor and Francis Group) ch 13
[30] Menapace R, Findl O and Kriechbaum K *et al* 2007 Accommodating intraocular lenses: a critical review of present and future concepts *Graefe's Arch. Clin. Exp. Ophthalmol.* **245** 473–89
[31] Uthoff D, Gulati A, Hepper D and Holland D 2007 Potentially accommodating 1CU intraocular lens: 1-year results in 553 eyes and literature review *J. Refract. Surg.* **23** 159–71
[32] https://bausch.com/ecp/our-products/cataract-surgery/lens-systems/crystalens-ao
[33] https://bausch.com/ecp/our-products/cataract-surgery/lens-systems/trulign-toric-iol
[34] DONG Z, Wang N L and Li J H 2010 Vision, subjective accommodation and lens mobility after TetraFlex accommodative intraocular lens implantation *Chin. Med. J.* **123** 2221–4
[35] Cleary G, Spalton D J and Marshall J 2010 Pilot study of new focus-shift accommodating intraocular lens *J. Cataract Refract. Surg.* **36** 762–70
[36] Preussner P R, Wahl J, Gerl R, Kreiner C and Serester A 2001 Akkommodatives linsenimplantat [accommodative lens implant] *Ophthalmologe* **98** 97–102 (in German)
[37] Menapace R 2006 Routine posterior optic buttonholing for eradication of posterior capsule opacification in adults: report of 500 consecutive cases *J. Cataract Refract. Surg.* **32** 929–43
[38] Alió J L, Plaza-Puche A B, Javaloy J and Ayala M J 2012 Comparison of the visual and intraocular optical performance of a refractive multifocal IOL with rotational asymmetry and an apodized diffractive multifocal IOL *J. Refract. Surg.* **28** 100–5
[39] Bohórquez V and Alarcon R 2010 Long-term reading performance in patients with bilateral dual-optic accommodating intraocular lenses *J. Cataract Refract. Surg.* **36** 1880–6
[40] Ossma I L, Galvis A, Vargas L G, Trager M J, Vagefi M R and McLeod S D 2007 Synchrony dual-optic accommodating intraocular lens. Part 2: pilot clinical evaluation *J. Cataract Refract. Surg.* **33** 47–52
[41] McLeod S D, Vargas L G, Portney V and Ting A 2007 Synchrony dual-optic accommodating intraocular lens. Part 1: optical and biomechanical principles and design considerations *J. Cataract Refract. Surg.* **33** 37–46
[42] Sergienko N 2008 Evaluation of the synchrony dual-optic accommodating intraocular lens *J. Cataract Refract. Surg.* **34** 8–9 author reply 9

[43] McLeod S D, Trager M J, Vargas L G and Portney V 2008 Reply: evaluation of the synchrony dual-optic accommodating intraocular lens *J. Cataract Refract. Surg.* **34** 9

[44] Tomás-Juan J and Murueta-Goyena Larrañaga A 2015 Axial movement of the dual-optic accommodating intraocular lens for the correction of the presbyopia: optical performance and clinical outcomes *J. Optom.* **8** 67–76

[45] Alvarez L W 1965 Two-element variable-power spherical lens *Patent* US3305294A

[46] Simonov A N, Vdovin G and Rombach M C 2006 Cubic optical elements for an accommodative intraocular lens *Opt. Express* **14** 7757–75

[47] Alió J L, Simonov A, Plaza-Puche A B, Angelov A, Angelov Y, van Lawick W and Rombach M 2016 Visual outcomes and accommodative response of the lumina accommodative intraocular lens *Am. J. Ophthalmol.* **164** 37–48

[48] Pepose J S, Burke J S and Qazi M A 2017 Accommodating intraocular lenses *Asia-Pacific J. Ophthalmol.* **6** 350–57

[49] Kohl J C, Werner L, Ford J R, Cole S C, Vasavada S A, Gardiner G L, Noristani R and Mamalis N 2014 Long-term uveal and capsular biocompatibility of a new accommodating intraocular lens *J. Cataract Refract. Surg.* **40** 2113–9

[50] Findl O, Kiss B and Petternel V *et al* 2003 Intraocular lens movement caused by ciliary muscle contraction *J. Cataract Refract. Surg.* **29** 669–76

[51] Zhang D Y, Lien V, Berdichevsky Y, Choi J and Lo Y H 2003 Fluidic adaptive lens with high focal length tunability *Appl. Phys. Lett.* **82** 3171–72

[52] Ben-Nun J and Alió J L 2005 Feasibility and development of a high-power real accommodating intraocular lens *J. Cataract Refract. Surg.* **31** 1802–8

[53] Jacob S 2020 Next-Generation IOLS https://eurotimes.org/next-generation-iols/

[54] Qiao W, Tsai F S, Cho S H, Yan H and Lo Y 2009 Fluidic intraocular lens with a large accommodation range *IEEE Photon. Technol. Lett.* **21** 304–6

[55] Portaliou D M, Kymionis G D, Panagopoulou S I, Kalyvianaki M I, Grentzelos M A and Pallikaris I G 2011 Long-term results of phakic refractive lens implantation in eyes with high myopia *J. Refract. Surg.* **27** 787–91

[56] Cumming J S, Slade S G and Chayet A 2001 Clinical evaluation of the model AT-45 silicone accommodating intraocular lens: results of feasibility and the initial phase of a food and drug administration clinical trial *Ophthalmology* **108** 2005–9 discussion 10

[57] Fine H I, Hoffman R S and Packer M 2007 Accommodating IOLs: Current Technology, Limitations, and Future Designs. https://aao.org/current-insight/accommodating-iols-current-technology-limitations-

[58] Liao X and Wilcox C D 2004 Method of manufacturing ophthalmic lenses made from hydrophobic acrylic polymers *Patent* WO2004029675A1

[59] IOL Technology on the Horizon 2010 *Review of Optometry* **147** SS15 https://link.gale.com/apps/doc/A219833998/AONE?u=capes&sid=bookmark-AONE&xid=667654cb

[60] Nishi O and Nishi K 1998 Accommodation amplitude after lens refilling with injectable silicone by sealing the capsule with a plug in primates *Arch. Ophthalmol.* **116** 1358–61

[61] Nishi O, Nishi K, Mano C, Ichihara M and Honda T 1997 Controlling the capsular shape in lens refilling *Archiv. Ophthalmol.* **115** 507–10

[62] Nishi O, Nishi Y, Chang S and Nishi K 2014 Accommodation amplitudes after an accommodating intraocular lens refilling procedure: *in vivo* update *J. Cataract Refract. Surg.* **40** 295–305

[63] Nishi Y, Mireskandari K, Khaw P and Findl O 2009 Lens refilling to restore accommodation *J. Cataract Refract. Surg.* **35** 374–82
[64] Kooijman L *et al* 2021 Is accommodation a confounder in pupillometry research? *Biol. Psychol.* **160** 108046
[65] Schneeberger N Electronic Autofocal Intraocular Lens (AF-IOL). https://helbling.ch/reference/electronic-autofocal-intraocular-lens-af-iol/?lang=en
[66] Donnenfeld E D 2013 An 'Autofocal' Accommodating IOL. https://crstoday.com/articles/2013-jun/an-autofocal-accommodating-iol/
[67] Fehr J N, Doll W and Schnell U 2013 Ophthalmic devices and methods with application-specific integrated circuits *Patent* CA2803128A1
[68] Doll W, Fehr J N, Gupta A, Schnell U and Trip R 2016 Systems and methods for power management of implantable ophthalmic devices *Patent* AU2013243919B2
[69] Boysset M 2020 Article of clothing and method for controlling cells using such an article of clothing *Patent* EP3923814A1
[70] The Ophthalmologist, The New Generation: Electronic IOLs 2019 https://theophthalmologist.com/subspecialties/the-new-generation-electronic-iols
[71] Simonov A N, Vdovin G and Loktev M 2007 Liquid-crystal intraocular adaptive lens with wireless control *Opt. Express* **15** 7468–78
[72] De Lima Monteiro D W, Ferreira Jr A I, Honorato F P, Loktev M, Soloviev O and Vdovin G 2005 Y-nano X-micro technologies: nanometric optical control *TechConnect Briefs* **3** 475–8
[73] Lin H C, Chen M S and Lin Y H 2011 A review of electrically tunable focusing liquid crystal lenses *Trans. Elec. Electron. Mater.* **12** 234–40
[74] Ahmed S N 2015 Position-sensitive detection and imaging *Physics and Engineering of Radiation Detection* 2nd edn (Amsterdam: Elsevier) pp 435–75
[75] ISO 11979-2:1999(E) First edition 1999-12-15 – Ophthalmic implants – Intraocular Lenses – Part 2: Optical properties and test methods 1999
[76] Joung Y H 2013 Development of implantable medical devices: from an engineering perspective *Int. Neurourol. J.* **17** 98–106
[77] Garner L J, Wilson L N, Lagoudas D C and Rediniotis O K 2000 Development of a shape memory alloy actuated biomimetic vehicle *Smart Mater. Struct.* **9** 673
[78] Miranda D O 2017 Desenvolvimento e estudo do efeito de diferentes nanoestruturas no desempenho de dispositivos baseados em hidrogéis contendo polímeros conjugados, UFMG
[79] Dragan E S 2014 Design and applications of interpenetrating polymer network hydrogels. A review *Chem. Eng. J.* **243** 572–90
[80] Carpi F and Smela E 2009 *Biomedical Applications of Electroactive Polymer Actuators.* (New York: Wiley)
[81] Eftekhari A 2010 *Nanostructured Conductive Polymers* (Cleveland, OH: Wiley)
[82] Soni S N, Bajpai J and Bajpai A K 2012 Investigation of electroactive behaviour of polyaniline containing polyelectrolyte nanocomposite membranes *Nanosci. Methods* **1** 164–82
[83] Miranda D O, Dorneles M F and Oréfice R L 2020 One-step process for the preparation of fast-response soft actuators based on electrospun hybrid hydrogel nanofibers obtained by reactive electrospinning with *in situ* synthesis of conjugated polymers *Polymer* **200** 122590
[84] Wang T, Farajollahi M, Choi Y S, Lin I T, Marshall J E, Thompson N M, Kar-Narayan S, Madden J D and Smoukov S K 2016 Electroactive polymers for sensing *Interface Focus* **6** 20160026

[85] Choi S T, Lee J Y, Kwon J O, Lee S and Kim W 2011 Varifocal liquid-filled microlens operated by an electroactive polymer actuator *Opt. Lett.* **36** 1920–2

[86] Kim J J, Liu H, Ousati A A and Jiang H 2020 Biologically inspired artificial eyes and photonics *Rep. Prog. Phys.* **83** 047101

[87] Rebello K J 2004 Applications of MEMS in surgery *Proc. IEEE* **92** 43–55

[88] Bogue R 2007 MEMS sensors: past, present and future *Sensor Rev.* **27** 7–13

[89] Dittrich P and Manz A 2006 Lab-on-a-chip: microfluidics in drug discovery *Nat. Rev. Drug Discov.* **5** 210–8

[90] Deboer C M, Lee J K, Wheelan B P, Cable C, Shi W, Tai Y C and Humayun M S 2016 Biomimetic accommodating intraocular lens using a valved deformable liquid balloon *IEEE Trans. Biomed. Eng.* **63** 1129–35

[91] Zhao Y, Nandra M S and Tai Y C 2011 A MEMS intraocular origami coil *2011 16th Int. Solid-State Sensors, Actuators and Microsystems Conf. (Beijing)* pp 2172–5

[92] Moura T, Tsukamoto T, Tanaka S, Filgueiras D and De Lima Monteiro D W 2014 Design of a dual lens of a dual lens system for a micromachined optical setup optical setup optical setup *29th Symp. on Microelectronics Technology and Devices (SBMicro)* pp 1–4

[93] Moura T, Tsukamoto T, De Lima Monteiro D W and Tanaka S 2015 Ring-shape SMA micro actuator with parylene retention spring for low power consumption, large displacement linear actuation *TRANSDUCERS 2015 18th Int. SolidState Sensors, Actuators and Microsystems Conf., 2015, Anchorage. 2015 Transducers* vol W2P pp 2148–51

[94] Moura T, Tsukamoto T, De Lima Monteiro D W and Tanaka S 2016 Hybrid MEMS-SMA structure for intraocular lenses *Sens. Actuators. A, Phys.* **243** 15–24

[95] Wang X 2007 *Crystallization and Martensitic Transformation Behavior of NiTi Shape Memory Alloy Thin Films* (Cambridge, MA: Harvard University)

[96] Bellouard Y 2008 Shape memory alloys for microsystems: a review from a material research perspective *Mater. Sci. Eng.* A **481–482** 582–89

[97] Moura T D O 2016 Dispositivo MEMS baseado em silicio, parileno e molas eletroativas para acomodacao optica *Thesis* PPGEE-UFMG

[98] Buckhurst P J, Wolffsohn J S, Davies L N and Naroo S A 2010 Surgical correction of astigmatism during cataract surgery *Clin. Exp. Optom.* **93** 409–18

[99] Kong H S, Glass J T and Davis R F 1986 Epitaxial growth of β-SiC thin films on 6H α-SiC substrates via chemical vapor deposition *Appl. Phys. Lett.* **49** 1074

[100] Meyerson B S 1994 Method and apparatus for low temperature, low pressure chemical vapor deposition of epitaxial silicon layers *Patent* US5906680A

[101] Shimabuko H H, Ramírez J C and Saito L A M 2020 Graphene-based capacitive structures fabrication for optical modulation devices *Frontiers in Optics/Laser Science* ed B Lee, C Mazzali, K Corwin and R Jason Jones *OSA Technical Digest (Optical Society of America)* paper FTu6B.8

Chapter 14

Intraocular optical spectroscopy: a proposal for Alzheimer's disease early diagnosis

Ado Jorio, Alexandre S Barbosa, Emerson A Fonseca, Lucas Lafeta, Leandro M Malard, Rafael P Vieira and Renan Cunha

The ocular media transparency provides a privileged condition for the search of retinal biomarkers of neurodegeneration, such as the amyloid-β peptide aggregation in Alzheimer's disease (AD). AD is a disabling, progressive neurodegenerative condition that can impact the vast majority of the aging world population. Nowadays its diagnosis is limited to tests that are only effective when the disease is already in an advanced later stage. In this chapter, we describe our initial efforts to develop an optical spectroscopy-based protocol for the early diagnosis of AD, by identifying the molecular fingerprints of amyloid-β peptide aggregation in the retina. Different optical spectroscopy and imaging methods are presented, with particular emphasis on Raman spectroscopy in brain tissues. This optical technique does not require the use of specific unnatural markers for molecular identification, so that various biomarkers, like proteins, lipids, nucleic acids, macromolecules can be imaged, therefore enabling a complete biochemical analysis. The achievements in *ex vivo* brain tissues and some technical aspects related to *in vivo* retinal spectroscopy are discussed, as well as the potential for a more general platform to study the eye as well as retinal and neurodegenerative diseases.

14.1 Introduction

The transparency of the ocular media enables the introduction of the 'intraocular spectroscopy' concept as a means to study not only the eye itself, including the cornea, lens, anterior chamber and vitreous body, but also neurodegenerative diseases, considering the retina and optic nerve extended from the diencephalon during the embryonic stage. The retina is a tissue composed by neuronal cells connected to the central nervous system, the ganglion cell layer composed by projections of neurons whose axons form the optic nerve, making the direct

doi:10.1088/978-0-7503-3263-7ch14

connection with the brain. The retina has a high density of synapses with a set of neurons working in fine-tuning to transduce the visual input to the lateral geniculate nucleus in the thalamus and the superior colliculus in the midbrain. Therefore, the retina reflects structural changes that occur in neural tissues [1]. This chapter describes our efforts to develop a diagnosis method able to detect the early stages of Alzheimer's disease, which is already well studied in *ex vivo* brain tissues, and it also presents the first steps of the method to be in future applied in *in vivo* intraocular optical spectroscopy.

The increasing need for Alzheimer's disease (AD) biomarkers relies on the fact that, in the last decades, there has been a significant extension in the life expectancy of the population, which brings along a higher incidence of age-related diseases. Dementia is a final outcome of a group of neurodegenerative disorders and is already the fifth leading cause of death worldwide in people aged 65 and older. Currently, AD represents 60% to 80% of dementia cases, characterized by progressive cognitive impairment, which causes difficulty in memory, language, and other abilities that affect daily routine activities. The recent research proposes AD as a *continuum* pathophysiology process, starting with molecular alterations evolving to mild cognitive impairment, culminating in the severe stage of the disease, and finally, death [2]. The symptoms of AD are, therefore, detected 15–20 years after the beginning of the neuropathological process [3], so that the diagnosis generally comes when the modifying phase of AD is in an advanced stage. This picture complicates the development of new treatments, stressing the importance of an early diagnosis method that can be broadly utilized.

New approaches for earlier interventions, based on identifying biochemical hallmarks *in vivo* to pave the way to an ideal early diagnostic context, include cerebrospinal fluid analysis and neuroimaging techniques, such as magnetic resonance imaging (MRI) and positron emission tomography (PET) scan. However, these techniques exhibit different limitations that forbid their use as widespread routine tests. Reduced levels of $A\beta42$ and $A\beta42/A\beta40$ ratio in cerebrospinal fluid are consistent with fully established AD but lack sensitivity in early AD diagnosis. The main structural findings in MRI are hippocampal atrophy, medial temporal lobe atrophy and diffuse cortical loss, however, such features also presented low specificity in early AD. A more detailed imaging of subcortical microstructure in AD has been obtained with diffusion MRI (dMRI), which estimates anisotropy and diffusivity based on the motion of molecules of water. dMRI provides evidence of distinct patterns of neural tracts involvement in AD. Still, although dMRI is a promising technique, its resolution hampers its use in AD early diagnosis. Similarly, amyloid loading obtained with PET techniques is inconsistent in early AD. Contrastingly, due to the easy access of the retina for imaging, it presents itself as the window to the brain, and our proposal is to monitor the incidence and evolution of AD using highly sensitive and highly specific intraocular Raman spectroscopy, starting an effort for early diagnosis methods via intraocular examination.

To address intraocular spectroscopy and its application to the early diagnosis of AD, this chapter is organized as follows: section 14.2 introduces the fundamental aspects of retinal spectroscopy, including some principles of optical spectroscopy, with a focus on Raman (section 14.2.1) and its applications in biomedicine (section 14.2.2), also addressing the general aspects of intraocular spectroscopy (section 14.3.3); section 14.3 presents the applied aspects of retinal Raman spectroscopy. It includes a short discussion of the biochemical aspects of the AD to structure the bases for the main results (section 14.3.1), which are the Raman spectral imaging of amyloid plaques (section 14.3.2), ending with an early-stage practical implementation of retinal spectroscopy (section 14.3.3). Section 14.4 delves into intraocular spectroscopy as a platform for studying structures of the anterior segment of the eye and vitreous (section 14.4.1) and followed by the main retinal findings obtained in ophthalmologic and neurologic conditions (section 14.4.2); in section 14.5 we present our perspectives for the evolution of this work.

14.2 Fundamental aspects of retinal spectroscopy

One of the pathological characteristics of AD is extracellular depositions of amyloid-β peptide (Aβ), forming Aβ plaques associated with neurofibrillary tangles, which are, to date, the main research targets in the field, potentially a marker for diagnosis. The deposition of Aβ takes years to happen, and it is cumulative. Recent works report visual alteration in AD patients and well-established changes in the retina, such as thinning of the nerve fiber layer, narrowing of blood vessels, and reduction of blood flow, in addition to the presence of amyloid plaque and neurofibrillary tangles [4]. The presence of amyloid plaques in human retina *in vivo*, and *post-mortem* with a predominance of amyloid plaques with dimensions of 5–15 μm, mostly in the ganglion cell layer, vessel walls and perivascular regions has been shown (see figure 14.1) [5]. This set of results reinforces the argument that retina examination can be a way to early diagnosis and to monitor the evolution of the AD. In fact, it was found that Aβ deposition in the retina occurs months before the appearance of plaques in the brain in a transgenic mouse model of AD [6]. Approaches such as *in vivo* using Aβ fluorescent markers on the retina [5, 6], tissue reflectance spectrum analysis [7], and fluorescence life-time imaging ophthalmoscope [8] have already started this promising race towards early diagnosis of AD through retinal examination.

14.2.1 Light–matter interaction and Raman spectroscopy

We are interested in a tool capable of providing a broadly accessible, high-resolution, specific, rapid, and clinically safe identification of a natural marker, i.e., a molecular fingerprint of the AD incidence and evolution, more specifically, the identification of Aβ plaques. The main purpose of this section is to show how certain types of light–matter interaction satisfy these criteria and allow us to explore AD biomarkers.

When light reaches matter, transmission, absorption, and reflection take place. However, this is a simple view of different phenomena, including photoluminescence (also broadly known as fluorescence or phosphorescence) and light scattering. Photoluminescence is a process in which light with a color corresponding to a given higher frequency is absorbed, and the material emits light back at lower frequencies

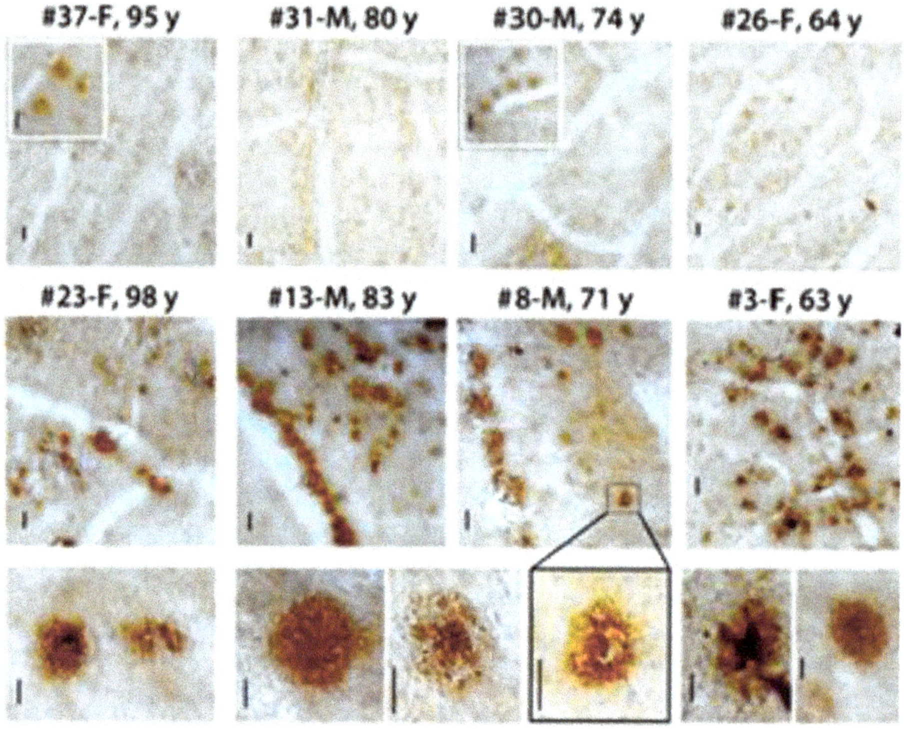

Figure 14.1. Amyloid-β (Aβ) aggregation in post-mortem retinas from AD patients. On the top, retinal micrographs from controls, matched with AD patients according to sex and age. In the middle, amyloid plaques were labeled with specific antibodies and highlighted with peroxidase. On the bottom, high-magnification images of Aβ aggregates. Scale bar: 20 μm (top and middle) and 10 μm (bottom). Adapted from [5] (open access journal) with permission from *The American Society for Clinical Investigation.*

(red-shifted) [9]. For this process to take place, the incident light absorption promotes electrons from their fundamental states in the molecule to high energy excited states; the electrons decay to the lower excited state by transferring energy to the material, and from this lowest excited level, the electrons decay back to the fundamental state, emitting light. The time delay between absorption and emission varies significantly from one material to the other, and it is frequently utilized to classify the type of photoluminescence (e.g., fluorescence: delays from pico- to nanoseconds; phosphorescence: delays from milliseconds to hours). This process can be induced by the absorption of one single photon or more. The latter case is regarded as a nonlinear optical process, since the emission does not scale linearly with the incoming power, but quadratically in the case of two-photon photoluminescence, cubically in the case of three-photon photoluminescence, and so on. This nonlinear effect is widely applied in microscopy since it allows higher resolution imaging, optical sectioning and background suppression. The two-photon excitation fluorescence (TPEF) microscopy technique will be of particular interest in section 14.3.4, as a practical implementation of intraocular imaging. Furthermore, since photoluminescence involves electronic transitions, the energy of the incident light must be compatible with the energy of these transitions. Usually, photoluminescence-based diagnosis methods use external compatible highly fluorescent molecules (markers) to identify specific molecules that are evidenced by the specific marker. However, some biological tissues are formed by biomolecules that can fluoresce (autofluorescence), so that external markers are not necessary, and the acquisition of the emission spectrum could, in principle, allow the identification of these biomolecules. Autofluorescence is important to dismiss the external agent, reducing errors, cost and health risk.

Light scattering is a process in which light–matter interaction causes a change in direction, and possibly also in the frequency of incident light. The process is regarded as an elastic scattering if the scattered frequency remains unchanged, and as an inelastic scattering otherwise [10–13]. The fundamental difference between scattering and photoluminescence resides in the fact that the entire scattering process is nearly instantaneous, with the time delay between absorption and emission in the order of femtoseconds. The incoming light turns into the outgoing photons coherently, without the incoherent decay of electrons from higher to lower levels. Consequently, the scattered light is very well-defined in energy, providing a much more accurate spectral fingerprint than photoluminescence, with sharp peaks compared to broad energy emissions. This is a key feature for applying spectroscopy in AD, as the acquisition of the scattering spectrum allows highly reliable identification of biomolecules. Here we focus on Raman scattering by molecules, a technique that provides very accurate spectral fingerprints with excellent chemical specificity, based on the inelastic scattering of light resulting from the coherent exchange of energy between the photons and the quanta of molecular vibrations [14, 15]. A set of vibrational mode frequencies provides specific indication of the elemental composition, chemical bonds, physical molecular structure and environmental interferences.

Microscopically, the Raman scattering takes place as follows: when a monochromatic light field (e.g., a laser, represented in figure 14.2(a) by the incident green arrow) interacts with a molecule (figure 14.2(b)), the electric field of light displaces the molecular electric charges, orienting existing dipoles or inducing new ones. In such a coupling, light predominantly undergoes elastic scattering (Rayleigh scattering, scattered green arrow in figure 14.2(a)). However, if the polarizability (ability to be polarized) of the molecule changes along with the molecular vibration, light can undergo inelastic spontaneous Raman scattering (SpRS). Figure 14.2(c) shows the energy diagram for the three possible scatterings: light loses energy and is scattered at the red-shifted Stokes frequency ω_S (red arrow in figure 14.2(a)), leaving the molecule in an excited vibrational state; light gains energy and is scattered at the blue-shifted anti-Stokes frequency ω_{aS} (violet arrow in figure 14.2(a)), leaving the molecule in the ground vibrational state, subjected to the condition that it was initially in an excited state—these are the Stokes (*S*) and anti-Stokes (*aS*) Raman scattering processes, respectively; lastly, light is emitted at the same frequency of the incoming pump light ω_P, and this process is named Rayleigh (elastic) scattering. Figure 14.2(d) shows schematically the intensity of each of these effects, which is regarded as a measure of the probability for each interaction. An interaction in which no net energy exchange

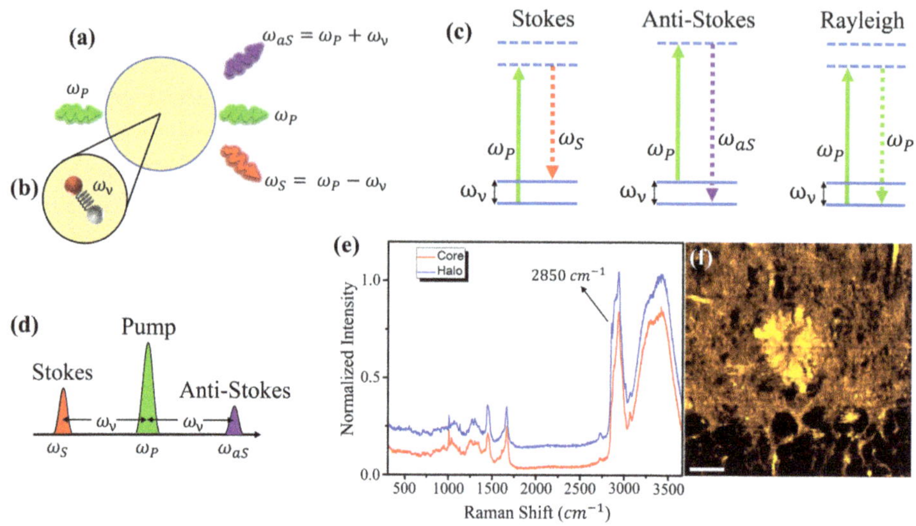

Figure 14.2. Pictorial representation and characteristics of spontaneous Raman scattering. (a and b) Light interacts with a molecule, undergoing spontaneous Raman scattering, characterized by the production of two new frequencies: Stokes (red arrow) and anti-Stokes (violet arrow). It also undergoes Rayleigh scattering (green arrow); (c) energy diagrams of the interactions represented in (a); (d) schematic representation of the intensities (and, therefore, event probability) of each effect represented in (a); (e) Raman spectra of an Aβ plaque (at the core and at the surrounding lipidic halo) and identification of the frequency used to acquire the hyperspectral Raman image of an Aβ plaque as shown in (f) based on the intensity of the scattered light at Stokes Raman shift 2850 cm^{-1}, which highlights the lipidic halo surrounding the Aβ plaque (at the center, not highlighted in this image), reproduced from [16] with permission from the *Royal Society of Chemistry*. Scale bar 20 μm.

occurs (Rayleigh scattering) is much more likely than an interaction in which light transfers energy to the molecule (Stokes Raman scattering) and even more likely than an interaction in which the molecule transfers energy to light (anti-Stokes Raman scattering). In practical terms, this means that the light will be emitted mostly by Rayleigh scattering, and the most intense response of the material will be a beam with the same frequency as the incident light. As will be discussed in more detail in the instrumentation described in section 14.3.1, when necessary, this problem is overcome with the addition of a filter blocking the frequency of the incident light in the detection.

Figure 14.2(e) shows the Stokes Raman spectrum of an amyloid plaque. Raman spectra plot the scattered light intensity as a function of the Raman shift, given in cm^{-1}, obtained using the formula $\omega_{S,aS} = 10^7(1/\lambda_{S,aS} - 1/\lambda_P)$, where λ_P and $\lambda_{S,aS}$ are the wavelengths of the incident (pump laser) and scattered (*S* or *aS*) light, respectively, in meters. This awkward unit cm^{-1} is actually very appropriate for vibrational spectroscopy, since most vibrations lie in the 10–4000 cm^{-1} spectral range. To work with Hz, for example, one would carry uncomfortable numbers, considering 1 $cm^{-1} = 2.9979\ldots \times 10^{10}$ Hz. Furthermore, usually, spectrometers have accuracy in the order of 0.1–10 cm^{-1}, again, making this unit appropriate for handling the numbers. By analyzing the vibrational modes of known molecules, it is possible to extract information about the chemical distribution and structure of the material, as well as to identify spectroscopic fingerprints that can serve as photonic biomarkers of AD. In addition, Raman spectroscopy (light as a function of Raman shift) can be combined with confocal microscopy (light as a function of position) to generate hyperspectral images, such as the lipid-based image of the Aβ plaque shown in figure 14.2(f). This image is built based on the intensity of the spectral peak at 2850 cm^{-1}, which is the frequency of a specific vibrational mode present in lipids. This demonstrates the informational capacity provided by SpRS, where we can access not only spectral biochemical information but also identify structural changes in a hyperspectral image. However, despite SpRS providing a highly accurate spectral fingerprint for molecular specific imaging, its most serious drawback is the low intensity of the signal. In a typical case, 1 in 10^8 photons of the incident radiation undergo SpRS [15], so that the data for spectral maps like the one in figure 14.1(f) take hours to be acquired. This has to be overcome to make this technique applicable for diagnosis proposals.

To get around the low intensity of SpRS, it is possible to implement a coherent excitation that amplifies the Raman interaction [14]. Consider the situation in which two light fields at ω_P and ω_S, and not just one at ω_P, interact with the molecule, as in figures 14.3(a) and (b). If the fields are spatially and temporally overlapped, and if the frequency difference ($\omega_P - \omega_S$) matches (is resonant with) a vibrational mode frequency ($\omega_P - \omega_S = \omega_\nu$), where ω_ν is a natural vibration frequency of the molecule in cm^{-1}, then the interaction generates highly enhanced Raman-based effects. Figure 14.3(c) shows the energy diagram of the process called *coherent anti-Stokes Raman scattering* (CARS), where two photons of frequency ω_P are absorbed while one incoming photon of energy $\omega_S = \omega_P - \omega_\nu$ coherently excite the ω_ν-vibrational mode, thus generating a high-frequency output at the anti-Stokes frequency $\omega_{CARS} = 2\omega_P - \omega_S$.

Differently, stimulated Raman scattering (SRS) is an exchange of energy between the incident fields at ω_P and ω_S, mediated by the exchange of energy with the material at ω_ν. As the diagram in figure 14.3(d) shows, this dissipative effect is characterized by the transformation of ω_P into ω_S. The main difference between SRS and the Stokes scattering in figure 14.2(c) is that the presence of a strong second field at ω_S, absent in the SpRS, stimulates the conversion of ω_P into ω_S (represented schematically here by a solid-line down-arrow in figure 14.3(d), rather than the dotted-line in figure 14.2(c); in other words, solid-line for stimulated and dotted-line for spontaneous decay), while the energy difference proportional to ($\omega_P - \omega_S$) is transferred to the material. CARS and SRS can be several orders of magnitude more likely than SpRS [12, 14], thus overcoming the main intensity limitation of Raman scattering. This enhancement is reflected in figures 14.3(e) and (f), where a region is imaged with CARS and SRS within a few seconds, in drastic contrast to the image of the same region acquired with SpRS, in figure 14.2(f), which took 20 h. Such features make CARS and SRS ultrafast Raman-based techniques that can be implemented with near-infrared excitation lasers, thus suitable for intraocular diagnostic applications.

Although the CARS was prepared to probe the same vibrational frequency associated with lipids (2850 cm^{-1}) in the Raman image of figure 14.2(f), the predominant proteic core of the plaque in figure 14.3(e), absent in figure 14.2(f), appears illuminated as a consequence of the non-resonant background. Such a

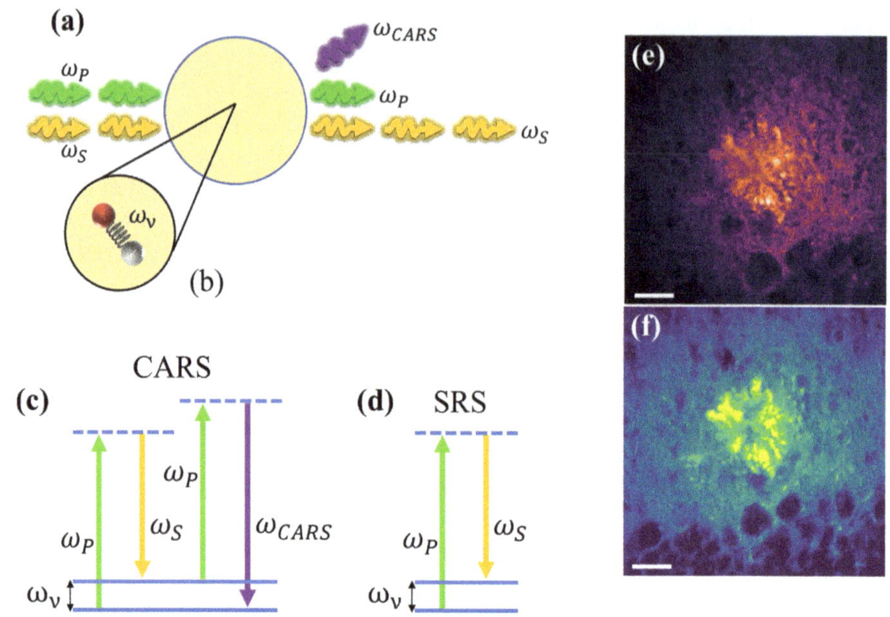

Figure 14.3. Pictorial representation and characteristics of the coherent Raman scattering. (a) and (b) The light of two different frequencies interacts with a molecule, undergoing coherent Raman scattering, characterized by the production of new effects: CARS (violet arrow) and SRS (increase in ω_S light intensity, yellow arrow and decrease in ω_P light intensity, green arrow); (c) and (d): energy diagrams of the interactions represented in figure 14.3(a); (e) CARS and (f) SRS images of the same region in figure 14.2(f), reproduced from [16] with permission from the *Royal Society of Chemistry*. Scale bar 20 μm.

background is a result of nonlinear competing processes at the same frequency, causing interferences that distort CARS intensity, preventing a perfect match with the natural SpRS. Additionally, the intensity in a CARS image can also be affected by one and two-photon fluorescence of autofluorescent molecules, which further adds to the difficulty of making a quantitative analysis from the CARS image.

On the other hand, the SRS signal occurs at the same frequency as the incident fields, so both fields are present in the detection, ω_P and ω_S [17]. The SRS signal is often extracted from the Rayleigh signal employing modulation transfer and lock-in amplification detection schemes. As a consequence, differently from CARS, the signal measured in the SRS is proportional to the number of scatterers [17], which allows a simpler quantitative interpretation of an SRS image, i.e., the SRS image is typically a trustworthy chemical map of the sample, where the intensity is linearly related to the molecular concentration. Furthermore, the non-resonant background present in CARS is not observed in SRS, which shares virtually an identical spectrum to that of SpRS, both identifying spectroscopic fingerprints that can serve as photonic biomarkers.

Achieving the criteria for an adequate diagnostic tool, label-free optical spectroscopic techniques are promising candidates. While multiphoton photoluminescence offers high-resolution and speed, it cannot retrieve chemical information. On the other hand, Raman spectroscopy allows the fingerprint identification of materials without extrinsic labels through its chemically specific nature. Coherent Raman techniques, such as CARS and SRS, allow for a gain of orders of magnitude in the probability of the Raman effect, making Raman spectro-microscopy a high-resolution, ultrafast, non-destructive, and non-invasive technique, suitable for biological and medical applications.

14.2.2 Raman spectroscopy application in biomedicine

The application of Raman spectroscopy in biology and medicine is an important relatively recent effort, made possible by the technological evolution of lasers, optics and detectors. It is evolving rapidly due to the high degree of specificity of the technique, and its non-destructive and non-invasive nature, as it only requires a safe beam of light near the visible range. Biological tissues are composed of a collection of different molecules exhibiting complex structures, which can be revealed by SpRS. In addition, the SpRS does not require prior processing of the tissue (or cell), such as fixation in formaldehyde, paraformaldehyde or glutaraldehyde [18]. There is a vast literature on the fields of biology and medicine utilizing SpRS *in vitro*, *ex vivo* and *in vivo*, for the diagnosis and the study of diseases in several tissues [19]. In cell biology studies, SpRS allows imaging of cells, intracellular structures and molecules, without the need for specific molecular markers, like in fluorescence methods.

For instance, SpRS has been widely used to identify tumor tissue, both *ex vivo* and *in vivo* [20]. All the biochemistry changes result in a spectral signature that makes it possible to distinguish, with satisfactory precision and spatial resolution, regions with and without tumor cells. The Raman-based fiber optic probe technique enables the detection of tumors *in situ* in real time in patients, giving rise to different proposals for *in vivo* Raman-based diagnosis for tumors [20]. Jermyn *et al* used a fiber optic probe attached to the SpRS to identify glioblastoma during surgical resection of brain tumor. The authors report an accuracy of 92% and specificity of 91%. The authors claim that the Raman probe was able to detect the presence of only 17 cancer cells/0.0625 mm^2 [21], showing its applicability as a highly sensitive probe.

The use of SpRS has gained a new phase with CARS and SRS. The coherent and stimulated Raman techniques, in addition to label-free and high-resolution imaging acquisition, adds the possibility of ultrafast acquisition images of cells and tissues [22]. Both techniques use infrared, which has a greater capacity to penetrate biological tissues and can reach down to 3 cm, depending on the wavelength and tissue, making the technique extremely promising in the use of biological studies and biomedical applications [23]. It is possible to conduct diffusion studies of chemical compounds in tissues and cells, to observe blood flow in non-invasive microscopy, to perform histological analysis of fresh tumor tissue, and a series of other approaches in the study of cell biology, including complex studies such as regulation of energy balance [24].

14.2.3 General aspects of intraocular spectroscopy

Several groups have reported high quality retina images using fluorescence. Fluorescence imaging is already used in commercial systems such as scanning laser ophthalmoscopy (SLO). These systems use a set of high-speed scanning mirrors in order to focus the laser at different positions in the retina, and the backscattered fluorescence is detected by a combination of appropriate band pass filters, detected by photomultiplier tubes or avalanche photodiodes. Furthermore, the system can be combined with adaptive optics in order to correct aberrations inherent to the eye, to make high-resolution images of the living retina. Such SLO systems are very similar to conventional scanning laser microscopy setups commonly used in biology, physics and other fields. Therefore, the modification of the SLO setup allows different optical microscopy modalities to be employed. For example, two-photon excitation fluorescence has been developed to image and study retina with different advantages compared to one-photon fluorescence [25–27]. As the two-photon microscopy is based on a nonlinear optical effect, it offers higher spatial resolution, including depth resolution. Moreover, the use of a near-infrared laser is safer to the retina tissue, since the tissue is almost transparent for these wavelengths, and allows imaging of the retina with three-dimensional resolution. Different groups have been able to demonstrate high quality retina imaging by two-photon microscopy, with and without adaptive optics [28, 29], or with eye tracking software. Figure 14.4(a) shows a typical retina image made by one- and two-photon excitation fluorescence, for comparison [25]. Both images show a high-quality image of a mouse retina, with

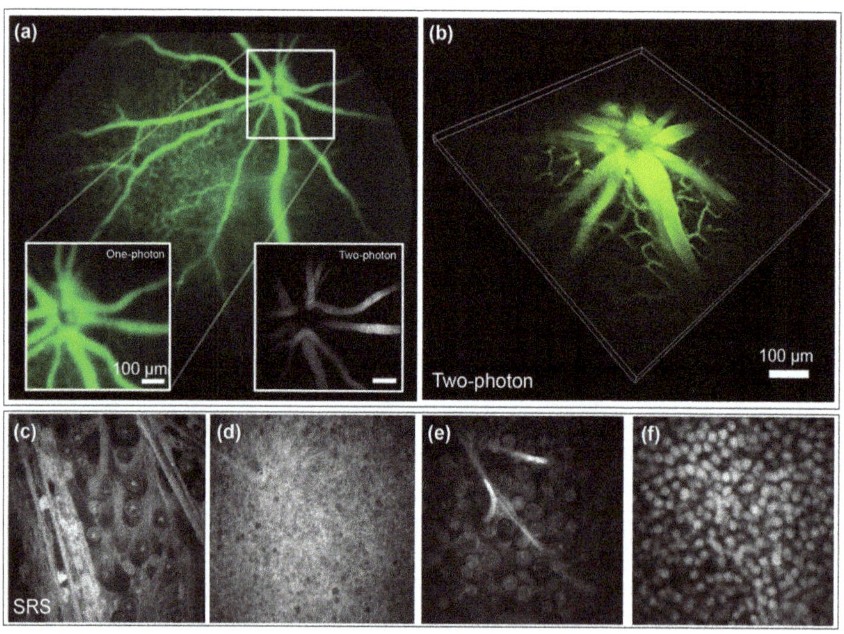

Figure 14.4. (a) Main panel: One-photon angiogram fluorescence image of *in vivo* mouse retina. Lower insets: zoom made with one- and two-photon fluorescence for comparison. (b) Three-dimensional image performed by two-photon angiogram fluorescence. Adapted from [25] (open access journal) with permission from the *Nature Publishing Group*. (c)–(f) SRS images of *in vitro* mouse retina taken at different layers: ganglion cell layer, inner plexiform layer, inner nuclear layer and outer nuclear layer, respectively. Adapted from [27] (open access journal) with permission from *The Optical Society, OSA*.

the advantage of the two-photon image being able to reconstruct a three-dimensional image by varying the focus depth, as shown in figure 14.4(b).

However, as already mentioned before, the use of fluorescence microscopy is limited due to the use of fluorescent molecular probes, hence progress has been made by using Raman spectroscopy, in special CARS and SRS to study the retina. CARS microscopy [26] has been used to image mouse retina with high spatial and depth resolution along the different layers of the retina. SRS has also been shown to be able to image mouse retinas [27] in different layers, as shown in figures 14.4(c)–(f). The SRS was compared to fluorescence images confirming that the stimulated Raman technique is adequate. Therefore, SLO systems with the modifications made to use pulsed near-infrared lasers, as shown with the implementation of two-photon excitation fluorescence, can be further adapted for CARS or SRS microscopies. The greatest difference is that these two stimulated Raman techniques use two different laser wavelengths that need to be overlapped in time and space. Clearly, the instrumentation for experiments of CARS and SRS in microscope setups is sufficiently developed to be adapted in the SLO systems.

In fact, Raman has been applied to ophthalmology for quite some time, for example for bovine lens protein, collagen of intact feline corneas, *in vivo* intraocular Raman spectroscopy in rabbits, and in humans [30]. Other studies performed *in vivo* intraocular Raman in humans: Evans *et al* used OCT microscope for *ex vivo* retina imaging combined with Raman spectroscopy [31]; in 2013, Masihzadeh *et al* applied CARS microscopy for label-free *ex vivo* retina imaging [32] and cornea [33]; in 2015, He *et al* used SRS to acquire high resolution images of mice retina (*ex vivo*) [34]; Berstein *et al* studied the carotenoid levels in retina by the use of resonant intraocular Raman spectroscopy in animals [35] and human eyes [36]. The drawback of their approach is that they are limited in studying only biomolecules whose absorption peaks match the laser energy utilized. In the case of proteins, such as amyloid-β, the absorption maximum is in the UV range, therefore it is not feasible to use resonance Raman scattering. More recently, it was reported the feasibility of using non-resonant Raman spectroscopy with power levels in accordance with safety regulations [37, 38], therefore enabling its use in the living retina. One of the key aspects of this work was to use novel high throughput spectrographs and high quality dichroic and long band pass filters for detecting the Stokes part of the Raman spectra.

14.3 Technical aspects of retinal spectroscopy

14.3.1 Instrumentation for intraocular spectroscopy

In section 14.2 we discussed the fundamentals of optical effects suitable to design diagnostic applications, specifically in the study of Alzheimer's disease and the advantages of Raman spectroscopy. In this section, we address the instrumental aspects more directly, discussing the experimental implementation of each effect. Figure 14.5 shows a setup designed to implement all the already mentioned optical spectroscopic techniques: spontaneous Raman spectroscopy (SpRS) and imaging, two-photon excitation fluorescence (TPEF), coherent anti-Stokes Raman spectroscopy (CARS) and stimulated Raman spectroscopy (SRS).

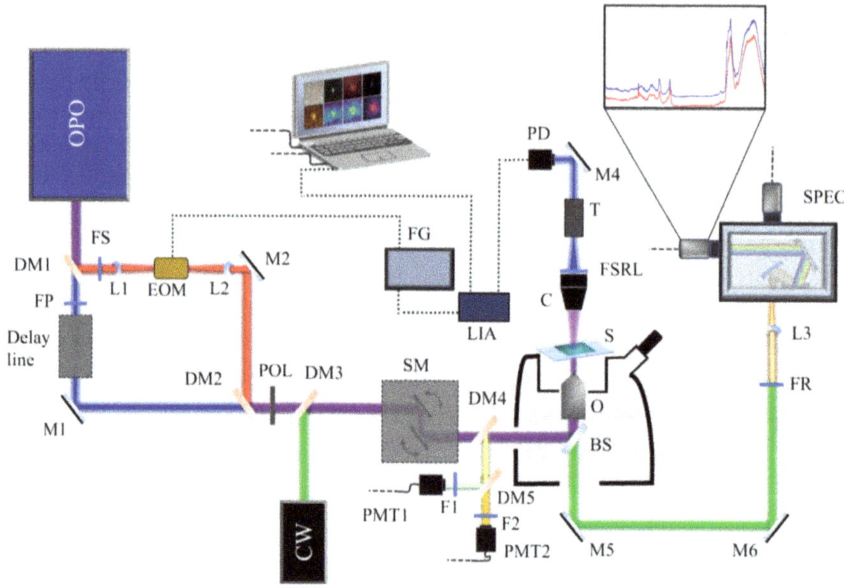

Figure 14.5. Experimental setup capable of implementing photoluminescence, TPEF, SpRS, CARS and SRS measurements (more information in [16]).

For SpRS, a continuous-wave (CW) laser is generally used for the excitation. The excitation beam is directed to the scanning mirrors SM through the dichroic mirror DM3, although a sample scanning stage can be alternatively used to generate the spectral imaging. The part reflected by the beam splitter BS is focused on the sample by the objective O and the backscattered response signal is transmitted by the BS and directed to the spectrometer SPEC through the mirrors M5 and M6. As mentioned in section 14.2.1, most of this light–matter interaction produces Rayleigh scattering at the same wavelength as the incident beam. In an ideal case, the blocking filter FR ensures that only the Raman signal is transmitted, eliminating the Rayleigh scattering. The lens L3 focuses the Raman signal on the spectrometer SPEC, which allows the spectrum to be recorded.

For TPEF microscopy, higher input excitation is needed, and a femtosecond pulsed laser is used. Pulsed lasers generate high excitation powers compressed in time, important for enhancing the nonlinear effect while keeping the average power low enough to protect the tissue against burning. An optical parametric oscillator (OPO) can provide such a laser beam with tunable wavelength. In figure 14.5, the excitation beam is transmitted through the dichroic mirrors DM1 and DM2 (following the path of the purple beam) and it is focused in the sample (S) by the objective O, which must be of high numerical aperture to achieve high spatial resolution. The sample imaging is performed through scanning the mirrors SM. The dichroic mirrors DM4 and DM5 direct the reflected signal generated by the material to the photomultiplier PMT1. Bandpass filter F1 can be inserted before the PMT1, allowing an unambiguous measurement of the effect. As discussed in section 14.2.1, photoluminescence is generated in broad emissions. This causes the different signals produced by different biomolecules in a biological tissue to overlap, increasing the difficulty of assigning a biochemical origin to each signal arriving at the PMT1, even with band pass filters. Consequently, although TPEF microscopy allows for label-free, rapid imaging, it is limited for fingerprint identification of photonic biomarkers; CARS and SRS, Raman-based nonlinear effects are best suited to accomplish this task.

For the implementation of CARS and SRS, the optical parametric oscillator OPO providing two picosecond pulsed lasers is more suitable than using continuous (CW) lasers. In this case, the OPO must also provide at least two different wavelengths, in which one is tunable, so that the frequency difference between the beams is resonant with different vibrational modes of the material to be probed. Figure 14.5 shows two such excitation beams: a tunable pump beam (blue) and a fixed Stokes beam (red). Tuning the wavelength of the pump beam controls the resonance with the vibrational mode. The dichroic mirror DM1 splits the beams in a Mach–Zehnder configuration where the blocking filters FP and FS ensure no contamination of the pump beam in the arm of the Stokes beam and vice-versa. The electro-optical modulator EOM connected to the function generator FG produces a high-frequency ($\geqslant$10 MHz) polarization-modulation in the Stokes beam. The dichroic mirror DM2 recombines both beams in a collinear configuration. Temporal synchronization is achieved through a delay line. The polarizer POL after the recombination of the beams transforms the EOM polarization-modulation into amplitude-modulation and ensures that both the unmodulated (pump) and modulated (Stokes) beams have the same polarization state. This modulation process is important for the SRS. Since the SRS signal occurs at the same wavelength as the Rayleigh scattering (see section 14.2.1), the latter cannot be eliminated by an optical filter, as in SpRS and CARS. In this case, an electronic filter through this modulation process is necessary. The scanning mirrors SM allow the beams to map the sample when focused by the high numerical aperture apochromatic objective O in the microscope. The backscattered CARS signal is then directed to the photomultiplier PMT2 by the dichroic mirrors DM4 and DM5. A band pass filter eliminates signal contamination by TPEF. The forward SRS signal is collected by the high numerical aperture collimator C and the blocking filter FSRL eliminates the (modulated) Stokes component of the beam. It is important to note that here we are schematically representing the forward SRS, but depending on the application, e.g., collecting signals from the retina, the backward SRS signal can also be detected. The telescope T reduces the beam movement on the detection area of the photodiode PD, where the SRS signal is extracted with the help of the lock-in amplifier LIA coupled to the PD, and connected to the EOM via the synchronization port of the FG. This synchronization allows that only the pump component modulated after interaction with the Stokes beam is amplified at the detection stage. The pump component that remained unmodulated did not simultaneously interact with the sample and, therefore, it is the Rayleigh scattering. This setup allows the SRS signal to reach a high signal-to-noise ratio and produces label-free, high-resolution and ultrafast images of Aβ plaques [16].

14.3.2 Biochemistry of Alzheimer's disease

Due to the high molecular selectivity, Raman spectroscopy is a powerful tool to probe and distinguish different biological macromolecules, such as lipids, proteins and nucleic acids, enabling optical images with chemical specificity [24]. Therefore, to develop a Raman spectroscopy-based diagnosis method for a specific disease, we have to establish the biochemical aspects related to its development. The molecular basis of AD comprises a plethora of biochemical and cellular events such as protein misfolding, hyperphosphorylation, oxidative stress, metal ions imbalance and neuroinflammation [39]. In combination, these events culminate in neuronal death and brain atrophy. Among these factors, amyloid-β (Aβ) peptide misfolding is one of the most studied processes.

Considering the amyloid hypothesis, the Aβ formation, misfolding and aggregation are crucial steps to the possible causes of the disease. Aβ is formed from the amyloid precursor peptide (APP) which is expressed in cells of the central nervous system and suffers proteolytic cleavage. From this point, Aβ monomers can form soluble oligomers, fibers and large aggregates in extracellular medium [40]. Due to the relevance of Aβ aggregation as a biochemical hallmark in AD and the possibility to characterize these events in the first years, the interest in developing early-stage diagnostic methods based on non-invasive spectroscopic techniques have significantly risen in recent years. In 2018, Palombo *et al* presented an unequivocally hyperspectral Raman image of amyloid plaque [41]. In the same year, Ji *et al* presented an SRS image of amyloid plaque in tissue from transgenic mice [42]. Peptides and proteins have some vibrational modes representative of their secondary structures that can be alpha helix (α-helix) and beta-sheet (β-sheet), the main one being amide I, which provides useful spectral indicators of structural changes. The amyloid plaques are formed by Aβ protein in a compact structure rich in the conformation of the β-sheet (amyloid fibrils), whose Raman vibration is characterized by the ~1675 cm^{-1} frequency shift. In 2019, our group made an in-depth Raman characterization of amyloid plaques in transgenic mice brains for AD, demonstrating different vibrations for identifying Aβ plaques by Raman spectroscopy [43], as discussed in the next section.

14.3.3 Raman spectra of amyloid plaques

Figure 14.6 shows a set of hyperspectral Raman imaging performed in brain slices of 12 months old transgenic mice with 0.5 μm step-size spatial resolution. In the spectral image, the traditional Aβ sheet vibration modes amide I (centered at 1675 cm^{-1})

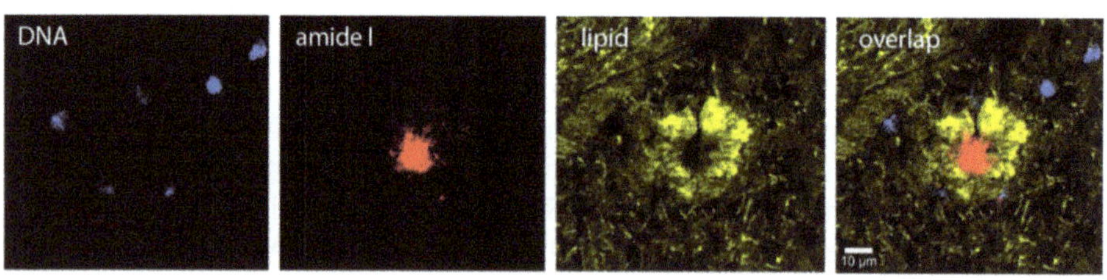

Figure 14.6. 0.5 μm step-size resolution hyperspectral imaging of amyloid plaque. Hyperspectral post-processing data selecting filters for DNA frequency (centered at 791 cm^{-1}, leftmost panel), amide I (centered at 1675 cm^{-1}, middle-left panel) and CH_2 symmetric stretching of the lipid (centered at 2854 cm^{-1}, middle-right panel). The images are artificially colored, and the rightmost panel shows their overlap. Adapted from [44] with permission from the *Royal Society of Chemistry*.

evidences the amyloid plaques core (in red, middle-left panel). The CH_2 symmetric stretching of the lipid (centered at 2854 cm^{-1}) shows the limits of the plaque formed by a lipid-rich halo structure, probably related with neurodegeneration and neuro-inflammation (in yellow, middle-right panel). DNA frequency (centered at 791 cm^{-1}) shows the cell nuclei (in blue, leftmost panel) spread around the lipid halo. The overlap figure (rightmost panel) shows the plaque structure, with a lipid accumulation forming the halo structure surrounding the dense-core plaques.

Considering it is possible to use Raman spectroscopy to identify the presence of the AD marker, i.e., the Aβ, protocols have to be developed to enable the identification of the disease development stage. One possibility is the obvious analysis of the amount of Aβ plaques, which increases with the evolution of the AD. Another possibility, which is indeed more interesting, is the analysis of the structural evolution of the plaques. Figure 14.7 shows the hyperspectral imaging of the amyloid plaque in brain slices of a younger (6 months) transgenic mouse. The plaque image in figure 14.7 was based on the signal from a different vibration, i.e., the amide III, demonstrating that other frequencies can be used to identify plaque in the tissue. Anyway, notice the image, itself, is not significantly different from the one shown in figure 14.6 from the 12 months old transgenic mouse.

The multi-peak Raman spectral analysis of biological materials is rather complex due to the large number of vibrational modes in this complex peptide. Principal component analysis (PCA) is a first-choice technique in multivariate analyses, performed with the aim to reduce dataset dimensionality. PCA is essential to proper investigation of data structure and provides the base picture to further techniques of hyperspectral imaging. In this procedure, principal components (PCs) replace primary variables, capturing relevant information in the dataset. The first PC (PC1) is geometrically defined as the axis of maximum variance of data distribution along the coordinates of primary variables, capturing maximum information of the original dataset. Subsequent PCs capture the remaining orthogonal variance of data in each step. Figure 14.8 shows the values (error bars are variance) of PC1 for specific Raman spectral regions, demonstrating that PCA can easily differentiate the spectral features between Wild Type (WT, black data) and Transgenic (Tg, red and blue data) mice, and it also suggests an evolution from Tg mice with 6 (red data) and 12 months (blue data) old [44]. Therefore, changes in the brain tissue of young adult and ageing transgenic mice can be identified by the Raman hyperspectral fingerprint.

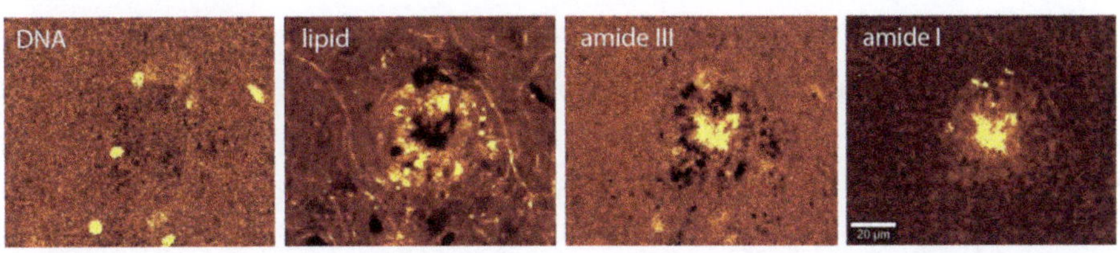

Figure 14.7. 0.5 μm step-size resolution hyperspectral imaging of amyloid plaque. Analysis in WITec software selecting filters for frequencies of DNA (centered at 791 cm^{-1}), lipid (centered at 2854 cm^{-1}), amide III (centered at 1233 cm^{-1}), and amide I (centered at 1675 cm^{-1}). Adapted from [43] with permission from the *Royal Society of Chemistry*.

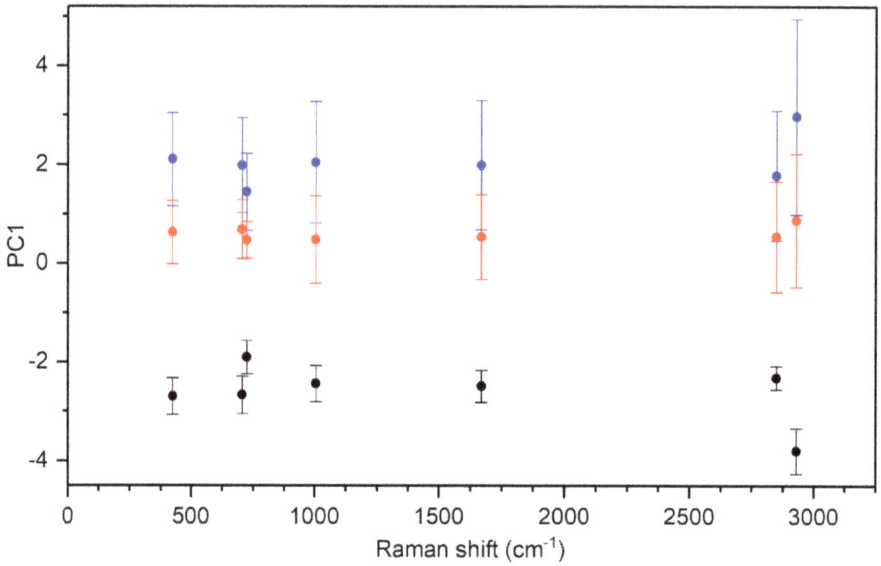

Figure 14.8. Differentiation among amyloid plaque rich brain tissues from WT, young and ageing Alzheimer's Tg mice, 6 and 12 months old, based on PCA at different Raman spectral regions. The first principal component (PC1) shows a clear trend from the WT (black symbols) and Tg mice of 6 (red) and 12 months old (blue). The error bars indicate the PC1 variance. Adapted from [44] with permission from the *Royal Society of Chemistry*.

PCA-based hyperspectral imaging can also be performed with score maps and cluster maps (not shown).

Finally, as already pointed out, although the spontaneous Raman hyperspectral image is powerful for characterizing and identifying amyloid plaques, as shown in figures 14.6–14.8, it is not an appropriate technique for *in vivo* application. SRS, as well as CARS, is the technique for label-free imaging amyloid plaque to be applied to *in vivo* diagnosis. Figure 14.9 shows the SRS amyloid plaques images by 1670 cm^{-1} (amide I frequency), 2850 cm^{-1} (CH_2 stretching of the total lipid), 2880 cm^{-1} (CH_2 stretching of the total lipid and protein), and 2930 cm^{-1} (CH_3 stretching of the total lipid and protein), which is present in biological tissues [44]. Differences in brightness are related to the rich biochemical structure of the lipidic halo. A fast SRS based imaging can, therefore, appropriately identify and characterize the different aspects of the Aβ plaques accordingly.

14.3.4 Proof of concept for intraocular spectroscopy

To implement a platform for diagnosing AD *in vivo*, in addition to the detection of the presence of Aβ in the tissue and its structure characterization, as described in section 14.3.3, it is necessary to experimentally validate the techniques in optical models that simulate the geometry of the human eye. These tests aim to evaluate the quality of the images and the safety of the equipment for the patient.

As shown in figure 14.10(a), the human eye is complex from the point of view of optics [45]. However, to address the feasibility of imaging Aβ plaques subjected to

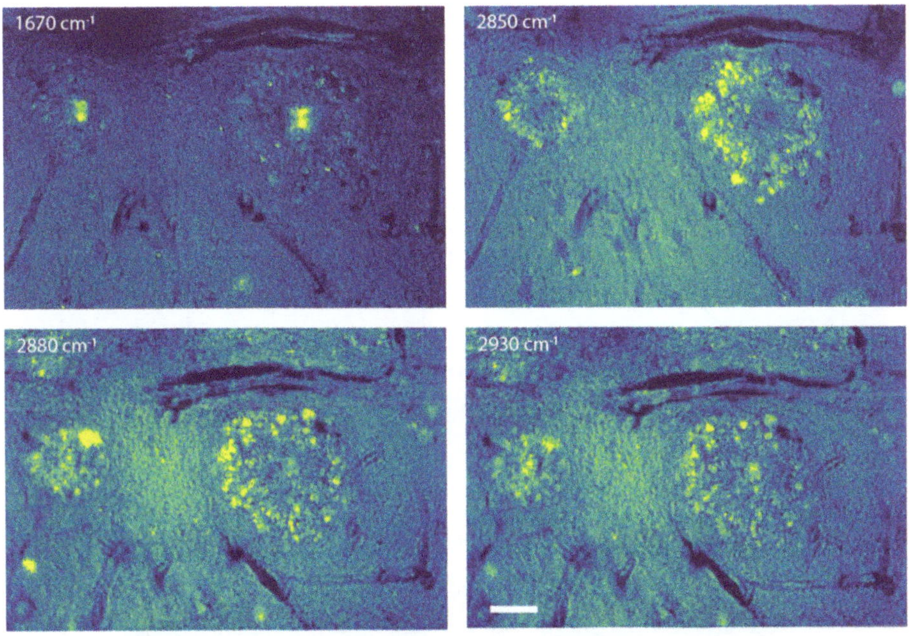

Figure 14.9. Stimulated Raman spectroscopy (SRS) imaging of amyloid plaques in mouse brain tissue. Aβ core (1670 cm^{-1}) on the top-left panel and lipid-rich halo distribution (2850 cm^{-1}) on the top-right panel, CH_2 stretching of the total lipid and protein (2880 cm^{-1}) on the bottom-left image, and CH_3 stretching of the total lipid and protein (2930 cm^{-1}) on the bottom-right image. Scale bar 30 μm.

the optical limitations of the eye, for simplicity, we simulate the required geometry using a biconvex lens of 25 mm focal length, equivalent to approximately 40 diopters to measure different reference materials in a reduced eye model, as shown in figure 14.10(b).

To validate the method, polystyrene microspheres and the brain tissue of mice are used as reference materials. The polystyrene microspheres are homogeneous in size and shape, easy to manipulate, and not perishable as biological tissues. Also, it presents an intense CARS signal when stimulating the Raman peak at ~3057 cm^{-1}, close to relevant Raman features of biological materials. The brain tissue used is taken from genetically modified double Tg mice APP: PS1 to develop AD, the frequency of interest for our proof of concept here is at 2930 cm^{-1}, although other frequencies (2850 cm^{-1} and 2880 cm^{-1}) are also relevant (see section 14.3.3) [44]. These frequencies are associated with lipids and protein Raman peaks. In the case of polystyrene, the sample is deposited directly in the sample holder that simulates the retina, which is attached to the lens holder. For the brain tissues, they are placed between two coverslips and sealed in order to prevent the sample dehydration, i.e., in this case, the rounded sample holder was not used.

Figure 14.11 shows TPEF and CARS images performed in the eye model for polystyrene (a and b) and the mice's brain (c–f). The results show that, considering the realistic geometry defined by this eye model, it is possible to obtain images with good

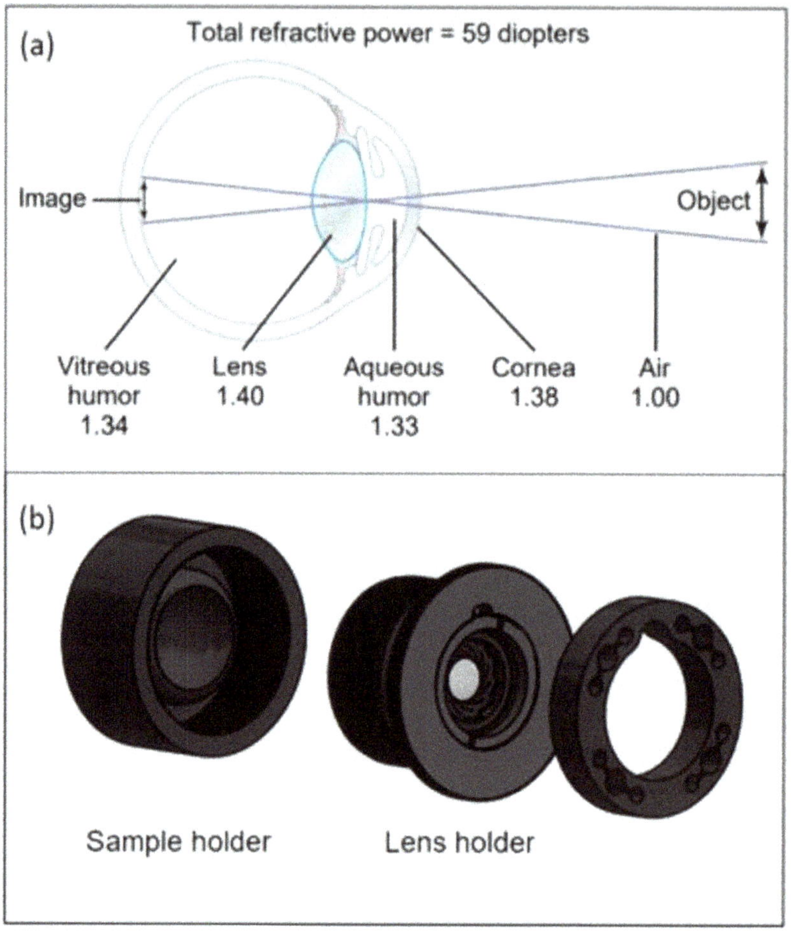

Figure 14.10. (a) Simplified illustration of the human eye's optics with its main structures and their respective refractive indices. Adapted from [45] with permission from *Elsevier*. (b) Schematics of our simplified eye model consisting of a curved sample holder that simulates the curvature of the retina and a lens holder with an iris.

quality and resolution for CARS in polystyrene (a and b). For the tissue, we obtained a good result for the TPFE (c and d), where it is possible to identify the amyloid plaques in the brain samples. However, for the CARS measurements in the brain (e and f), we obtained a strong signal coming from the tissue, masking the resonant CARS signal coming from the amyloid plaques. Improvements must be considered to enhance the quality of the CARS images in this setup, such as avoiding chromatic aberration introduced by the lenses in the system, and to implement SRS in the backscattering configuration. Since the human eyes have chromatic aberration, especially for wavelengths in the near-infrared, it is probably necessary to introduce corrective lenses for the different wavelengths in the system, in order to perform *in vivo* tests [46]. Although improvements are indeed necessary, they are usual in this type of spectroscopy and we consider this proof of concept satisfactory.

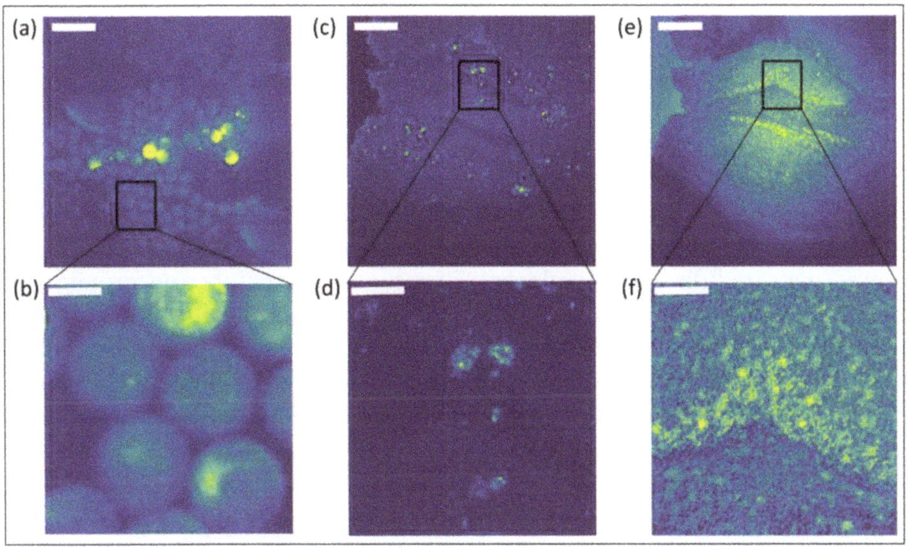

Figure 14.11. Measurements performed with the simplified eye model: (a) CARS measurement of polystyrene sphere of 90 μm in diameter, using 1064 nm and 803 nm; (b) an enlarged image from (a). (c) TPEF measurements in a 12-month-old transgenic mouse brain tissue using a picosecond pulsed laser at 830 nm; (d) an enlarged image of the amyloid plaques in (c). (e) CARS measurement of the same tissue using 1064 nm and 811 nm laser beams, in resonance with CH_3 stretching of the total lipid and protein (2930 cm^{-1}); (f) an enlarged image of the region of the amyloid plaques. Scale bars indicate 270 μm in the overview upper images, and 45 μm in enlarged lower images.

14.4 The intraocular spectroscopy as a platform

Up to this point, our focus has been on the protocol for early diagnosis of AD. However, the intraocular spectroscopy has potential to be a platform that can be utilized generally to study the eye, retinal and neurodegenerative diseases, following similar procedures. In this section we give some relevant examples that can be used to turn the specific technique into a generic platform.

14.4.1 Cornea, lens, anterior chamber, and vitreous body

14.4.1.1 Corneal edema and haze

Raman spectroscopy provides a biochemical in-depth assessment of anterior ocular structures. Corneal haze results from degradation of the lattice composed by collagen I and V and stabilized by core sulfated proteoglycans and linear side chains of disaccharides and sulphate esters. This lattice generates amides I and III consistent with the α helix on Raman spectroscopy [47]. Lattice degeneration is associated with a red-shifted amide I and a blue-shifted amide III, consistent with a change from α-helices into β-sheets, as well as reduced peaks of sulphated proteoglycans at 1000 cm^{-1} and 1248 cm^{-1}. In addition, stromal hydration is assessed by the ratio of OH from water (at 3100–3700 cm^{-1}) to CH from collagen (at 2850–3030 cm^{-1}) [48, 49]. Raman spectroscopy can provide information aiming at the identification of infectious agents [49, 50], therapeutic response in infectious keratitis [50], and transport of ocular drugs through epithelium and stroma [51].

14.4.1.2 Lens opacification

Disturbance of the highly packed lens proteins impairs lens transparency. Age-related nuclear dehydration produces a decreased ratio of OH stretching to CH stretching, in addition to increased peaks from disulfide bonds (~510 cm^{-1}), whereas a significant nuclear cataract is related to deamidation and glycation of lens proteins, increased disulfide bonds, loss of secondary structure and protein aggregation. In consequence, a blue-shifted amide I and red-shifted amide III are disclosed, consistent with a change from α helices to β sheets [52]. These unfolded nuclear proteins are more prone to oxidation, which triggers nuclear opacification [53]. Conversely, unfolded cortical proteins increase lens membrane permeability, producing cortical hydration and opacification [54]. High peaks from water [52] are early findings in cortical cataracts. Further, peaks from macular pigments, especially lutein and zeaxanthin at 1525 cm^{-1}, obtained in pseudophakic patients, provide objective parameters for a better choice of intraocular lens [55].

14.4.1.3 Anterior chamber and vitreous body

Raman spectra obtained from aqueous present peaks from water, solutes, amino acids, and proteins. Among these, usually the most relevant peaks are related to water (~3410 cm^{-1} and a shoulder at ~3250 cm^{-1} from symmetric and asymmetric OH stretch respectively) and proteins (~1238 cm^{-1} from amide III, ~1450 cm^{-1} from CH bending and both 2890 cm^{-1} and 2945 cm^{-1} from CH stretching) [56]. Changes in water peak intensities, as well as ratios of water to CH and amides provide spectral data regarding protein concentration in aqueous and blood-aqueous barrier permeability. Protein analysis from aqueous and vitreous can be performed with Raman spectroscopy, providing a tool for the differential diagnosis of intraocular inflammation [57]. Experimental models of aseptic vitritis and infectious endophthalmitis were consistent with distinct spectral findings. Whereas both conditions present peaks from amide III (at 1258 cm^{-1} and 1339 cm^{-1}) and CH_2 (at 1451 cm^{-1}), those in infectious endophthalmitis were more intense and associated with significant peaks of the amino acid phenylalanine (at 1004 cm^{-1}) and a significant blue-shifted amide I (at 1668 cm^{-1}) [57]. Raman spectroscopy can be used in bioavailability studies of topical and intravitreal drugs [51].

14.4.2 Retinal and neurodegenerative diseases

14.4.2.1 Retinal dystrophies

Rhodopsin is composed of a rod opsin and a transmembrane multidomain retinal chromophore [58]. The photon-induced isomerization of 11-cis into 11-all trans retinal is the starting event of phototransduction. Peaks at 971 cm^{-1}, 1216 cm^{-1}, and 1237 cm^{-1} are obtained from chromophore, whereas an intense peak at ~1524 cm^{-1} is obtained from the 11-cis retinal isomer, until light adaptation and bleaching [59]. Structural changes in opsin are related to progressive rod degeneration, as in autosomal dominant retinitis pigmentosa [60], whereas structural changes in the chromophore produce stabilization in the 11-all trans isomer and light desensitization. The lack of 11-cis retinal is related to congenital blindness, as in Leber congenital amaurosis [59].

14.4.2.2 Age-related macular degeneration

Depletion of macular pigments and damage in retinal pigment epithelium (RPE) and Bruch's membrane are well-established risk factors for age-related macular degeneration (ARMD). Among macular xanthophylls, the most important are lutein and zeaxanthin, localized in the Henle's layer of parafoveal and foveal regions, respectively [60]. Both absorb short wavelength light (the blue–green light between 400 and 500 nm) and limit macular photochemical damage. Lutein and zeaxanthin produce intense peaks from C=C stretching (at 1525 cm^{-1}) and C–C stretching (at 1159 cm^{-1}) in addition to a moderate peak from $C–CH_3$ rocking (at 1008 cm^{-1}). Among these, the peak at 1525 cm^{-1} is used as a signature of the whole macular pigments [60]. Similar backbone and terminal OH groups justify similar Raman spectra [61].

Due to the high metabolic activity of photoreceptors and RPE, the RPE-Bruch's membrane complex presents a glucose-enriched, high oxygenated microenvironment, which is particularly susceptible to oxidative damage [62]. Experimental studies with human cultured RPE exposed to oxidation revealed higher 812/832 and 1254/1650 peak ratios, suggesting that the breathing mode of tyrosine and amide I backbone of RPE proteins are especially sensitive to oxidation [62]. Further, age-related deposits of advanced glycation and lipoxidation end-products in Bruch's membrane present significant association with protein remodeling of extracellular matrix and RPE dysfunction [62].

14.4.2.3 Glaucoma

Chronic open-angle glaucoma is a major age-related neurodegenerative condition involving primarily the retinal ganglion cell layer and characterized by transsynaptic ascending degeneration to lateral geniculate nucleus, optic radiations, primary visual cortex and higher-order visual centers [63]. Evidence suggests that geniculate and post-geniculate involvement in glaucoma occurs before significant RGC loss [64], disclosing a pattern of transsynaptic degeneration similar to chronic neurodegenerative conditions such as Alzheimer's disease (AD) and Parkinson's disease (PD). Conversely, preferential damage in primary visual cortex and high-order visual processing centers occurs in a subgroup of patients with both AD and PD [65].

In addition to a common pattern of transsynaptic degeneration, studies demonstrate that amyloid-β and α-synuclein, the main biomarkers of AD and PD respectively, are both involved in ganglion cell apoptosis. Glaucoma and AD involve primarily the magnocellular visual paths, and a significant subgroup of patients with AD fulfill diagnostic criteria of glaucoma [66]. Besides, Aβ was identified in the lateral geniculate nucleus and primary visual cortex in experimental glaucoma, consistent with an Aβ-related, ascending transsynaptic degeneration [64]. A study using Raman spectroscopy and PCA was capable to differentiate and classify differences between dog's glaucomatous and healthy ganglion cells [67].

14.4.2.4 Parkinson's disease

Parkinson's disease (PD) is the major neurodegenerative condition primarily related to motor disability. The hallmarks of PD are the Lewy bodies, composed by α-synuclein fibrils assembled into β-sheets aggregates in dopamine circuits of substantia nigra and striatum [68]. Visual impairment is a major non-motor manifestation of PD, produced by α-synuclein aggregation in dopaminergic circuits of inner and outer plexiform layers [69]. Consequently, inputs to ganglion cells are suppressed, with impairment of color discrimination, contrast sensitivity and motion perception [70]. As in AD, α-synuclein aggregation is related to Müller's cell activation, neuroinflammation and oxidative damage of photoreceptor and ganglion cells [69]. However, in PD parvocellular ganglion cells are primarily affected.

The aggregation of α-synuclein is associated with a narrowed and red-shifted amide I, a blue-shifted amide III and an identifiable amide II. Distinct spectra are obtained in function of microenvironment factors, such as oxidative stress and pH, which modulate the kinetics of α-synuclein aggregation [71]. The narrower amide I peak at acidic media suggests a faster aggregation with a more ordered conformation, whereas the broader amide I at neutral pH is consistent with longer times of aggregation and higher polymorphism [71]. Such spectral differences provide tools to a better understanding of the microenvironment factors related to α-synuclein-related neurodegeneration.

14.5 Summary and perspectives

Here we propose the use of intraocular Raman spectroscopy as a protocol for the early diagnosis of AD. We believe we have provided the initial background evidence for the viability of the method, which should now evolve to studies in humans. The urgent need for new biomarkers to AD is related to the limitation of current diagnosis approaches. The standard protocol is based on a variety of procedures, including the Mini-Mental Stage Exam, mostly based on the clinical evaluation of signs and symptoms. In general, these procedures are conducted when cognitive impairment is already installed. Current pharmacological therapy in AD is unable to significantly modify the progressive neurodegeneration that characterizes this condition, and the lack of an early-stage diagnosis decreases the chances of developing new disease-modifying therapies. Therefore, the major aim in AD investigation must be the search for novel biomarkers to be used in early diagnosis.

Available methods to an early diagnosis have risen as promising tools in AD biomarkers' characterization. To date, however, they have failed as a broadly accessible routine test in many aspects. Lumbar puncture to collect cerebrospinal fluid analysis requires specialized technicians, and it is an invasive procedure which exhibits current limitations on accuracy, still inconsistent. As regards neuroimaging fields and related spectroscopic techniques, the development of an efficient, safe and non-invasive protocol is challenging. Despite the recent advances in MRI, this technique is expensive, unsuitable for patients with claustrophobia, and only provides information about morphological changes which are related to already installed neurodegeneration stages of AD. Methods which apply radiotracers, such as ^{18}F-containing molecules in PET scan, allow adequate detection of biochemical alterations by binding specific regions linked to AD protein aggregation and can be associated with MRI. The disadvantages of PET concern costs and the fact of exposing the patient to gamma radiation, even if at low levels.

The above-mentioned techniques and their limitations reinforce the ongoing search for developing financially affordable and less invasive first-screen options aiming at early AD diagnosis. In this context, targeting the retina as a neuronal structure and using Raman spectroscopy at excitation frequencies on the near-infrared and visible ranges are promising approaches. The development of compounds which selectively bind to protein aggregates and display fluorescent properties are also alternatives, however, they imply accurate pharmacokinetics and toxicological studies for safety.

Thus, the fast, accessible and non-invasive analysis of biochemical alterations in the retina emerged as a promising approach in early-stage diagnosis, applying not only to AD, but to any retinal and neurodegenerative disorders. The ability to early diagnose such conditions is a first step required to the development of effective therapies, and here we demonstrated the validity of Raman spectroscopy-based technique and protocol that should now proceed to be tested in humans.

Chapter highlights

- Intraocular spectroscopy for early diagnosis of retinal and neurodegenerative diseases.
- Early diagnosis protocol for Alzheimer's disease based on biochemical hallmarks.
- Raman spectroscopy as a fingerprint for molecular disorder.
- Raman spectroscopy imaging evidences of amyloid-β plaques' internal structure.

References

[1] London A, Benhar I and Schwartz M 2013 *Nature. Rev. Neurol.* **9** 44–53

[2] Jack C R Jr *et al* 2018 *Alzheimer's Dementia* **14** 535–62

[3] Alzheimer's Association 2019 *Alzheimer's Dementia* **15** 321–87

[4] Guo L, Duggan J and Cordeiro M 2010 *Curr. Alzheimer Res.* **7** 3–14

[5] Koronyo Y *et al* 2017 *JCI Insight* **2** 1–19

[6] Koronyo-hamaoui M *et al* 2011 *Neuroimage* **54** S204–17

[7] More S S, Beach J M, McClelland C, Mokhtarzadeh A and Vince R 2019 *ACS Chem. Neurosci.* **10** 4492–501

[8] Jentsch S *et al* 2015 *Acta Ophthalmol.* **93** e241–7

[9] Dresselhaus M S, Dresselhaus G, Cronin S B and Souza Filho A G 2018 *Solid State Properties: From Bulk to Nano* (Berlin: Springer)

[10] Shen Y R 1984 *The Principles of Nonlinear Optics* (New York: Wiley)

[11] Fox M 2011 *Optical Properties of Solids (Oxford Master Series in Physics, 2nd edn)* (Oxford: Oxford University Press)

[12] Boyd R 2008 *Nonlinear Optics* (London: Academic)

[13] Saleh B E A and Teich M C 2007 *Fundamental of Photonics* 2nd edn (New York: Wiley)

[14] Rigneault H and Pascal B 2018 *APL Photon.* **3** 091101
[15] Jones R R, Hooper D C, Zhang L, Wolverson D and Valev V K 2019 *Nanoscale Res. Lett.* **14** 231
[16] Cunha R *et al* 2021 *Analyst* **146** 2945–54
[17] Min W, Freudiger C W, Lu S and Xie X S 2011 *Annu. Rev. Phys. Chem.* **62** 507–30
[18] Butler H J *et al* 2016 *Nat. Protoc.* **11** 664–87
[19] Talari A C S, Movasaghi Z, Rehman S and Rehman I 2015 *Appl. Spectrosc. Rev.* **50** 46–111
[20] Jermyn M *et al* 2016 *Phys. Med. Biol.* **61** R370–400
[21] Jermyn M *et al* 2015 *Sci. Transl. Med.* **7** 93180D
[22] Folick A, Min W and Wang M C 2011 *Curr. Opin. Genet. Dev.* **21** 585–90
[23] Henderson T A and Morries L D 2015 *Neuropsychiatr. Dis. Treat.* **11** 2191–208
[24] de Almeida Barros A G *et al* 2014 *PLoS One* **9** e85874
[25] Bar-Noam A, Farah N and Shoham S L 2016 *Sci. Appl.* **5** e16007
[26] Masihzadeh O *et al* 2013 *Invest. Ophthalmol. Vis. Sci.* **54** 3094
[27] Sicong H *et al* 2015 *Biomed. Opt. Express* **6.3** 1055–66
[28] Gray D C *et al* 2006 *Opt. Express* **14** 7144–58
[29] Qin Z *et al* 2020 *Light: Sci. Appl.* **9** 79
[30] Erckens R J *et al* 2001 *Lasers Med. Sci.* **16.4** 236–52
[31] Evans J W *et al* 2009 *J. Biophotonics.* **2** 398–406
[32] Masihzadeh O *et al* 2013 *Invest. Ophthalmol. Vis. Sci.* **5** 3094–101
[33] Ammar D A *et al* 2013 *Invest. Ophthalmol. Vis. Sci.* **54.8** 5258–65
[34] He S *et al* 2015 *Biomed. Opt. Express* **6.3** 1055
[35] Bernstein P S *et al* 1998 *Invest. Ophthalmol. Vis. Sci.* **39.11** 2003–11
[36] Ermakov I V *et al* 2001 *Opt. Lett.* **26** 202–4
[37] Stiebing C *et al* 2019 *Neurophoton.* **6.4** 041106
[38] Hunter J J *et al* 2012 *Prog. Retin. Eye Res.* **31.1** 28–42
[39] Querfurth H W and LaFerla F M 2010 *N. Engl. J. Med.* **362.4** 329–44
[40] Selkoe D J and Hardy J 2016 *EMBO Mol. Med.* **8.6** 595–608
[41] Palombo F 2018 *Analyst* **143** 850–7
[42] Ji M *et al* 2018 *Sci. Adv.* **4** 11
[43] Fonseca E A *et al* 2019 *Analyst* **144** 7049–56
[44] Fonseca E A *et al* 2021 *Analyst* **146** 6014–25
[45] Hall J E and Hall M E 2020 *Textbook of Medical Physiology* 14th edn (Amsterdam: Elsevier)
[46] Fernández E J *et al* 2006 *Opt. Express* **14.13** 6213–25
[47] Murab S and Ghosh S 2016 *Biomed. Mater.* **11.6** *065005*
[48] Bauer N J C *et al* 1998 *Invest. Ophthalmol. Vis. Sci.* **39** 831–35
[49] Stöckel S *et al* 2016 *J. Raman Spectrosc.* **47** 89–109
[50] Rusciano G *et al* 2013 *PLoS One* **8.8** 1–8
[51] Bauer N J C *et al* 1999 *J. Ocul. Pharmacol. Ther.* **15.2** 123–34
[52] Huang C C and Chen W 2018 *Appl. Spectrosc. Rev.* **53.9** 689–702
[53] Michael R and Bron A J 2011 *Philos. Trans. R. Soc. Lond. B. Biol. Sci.* **366** 1278–92
[54] Manoharan R, Wang Y Y and Feld M S 1996 *Spectrochim. Acta A* **52** 215–49
[55] Obana A *et al* 2011 *Ophthalmology* **118** 1852–8
[56] Carey D M and Korenowski G M 1998 *J. Chem. Phys.* **108.7** 2669–76
[57] Rossi E E *et al* 2012 *J. Photochem. Photobiol.* B **107.1** 73–8

[58] Sakmar P T *et al* 2002 *Annu. Rev. Biophys. Biomol. Struct.* **31** 443–84
[59] Park P S H 2014 *Adv. Pharmacol.* **70** 1–36
[60] Bernstein P S *et al* 2004 *Arch. Biochem. Biophys.* **430** 163–9
[61] Gellerman W *et al* 2002 *J. Opt. Soc. Am.* A **19.6** 1172–86
[62] Beattie J R *et al* 2010 *FASEB J.* **24** 4816–24
[63] Ramirez A I *et al* 2017 *Front. Aging Neurosci.* **9.214** 1–21
[64] Yan Z *et al* 2017 *Invest. Ophthalmol. Vis. Sci.* **58** 5434–43
[65] Crutch S A *et al* 2012 *Lancet Neurol.* **11** 170–8
[66] Kusne Y *et al* 2017 *Acta Ophthalmol.* **95** 668–76
[67] Wang Q *et al* 2011 *J. Biomed. Opt.* **16.10** 107006
[68] Walker Z *et al* 2015 *Lancet* **386** 1683–97
[69] La Morgia C *et al* 2017 *Front. Neurol.* **8** 710
[70] Mammadova N *et al* 2019 *Neurobiol. Dis.* **121** 1–16
[71] Flynn J D *et al* 2018 *J. Biol. Chem.* **293.3** 767–76

IOP Publishing

Advances in Ophthalmic Optics Technology

Davies William de Lima Monteiro and Bruno Lovaglio Cançado Trindade

Chapter 15

Predicting cataracts through automatic image analysis and classification

Monica Mitiko Soares Matsumoto, Mariana Vaz Goulart
Marcus Henrique Victor Jr and Paulo Schor

Cataracts is the most common cause of reversible blindness in the world, it is estimated that 95 million people have vision impairment due to it [1]. In many cases cataracts can lead to a complete loss of vision capacity. When the patient's vision is compromised, cataract surgery is performed, and the surgery consists of the replacement of the cloudy natural lens with an artificial clear intraocular lens (IOL).

There are a couple of imaging techniques for cataract diagnosis. Those examinations can assess the opacity of the anterior portion of the eye, especially the lens. The images can be automatically analyzed. Visual cues such as affected area and opacity heterogeneity guide the diagnosis. The pattern in the images is usually complex, and for some clinical applications automatic diagnosis can outperform experienced health professionals.

In this chapter, we address automatic cataract detection. First, we give an overview of eye lens physiology and clinical cataract detection. After that, we introduce image processing techniques relevant to most imaging modalities. Then, we focus on image analysis and classification. Extracting appropriate features of the image is a way to convey ad hoc clinical evaluation. We subsequently outline some artificial intelligence techniques commonly used for the classification task.

The chapter covers clinical cataract detection, classification and challenges. Foremost, we present a thorough literature review of automatic cataract classification. We cover methodologies, results and discussions about computerized automatic diagnosis.

15.1 An introduction to cataracts—eye-lens physiology and opacification

As previously mentioned in chapter 2, the crystalline lens embryologically develops from an invagination of the ectoderm. Recent studies postulate that external pressure is a major force to promote such 3D structure [2].

doi:10.1088/978-0-7503-3263-7ch15

From the lens development and like any lens, it is pristine and has very smooth surfaces. As a special lens, its infant content is usually soft—a very liquid gel. Therefore, the lens can deform under a slight stress force such the one exerted by the zonule fibers, linked to the ciliary muscle.

The role of age, sun exposure and microenvironmental oxidative stress is well known in the pathogenesis of lens structural changes [3]. The continuous epithelium cell migration and central packing of the lens relates to nucleus sclerosis and further hardening of the lens.

Currently, the concept of cataracts refers to a non-functional lens. Recently, a broader concept was created, named dysfunctional lens syndrome (DLS), to define the ongoing process that alters the crystalline lens to a hard, deformed and eventually opaque crystalline lens [4]. It is a pre-cataracts stage where the lens is clinically normal, though with imperceptible transparency changes that start affecting diffraction and scattering of the light.

Due to advances in intervention techniques and implants, this concept of non-opaque-cataracts or dysfunctional lens, is useful to understand and restore vision of younger patients. The novelties include new fluidics control surgery that lowers the intervention risk (by increasing safety) and have superior efficacy of multifocal and extension of the depth of focus in IOL.

In clinical practice, physicians put together symptoms and signs to define a disease and that considers several other factors such as functional and emotional status of the person to decide on an intervention. And although the classic opaque cataracts concept fulfills the surgical requirements, DLS is a broader concept that still lacks an assertive clinical definition and objective diagnostic devices such as imaging, biomechanical and optical diagnostic frameworks.

We will refer in this chapter to cataracts as the classic opacification of the lens, seen through optical diagnostic equipment, but attention will be paid to the alternative devices to be used in order to clarify and follow up the dysfunctional lenses.

15.2 Cataract detection clinical review

Cataracts can be typically classified into three types according to their cause: senile cataracts, congenital cataracts and secondary cataracts. The congenital cataracts are a condition present since birth. Secondary cataracts can occur due to physical trauma to the eyes or can be induced by the continuous use of certain medications. Senile cataracts are age-related and are the most common type of the disease in adults. Senile cataracts can be divided into three types: nuclear, cortical and posterior subcapsular cataracts (PSC) [1]. In nuclear cataracts, opacity begins at the center of the lens and spreads to the surface, cortical cataracts begin at the outer edge of the lens and grow towards the center, and PSC cataracts occur at the posterior surface of the lens [5]. The three types of senile cataracts can be seen in figure 15.1 in cross-section.

In the conventional method for cataract analysis and diagnosis, the ophthalmologist visualizes the structures of the eye using a device called biomicroscope or slit lamp. This equipment illuminates the eye with different angles to visualize opacities of the lens.

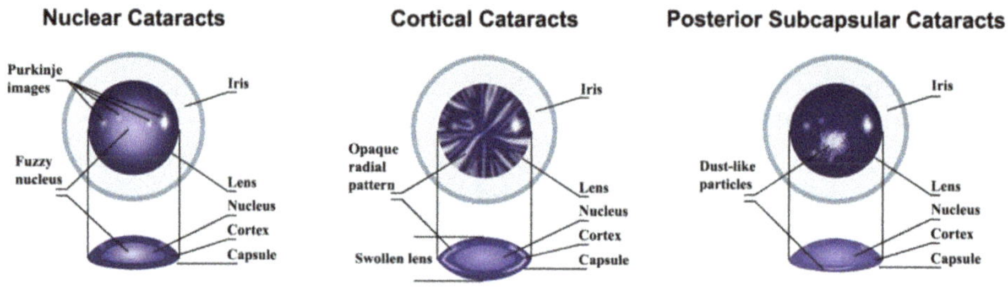

Figure 15.1. Appearance of lens opacity in different cataracts types.

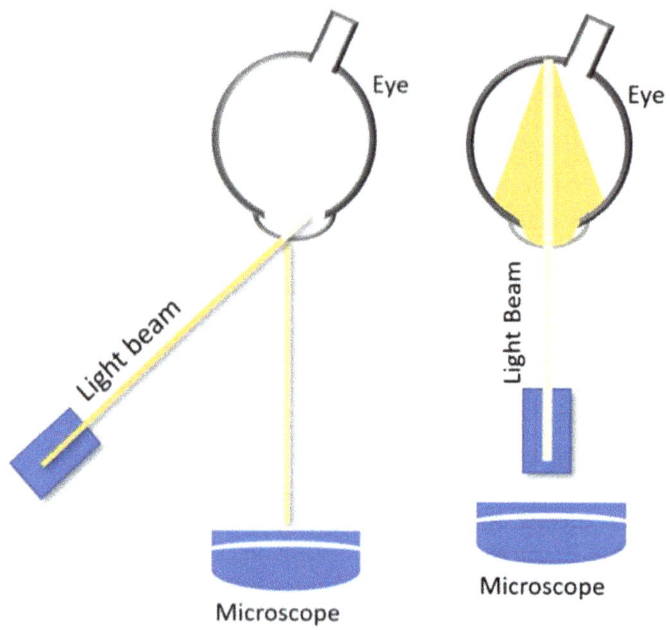

Figure 15.2. On the left, illustration of the direct slit lamp illumination to assess nuclear cataract. The yellow represents the light beam impinging the eye and the blue at the bottom represents the part of the biomicroscope in which the image is seen by the examiner. On the right, illustration of retro illumination to assess cortical cataract and PSC. The yellow represents the light beam from the biomicroscope reaching the posterior layers of the eye and scattering back.

Direct slit lamp light, shown in figure 15.2, gives a quick assessment of nuclear cataracts (NCs). Figure 15.3 depicts an image obtained with this type of illumination, in (a) the visualized structures are presented, in (b) a sample of an eye with no sight-threatening cataracts and in (c) an eye with nuclear cataracts.

Under retro illumination, the light reaches the posterior layers of the eye and the red reflex occurs because of the blood filled choriocapillaris layer. This way, light goes through the transparent structures and the opaque structures become dark areas in the image. The retro illumination helps to evaluate cortical cataracts and PSC [5]. The images are similar to the opacity seen in figure 15.1.

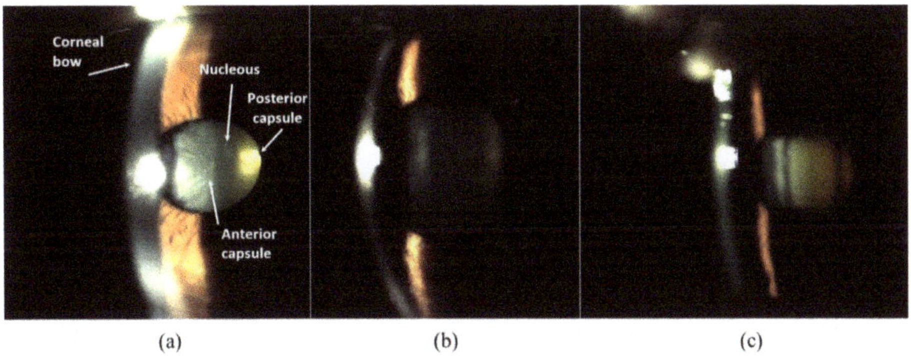

Figure 15.3. Ocular structures visualized by microscopy. (a) Main ocular parts visualized: cornea, lens nucleus, posterior capsule, and anterior capsule, (b) eye with no sight-threatening cataract and (c) eye with nuclear cataract.

Another examination that is not specific for cataract detection, but can indirectly detect it, is the eye fundus examination. The lens opacity is detected by looking at the retina in the fundus images. If the image is clear, there is probably no opacity, but if the image is difficult to see, blurred or not possible, there is opacity in the anterior region of the globe.

After assessing whether the patient has cataract or not, doctors can evaluate the lens opacity using a classification system (Lens Opacities Classification System III—LOCSIII). LOCSIII is an improved LOCSII-derived system for grading senile cataracts in slit lamp (direct lighting) and retro illumination images [6]. The classification system has reference images of different levels of nuclear, cortical and posterior subcapsular cataracts. These images are used as an evaluation guideline to establish the degree of the cataract.

15.3 Digital medical image and processing

In this section, we will address how medical images are computationally processed and utilized. The different imaging modalities produce digital images that can be processed by algorithms. We will outline a common pipeline to process the image and make it ready to be quantified. The following subsections describe definitions and methods to process and analyze such images. We will use an infrared image of the eye as an example that can be used to classify and quantify the opacification through the pupil.

15.3.1 Introduction to medical imaging and processing

Imaging techniques are consolidated as one of the main medical examinations. Through imaging, physicians may evaluate the anatomy and functional disorders of the patient. Different image modalities are used, for instance, to identify tissue changes, cancer staging and bone damage, to assess lung and cardiac function, to plan for surgeries and to measure body structures.

The most common modalities in the clinical evaluation are computed tomography (CT), x-ray, magnetic resonance imaging (MRI), optical coherence tomography

(OCT), ultrasound (US), visible light photography and other light spectrum-based modalities (e.g., infrared IR).

In this chapter, we will center our attention on the acquired digital image, which can be enhanced and processed in order to automatically retrieve the desired information.

15.3.2 Digital image

A scene can be spatially sampled by sensors and its local intensity can be quantized, determining a grayscale image. Generally, the intensity is represented by an integer between 0 and $2^N - 1$, where N is the number of levels supported by the imaging system. For example, in an 8-bit system, the intensity values will be from 0 to 255.

Generally, each element recorded by the acquisition system represents a fraction of the imaged object. In a 2D image, this element is called a picture element (pixel). When there is a depth associated with it, the element is represented as a volume element (voxel), e.g., 3D reconstruction. A spel is a name given to a general element with any dimension that associates with a space element. Thus, each spel stores the intensity value for the imaging modality used and can have, for instance, 4D (if intensity changes with time) or 5D (if space element shape changes with time).

A spel can also contain different captured attributes. In the context of photographs, instead of shades of gray (just one attribute), we can have color images (containing three attributes) in which an array of different color channels represents a pixel intensity. Thus, each modality may have its various attributes recorded for further processing and analysis.

For the sake of simplicity, we will refer to any space element as a pixel in this chapter.

15.3.3 Histogram

The histogram is an initial characterization of the image and can be used to improve its visualization. It consists of a graph in which the possible intensity values are on the horizontal axis (bins). Their heights (vertical axis) contain the number of pixels of each intensity value.

Figure 15.4 depicts a predominantly dark image with 47 580 pixels whose distribution has a high peak at a low intensity value. Both eyes are brighter than the vicinity, which is confirmed by a much smaller peak at a higher intensity value in the histogram.

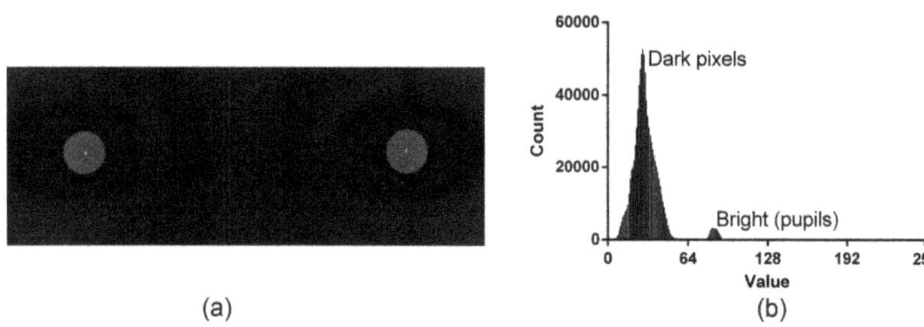

Figure 15.4. Example of a frontal infrared picture of the eyes (a) retro illumination image with 366 × 130 pixels (47 580 pixels), (b) its histogram with two peaks: 'Dark' corresponding to the predominantly dark image and 'Bright' corresponding to both pupils.

The histogram building process involves reading the image pixel by pixel and counting the occurrence of each intensity value. Although the histogram represents the number of events of a given value, it does not specify where it occurred in the image. Thus, the spatial information is lost in the histogram.

The sum of all heights in the histogram is equal to the total pixels of the image. The normalized histogram can be obtained by dividing the heights by the total number of pixels. In statistical terms, the normalized histogram is the probability density function (PDF) of the digital image, indicating the probability (between 0 and 1) of observing a pixel intensity value in the image. The normalized histogram's sum is the cumulative distribution function (CDF), which indicates the probability of a pixel having a value equal to or less than a given value. CDF grows from 0 to 1, as PDF always assumes non-negative values. The CDF of a particular value is obtained by adding the PDF values from 0 to the desired value. For instance, from figure 15.5, around 10 000 pixels have values equal to or less than intensity 23; and approximately 46 000 pixels have values equal to or less than 80. Such cumulative counts also show how the assessed image appears dark due to its large count of low-value pixels.

In an N-pixel image, a_i and h_i are the possible intensity values and the corresponding sum of occurrences, respectively. We can define the average pixel intensity value a as equation (15.1):

$$\underline{a} = \frac{\sum_{i=1}^{N} a_i h_i}{\sum_{i=1}^{N} h_i} \tag{15.1}$$

For instance, in an 8-bit image, the possible pixel values are between 0 and 255, if a is significantly smaller or greater than 128 the image will be mostly dark or bright, respectively. Figure 15.6 depicts a bright and a dark version of the same image.

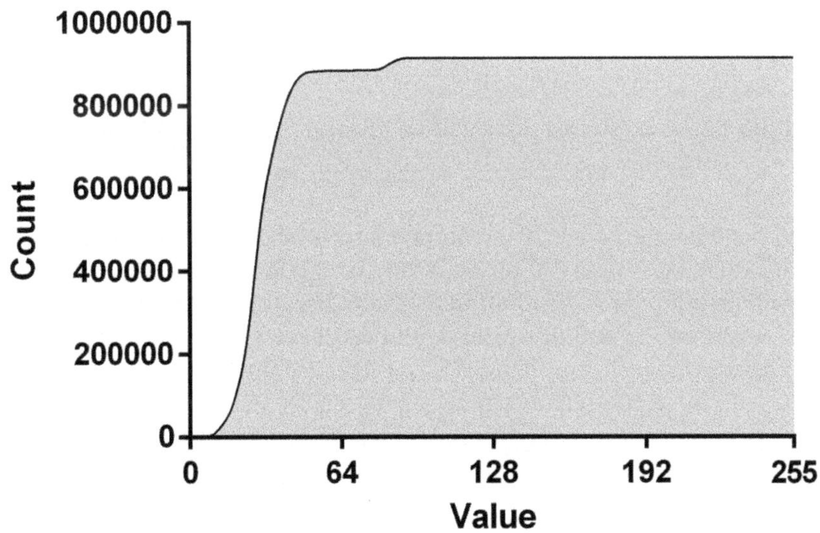

Figure 15.5. Cumulative distribution function related to figure 15.4. Notice the final count corresponding to the total number of pixels in the image.

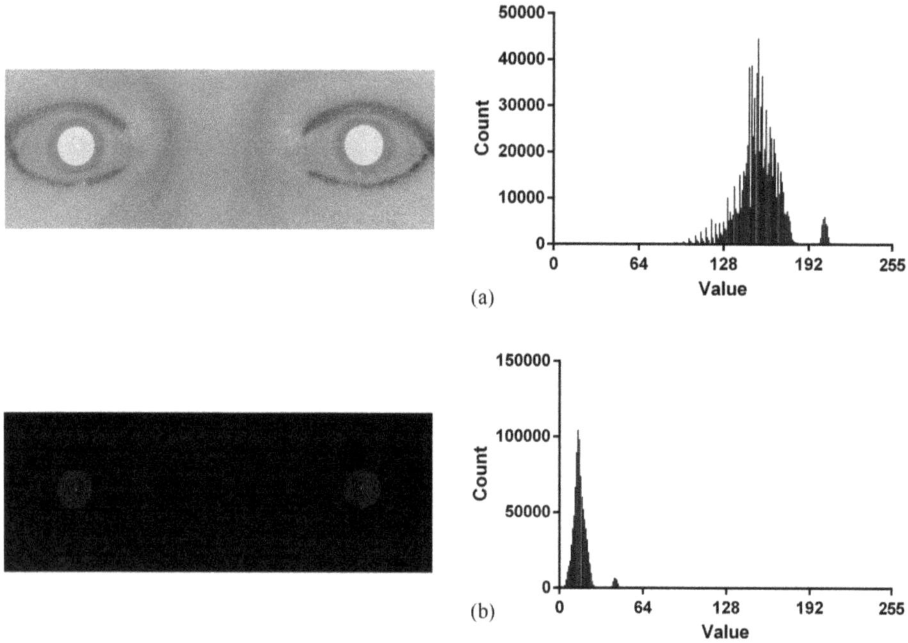

Figure 15.6. Overall bright (a) and dark (b) images with their respective histograms on the right.

15.3.4 Dynamic range

The dynamic range (DR) is directly related to the difference between the maximum (a_{max}) and minimum (a_{min}) intensity values of the pixels found in an image. It can be expressed as equation (15.2):

$$DR = 20\log_{10}(a_{max} - a_{min}) \tag{15.2}$$

For example, in a 12-bit CT image (values between −1000 and 3095), the DR will be given by equation (15.3):

$$DR = 20\log_{10}(3095 - (-1000)) = 72 \text{ dB} \tag{15.3}$$

In a 10-bit fluoroscopy image, it will be given by equation (15.4):

$$DR = 20\log_{10}(1024) = 60 \text{ dB} \tag{15.4}$$

Ideally, the dynamic range of the observed scene should be equal to the dynamic range of the imaging system (2^N for an N-bit system). Suppose the dynamic range of the scene under study is wider than that of the detector. In that case, there will be underflow, overflow, or both for pixels with borderline values, and such information will be irreversibly lost. In figure 15.7, there are four hypothetical histogram situations, with different dynamic ranges between scene and detector.

15.3.5 Signal-to-noise ratio

The signal-to-noise ratio (SNR) of an image can be estimated using the image itself. The signal (or average intensity) of the image is the square of the image's average

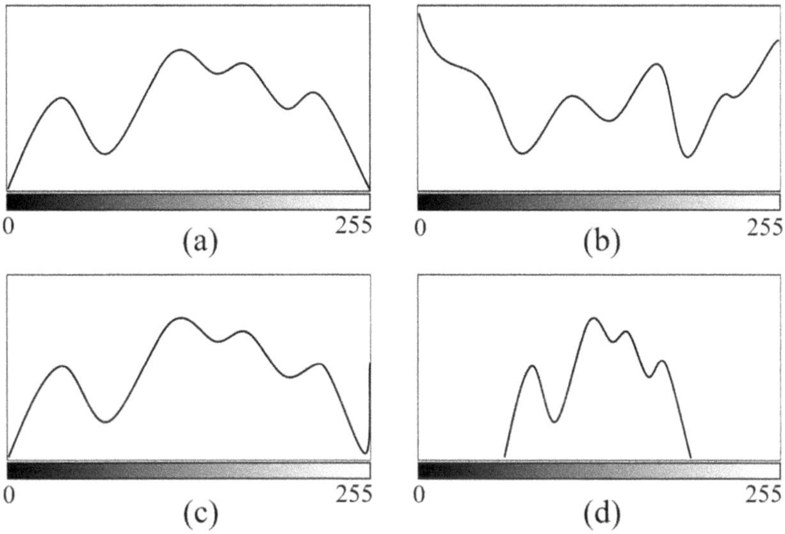

Figure 15.7. The histograms display: (a) matching dynamic ranges of the scene and the detector, (b) scene's dynamic range greater than that of the detector (under and overflow of minimum and maximum intensity boundaries), (c) matching between dynamic ranges of the scene and the detector, but with a saturation of high pixel values (possibly due to increased exposure time), (d) dynamic range of the scene below that of the detector.

value ($\underline{a}^2$). The noise is the variance of the pixel values (σ_a^2), being taken in a region of the image whose tone is approximately constant, such as its background. This restriction is necessary so that signal variation is not confused with noise. Thus, the SNR can be calculated as equation (15.5):

$$\mathrm{SNR} = 10\log_{10}\left(\frac{\underline{a}^2}{\sigma_a^2}\right) = 20\log_{10}\left(\frac{\underline{a}}{\sigma_a}\right) \tag{15.5}$$

Similarly, the contrast-to-noise ratio (CNR) can be defined as a measure of the relationship between the difference in intensity between the average foreground ($\underline{a_{FG}}$) and the average background ($\underline{a_{BG}}$), and the noise variance (σ_a^2), as equation (15.6):

$$\mathrm{CNR} = 20\log_{10}\left(\frac{\underline{a_{FG}} - \underline{a_{BG}}}{\sigma_a}\right) \tag{15.6}$$

15.3.6 Histogram stretch

An image with a dynamic range lower than the entire available dynamic range of the imaging system can be displayed in the full range if its histogram is stretched. Figure 15.8 depicts the original histogram (solid line) and its stretched version (dashed line).

An immediate solution is a linear stretching, preserving the shape of the histogram. Thus, the minimum a_{min} value is mapped to 0 and the maximum a_{max} value to 255 (in an 8-bit system, $N = 8$), then an arbitrary intermediate value a_i is mapped as g_i as given by equation (15.7):

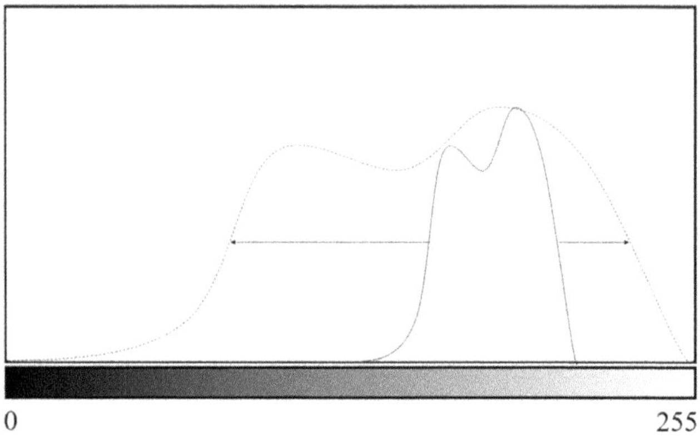

Figure 15.8. The original histogram (solid line) is stretched (dashed line) for occupying the entire dynamic range of the detection system.

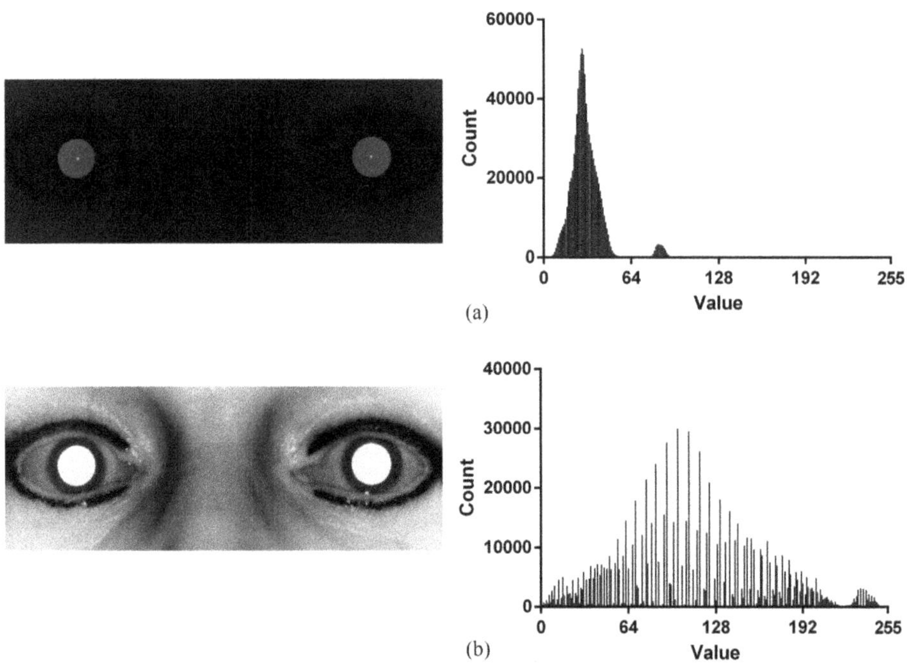

Figure 15.9. (a) Original image and its stretched version. (b) The respective histograms show how the dynamic range is fulfilled.

$$g_i = \left(\frac{2^N - 1}{a_{\max} - a_{\min}} \right) a_i \tag{15.7}$$

In this transformation, the number of bins is conserved, resulting in empty bin appearance between the originally transformed bins. In figure 15.9, there is an example of original and histogram stretched images. Notice how the stretched version offers many more details that were hidden in the original image because it improves image contrast.

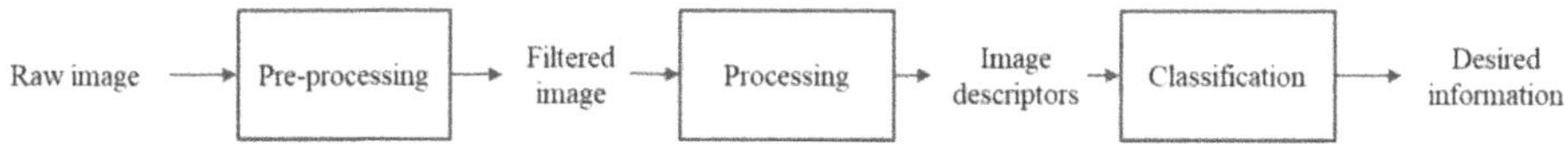

Figure 15.10. Block diagram of an image processing and analysis pipeline.

15.3.7 General pipeline of image processing

The analysis of the images follows a general pipeline of processing, from the original image to the desirable outcome. We will outline the main steps and tools used. Nevertheless, classical image processing is ad hoc and the computational approach is tailored to the image specificities inherent to the image acquisition. Additionally, the process is wired to the goal of the analysis; for instance, automatically classifying the cataracts in the LOCS III definitions.

The main steps will be divided, for convenience, as following: pre-processing, processing and classification (figure 15.10). We will define them and display the steps. Note that the steps may or not be used, and additional procedures may also be inserted.

Pre-processing: a preparation step to enhance the image, before it goes properly to a processing step. During acquisition, images are prone to several artifacts and noise. As an effect, for instance, images can get blurred, there are fluctuations in the intensity values, shadows (US, CT) or an underlying intensity gradient (MRI). In ophthalmology, imaging sources are usually optical. The optical path has deviations coming from lens imperfections, diffraction, scattering or from the lens extent limitations. Another source of noise is the thermal variations of the image sensors. The photograph exposition time can also lead to image noise.

Another commonly used modality in ophthalmology is OCT, a light-based technique that can quantify the layers of the retina. Due to its nature, when light penetrates the tissue, there is also noise coming from diffraction, scattering and intensity attenuation. It has depth information, which is different from a photograph. The noise comes from diffraction of different tissue layers, and this undesired noise is called speckle. As mentioned in section 15.3.2, images have a range of intensities which are integer values from 0 to $2^N - 1$, and this quantization in N bits can lead to another source of error, when converting the continuous values to integers.

To deal with noise and restore or enhance a desired region, we may apply image filters, mainly represented in the spatial or frequency domains. A filter might be designed to reduce noise or enhance borders. Also, other image operations can be applied to restore an image based on the knowledge of the physical properties of the imaging equipment. Most of the filters are usually applied with a convolution operation. The convolution is a specific mathematical function that replaces a pixel intensity based on the regional pixels in the image and a second function in the frequency-spatial domain. The function is applied to the whole image, considering this local operation as a sliding window. The digital convolution operation is defined in equation (15.8):

$$f'(x, y) = \sum_{s=-a}^{a} \sum_{t=-b}^{b} w(s, t) f(x - s, y - t) \tag{15.8}$$

$f'(x, y)$ is the filtered image of pixel with (x,y) coordinates, $f(x,y)$ is the original image pixel intensity, $w(x,y)$ is the convolution kernel, and the convolution dimensions are $2a + 1$ and $2b + 1$.

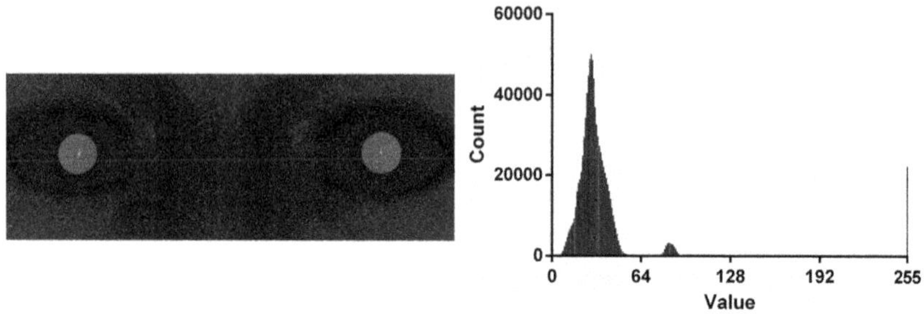

Figure 15.11. Example of IR pupil image with salt and pepper noise (left) and corresponding histogram (right).

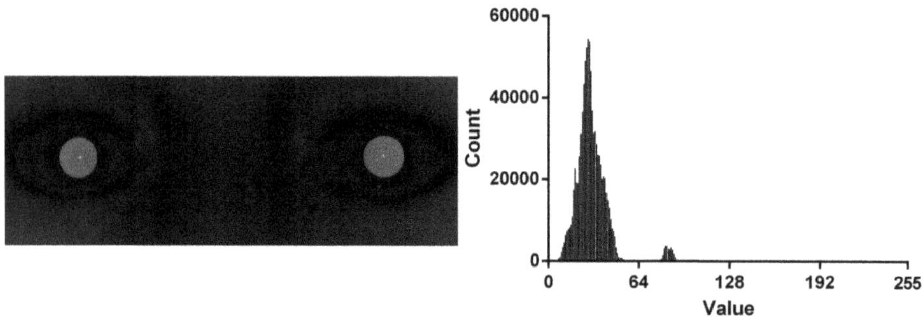

Figure 15.12. Median filter on previous image, with a radius of 2 pixels.

When noise can be statistically modelled, another approach is to use statistical operations that suppress noise, such as mean, median, minimum and maximum in a region. In figure 15.11, we can observe a salt and pepper noise, which is a random saturation of some pixels, either too dark or too bright (observe the extremes in the histogram). Regionally, an individual pixel can be replaced by its local median and have a value similar to what it is expected to feature. It is different than using the mean because the extreme values of the salt and pepper noise deviate from the mean value of the region, then a percentile distribution is the best approach. Figure 15.12 depicts figure 15.11 filtered by a median filter with radius 2, reducing the salt and pepper noise.

Processing: this is the step to assess the desired region of interest (ROI) of the image and obtain image descriptors. One example is to quantify the distance between the layers of the retina from OCT examination. For that goal, the surfaces of each tissue need to be delineated, see figure 15.13. Then, average values of distances between layers are calculated.

Another segmentation example is to quantify cataract extension to what percentage of the pupil is opacified. For that, the pupil needs to be segmented and the opacification within the pupil segmented and quantified. In section 15.4, the entire process for cataract opacification analysis is explored in detail.

For segmentation purpose, images are split into foreground and background regions. The object of interest is called foreground or region of interest (ROI) and the remaining image area is the background. The ROI represents the geometric information of the object and it can be a set of binary or fuzzy pixels. Therefore, the

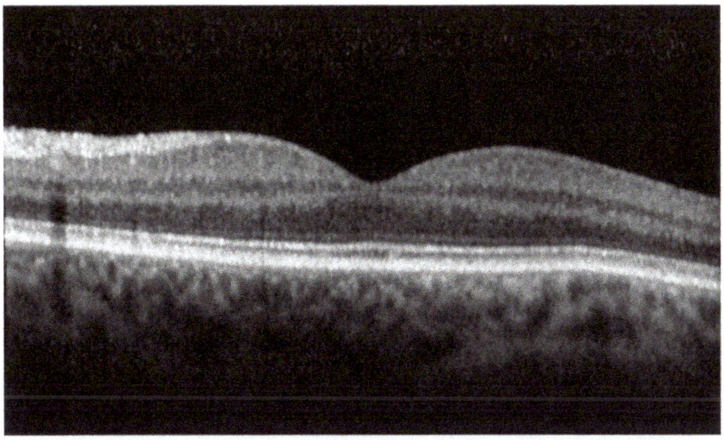

(a)

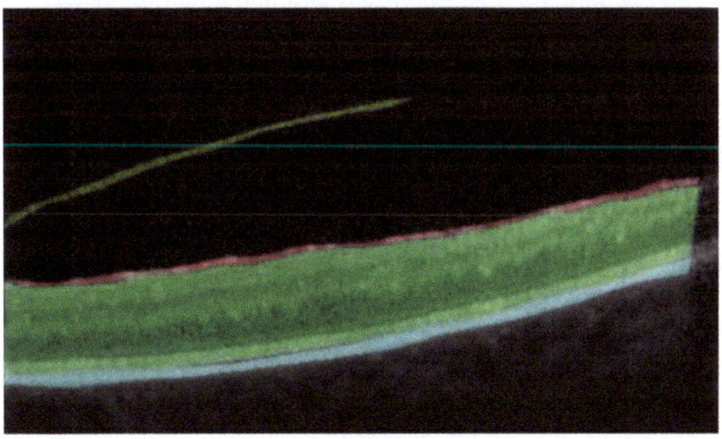

(b)

Figure 15.13. OCT of retina (a) and OCT image with overlay of different tissue segmentations (b). Courtesy: Thiago Martins/EPM-Unifesp.

first step is to segment or delineate the ROI. As an example, in figure 15.9, the pupil ROI is the foreground to be segmented and the remaining pixels are the background.

There are different perspectives of the types of techniques for segmentation, according to:

- the level of interaction with the user: manual, semi-automatic or automatic;
- the geometric descriptor: region based or contour based;
- the understanding of connectedness: similarity or continuity [7]; or
- the segmentation model: purely image based, shape based or hybrid [8].

For instance, a segmentation technique is the intensity threshold and in section 15.5 we have explored the optimal thresholding based on the intensity proposed by Otsu [9]. Some examples of region based segmentation applicable to ophthalmology are region growing [10], fuzzy connectedness [11] and watershed [12]. An example of contour based techniques is the active shape model called snakes [13].

Furthermore, depending on the medical application, there is a processing step called registration. It is used to compare images, in order to have the same patient pre- and post-intervention comparison. Another analysis is of different patients, when comparing similar regions, such as brain region analysis. For the same patient image comparison, simple matrix tools might be useful, such as rotation, translation and rescaling. For objects like the brain that have different shapes and connections, a more flexible structure is needed to match pixels between images. An established method is the template matching or ATLAS based [14], which is based on manual segmentation of a diverse set of individuals. Another recent method is the diffeomorphic transformation [15].

When we have the proper ROI segmented and registered, we can finally extract the features of the image, also called image descriptors. The features are used for:

- quantification: size in metric system of a region or from surfaces, e.g., area, perimeter, length, volume, minimum and maximum diameter;
- compact representation: geometric descriptor such as polygon approximation, Moore neighbor tracing, chain codes and topology skeleton representation;
- function analysis: time dependent quantification, e.g., blood perfusion in the brain (functional MRI) and pupil light reflex (PLR) analysis;
- radiomics: features related to pixel distribution and spatial characteristics that are not easily grasped by human eyes. For instance, features may be related to texture, statistics, invariant moments, shape, scale invariant image descriptor and entropy (information theory).

Classification: A computer automated system has the ultimate goal to yield relevant information for clinical decisions. Classification can be understood as the step to automatically categorize ROIs for diagnosis, or group similar images to cluster and display for human analysis. The output is categorized as binary (e.g., benign or malignant tumor), multilabel (e.g., PSC, cortical and nuclear cataracts) or fuzzy (continuous output from 0 to 1, similar to a class probability).

Supervised classifiers need at least a training and a testing set, in order to tune the model to label an unknown sample. In ophthalmic images, an expert will evaluate the training set and manually label each sample. This is a time-consuming task that might impact in the number of images labeled and can affect statistically the results of the automatic classifier. Some examples of classical classifiers [16] are neural network, k-nearest neighbor (KNN), support vector machine (SVM) and AdaBoost [17]. A recent classifier that has been a state-of-art method for images is convolutional neural networks (CNN)/deep learning [18], the drawback is that it needs a massive number of images to work properly.

15.4 Automatic cataract detection review

In this section, we present a review of methods that analyze eye images and classify cataracts. Different approaches to the processing and classification of images are described, especially for retro illuminated, direct illuminated and eye fundus images. We have also described other correlated approaches of classification of other eye

Table 15.1. Cross evaluation table of proposed diagnostic method and ground truth.

		Ground truth	
		Positive	Negative
Proposed method	Positive	TP	FP
	Negative	FN	TN

diseases. In section 15.6 we describe in detail a method proposed by us for cataract analysis of IR based images of the posterior pole of the eye for cataract detection.

In this section we refer to evaluation metrics as a cross evaluation of a diagnostic method and ground truth. The evaluation considers the distribution of the number of true positives (TPs), false positives (FPs), true negatives (TNs) and false negatives (FNs). The cross evaluation table is shown on table 15.1.

In this subsection, we review the literature results and use the metrics as the following definitions:

- True positives (TP): outcome correctly classified as positive;
- True negatives (TN): outcome correctly classified as negative;
- False positives (FP): outcome incorrectly classified as positive;
- False negatives (FN): outcome incorrectly classified as negative.

From the above definitions, we can derive some performance metrics:

- Sensitivity is the ratio between TP and the total number of positives (TP + FN);
- Specificity is the ratio between TN and the total number of negatives (TN + FP);
- Precision is the ratio between TP and TP + FP;
- Accuracy is the ratio between correct predictions (TP + TN) over all number of data (TP + TN + FP + FN).

15.4.1 Retro illuminated images

There are several research groups that have analyzed retro illuminated images of eyes, captured by an ocular biomicroscope [19–24]. These images are similar to the ones used in this chapter by the authors, because they also reflect light in the posterior pole of the eye. The difference is that, in these papers, the images are captured using visible light.

The authors [24] conducted the research using 607 images specifically to classify cortical opacity. The detection of the pupil region of interest was based on envelope detection and ellipse adjustment. The detection of cortical opacity was performed using the polar coordinate transform of the ROI followed by a Sobel vertical edge detector. The Sobel detector was also used to remove PSC-type cataract opacity. The method assigned larger weight to opacities close to the border of the pupil. In total, 102 images were tested for cortical opacity detection and the accuracy was 86.3%.

CHOW *et al* [21] analyzed 725 retro illumination images to detect cortical cataracts and PSC with a method based on texture and intensity analysis. To detect opacity, a texture analysis was the first step, dividing the images between clear lenses (with mild or no cataracts) and opaque lenses (with moderate or severe cataracts).

In clear lenses, the detection of opacity was obtained with a local entropy filter. In opaque lenses, a local entropy filter was used and also a threshold to select the ROI. The binary OR operator was used between these two results to define the final result. The accuracy of the method was 42.2% for cataract detection.

In another investigation, the researchers [20] have detected the ROI with a Canny and Laplacian edge detector, the detected edges were approximated to an ellipse. To detect cortical opacity, the image was converted to polar coordinates and a local threshold was applied to differentiate between darker areas and edge detection in horizontal and vertical directions. Angular opacities were subtracted from radial opacities to keep only cortical opacities. This is another approach where the authors specify the type of cataracts to be classified. Then, a region growing algorithm is performed to detect cortical opacities. The accuracy was 89.3%.

In these first three references, the methods are based on ad hoc differences between the types of cataracts and the output is the classification of specific cataracts.

SHAHEEN and AKRAM [19] developed a program and a user interface to receive images of ocular biomicroscopy with direct illumination or retro illumination, the method classified and defined the degree of cataracts using comparison with an image database with a classification system. The accuracy of nuclear cataracts graduation results was 94%. In this case, the program requires the user to inform the type of input image, e.g., retro illumination, which limits its use by a lay person.

Another method developed with retro illumination slit lamp images was developed by SIGIT *et al* [25], the researchers used Hough circle transform to segment the pupil ROI. The extracted features were average intensity and uniformity and the classification was performed using a neural network (single perceptron). The method was tested in a 30-image set and the accuracy was 96.6%.

FUADAH *et al* [26] used texture analysis in the retro illumination images, applying a gray level co-occurrence matrix (Gray Level Co-occurrence Matrix—GLCM) and other metrics such as contrast, dissimilarity, uniformity, correlation and homogeneity features. The classification is performed using KNN. The accuracy obtained in the classification was 97.5%. The article also briefly mentions that the application was implemented to acquire images and test them directly on a smartphone.

In the work of JIANG *et al* [27], the authors proposed a computer-aided method for the evaluation of pediatric cataracts and posterior capsular opacity (PCO) in retro illumination images. The method approaches the processing step with a double Canny edge detector and a Hough transform for segmentation of the pupils, followed by a deep residual CNN (deep residual CNN) with 50 layers, combined with a cost-sensitive method for extracting high level features and classification of images. The study analyzed 2705 images, of which 735 were positive for PCO and 1970 negative, all images were labeled according to the diagnosis of three ophthalmologists. From the dataset, 4/5 of the images were used for training and 1/5 for testing. The work showed satisfactory results, with average accuracy: 92.24%, specificity: 93.19%, sensitivity: 89.66% and area under the ROC curve (AUC) 97.11%.

Among these last four references, the segmentation technique was the edge detection with the Hough transform. These references also had extracted features that were used in our approach, described in detail in section 15.5.

15.4.2 Direct illuminated images

In the references [19, 20, 28, 29], the authors used slit lamp images with direct illumination for cataract nuclear detection. The processing method used in [28] and [20] to detect the lens contour (ROI) was the active shape model (ASM). The degree of cataracts was predicted using a training set with previous knowledge of the degree, the classification technique was SVM. The method of Li *et al* [28] correctly classified 95% (accuracy) of the database. The approach of Li *et al* [20] had a classification accuracy of 72% of the images with less than 0.5 error in the cataract degree.

15.4.3 Eye fundus images

The approaches by [30–32] used fundus images to analyze lens opacities, sample in figure 15.8 (section 15.2). The type of image analyzed is different from previous examinations, but the final goal is the same—to classify the presence or absence of cataracts. In [30], the ROI of the image was divided into sub regions, the block containing the optical disc was where the cataracts clearly appeared and was therefore treated as the most relevant sub region. The classification technique was SVM and the accuracy for classification in four different cataract classes was 87.52%. In [31] the two classes classification (SVM) had accuracy of 91.11%. In [32], the feature extraction and classification approach was deep CNN and the accuracy was 93.52%.

15.4.4 Classification of other eye diseases

The researchers [33] approached the automatic detection of diabetic retinopathy and diabetic macular edema in fundus images with a deep learning method. The neural network was trained using 128 175 images, which were evaluated and classified by 54 ophthalmologists. Using a database of 9963 images for training, the results had a sensitivity of 97.5% and specificity of 93.4%.

This is an important reference, because in 2018, it was the first medical equipment that used artificial intelligence to detect diabetic retinopathy in adults to be approved by the US Food and Drug Administration (FDA) [34]. This fact demonstrates that the development of artificial intelligence in medical equipment is growing and it can be done reliably. In the near future, new equipment and algorithms like those demonstrated in the next section may have real applications in patient diagnosis.

15.4.5 Analysis of computer-aided diagnosis (CAD) methods for eye diseases

In the work by Zhang *et al* [5], a survey on CAD for ocular diseases reviewed CAD methods for eye diseases and different types of data, the authors showed the considerable progress of these techniques in recent years. The research considered three types of data: clinical tests, image and genetic based, highlighting the image database as the most important in the diagnosis of eye diseases. Among the diseases with CAD studies, there were diabetic retinopathy, age-related macular degeneration (AMD), glaucoma and cataracts. Among imaging examinations surveyed were fundus images, OCT, slit lamp images and Heidelberg retinal tomography (HRT). The first observation about the progress of CAD was about the trend of the creation of automatic systems that learn from the dataset, from semi-automatic and expert dependent to a completely automatic system. The second observation was the trend of using heterogeneous data sources, i.e., more than one type of data to make the diagnosis. For example, heterogeneous data of imaging combined with clinical or genetic data.

15.4.6 Summary of methods for automatic cataract detection

In tables 15.2 and 15.3, we present a comparison between the main references cited in the previous subsections. In table 15.2, we show the automatic system input, the number of images used in the study, the type of cataracts classified and the segmentation method. In addition, we describe the main features extracted, the classification and accuracy obtained in the referred approaches in table 15.3. Notice the difference between approaches, we have highlighted the ones with retro illumination type of imaging. There is a large variation in the number of images in each study, as the acquisition, specialist review and manual segmentation are tasks that are time intensive. Nevertheless, all methods have the pipeline process described in the previous section (15.3): pre-processing, processing and classification.

15.5 Detection and classification of cataracts in infrared retro illuminated image

In this subsection we describe a full processing of retro illuminated images based on infrared camera, recorded with incident IR light (850 nm). The images capture the opacity present in the retro light path; hence it can be used for cataract detection and classification. In section 15.3, we have split the main image processing steps. In this subsection, pre-processing is described in section 15.5.1 (color conversion), processing in section 15.5.2 (segmentation) and section 15.5.3 (feature extraction), and classification in section 15.5.4.

Table 15.2. Description of input image type, number of input images in the study, type of classified cataracts and segmentation method of the main references.

Reference	Image type	Number of images	Cataract	Segmentation
LI *et al* [24]	Retro illumination with visible light	102	Cortical	Convex Envelope Analysis + Ellipse fitting
GAO *et al* [22]	Retro illumination with visible light	4545	Cortical and Posterior Subcapsular	Edge detection + Ellipse Fitting
LI *et al* [20]	Retro illumination with visible light	300	Cortical	Canny and Laplacian edge detector + Ellipse Fitting
CHOW *et al* [21]	Retro illumination with visible light	725	Cortical and Posterior Subcapsular	Canny and Laplacian edge detector + Envelope detector + Ellipse fitting

Table 15.3. Description of extracted features, classification methods and accuracy of the main references.

Reference	Extracted features	Classification	Accuracy
LI *et al* [24]	Polar transform + Radial edge detection—Sobel Vertical and horizontal edge detection	Not described	86.30%
GAO *et al* [22]	Mean, standard deviation, asymmetry, flattening and local entropy	Linear discriminant analysis (LDA)	87.80%
LI *et al* [20]	Polar transform + Local threshold and detection of radial and angular edge + growth of opacity region	SVM	89.30%
CHOW *et al* [21]	Local entropy filter for texture detection + global threshold	Not described	42.20%

15.5.1 Color to grayscale conversion

The camera, with no IR filter, records the scene as colored images. For opacity analysis, the color content is not crucial to distinguish between dark and light regions. Thus, the colored pixel values are converted into grayscale values using the weighted sum method described in equation (15.9) [35]. The new grayscale value (Y) is obtained by linearly transforming the color channel values red (R), green (G), and blue (B). In figure 15.14, there is an example of conversion.

$$Y = 0.299 \times R + 0.587 \times G + 0.114 \times B \tag{15.9}$$

15.5.2 Pupil segmentation

For the evaluation of opacity, the ROI is the pupil internal area. The pupil is the region where the posterior reflection emerges, and the captured optical path highlights the opacities that characterize cataracts. We need to filter the image to find the pupil edges and segment it to define the ROI.

There are several methods for detecting edges and patterns in images. In general, edges are detected by a gradient difference in pixel intensity, which reveals the boundaries of an object. In our case, however, a gradient-based detection method would not work, because the lens opacity can affect the pupil perimeter, see example in figure 15.15. For this reason, in addition to edge detection, we explore the circular shape of the pupils. The ROI detection approach has two steps, the gradient-based edge detection and the circular Hough transformation.

The process consists of gradient identification and subsequent application of the circle Hough transform. The first operation is the edge detection, which is performed using the Canny method [7]. We can find the image intensity gradient by applying the Sobel operator in horizontal (x-axis) and vertical (y-axis) directions, using the 3×3 convolution matrices shown in equations (15.10) and (15.11):

$$G_x = [-1 \;\; 0 \;\; +1 \;\; -2 \;\; 0 \;\; +2 \;\; -1 \;\; 0 \;\; +1\,] \tag{15.10}$$

$$G_y = [-1 \;\; -2 \;\; -1 \;\; 0 \;\; 0 \;\; 0 \;\; +1 \;\; +2 \;\; +1\,] \tag{15.11}$$

(a)

(b)

Figure 15.14. Color (a) to grayscale (b) conversion of a retro illuminated image recorded by an infrared camera.

Thus, for each pixel, we obtain the magnitude and phase of the resulting gradient, using equations (15.12) and (15.13):

$$G = \sqrt{\left(G_x^2 + G_y^2\right)} \tag{15.12}$$

$$\theta = \arctan\left(\frac{G_y}{G_x}\right) \tag{15.13}$$

After the Sobel filter, there is a refinement process. First, a non-maximum suppression step is performed, it consists of finding all local gradient maxima (in the direction of the gradient), as in figure 15.16 and then eliminating all pixels below that value. This operation results in thinner edges, because it removes the smaller values.

After non-maximum suppression, a two-level threshold hysteresis is performed with the following rules:

- All gradient values greater than the maximum threshold ($t_{\max}$) are considered edges;
- All gradient values less than the minimum threshold ($t_{\min}$) are not considered as edges;
- The gradient values between $t_{\min}$ and $t_{\max}$ are only considered edges if they are connected to pixels (neighbors) that are greater than $t_{\max}$, considered for sure as edges.

Figure 15.15. Gradient-based edge detection in an image of eyes with opacity. The white line over the image represents the detected edges. Notice how opacities corrupt the detected edges in pupil's edges.

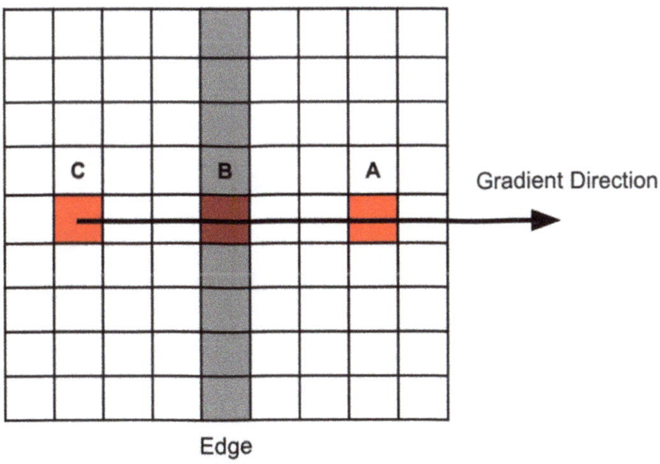

Figure 15.16. Pixels analyzed in non-maximum suppression. Instead of having a thicker edge (pixels A, B, and C), the lower gradient values with the same direction (A and C) are excluded from the edge, remaining only the maximum gradient value **B**.

After this operation, small edge lines and noise are reduced.

As the pupils have a circular shape, the Hough transform was used to identify circular regions. A circle of radius R and center at (a, b) Euclidean coordinates can be described by equations (15.14) and (15.15) parameterized by the angle (θ) within the interval $[0, 2\pi]$:

$$x(\theta) = a + R \times \cos\cos(\theta) \tag{15.14}$$

$$y(\theta) = b + R \times \sin\sin(\theta) \tag{15.15}$$

In the algorithm, the edge pixels are found in the previous steps to find the parameters (a, b) of the image circles. For each pixel on the edge, a center and a circle of a certain radius are created, as shown in figure 15.17. These new circles are stored in an accumulation matrix. After analyzing all the points on the edges, the point of greatest accumulation corresponds to the circle's center in the original image [36].

Figure 15.18 is an example of the original-colored image (a), the edge detection (b), and the result of the circles detected by the Hough transform drawn on the image (c).

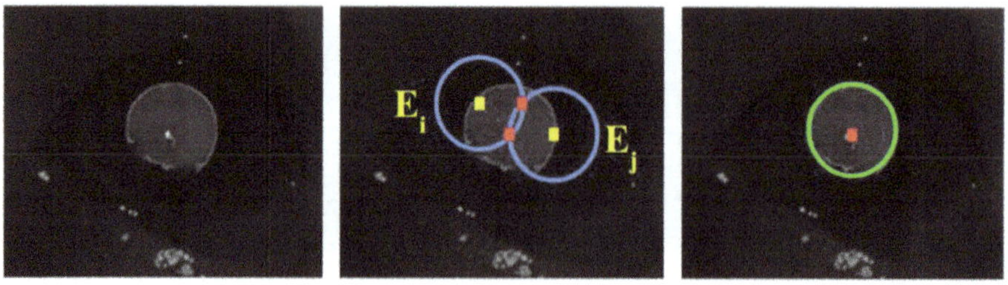

Figure 15.17. Example of pupil image with edges shown with white pixels on the left. The central image has two edge pixels (E_i and E_j, in yellow) and the corresponding circle locus of circle centers in blue that contain that edge pixel. On the right, the pixel in red is considered the final center of the Hough transform with radius shown in green.

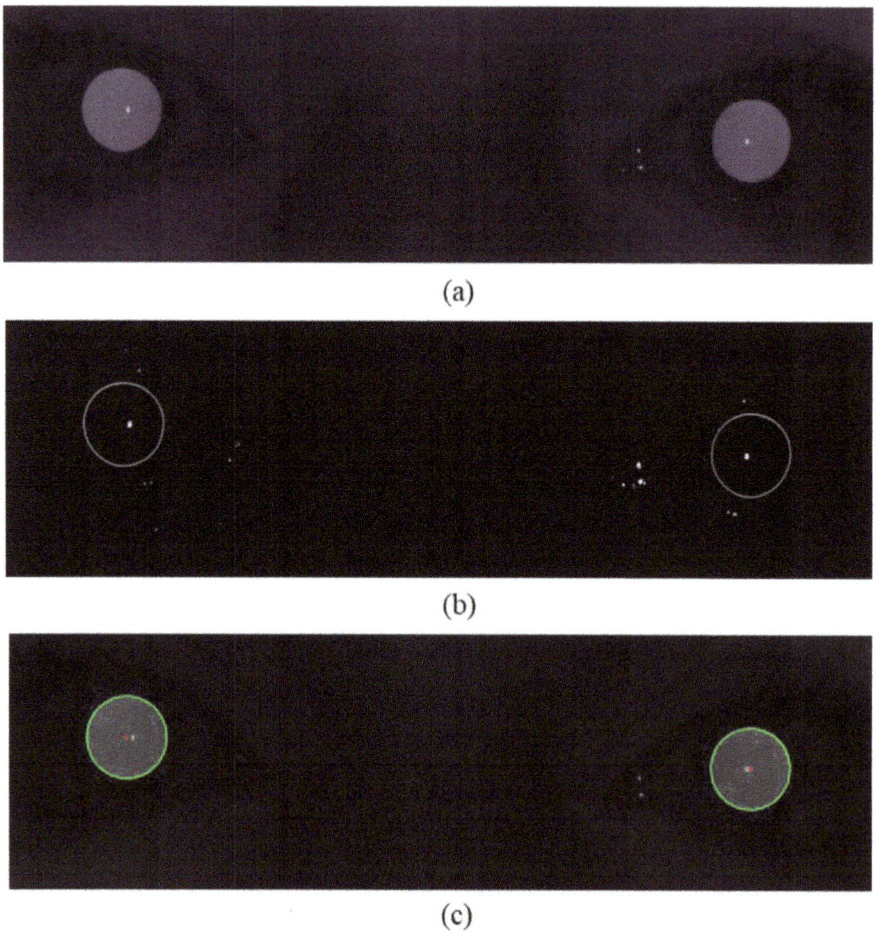

Figure 15.18. Example of segmentation processing: (a) original-colored image, (b) Canny filter detected edges, (c) image with the circles detected by the Hough method (represented in green) and the center of the circle (represented as a red dot).

15.5.3 Feature extraction

Features are measurements that characterize each image ROI and provide quantitative information for differentiating one from another. They are the data received by the classifier to separate the images into the desired groups. It is important to know the characteristics of the images and understand what aspects the specialists take into consideration to diagnose and translate these aspects into quantitative data.

When clinically assessing cataracts, specialists may consider the following findings:

- extension of cloudiness of the crystalline lens;
- location of the opaque area;
- the image pattern of the non-functional area; and
- symmetry of the abnormal findings as the lack of symmetry is related to the disease.

In an automated device, the findings may be, respectively, translated into ad hoc features such as:

- percentage of opaque region;
- pattern of opaque location (e.g. nuclear, radial or diffuse);
- texture features; and
- similarity measurements between pupils.

We describe two ad hoc features, the percentage of opacity (based on Otsu threshold) and texture features (first-order features and Hu's invariant moments)

Otsu threshold: In general, when dealing with image segmentation of two regions (e.g., foreground and background), the histogram is expected to be bimodal. When noisy, the image can be pre-processed. There is an optimal way to find the intensity threshold to separate both modes, the Otsu threshold method. Pixels with values are above the Otsu threshold are considered foreground (binary 1), and those values below the threshold are considered background (binary 0). Finally, we obtain a binary image containing foreground and background corresponding to the desired segmentation.

To find the optimal threshold value, the Otsu method iterates through all possible threshold values, calculating the variance for each side created in the histogram, to find the value where the weighted sum of the two variances is minimal [9].

With the Otsu method, we were able to divide the light and dark pupil areas, where the dark area represents regions where the beams of the luminous reflection could not go through the lens, see figure 15.19. A challenge encountered when using the Otsu threshold method to detect opacity is that unwanted opacities of origin were also detected. In astigmatism, the darker regions are not related to opacity in the lens but due to refraction, see figure 15.19 (middle line).

Opacity ratio: To quantify the opacity determined using the Otsu threshold method and use it as an attribute for classification, the opacity ratio (OR) was established as shown in equation (15.16):

$$\text{Opacity ratio} = \frac{N_{\text{o}}}{N}, \tag{15.16}$$

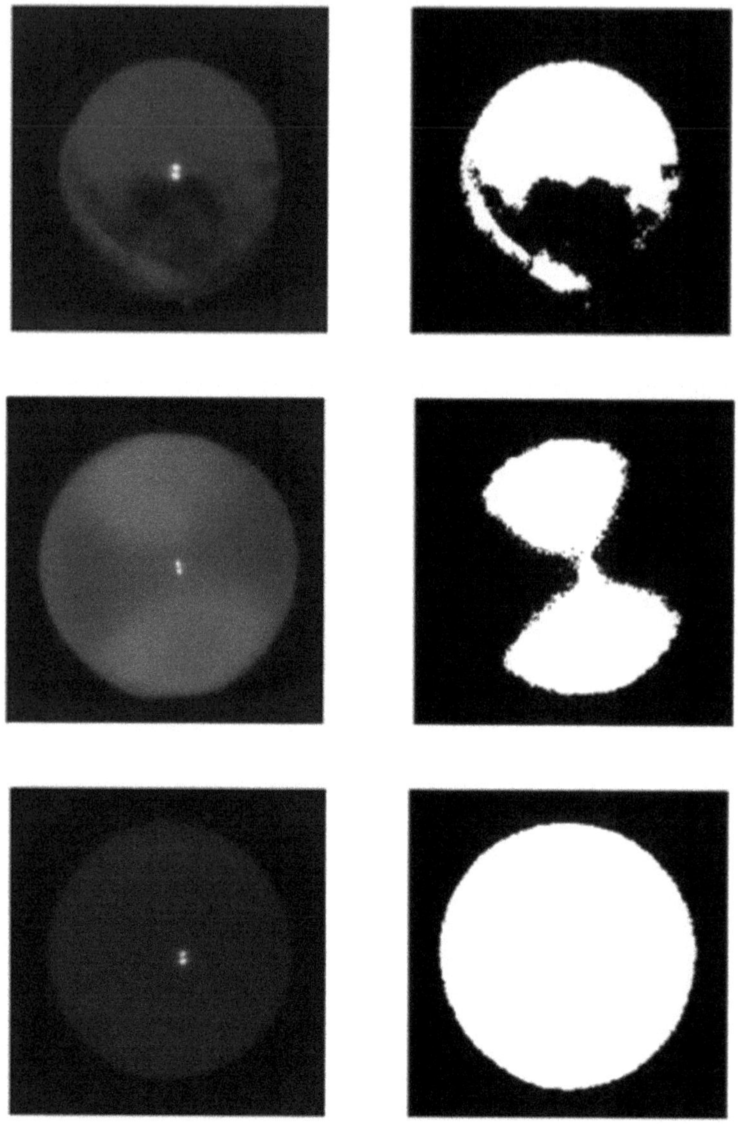

Figure 15.19. Example of Otsu threshold method: original images (left) and binary images (right). From top to bottom: eye with opacities, an eye with astigmatism, and eye without opacities. Notice the astigmatism case mislabeled as opacification using the Otsu threshold method.

where N_o is the number of opaque pixels (black), N_c is the number of light pixels (white), and N is the total number of pixels in the ROI ($N_o + N_c$).

In eyes with opacity, the expected opacity ratio is high, and in eyes with clear lenses (without opacity), the expected ratio is low. Although the opacity ratio is a simple attribute, it describes well the difference between opaque and non-opaque images and it is a one-variable form of classifying the images quantitatively.

First-order features: The first-order statistical features describe the distribution of pixel intensities within the ROI, as eyes with clear lenses and eyes with opaque lenses show different distributions.

Following, four first-order features are presented to be used for classification: entropy, mean absolute deviation, variance, and uniformity. As the image illumination levels, which reflects the pixel intensity value, was not normalized during the experiments, we are interested in features that are not dependent on illumination conditions. Equations (15.17)–(15.20) describe each feature [37]:

- Entropy quantifies the uncertainty or randomness of the image values:

$$\text{Entropy} = -\sum_{i=1}^{N_g} p(i)\log_2(p(i) + \epsilon) \tag{15.17}$$

where ϵ is a positive, small, arbitrary value.

- Mean absolute deviation (MAD) is the average distance of all intensity values from the average image value ($\underline{X}$):

$$\text{MAD} = \frac{1}{N_p}\sum_{i=1}^{N_p} |X(i) - \underline{X}| \tag{15.18}$$

- Variance measures the amount of variation or dispersion from the average value ($\underline{X}$):

$$\text{Variance} = \frac{1}{N_p}\sum_{i=1}^{N_p} (X(i) - \underline{X})^2 \tag{15.19}$$

- Uniformity is the sum of the squares of each intensity value of the probability density function. It is a measure of image homogeneity:

$$\text{Uniformity} = \sum_{i=1}^{N_g} p(i)^2 \tag{15.20}$$

The terms of the first-order features are given by:

o. N_p: the number of pixels contained in the ROI.
p. N_g: number of allowed intensity levels, in our case it is equal to 255 (8-bit resolution).
q. $X(i)$: pixel intensity value.
r. $P(i)$: intensity frequency distribution function with Ng discrete intensity levels.
s. $p(i)$: probability density function $= \frac{P(i)}{Np}$, $p(i)$ is the same as PDF explained in section 15.3.3.

Hu's invariant moments: Image moments are statistical parameters to measure the distribution of the pixels and intensity locations. A set of seven invariant moments, called Hu moments [38], are also calculated as features for each image, see equations (15.21)–(15.27). These seven moments are invariant for translation, scale, rotation, and mirroring. It is expected that they would present close values for similar images, such as eyes without opacities or eyes with astigmatism, invariant to differences in scale or rotation in the images. Thus, helping the classifier to classify cataract and non-cataract images.

$$hu_1 = \eta_{20} + \eta_{02} \tag{15.21}$$

$$hu_2 = (\eta_{20} - \eta_{02})^2 + 4\eta_{11}^2 \tag{15.22}$$

$$hu_3 = (\eta_{30} - 3\eta_{12})^2 + (3\eta_{21} - \eta_{03})^2 \tag{15.23}$$

$$hu_4 = (\eta_{30} + \eta_{12})^2 + (\eta_{21} + \eta_{03})^2 \tag{15.24}$$

$$\begin{aligned} hu_5 = {} & (\eta_{30} - 3\eta_{12})(\eta_{30} + \eta_{12})[(\eta_{30} + \eta_{12})^2 - 3(\eta_{21} + \eta_{03})^2] \\ & + (3\eta_{21} - \eta_{03})(\eta_{21} + \eta_{03})[3(\eta_{30} + \eta_{12})^2 - (\eta_{21} + \eta_{03})^2] \end{aligned} \tag{15.25}$$

$$hu_6 = (\eta_{20} - \eta_{02})[(\eta_{30} + \eta_{12})^2 - (\eta_{21} + \eta_{03})^2] + 4\eta_{11}(\eta_{30} + \eta_{12})(\eta_{21} + \eta_{03}) \tag{15.26}$$

$$\begin{aligned} hu_7 = {} & (3\eta_{21} - \eta_{03})(\eta_{30} + \eta_{12})[(\eta_{30} + \eta_{12})^2 - 3(\eta_{21} + \eta_{03})^2] \\ & + (3\eta_{12} - \eta_{30})(\eta_{21} + \eta_{03})[3(\eta_{30} + \eta_{12})^2 - (\eta_{21} + \eta_{03})^2] \end{aligned} \tag{15.27}$$

where the 2D moments of order $(p+q)$ of a $M \times N$ digital image $f(x,y)$ are defined as follow in equations (15.28)–(15.32) [7]:

$$m_{pq} = \sum_{x=0}^{M-1} \sum_{y=0}^{N-1} x^p y^q f(x, y), \quad p, q \in N \tag{15.28}$$

$$\mu_{pq} = \sum_{x=0}^{M-1} \sum_{y=0}^{N-1} (x - \underline{x})^p . (y - \underline{y})^q f(x, y) \tag{15.29}$$

$$\underline{x} = \frac{m_{10}}{m_{00}} \tag{15.30}$$

$$\underline{y} = \frac{m_{01}}{m_{00}} \tag{15.31}$$

$$\eta_{pq} = \mu_{pq}\left(\mu_{00}^{\left(\frac{p+q}{2}\right)+1}\right)^{-1}, \quad p + q = 2, 3, \ldots \tag{15.32}$$

15.5.4 Classification

The image classification can be performed using several approaches using the extracted features as inputs for such algorithms. We will present two machine learning-based methods: the KNN and the SVM.

K-nearest neighbors: The KNN classification algorithm classifies elements according to the label of their closest neighbors. K is an integer and positive number, and in general odd, that defines the number of considered neighbors. The rating can come from a vote or a weight metric regarding the proximity between labeled neighbors and the new test sample. In this algorithm, training and testing are a single step, and for each new sample, it is necessary to perform all measurements again, which may make the method slow for large data sets.

We can combine all features in different feature-sets and test the KNN algorithm for different K values, e.g., $K = 1, 3, 5, 7, 11$. The training and test strategy might be to separate the two eyes of the same patient and train with the whole dataset except by two images: one with cataracts and another without (leave-two-out). These two images are posteriorly used to test the trained classifier.

Support vector machine: SVM algorithms search for an optimal hyperplane separating two classes, e.g., images with or without opacities. To solve nonlinear problems, we apply nonlinear transformations (kernel functions) for changing the input space to the feature space. The optimal hyperplane separating the two classes follows equation (15.33):

$$\Omega(x^T)w + b = 0 \tag{15.33}$$

where x is the input data, Ω is the kernel function, w represents weights, and b is the bias. Commonly used kernel functions in SVM problems are linear, polynomial, gaussian, and sigmoid functions.

Classifier assessment: After obtaining the classifier, we need to compare the outcomes objectively. We can use performance metrics to compare the effectiveness of each classifier such as sensitivity, specificity, precision and accuracy, defined in the beginning of section 15.4.

By comparing the performance metrics among different classifiers, it is possible to quantitatively choose one to be the best classifier and to progressively improve a classifier by tuning its parameters, thus reaching even better outcomes from the same classifier.

15.6 Summary

In this chapter, we have addressed the main topics for cataract detection with a CAD system. It is important to reassert the relevance of automated image analysis, which can improve clinical eye care in a multitude of aspects. The imaging equipment can be used for screening and diagnosis. For instance, with the examinations automatically analyzed, physicians can focus on patient cataract treatment and spend more quality time with the patient. Additionally, a negative cataract diagnosis lets patients drop the clinical office queue. In places where patients have difficult access to healthcare, cataract imaging can be a tool for screening and early detection of this disease. The positive consequences are that more people can be referred to a cataract specialist and have early cataracts follow up.

The first important step to detect and analyze cataract images was the study and understanding the images from a clinical point of view. Understanding the characteristics present in the eyes with cataracts and different pathologies, such as astigmatism or any other pure optical light deviation, and their results in the image. With that in mind, we can choose features that could bring appropriate, relevant descriptors for classification. It was also essential to understand the motivation of the project, how important it is to develop a screening tool that will not completely replace medical care but will be a step to improve eye care.

Although we presented several methods to detect and classify cataracts, there is a common framework to process the images, see figure 15.10, and the framework is as follows. After raw image acquisition, there are pre-processing steps on the digital image, then processing to segment the ROI and analyze the lens opacity. There are different methods to detect the pupil, based on edge detection or shape.

Within the pupil, cataract opacity is detected and classified, e.g., through intensity variations (statistics), texture, and invariant moment. As we have discussed, the final performance of the algorithm also depends on the quality of the training set, which is used to tune the classifier mathematically. Further studies of the dysfunctional lenses benefit from using more extensive databases, consequently safer and more effective outcomes are delivered to the population. And some automatic methods can even perform equally or better than a specialized physician or other healthcare provider. For screening purposes, the sensitivity of the method is the most important metric because the false negative (FN) rate should be as low as possible.

There are recent trends of incorporating as much patient data as possible to improve diagnosis and patient care. Another trend is that FDA-approved medical devices can drive examination and analyze patient images through artificial intelligence software embedded in the device. Therefore, automatic methods are welcome to assist physicians, improve the healthcare framework and reduce healthcare costs.

Chapter highlights

- Introduction to lens physiology and cataract opacities.
- Overview of image acquisition, different imaging modalities and inherent noises.
- Understanding of a pipeline from raw image processing, cataract feature detection to automated classification.
- Review of cataract image detection and analysis methods in literature.
- Detailed processing of infrared retro illuminated images and cataract analysis.

References

[1] Liu Y C, Wilkins M, Kim T, Malyugin B and Mehta J S 2017 Cataracts *Lancet* **390** 600–12

[2] Hosseini H S and Taber L A 2018 How mechanical forces shape the developing eye *Prog. Biophys. Mol. Biol.* **137** 25–36

[3] Vinson J A 2006 Oxidative stress in cataracts *Pathophysiology* **13** 151–62

[4] Waring G O and Rocha K M 2018 Characterization of the dysfunctional lens syndrome and a review of the literature *Curr. Ophthalmol. Rep.* **6** 249–55

[5] Zhang Z, Srivastava R, Liu H, Chen X, Duan L and Kee Wong D W *et al* 2014 A survey on computer aided diagnosis for ocular diseases *BMC Med. Inf. Decis. Making* **14** 80

[6] Chylack L T Jr, Wolfe J K, Singer D M, Leske M C, Bullimore M A and Bailey I L *et al* 1993 The lens opacities classification system III. The Longitudinal Study of Cataract Study Group *Arch. Ophthalmol.* **111** 831–6

[7] Gonzalez R C and Woods R E 2008 *Digital Image Processing* (London/Englewood Cliffs, NJ: Pearson/Prentice-Hall)

[8] Chen X, Udupa J K, Alavi A and Torigian D A 2013 GC-ASM: synergistic integration of graph-cut and active shape model strategies for medical image segmentation *Comput. Vis. Image Underst.* **117** 513–24

[9] Otsu N 1979 A threshold selection method from gray-level histograms *IEEE Trans. Syst. Man Cybern.* **9** 62–6

[10] Qian Zhao Y, Hong Wang X, Fang Wang X and Shih F Y 2014 Retinal vessels segmentation based on level set and region growing *Pattern Recognit.* **47** 2437–46

[11] Herman G T and Carvalho B M 2001 Multiseeded segmentation using fuzzy connectedness *IEEE Trans. Pattern Anal. Mach. Intell.* **23** 460–74

[12] DeBuc D C 2011 *A Review of Algorithms for Segmentation of Retinal Image Data Using Optical Coherence Tomography* (London: IntechOpen)

[13] Kass M, Witkin A and Terzopoulos D 1988 Snakes: active contour models *Int. J. Comput. Vision* **1** 321–31

[14] Aljabar P, Heckemann R A, Hammers A, Hajnal J V and Rueckert D 2009 Multi-atlas based segmentation of brain images: atlas selection and its effect on accuracy *Neuroimage* **46** 726–38

[15] Khan A R, Wang L and Beg M F 2008 FreeSurfer-initiated fully-automated subcortical brain segmentation in MRI using large deformation diffeomorphic metric mapping *Neuroimage* **41** 735–46

[16] Bishop C M and Nasrabadi N M 2006 *Pattern Recognition and Machine Learning* (New York: Springer)

[17] Schapire R E and Singer Y 1999 Improved boosting algorithms using confidence-rated predictions *Mach. Learn.* **37** 297–336

[18] Goodfellow IB Y and Courville A 2016 *Deep Learning* (Cambridge, MA: MIT Press)

[19] Shaheen I and Akram M U 2017 An integrated framework for clinical grading of cataract *2017 1st Int. Conf. on Next Generation Computing Applications (NextComp) (19–21 July 2017)*

[20] Li H, Lim J, Liu J, Wong D, Tan N M and Lu S *et al* 2009 Computerized systems for cataract grading *2009 2nd Int. Conf. on Biomedical Engineering and Informatics* pp 1–4

[21] Chow Y C, Gao X, Li H, Lim J H, Sun Y and Wong T Y 2011 Automatic detection of cortical and PSC cataracts using texture and intensity analysis on retro-illumination lens images *Annu. Int. Conf. IEEE Eng. Med. Biol. Soc.* **2011** 5044–7

[22] Gao X, Li H, Lim J H and Wong T Y 2011 Computer-aided cataract detection using enhanced texture features on retro-illumination lens images *18th IEEE Int. Conf. on Image Processing (11–14 September 2011)*

[23] Li H, Lim J H, Liu J, Wong D W K, Foo Y and Sun Y 2010 Automatic detection of posterior subcapsular cataract opacity for cataract screening *Annual Int. Conf. of the IEEE Engineering in Medicine and Biology (31 August–4 September 2010)*

[24] Huiqi L, Liling K, Joo Hwee L, Jiang L, Damon Wing Kee W and Wong T Y 2008 Automatic opacity detection in retro-illumination images for cortical cataract diagnosis *2008 IEEE Int. Conf. on Multimedia and Expo (23 June–26 April 2008)*

[25] Sigit R, Kom M, Satmoko M B, Basuki D K and Si S 2018 Classification of cataract slit-lamp image based on machine learning *2018 Int. Seminar on Application for Technology of Information and Communication* pp 597–602

[26] Fuadah Y, Setiawan A and Mengko T 2015 Mobile cataract detection using optimal combination of statistical texture analysis *2015 4th Int. Conf. Instrumentation, Communications, Information Technology, and Biomedical Engineering (ICICI-BME)* pp 232–6

[27] Jiang J, Liu X, Zhang K, Long E, Wang L and Li W *et al* 2017 Automatic diagnosis of imbalanced ophthalmic images using a cost-sensitive deep convolutional neural network *Biomed. Eng. Online* **16** 132

[28] Li H, Lim J H, Liu J, Mitchell P, Tan A G, Wang J J and Wong T Y 2010 A computer-aided diagnosis system of nuclear cataract *IEEE Trans. Biomed. Eng.* **57** 1690–8

[29] Huang W, Chan K L, Li H, Lim J H, Liu J and Wong T Y 2011 A computer assisted method for nuclear cataract grading from slit-lamp images using ranking *IEEE Trans. Med. Imaging* **30** 94–107

[30] Qiao Z, Zhang Q, Dong Y and Yang J 2017 Application of SVM based on genetic algorithm in classification of cataract fundus images *2017 IEEE Int. Conf. on Imaging Systems and Techniques (IST) (18–20 October 2017)*

[31] Harini V and Bhanumathi V 2016 Automatic cataract classification system *2016 Int. Conf. on Communication and Signal Processing (ICCSP) (6–8 April 2016)*

[32] Linglin Z, Jianqiang L, Li Z, He H, Bo L and Yang J 2017 Automatic cataract detection and grading using deep convolutional neural network *2017 IEEE 14th Int. Conf. on Networking, Sensing and Control (ICNSC) (16–18 May 2017)*

[33] Gulshan V, Peng L, Coram M, Stumpe M C, Wu D and Narayanaswamy A *et al* 2016 Development and validation of a deep learning algorithm for detection of diabetic retinopathy in retinal fundus photographs *JAMA* **316** 2402–10

[34] FDA FDA permits marketing of artificial intelligence-based device to detect certain diabetes-related eye problems. Available from: https://fda.gov/news-events/press-announcements/fda-permits-marketing-artificial-intelligence-based-device-detect-certain-diabetes-related-eye

[35] Gomes J and Velho L 1997 *Image Processing for Computer Graphics* (Berlin: Springer Science & Business Media)

[36] Kimme C, Ballard D and Sklansky J 1975 Finding circles by an array of accumulators *Commun. ACM* **18** 120–2

[37] van Griethuysen J J M, Fedorov A, Parmar C, Hosny A, Aucoin N and Narayan V *et al* 2017 Computational radiomics system to decode the radiographic phenotype *Cancer Res.* **77** e104

[38] Ming-Kuei H 1962 Visual pattern recognition by moment invariants *IRE Trans. Inf. Theory* **8** 179–87

www.ingramcontent.com/pod-product-compliance
Ingram Content Group UK Ltd.
Pitfield, Milton Keynes, MK11 3LW, UK
UKHW051907100926
8556IPUK00005B/796

9 780750 332613